Universitext

For other titles in this series, go to
www.springer.com/series/223

Jean Gallier

Discrete Mathematics

 Springer

Jean Gallier
University of Pennsylvania
Department of Computer and Information Science
3330 Walnut Street
Philadelphia, PA 19104
USA
jean@cis.upenn.edu

ISBN 978-1-4419-8046-5   e-ISBN 978-1-4419-8047-2
DOI 10.1007/978-1-4419-8047-2
Springer New York Dordrecht Heidelberg London

Mathematics Subject Classification (2010): 03-01, 03F03, 05-01, 05C21, 06-01, 11-01, 11A05, 68-01

Springer is part of Springer Science+Business Media (www.springer.com)

To my family, especially Anne and Mia, for their love and endurance

Preface

The curriculum of most undergraduate programs in computer science includes a course titled *Discrete Mathematics*. These days, given that many students who graduate with a degree in computer science end up with jobs where mathematical skills seem basically of no use,[1] one may ask why these students should take such a course. And if they do, what are the most basic notions that they should learn?

As to the first question, I strongly believe that *all* computer science students should take such a course and I will try justifying this assertion below.

The main reason is that, based on my experience of more than twenty-five years of teaching, I have found that the majority of the students find it very difficult to present an argument in a rigorous fashion. The notion of a proof is something very fuzzy for most students and even the need for the rigorous justification of a claim is not so clear to most of them. Yet, they will all write complex computer programs and it seems rather crucial that they should understand the basic issues of program correctness. It also seems rather crucial that they should possess some basic mathematical skills to analyze, even in a crude way, the complexity of the programs they will write. Don Knuth has argued these points more eloquently than I can in his beautiful book, *Concrete Mathematics*, and I do not elaborate on this any further.

On a scholarly level, I argue that some basic mathematical knowledge should be part of the scientific *culture* of any computer science student and more broadly, of any engineering student.

Now, if we believe that computer science students should have some basic mathematical knowledge, what should it be?

There is no simple answer. Indeed, students with an interest in algorithms and complexity will need some discrete mathematics such as combinatorics and graph theory but students interested in computer graphics or computer vision will need some geometry and some continuous mathematics. Students interested in databases will need to know some mathematical logic and students interested in computer architecture will need yet a different brand of mathematics. So, what's the common core?

---

[1] In fact, some people would even argue that such skills constitute a handicap!

As I said earlier, most students have a very fuzzy idea of what a proof is. This is actually true of most people. The reason is simple: it is quite difficult to define precisely what a proof is. To do this, one has to define precisely what are the "rules of mathematical reasoning" and this is a lot harder than it looks. Of course, defining and analyzing the notion of proof is a major goal of mathematical logic.

Having attempted some twenty years ago to "demystify" logic for computer scientists and being an incorrigible optimist, I still believe that there is great value in attempting to teach people the basic principles of mathematical reasoning in a precise but not overly formal manner. In these notes, I define the notion of proof as a certain kind of tree whose inner nodes respect certain proof rules presented in the style of a natural deduction system "a la Prawitz." Of course, this has been done before (e.g., in van Dalen [6]) but our presentation has more of a "computer science" flavor which should make it more easily digestible by our intended audience. Using such a proof system, it is easy to describe very clearly what is a proof by contradiction and to introduce the subtle notion of "constructive proof". We even question the "supremacy" of classical logic, making our students aware of the fact that there isn't just one logic, but different systems of logic, which often comes as a shock to them.

Having provided a firm foundation for the notion of proof, we proceed with a quick and informal review of the first seven axioms of Zermelo–Fraenkel set theory. Students are usually surprised to hear that axioms are needed to ensure such a thing as the existence of the union of two sets and I respond by stressing that one should always keep a healthy dose of skepticism in life.

What next? Again, my experience has been that most students do not have a clear idea of what a function is, even less of a partial function. Yet, computer programs may not terminate for all input, so the notion of partial function is crucial. Thus, we carefully define relations, functions, and partial functions and investigate some of their properties (being injective, surjective, bijective).

One of the major stumbling blocks for students is the notion of proof by induction and its cousin, the definition of functions by recursion. We spend quite a bit of time clarifying these concepts and we give a proof of the validity of the induction principle from the fact that the natural numbers are well ordered. We also discuss the pigeonhole principle and some basic facts about equinumerosity, without introducing cardinal numbers.

We introduce some elementary concepts of combinatorics in terms of counting problems. We introduce the binomial and multinomial coefficients and study some of their properties and we conclude with the inclusion–exclusion principle.

Next, we introduce partial orders, well-founded sets, and complete induction. This way, students become aware of the fact that the induction principle applies to sets with an ordering far more complex that the ordering on the natural numbers. As an application, we prove the unique prime factorization in $\mathbb{Z}$ and discuss gcds and versions of the Euclidean algorithm to compute gcds including the so-called extended Euclidean algorithm which relates to the Bezout identity.

Another extremely important concept is that of an equivalence relation and the related notion of a partition.

As applications of the material on elementary number theory presented in Section 5.4, in Section 5.8 we give an introduction to Fibonacci and Lucas numbers as well as Mersenne numbers and in Sections 5.9, 5.10, and 5.11, we present some basics of public key cryptography and the RSA system. These sections contain some beautiful material and they should be viewed as an incentive for the reader to take a deeper look into the fascinating and mysterious world of prime numbers and more generally, number theory. This material is also a gold mine of programming assignments and of problems involving proofs by induction.

We have included some material on lattices, Tarski's fixed point theorem, distributive lattices, Boolean algebras, and Heyting algebras. These topics are somewhat more advanced and can be omitted from the "core".

The last topic that we consider crucial is graph theory. We give a fairly complete presentation of the basic concepts of graph theory: directed and undirected graphs, paths, cycles, spanning trees, cocycles, cotrees, flows, and tensions, Eulerian and Hamiltonian cycles, matchings, coverings, and planar graphs. We also discuss the network flow problem and prove the max-flow min-cut theorem in an original way due to M. Sakarovitch.

These notes grew out of lectures I gave in 2005 while teaching CIS260, Mathematical Foundations of Computer Science. There is more material than can be covered in one semester and some choices have to be made regarding what to omit. Unfortunately, when I taught this course, I was unable to cover any graph theory. I also did not cover lattices and Boolean algebras.

Beause the notion of a graph is so fundamental in computer science (and elsewhere), I have restructured these notes by splitting the material on graphs into two parts and by including the introductory part on graphs (Chapter 3) before the introduction to combinatorics (Chapter 4). The only small inconvenience in doing so is that this causes a forward reference to the notion of an equivalence relation which only appears in Chapter 5. This is not a serious problem. In fact, this gives us a chance to introduce the important concept of an equivalence relation early on, without any proof, and then to revisit this notion more rigorously later on.

Some readers may be disappointed by the absence of an introduction to probability theory. There is no question that probability theory plays a crucial role in computing, for example, in the design of randomized algorithms and in the probabilistic analysis of algorithms. Our feeling is that to do justice to the subject would require too much space. Unfortunately, omitting probability theory is one of the tough choices that we decided to make in order to keep the manuscript of manageable size. Fortunately, probability and its applications to computing are presented in a beautiful book by Mitzenmacher and Upfal [4] so we don't feel too bad about our decision to omit these topics.

There are quite a few books covering discrete mathematics. According to my personal taste, I feel that two books complement and extend the material presented here particularly well: *Discrete Mathematics*, by Lovász, Pelikán, and Vesztergombi [3], a very elegant text at a slightly higher level but still very accessible, and *Discrete Mathematics*, by Graham, Knuth, and Patashnik [2], a great book at a significantly higher level.

My unconventional approach of starting with logic may not work for everybody, as some individuals find such material too abstract. It is possible to skip the chapter on logic and proceed directly with sets, functions, and so on. I admit that I have raised the bar perhaps higher than the average compared to other books on discrete maths. However, my experience when teaching CIS260 was that 70% of the students enjoyed the logic material, as it reminded them of programming. I hope this book will inspire and will be useful to motivated students.

A final word to the teacher regarding foundational issues: I tried to show that there is a natural progression starting from logic, next a precise statement of the axioms of set theory, and then to basic objects such as the natural numbers, functions, graphs, trees, and the like. I tried to be as rigorous and honest as possible regarding some of the logical difficulties that one encounters along the way but I decided to avoid some of the most subtle issues, in particular a rigorous definition of the notion of cardinal number and a detailed discussion of the axiom of choice. Rather than giving a flawed definition of a cardinal in terms of the equivalence class of all sets equinumerous to a set, which *is not* a set, I only defined the notions of domination and equinumerosity. Also, I stated precisely two versions of the axiom of choice, one of which (the graph version) comes up naturally when seeking a right inverse to a surjection, but I did not attempt to state and prove the equivalence of this formulation with other formulations of the axiom of choice (such as Zermelo's well-ordering theorem). Such foundational issues are beyond the scope of this book; they belong to a course on set theory and are treated extensively in texts such as Enderton [1] and Suppes [5].

Acknowledgments: I would like to thank Mickey Brautbar, Kostas Daniilidis, Max Mintz, Joseph Pacheco, Steve Shatz, Jianbo Shi, Marcelo Siqueira, and Val Tannen for their advice, encouragement, and inspiration.

References

1. Herbert B. Enderton. *Elements of Set Theory*. New York: Academic Press, first edition, 1977.
2. Ronald L. Graham, Donald E. Knuth, and Oren Patashnik. *Concrete Mathematics: A Foundation For Computer Science*. Reading, MA: Addison Wesley, second edition, 1994.
3. L. Lovász, J. Pelikán, and K. Vesztergombi. *Discrete Mathematics. Elementary and Beyond*. Undergraduate Texts in Mathematics. New York: Springer, first edition, 2003.
4. Michael Mitzenmacher and Eli Upfal. *Probability and Computing. Randomized Algorithms and Probabilistic Analysis*. Cambridge, UK: Cambridge University Press, first edition, 2005.
5. Patrick Suppes. *Axiomatic Set Theory*. New York: Dover, first edition, 1972.
6. D. van Dalen. *Logic and Structure*. Universitext. New York: Springer Verlag, second edition, 1980.

Philadelphia, November 2010   *Jean Gallier*

Contents

Chapter 1
Mathematical Reasoning, Proof Principles, and Logic

1.1 Introduction

One of the main goals of this book is to learn how to

construct and read mathematical proofs.

Why?

1. Computer scientists and engineers write *programs* and build *systems*.
2. It is very important to have *rigorous methods* to check that these programs and systems behave as expected (are *correct*, have *no bugs*).
3. It is also important to have methods to *analyze the complexity* of programs (*time/space complexity*).

More generally, it is crucial to have a firm grasp of the *basic reasoning principles and rules of logic*. This leads to the question:

What is a proof.

There is no short answer to this question. However, it seems fair to say that a proof is some kind of *deduction (derivation)* that proceeds from a set of *hypotheses (premises, axioms)* in order to derive a *conclusion*, using some *logical rules*.

A first important observation is that there are different *degrees of formality* of proofs.

1. Proofs can be very *informal*, using a set of loosely defined logical rules, possibly omitting steps and premises.
2. Proofs can be *completely formal*, using a very clearly defined set of rules and premises. Such proofs are usually processed or produced by programs called *proof checkers* and *theorem provers*.

Thus, a human prover evolves in a *spectrum of formality*.

J. Gallier, *Discrete Mathematics*, Universitext,
DOI 10.1007/978-1-4419-8047-2_1, © Springer Science+Business Media, LLC 2011

It should be said that *it is practically impossible to write formal proofs.* This is because it would be extremely tedious and time-consuming to write such proofs and these proofs would be huge and thus, very hard to read.

In principle, it is possible to write formalized proofs and sometimes it is desirable to do so if we want to have absolute confidence in a proof. For example, we would like to be sure that a flight-control system is not buggy so that a plane does not accidentally crash, that a program running a nuclear reactor will not malfunction, or that nuclear missiles will not be fired as a result of a buggy "alarm system".

Thus, it is very important to develop tools to assist us in constructing formal proofs or checking that formal proofs are correct and such systems do exist (examples: Isabelle, COQ, TPS, NUPRL, PVS, Twelf). However, 99.99% of us will not have the time or energy to write formal proofs.

Even if we never write formal proofs, it is important to understand clearly what are the rules of reasoning that we use when we construct informal proofs.

The goal of this chapter is to try answering the question, "What is a proof." We do so by formalizing the basic rules of reasoning that we use, most of the time subconsciously, in a certain kind of formalism known as a *natural deduction system.* We give a (very) quick introduction to *mathematical logic,* with a very deliberate *proof-theoretic* bent, that is, neglecting almost completely all semantic notions, except at a very intuitive level. We still feel that this approach is fruitful because the mechanical and rules-of-the-game flavor of proof systems is much more easily grasped than semantic concepts. In this approach, we follow Peter Andrews' motto [1]:

"To truth through proof."

We present various natural deduction systems due to Prawitz and Gentzen (in more modern notation), both in their intuitionistic and classical version. The adoption of natural deduction systems as proof systems makes it easy to question the validity of some of the inference rules, such as the *principle of proof by contradiction.* In brief, we try to explain to our readers the difference between *constructive* and *classical* (i.e., not necessarily constructive) proofs. In this respect, we plant the seed that there is a deep relationship between *constructive proofs* and the notion of *computation* (the "Curry–Howard isomorphism" or "formulae-as-types principle," see Section 1.11 and Howard [13]).

1.2 Inference Rules, Deductions, The Proof Systems $\mathcal{N}_m^{\Rightarrow}$ and $\mathcal{NG}_m^{\Rightarrow}$

In this section, we review some basic proof principles and attempt to clarify, at least informally, what constitutes a mathematical proof.

In order to define the notion of proof rigorously, we would have to define a formal language in which to express statements very precisely and we would have to set up a proof system in terms of axioms and proof rules (also called inference rules). We do not go into this as this would take too much time. Instead, we content ourselves with an intuitive idea of what a statement is and focus on stating as precisely as possible the rules of logic that are used in constructing proofs. Readers who really want to see a thorough (and rigorous) introduction to logic are referred to Gallier [4], van Dalen [23], or Huth and Ryan [14], a nice text with a computer science flavor. A beautiful exposition of logic (from a proof-theoretic point of view) is also given in Troelstra and Schwichtenberg [22], but at a more advanced level. Frank Pfenning has also written an excellent and more extensive introduction to constructive logic. This is available on the Web at

http://www.andrew.cmu.edu/course/15-317/handouts/logic.pdf

By the way, my book has been out of print for some time but you can get it free (as pdf files) from my logic website

http://www.cis.upenn.edu/˜jean/gbooks/logic.html

We also highly recommend the beautifully written little book by Timothy Gowers (Fields Medalist, 1998) [10] which, among other things, discusses the notion of proof in mathematics (as well as the necessity of formalizing proofs without going overboard).

In mathematics, we **prove statements.** Statements may be *atomic* or *compound*, that is, built up from simpler statements using *logical connectives*, such as *implication* (if–then), *conjunction* (and), *disjunction* (or), *negation* (not), and (existential or universal) *quantifiers*.

As examples of atomic statements, we have:

1. "A student is eager to learn."
2. "A students wants an A."
3. "An odd integer is never 0."
4. "The product of two odd integers is odd."

Atomic statements may also contain "variables" (standing for arbitrary objects). For example

1. human(x): "x is a human."
2. needs-to-drink(x): "x" needs to drink.

An example of a compound statement is

$$\text{human}(x) \Rightarrow \text{needs-to-drink}(x).$$

In the above statement, $\Rightarrow$ is the symbol used for logical implication. If we want to assert that every human needs to drink, we can write

$$\forall x(\text{human}(x) \Rightarrow \text{needs-to-drink}(x));$$

This is read: "For every x, if x is a human then x needs to drink."

If we want to assert that some human needs to drink we write

$$\exists x(\text{human}(x) \Rightarrow \text{needs-to-drink}(x));$$

This is read: "There is some x such that, if x is a human then x needs to drink."

We often denote statements (also called *propositions* or *(logical) formulae*) using letters, such as A, B, P, Q, and so on, typically upper-case letters (but sometimes Greek letters, φ, ψ, etc.).

If P and Q are statements, then their *conjunction* is denoted $P \wedge Q$ (say, P and Q), their *disjunction* denoted $P \vee Q$ (say, P or Q), their *implication* $P \Rightarrow Q$ or $P \supset Q$ (say, if P then Q). Some authors use the symbol $\rightarrow$ and write an implication as $P \rightarrow Q$. We do not like to use this notation because the symbol $\rightarrow$ is already used in the notation for functions ($f \colon A \rightarrow B$). We mostly use the symbol $\Rightarrow$.

We also have the atomic statements $\bot$ (*falsity*), which corresponds to **false** (think of it as the statement that is false no matter what), and the atomic statement $\top$ (*truth*), which corresponds to **true** (think of it as the statement that is always true). The constant $\bot$ is also called *falsum* or *absurdum*. It is a formalization of the notion of *absurdity* (a state in which contradictory facts hold). Then, it is convenient to define the *negation* of P as $P \Rightarrow \bot$ and to abbreviate it as $\neg P$ (or sometimes $\sim P$). Thus, $\neg P$ (say, not P) is just a shorthand for $P \Rightarrow \bot$.

This interpretation of negation may be confusing at first. The intuitive idea is that $\neg P = (P \Rightarrow \bot)$ is true if and only if P is not true because if both P and $P \Rightarrow \bot$ were true then we could conclude that $\bot$ is true, an absurdity, and if both P and $P \Rightarrow \bot$ were false then P would have to be both true and false, again, an absurdity. Actually, because we don't know what truth is, it is "safer" (and more constructive) to say that $\neg P$ is provable iff for every proof of P we can derive a contradiction (namely, $\bot$ is provable). In particular, P should not be provable. For example, $\neg(Q \wedge \neg Q)$ is provable (as we show later, because any proof of $Q \wedge \neg Q$ yields a proof of $\bot$). However, the fact that a proposition P is **not** provable does not imply that $\neg P$ **is** provable. There are plenty of propositions such that both P and $\neg P$ are not provable, such as $Q \Rightarrow R$, where Q and R are two unrelated propositions (with no common symbols).

Whenever necessary to avoid ambiguities, we add matching parentheses: $(P \wedge Q)$, $(P \vee Q)$, $(P \Rightarrow Q)$. For example, $P \vee Q \wedge R$ is ambiguous; it means either $(P \vee (Q \wedge R))$ or $((P \vee Q) \wedge R)$.

Another important logical operator is *equivalence*. If P and Q are statements, then their *equivalence*, denoted $P \equiv Q$ (or $P \Longleftrightarrow Q$), is an abbreviation for $(P \Rightarrow Q) \wedge (Q \Rightarrow P)$. We often say "$P$ if and only if Q" or even "P iff Q" for $P \equiv Q$. As we show shortly, to prove a logical equivalence $P \equiv Q$, we have to prove **both** implications $P \Rightarrow Q$ and $Q \Rightarrow P$.

An implication $P \Rightarrow Q$ should be understood as an if–then statement; that is, if P is true then Q is also true. A better interpretation is that any proof of $P \Rightarrow Q$ can be used to construct a proof of Q given any proof of P. As a consequence of this interpretation, we show later that if $\neg P$ is provable, then $P \Rightarrow Q$ is also provable (instantly) whether or not Q is provable. In such a situation, we often say that $P \Rightarrow Q$ is *vacuously provable*. For example, $(P \wedge \neg P) \Rightarrow Q$ is provable for any arbitrary Q (because if we assume that $P \wedge \neg P$ is provable, then we derive a contradiction, and then another rule of logic tells us that any proposition whatsoever is provable. However, we have to wait until Section 1.3 to see this).

Of course, there are problems with the above paragraph. What does truth have to do with all this? What do we mean when we say, "*P* is true"? What is the relationship between truth and provability?

These are actually deep (and tricky) questions whose answers are not so obvious. One of the major roles of logic is to clarify the notion of truth and its relationship to provability. We avoid these fundamental issues by dealing exclusively with the notion of proof. So, the big question is: what is a proof.

Typically, the statements that we prove depend on some set of *hypotheses*, also called *premises* (or *assumptions*). As we show shortly, this amounts to proving implications of the form

$$(P_1 \wedge P_2 \wedge \cdots \wedge P_n) \Rightarrow Q.$$

However, there are certain advantages in defining the notion of *proof* (or *deduction*) of a proposition from a set of premises. Sets of premises are usually denoted using upper-case Greek letters such as Γ or Δ.

Roughly speaking, a *deduction* of a proposition Q from a set of premises Γ is a finite labeled tree whose root is labeled with Q (the *conclusion*), whose leaves are labeled with premises from Γ (possibly with multiple occurrences), and such that every interior node corresponds to a given set of *proof rules* (or *inference rules*). Certain simple deduction trees are declared as obvious proofs, also called *axioms*.

There are many kinds of proof systems: Hilbert-style systems, natural-deduction systems, Gentzen sequents systems, and so on. We describe a so-called *natural deduction system* invented by G. Gentzen in the early 1930s (and thoroughly investigated by D. Prawitz in the mid 1960s).

Fig. 1.1 David Hilbert, 1862–1943 (left and middle), Gerhard Gentzen, 1909–1945 (middle right), and Dag Prawitz, 1936– (right)

The major advantage of this system is that it captures quite nicely the "natural" rules of reasoning that one uses when proving mathematical statements. This does not mean that it is easy to find proofs in such a system or that this system is indeed very intuitive. We begin with the inference rules for implication and first consider the following question.

How do we proceed to prove an implication, $A \Rightarrow B$?

The rule, called $\Rightarrow$-*intro*, is: *assume that A has already been proven and then prove B, making as many uses of A as needed.*

Let us give a simple example. The *odd numbers* are the numbers

$$1, 3, 5, 7, 9, 11, 13, \ldots.$$

Equivalently, a whole number n is odd iff it is of the form $2k + 1$, where $k = 0, 1, 2, 3, 4, 5, 6, \ldots$. Let us denote the fact that a number n is odd by $\mathrm{odd}(n)$. We would like to prove the implication

$$\mathrm{odd}(n) \Rightarrow \mathrm{odd}(n + 2).$$

Following the rule $\Rightarrow$-intro, we assume $\mathrm{odd}(n)$ (which means that we take as proven the fact that n is odd) and we try to conclude that $n + 2$ must be odd. However, to say that n is odd is to say that $n = 2k + 1$ for some whole number k. Now,

$$n + 2 = 2k + 1 + 2 = 2(k + 1) + 1,$$

which means that $n + 2$ is odd. (Here, $n = 2h + 1$, with $h = k + 1$, and $k + 1$ is a whole number because k is.)

Therefore, we proved that *if we assume* $\mathrm{odd}(n)$*, then we can conclude* $\mathrm{odd}(n + 2)$, and according to our rule for proving implications, we have indeed proved the proposition

$$\mathrm{odd}(n) \Rightarrow \mathrm{odd}(n + 2).$$

Note that the effect of rule $\Rightarrow$-intro is to *introduce* the premise $\mathrm{odd}(n)$, which was temporarily assumed, into the left-hand side (we also say *antecedent*) of the proposition $\mathrm{odd}(n) \Rightarrow \mathrm{odd}(n + 2)$. This is why this rule is called the *implication introduction*.

It should be noted that the above proof of the proposition $\mathrm{odd}(n) \Rightarrow \mathrm{odd}(n + 2)$ *does not depend* on any premises (other than the implicit fact that we are assuming n is a whole number). In particular, this proof does not depend on the premise, $\mathrm{odd}(n)$, which was assumed (became "active") during our subproof step. Thus, after having applied the rule $\Rightarrow$-intro, we should really make sure that the premise $\mathrm{odd}(n)$ which was made temporarily active is deactivated, or as we say, *discharged*. When we write informal proofs, we rarely (if ever) explicitly discharge premises when we apply the rule $\Rightarrow$-intro but if we want to be rigorous we really should.

For a second example, we wish to prove the proposition $P \Rightarrow (Q \Rightarrow P)$.

According to our rule, we assume P as a premise and we try to prove $Q \Rightarrow P$ assuming P. In order to prove $Q \Rightarrow P$, we assume Q as a new premise so the set of premises becomes $\{P, Q\}$, and then we try to prove P from P and Q. This time, it should be obvious that P is provable because we assumed both P and Q.

Indeed, the rule that P is always provable from any set of assumptions including P itself is one of the basic *axioms* of our logic (which means that it is a rule that requires no justification whatsover). So, we have obtained a proof of $P \Rightarrow (Q \Rightarrow P)$.

What is not entirely satisfactory about the above "proof" of $P \Rightarrow (Q \Rightarrow P)$ is that when the proof ends, the premises P and Q are still hanging around as "open" assumptions. However, a proof should not depend on any "open" assumptions and to rectify this problem we introduce a mechanism of "discharging" or "closing" premises, as we already suggested in our previous example.

What this means is that certain rules of our logic are required to discard (the usual terminology is "discharge") certain occurrences of premises so that the resulting proof does not depend on these premises.

Technically, there are various ways of implementing the discharging mechanism but they all involve some form of tagging (with a "new" variable). For example, the rule formalizing the process that we have just described to prove an implication, $A \Rightarrow B$, known as $\Rightarrow$-*introduction*, uses a tagging mechanism described precisely in Definition 1.1.

Now, the rule that we have just described is not sufficient to prove certain propositions that should be considered provable under the "standard" intuitive meaning of implication. For example, after a moment of thought, I think most people would want the proposition $P \Rightarrow ((P \Rightarrow Q) \Rightarrow Q)$ to be provable. If we follow the procedure that we have advocated, we assume both P and $P \Rightarrow Q$ and we try to prove Q. For this, we need a new rule, namely:

If P and $P \Rightarrow Q$ are both provable, then Q is provable.

The above rule is known as the $\Rightarrow$-*elimination rule* (or *modus ponens*) and it is formalized in tree-form in Definition 1.1.

We now make the above rules precise and for this, we represent proofs and deductions as certain kinds of trees and view the logical rules (inference rules) as tree-building rules. In the definition below, the expression Γ, P stands for the union of the multiset Γ and P. So, P may already belong to Γ. A picture such as

$$\Delta$$
$$\mathcal{D}$$
$$P$$

represents a deduction tree $\mathcal{D}$ whose root is labeled with P and whose leaves are labeled with propositions from the *multiset* Δ (a set possibly with multiple occurrences of its members). Some of the propositions in Δ may be tagged by variables. The list of untagged propositions in Δ is the list of *premises* of the deduction tree. We often use an abbreviated version of the above notation where we omit the deduction $\mathcal{D}$, and simply write

$$\Delta$$
$$P$$

For example, in the deduction tree below,

$$
\cfrac{\cfrac{P \Rightarrow (R \Rightarrow S) \qquad P}{R \Rightarrow S} \qquad \cfrac{Q \Rightarrow R \qquad \cfrac{P \Rightarrow Q \qquad P}{Q}}{R}}{S}
$$

no leaf is tagged, so the premises form the multiset

$$\Delta = \{P \Rightarrow (R \Rightarrow S), P, Q \Rightarrow R, P \Rightarrow Q, P\},$$

with two occurrences of P, and the conclusion is S.

As we saw in our earlier example, certain inferences rules have the effect that some of the original premises may be discarded; the traditional jargon is that some premises may be *discharged* (or *closed*). This is the case for the inference rule whose conclusion is an implication. When one or several occurrences of some proposition P are discharged by an inference rule, these occurrences (which label some leaves) are tagged with some new variable not already appearing in the deduction tree. If x is a new tag, the tagged occurrences of P are denoted P^x and we indicate the fact that premises were discharged by that inference by writing x immediately to the right of the inference bar. For example,

$$\frac{\dfrac{P^x, Q}{Q}}{P \Rightarrow Q} \quad x$$

is a deduction tree in which the premise P is discharged by the inference rule. This deduction tree only has Q as a premise, inasmuch as P is discharged.

What is the meaning of the horizontal bars? Actually, nothing really. Here, we are victims of an old habit in logic. Observe that there is always a single proposition immediately under a bar but there may be several propositions immediately above a bar. The intended meaning of the bar is that the proposition below it is obtained as the result of applying an inference rule to the propositions above it. For example, in

$$\frac{Q \Rightarrow R \qquad Q}{R}$$

the proposition R is the result of applying the $\Rightarrow$-elimination rule (see Definition 1.1 below) to the two premises $Q \Rightarrow R$ and Q. Thus, the use of the bar is just a convention used by logicians going back at least to the 1900s. Removing the bar everywhere would not change anything in our trees, except perhaps reduce their readability. Most logic books draw proof trees using bars to indicate inferences, therefore we also use bars in depicting our proof trees.

Because propositions do not arise from the vacuum but instead are built up from a set of atomic propositions using logical connectives (here, $\Rightarrow$), we assume the existence of an "official set of atomic propositions," or set of *propositional symbols*, $\mathbf{PS} = \{\mathbf{P}_1, \mathbf{P}_2, \mathbf{P}_3, \ldots\}$. So, for example, $\mathbf{P}_1 \Rightarrow \mathbf{P}_2$ and $\mathbf{P}_1 \Rightarrow (\mathbf{P}_2 \Rightarrow \mathbf{P}_1)$ are propositions. Typically, we use upper-case letters such as P, Q, R, S, A, B, C, and so on, to denote arbitrary propositions formed using atoms from $\mathbf{PS}$.

Definition 1.1. The axioms, inference rules, and deduction trees for *implicational logic* are defined as follows.

Axioms.

(i) Every one-node tree labeled with a single proposition P is a deduction tree for P with set of premises $\{P\}$.

(ii) The tree

$$\frac{\Gamma, P}{P}$$

is a deduction tree for P with multiset set of premises, $\Gamma \cup \{P\}$.

The above is a concise way of denoting a two-node tree with its leaf labeled with the multiset consisting of P and the propositions in Γ, each of these propositions (including P) having possibly multiple occurrences but at least one, and whose root is labeled with P. A more explicit form is

$$\frac{\overbrace{P_1, \cdots, P_1}^{k_1}, \cdots, \overbrace{P_i, \cdots, P_i}^{k_i}, \cdots, \overbrace{P_n, \cdots, P_n}^{k_n}}{P_i},$$

where $k_1, \ldots, k_n \geq 1$ and $n \geq 1$. This axiom says that we always have a deduction of P_i from any set of premises including P_i.

The $\Rightarrow$-*introduction rule*.

If $\mathcal{D}$ is a deduction tree for Q from the premises in $\Gamma \cup \{P\}$, then

$$\frac{\begin{array}{c} \Gamma, P^x \\ \mathcal{D} \\ Q \end{array}}{P \Rightarrow Q} \; x$$

is a deduction tree for $P \Rightarrow Q$ from Γ.

Note that this inference rule has the additional effect of discharging some occurrences of the premise P. These occurrences are tagged with a new variable x, and the tag x is also placed immediately to the right of the inference bar. This is a reminder that the deduction tree whose conclusion is $P \Rightarrow Q$ no longer has the occurrences of P labeled with x as premises.

The $\Rightarrow$-*elimination rule*.

If $\mathcal{D}_1$ is a deduction tree for $P \Rightarrow Q$ from the premises Γ and $\mathcal{D}_2$ is a deduction for P from the premises Δ, then

$$\frac{\begin{array}{cc} \begin{array}{c} \Gamma \\ \mathcal{D}_1 \\ P \Rightarrow Q \end{array} & \begin{array}{c} \Delta \\ \mathcal{D}_2 \\ P \end{array} \end{array}}{Q}$$

is a deduction tree for Q from the premises in $\Gamma \cup \Delta$. This rule is also known as *modus ponens*.

In the above axioms and rules, Γ or Δ may be empty; P, Q denote arbitrary propositions built up from the atoms in **PS**; and $\mathcal{D}, \mathcal{D}_1$, and $\mathcal{D}_2$ denote deductions, possibly a one-node tree.

A *deduction tree* is either a one-node tree labeled with a single proposition or a tree constructed using the above axioms and rules. A *proof tree* is a deduction tree such that *all its premises are discharged*. The above proof system is denoted $\mathcal{N}_m^{\Rightarrow}$ (here, the subscript m stands for *minimal*, referring to the fact that this a bare-bones logical system).

Observe that a proof tree has at least two nodes. A proof tree Π for a proposition P may be denoted

$$\Pi$$
$$P$$

with an empty set of premises (we don't display $\emptyset$ on top of Π). We tend to denote deductions by the letter $\mathcal{D}$ and proof trees by the letter Π, possibly subscripted.

In words, the $\Rightarrow$-introduction rule says that in order to prove an implication $P \Rightarrow Q$ from a set of premises Γ, we assume that P has already been proved, add P to the premises in Γ, and then prove Q from Γ and P. Once this is done, the premise P is deleted.

This rule formalizes the kind of reasoning that we all perform whenever we prove an implication statement. In that sense, it is a natural and familiar rule, except that we perhaps never stopped to think about what we are really doing. However, the business about discharging the premise P when we are through with our argument is a bit puzzling. Most people probably never carry out this "discharge step" consciously, but such a process does take place implicitly.

It might help to view the action of proving an implication $P \Rightarrow Q$ as the construction of a program that converts a proof of P into a proof of Q. Then, if we supply a proof of P as input to this program (the proof of $P \Rightarrow Q$), it will output a proof of Q. So, if we don't give the right kind of input to this program, for example, a "wrong proof" of P, we should not expect the program to return a proof of Q. However, this does not say that the program is incorrect; the program was designed to do the right thing only if it is given the right kind of input. From this functional point of view (also called constructive), if we take the simplistic view that P and Q assume the truth values **true** and **false**, we should not be shocked that if we give as input the value **false** (for P), then the truth value of the whole implication $P \Rightarrow Q$ is **true**. The program $P \Rightarrow Q$ is designed to produce the output value **true** (for Q) if it is given the input value **true** (for P). So, this program only goes wrong when, given the input **true** (for P), it returns the value **false** (for Q). In this erroneous case, $P \Rightarrow Q$ should indeed receive the value **false**. However, in all other cases, the program works correctly, even if it is given the wrong input (**false** for P).

For a concrete example, say P stands for the statement,
"Our candidate for president wins in Pennsylvania"
and Q stands for
"Our candidate is elected president."

Then, $P \Rightarrow Q$, asserts that *if* our candidate for president wins in Pennsylvania *then* our candidate is elected president.

If $P \Rightarrow Q$ holds, then if indeed our candidate for president wins in Pennsylvania then for sure our candidate will win the presidential election. However, if our candidate does not win in Pennsylvania, we can't predict what will happen. Our candidate may still win the presidential election but he may not.

If our candidate president does not win in Pennsylvania, our prediction is not proven false. In this case, the statement $P \Rightarrow Q$ should be regarded as holding, though perhaps uninteresting.

For one more example, let $\text{odd}(n)$ assert that n is an odd natural number and let $Q(n, a, b)$ assert that $a^n + b^n$ is divisible by $a + b$, where a, b are any given natural numbers. By divisible, we mean that we can find some natural number c, so that

$$a^n + b^n = (a + b)c.$$

Then, we claim that the implication $\text{odd}(n) \Rightarrow Q(n, a, b)$ is provable.

As usual, let us assume $\text{odd}(n)$, so that $n = 2k + 1$, where $k = 0, 1, 2, 3, \ldots$. But then, we can easily check that

$$a^{2k+1} + b^{2k+1} = (a + b) \left(\sum_{i=0}^{2k} (-1)^i a^{2k-i} b^i \right),$$

which shows that $a^{2k+1} + b^{2k+1}$ is divisible by $a + b$. Therefore, we proved the implication $\text{odd}(n) \Rightarrow Q(n, a, b)$.

If n is not odd, then the implication $\text{odd}(n) \Rightarrow Q(n, a, b)$ yields no information about the provablity of the statement $Q(n, a, b)$, and that is fine. Indeed, if n is even and $n \geq 2$, then in general, $a^n + b^n$ is not divisible by $a + b$, but this may happen for some special values of n, a, and b, for example: $n = 2$, $a = 2$, $b = 2$.

1. Only the leaves of a deduction tree may be discharged. Interior nodes, including the root, are *never* discharged.
2. Once a set of leaves labeled with some premise P marked with the label x has been discharged, none of these leaves can be discharged again. So, each label (say x) can only be used once. This corresponds to the fact that some leaves of our deduction trees get "killed off" (discharged).
3. A proof is a deduction tree whose leaves are *all discharged* (Γ is empty). This corresponds to the philosophy that if a proposition has been proved, then the validity of the proof should not depend on any assumptions that are still active. We may think of a deduction tree as an unfinished proof tree.
4. When constructing a proof tree, we have to be careful not to include (accidentally) extra premises that end up not being discharged. If this happens, we probably made a mistake and the redundant premises should be deleted. On the other hand, if we have a proof tree, we can always add extra premises to the leaves and create a new proof tree from the previous one by discharging all the new premises.
5. Beware, when we deduce that an implication $P \Rightarrow Q$ is provable, we **do not** prove that P **and** Q are provable; we only prove that **if** P is provable **then** Q is provable.

The $\Rightarrow$-elimination rule formalizes the use of *auxiliary lemmas*, a mechanism that we use all the time in making mathematical proofs. Think of $P \Rightarrow Q$ as a lemma that has already been established and belongs to some database of (useful) lemmas. This lemma says if I can prove P then I can prove Q. Now, suppose that we manage to give a proof of P. It follows from the $\Rightarrow$-elimination rule that Q is also provable.

Observe that in an introduction rule, the conclusion contains the logical connective associated with the rule, in this case, $\Rightarrow$; this justifies the terminology "introduction". On the other hand, in an elimination rule, the logical connective associated with the rule is gone (although it may still appear in Q). The other inference rules for $\wedge$, $\vee$, and the like, follow this pattern of introduction and elimination.

Examples of proof trees.

(a)

$$\cfrac{\cfrac{P^x}{P}}{P \Rightarrow P}\ x$$

So, $P \Rightarrow P$ is provable; this is the least we should expect from our proof system! Note that

$$\cfrac{P^x}{P \Rightarrow P}\ x$$

is also a valid proof tree for $P \Rightarrow P$, because the one-node tree labeled with P^x is a deduction tree.

(b)

$$\cfrac{\cfrac{\cfrac{(Q \Rightarrow R)^y \quad \cfrac{(P \Rightarrow Q)^z \quad P^x}{Q}}{\cfrac{R}{P \Rightarrow R}\ x}}{(Q \Rightarrow R) \Rightarrow (P \Rightarrow R)}\ y}{(P \Rightarrow Q) \Rightarrow ((Q \Rightarrow R) \Rightarrow (P \Rightarrow R))}\ z$$

In order to better appreciate the difference between a deduction tree and a proof tree, consider the following two examples.

1. The tree below is a deduction tree, beause two of its leaves are labeled with the premises $P \Rightarrow Q$ and $Q \Rightarrow R$, that have not been discharged yet. So, this tree represents a deduction of $P \Rightarrow R$ from the set of premises $\Gamma = \{P \Rightarrow Q, Q \Rightarrow R\}$ but it is *not a proof tree* because $\Gamma \neq \emptyset$. However, observe that the original premise P, labeled x, has been discharged.

$$\cfrac{\cfrac{Q \Rightarrow R \quad \cfrac{P \Rightarrow Q \quad P^x}{Q}}{R}}{P \Rightarrow R}\ x$$

2. The next tree was obtained from the previous one by applying the $\Rightarrow$-introduction rule which triggered the discharge of the premise $Q \Rightarrow R$ labeled y, which is no longer active. However, the premise $P \Rightarrow Q$ is still active (has not been discharged yet), so the tree below is a deduction tree of $(Q \Rightarrow R) \Rightarrow (P \Rightarrow R)$ from the set of premises $\Gamma = \{P \Rightarrow Q\}$. It is not yet a proof tree inasmuch as $\Gamma \neq \emptyset$.

$$\cfrac{\cfrac{(Q \Rightarrow R)^y \qquad \cfrac{P \Rightarrow Q \qquad P^x}{Q}}{\cfrac{R}{P \Rightarrow R}\; x}{(Q \Rightarrow R) \Rightarrow (P \Rightarrow R)}\; y$$

Finally, one more application of the $\Rightarrow$-introduction rule discharged the premise $P \Rightarrow Q$, at last, yielding the proof tree in (b).

(c) This example illustrates the fact that different proof trees may arise from the same set of premises $\{P, Q\}$: for example,

$$\cfrac{\cfrac{\cfrac{P^x, Q^y}{P}}{P \Rightarrow P}\; x}{Q \Rightarrow (P \Rightarrow P)}\; y$$

and

$$\cfrac{\cfrac{\cfrac{P^x, Q^y}{P}}{Q \Rightarrow P}\; y}{P \Rightarrow (Q \Rightarrow P)}\; x$$

Similarly, there are six proof trees with a conclusion of the form

$$A \Rightarrow (B \Rightarrow (C \Rightarrow P))$$

begining with the deduction

$$\frac{P^x, Q^y, R^z}{P}$$

corresponding to the six permutations of the premises P, Q, R.

Note that we would not have been able to construct the above proofs if Axiom (ii),

$$\frac{\Gamma, P}{P} \quad ,$$

were not available. We need a mechanism to "stuff" more premises into the leaves of our deduction trees in order to be able to discharge them later on. We may also view Axiom (ii) as a *weakening rule* whose purpose is to weaken a set of assumptions. Even though we are assuming all of the proposition in Γ and P, we only retain the assumption P. The necessity of allowing multisets of premises is illustrated by the following proof of the proposition $P \Rightarrow (P \Rightarrow (Q \Rightarrow (Q \Rightarrow (P \Rightarrow P))))$:

$$
\cfrac{\cfrac{\cfrac{\cfrac{\cfrac{\cfrac{P^u, P^v, P^y, Q^w, Q^x}{P}}{P \Rightarrow P} \; y}{Q \Rightarrow (P \Rightarrow P)} \; x}{Q \Rightarrow (Q \Rightarrow (P \Rightarrow P))} \; w}{P \Rightarrow (Q \Rightarrow (Q \Rightarrow (P \Rightarrow P)))} \; v}{P \Rightarrow (P \Rightarrow (Q \Rightarrow (Q \Rightarrow (P \Rightarrow P))))} \; u
$$

(d) In the next example, the two occurrences of A labeled x are discharged simultaneously.

$$
\cfrac{\cfrac{\cfrac{\cfrac{\cfrac{(A \Rightarrow (B \Rightarrow C))^z \quad A^x}{B \Rightarrow C} \qquad \cfrac{(A \Rightarrow B)^y \quad A^x}{B}}{C}}{A \Rightarrow C} \; x}{(A \Rightarrow B) \Rightarrow (A \Rightarrow C)} \; y}{\big(A \Rightarrow (B \Rightarrow C)\big) \Rightarrow \big((A \Rightarrow B) \Rightarrow (A \Rightarrow C)\big)} \; z
$$

(e) In contrast to Example (d), in the proof tree below the two occurrences of A are discharged separately. To this effect, they are labeled differently.

$$
\cfrac{\cfrac{\cfrac{\cfrac{\cfrac{\cfrac{(A \Rightarrow (B \Rightarrow C))^z \quad A^x}{B \Rightarrow C} \qquad \cfrac{(A \Rightarrow B)^y \quad A^t}{B}}{C}}{A \Rightarrow C} \; x}{(A \Rightarrow B) \Rightarrow (A \Rightarrow C)} \; y}{\big(A \Rightarrow (B \Rightarrow C)\big) \Rightarrow \big((A \Rightarrow B) \Rightarrow (A \Rightarrow C)\big)} \; z}{A \Rightarrow \Big(\big(A \Rightarrow (B \Rightarrow C)\big) \Rightarrow \big((A \Rightarrow B) \Rightarrow (A \Rightarrow C)\big)\Big)} \; t
$$

Remark: How do we find these proof trees? Well, we could try to enumerate all possible proof trees systematically and see if a proof of the desired conclusion turns up. Obviously, this is a very inefficient procedure and moreover, how do we know that all possible proof trees will be generated and how do we know that such a method will terminate after a finite number of steps (what if the proposition proposed as a conclusion of a proof is not provable)?

Finding an algorithm to decide whether a proposition is provable is a very difficult problem and, for sets of propositions with enough "expressive power" (such as propositions involving first-order quantifiers), it can be shown that there is **no** procedure that will give an answer in all cases and terminate in a finite number of steps for all possible input propositions. We come back to this point in Section 1.11. However, for the system $\mathcal{N}_m^{\Rightarrow}$, such a procedure exists but it is not easy to prove that it terminates in all cases and in fact, it can take a very long time.

What we did, and we strongly advise our readers to try it when they attempt to construct proof trees, is to construct the proof tree from the bottom up, starting from the proposition labeling the root, rather than top-down, that is, starting from the leaves. During this process, whenever we are trying to prove a proposition $P \Rightarrow Q$, we use the $\Rightarrow$-introduction rule backward, that is, we add P to the set of active premises and we try to prove Q from this new set of premises. At some point, we get stuck with an atomic proposition, say R. Call the resulting deduction $\mathcal{D}_{bu}$; note that R is the only active (undischarged) premise of $\mathcal{D}_{bu}$ and the node labeled R immediately below it plays a special role; we call it the special node of $\mathcal{D}_{bu}$. The trick is now to switch strategy and start building a proof tree top-down, starting from the leaves, using the $\Rightarrow$-elimination rule. If everything works out well, we get a deduction with root R, say $\mathcal{D}_{td}$, and then we glue this deduction $\mathcal{D}_{td}$ to the deduction $\mathcal{D}_{bu}$ in such a way that the root of $\mathcal{D}_{td}$ is identified with the special node of $\mathcal{D}_{bu}$ labeled R. We also have to make sure that all the discharged premises are linked to the correct instance of the $\Rightarrow$-introduction rule that caused them to be discharged. One of the difficulties is that during the bottom-up process, we don't know how many copies of a premise need to be discharged in a single step. We only find out how many copies of a premise need to be discharged during the top-down process.

Here is an illustration of this method for Example (d). At the end of the bottom-up process, we get the deduction tree $\mathcal{D}_{bu}$:

$$\frac{\dfrac{\dfrac{(A \Rightarrow (B \Rightarrow C))^z \quad (A \Rightarrow B)^y \quad A^x \quad C}{C}}{A \Rightarrow C}\ x}{(A \Rightarrow B) \Rightarrow (A \Rightarrow C)}\ y$$
$$(A \Rightarrow (B \Rightarrow C)) \Rightarrow ((A \Rightarrow B) \Rightarrow (A \Rightarrow C))\ z$$

At the end of the top-down process, we get the deduction tree $\mathcal{D}_{td}$:

$$\frac{\dfrac{A \Rightarrow (B \Rightarrow C) \quad A}{B \Rightarrow C} \quad \dfrac{A \Rightarrow B \quad A}{B}}{C}$$

Finally, after gluing $\mathcal{D}_{td}$ on top of $\mathcal{D}_{bu}$ (which has the correct number of premises to be discharged), we get our proof tree:

$$\cfrac{\cfrac{\cfrac{(A \Rightarrow (B \Rightarrow C))^z \qquad A^x}{B \Rightarrow C} \qquad \cfrac{(A \Rightarrow B)^y \qquad A^x}{B}}{\cfrac{C}{A \Rightarrow C} \; x}}{\cfrac{(A \Rightarrow B) \Rightarrow (A \Rightarrow C)}{\big(A \Rightarrow (B \Rightarrow C)\big) \Rightarrow \big((A \Rightarrow B) \Rightarrow (A \Rightarrow C)\big)} \; z} \; y$$

Let us return to the functional interpretation of implication by giving an example. The proposition $P \Rightarrow ((P \Rightarrow Q) \Rightarrow Q)$ has the following proof.

$$\cfrac{\cfrac{\cfrac{(P \Rightarrow Q)^x \qquad P^y}{Q}}{(P \Rightarrow Q) \Rightarrow Q} \; x}{P \Rightarrow ((P \Rightarrow Q) \Rightarrow Q)} \; y$$

Now, say P is the proposition $R \Rightarrow R$, which has the proof

$$\cfrac{\cfrac{R^z}{R}}{R \Rightarrow R} \; z$$

Using $\Rightarrow$-elimination, we obtain a proof of $((R \Rightarrow R) \Rightarrow Q) \Rightarrow Q$ from the proof of $(R \Rightarrow R) \Rightarrow (((R \Rightarrow R) \Rightarrow Q) \Rightarrow Q)$ and the proof of $R \Rightarrow R$:

$$\cfrac{\cfrac{\cfrac{((R \Rightarrow R) \Rightarrow Q)^x \qquad (R \Rightarrow R)^y}{Q}}{((R \Rightarrow R) \Rightarrow Q) \Rightarrow Q} \; x \qquad}{(R \Rightarrow R) \Rightarrow (((R \Rightarrow R) \Rightarrow Q) \Rightarrow Q)} \quad \cfrac{\cfrac{R^z}{R}}{R \Rightarrow R} \; z$$
$$\overline{\qquad\qquad ((R \Rightarrow R) \Rightarrow Q) \Rightarrow Q \qquad\qquad}$$

Note that the above proof is redundant. A more direct proof can be obtained as follows. Undo the last $\Rightarrow$-introduction in the proof of $(R \Rightarrow R) \Rightarrow (((R \Rightarrow R) \Rightarrow Q) \Rightarrow Q)$:

$$\cfrac{\cfrac{((R \Rightarrow R) \Rightarrow Q)^x \qquad R \Rightarrow R}{Q}}{((R \Rightarrow R) \Rightarrow Q) \Rightarrow Q} \; x$$

and then glue the proof of $R \Rightarrow R$ on top of the leaf $R \Rightarrow R$, obtaining the desired proof of $((R \Rightarrow R) \Rightarrow Q) \Rightarrow Q$:

$$
\cfrac{
  ((R \Rightarrow R) \Rightarrow Q)^x \qquad \cfrac{\cfrac{R^z}{R}}{R \Rightarrow R} \; z
}{
  \cfrac{Q}{((R \Rightarrow R) \Rightarrow Q) \Rightarrow Q} \; x
}
$$

In general, one has to exercise care with the label variables. It may be necessary to rename some of these variables to avoid clashes. What we have above is an example of *proof substitution* also called *proof normalization*. We come back to this topic in Section 1.11.

The process of discharging premises when constructing a deduction is admittedly a bit confusing. Part of the problem is that a deduction tree really represents the last of a sequence of stages (corresponding to the application of inference rules) during which the current set of "active" premises, that is, those premises that have not yet been discharged (closed, cancelled) evolves (in fact, shrinks). Some mechanism is needed to keep track of which premises are no longer active and this is what this business of labeling premises with variables achieves. Historically, this is the first mechanism that was invented. However, Gentzen (in the 1930s) came up with an alternative solution that is mathematically easier to handle. Moreover, it turns out that this notation is also better suited to computer implementations, if one wishes to implement an automated theorem prover.

The point is to keep a record of all undischarged assumptions at every stage of the deduction. Thus, a deduction is now a tree whose nodes are labeled with expressions of the form $\Gamma \to P$, called *sequents*, where P is a proposition, and Γ is a record of all undischarged assumptions at the stage of the deduction associated with this node.

During the construction of a deduction tree, it is necessary to discharge packets of assumptions consisting of one or more occurrences of the same proposition. To this effect, it is convenient to tag packets of assumptions with labels, in order to discharge the propositions in these packets in a single step. We use variables for the labels, and a packet labeled with x consisting of occurrences of the proposition P is written as $x: P$. Thus, in a sequent $\Gamma \to P$, the expression Γ is any finite set of the form $x_1: P_1, \ldots, x_m: P_m$, where the x_i are pairwise distinct (but the P_i need not be distinct). Given $\Gamma = x_1: P_1, \ldots, x_m: P_m$, the notation $\Gamma, x: P$ is only well defined when $x \neq x_i$ for all i, $1 \leq i \leq m$, in which case it denotes the set $x_1: P_1, \ldots, x_m: P_m, x: P$.

Using sequents, the axioms and rules of Definition 1.2 are now expressed as follows.

Definition 1.2. The axioms and inference rules of the system $\mathcal{NG}_m^{\Rightarrow}$ (*implicational logic, Gentzen-sequent style (the $\mathcal{G}$ in $\mathcal{NG}$ stands for Gentzen)*) are listed below:

$$\Gamma, x: P \to P \quad \text{(Axioms)}$$

$$\frac{\Gamma, x\colon P \to Q}{\Gamma \to P \Rightarrow Q} \quad (\Rightarrow\text{-}intro)$$

$$\frac{\Gamma \to P \Rightarrow Q \quad \Gamma \to P}{\Gamma \to Q} \quad (\Rightarrow\text{-}elim)$$

In an application of the rule ($\Rightarrow$-*intro*), observe that in the lower sequent, the proposition P (labeled x) is deleted from the list of premises occurring on the left-hand side of the arrow in the upper sequent. We say that the proposition P that appears as a hypothesis of the deduction is *discharged* (or *closed*). A *deduction tree* is either a one-node tree labeled with an axiom or a tree constructed using the above inference rules. A *proof tree* is a deduction tree whose conclusion is a sequent with an empty set of premises (a sequent of the form $\emptyset \to P$).

It is important to note that the ability to label packets consisting of occurrences of the same proposition with different labels is essential in order to be able to have control over which groups of packets of assumptions are discharged simultaneously. Equivalently, we could avoid tagging packets of assumptions with variables if we assume that in a sequent $\Gamma \to C$, the expression Γ, also called a *context*, is a *multiset* of propositions.

Let us display the proof tree for the second proof tree in Example (c) in our new Gentzen-sequent system. The orginal proof tree is

$$\cfrac{\cfrac{\cfrac{P^x, Q^y}{P}}{Q \Rightarrow P} \; y}{P \Rightarrow (Q \Rightarrow P)} \; x$$

and the corresponding proof tree in our new system is

$$\frac{\dfrac{x\colon P, y\colon Q \to P}{x\colon P \to Q \Rightarrow P}}{\to P \Rightarrow (Q \Rightarrow P)}$$

Below we show a proof of Example (d) given above in our new system. Let

$$\Gamma = x\colon A \Rightarrow (B \Rightarrow C), y\colon A \Rightarrow B, z\colon A.$$

$$\cfrac{\cfrac{\cfrac{\cfrac{\cfrac{\cfrac{\Gamma \to A \Rightarrow (B \Rightarrow C) \quad \Gamma \to A}{\Gamma \to B \Rightarrow C} \quad \cfrac{\Gamma \to A \Rightarrow B \quad \Gamma \to A}{\Gamma \to B}}{x\colon A \Rightarrow (B \Rightarrow C), y\colon A \Rightarrow B, z\colon A \to C}}{x\colon A \Rightarrow (B \Rightarrow C), y\colon A \Rightarrow B \to A \Rightarrow C}}{x\colon A \Rightarrow (B \Rightarrow C) \to (A \Rightarrow B) \Rightarrow (A \Rightarrow C)}}{\to (A \Rightarrow (B \Rightarrow C)) \Rightarrow ((A \Rightarrow B) \Rightarrow (A \Rightarrow C))}$$

Remark: An attentive reader will have surely noticed that the second version of the ⇒-elimination rule,

$$\frac{\Gamma \to P \Rightarrow Q \quad \Gamma \to P}{\Gamma \to Q} \quad (\Rightarrow\text{-}elim),$$

differs slightly from the first version given in Definition 1.1. Indeed, in Prawitz's style, the rule that matches exactly the ⇒-elim rule above is

$$
\begin{array}{cc}
\Gamma & \Gamma \\
\mathscr{D}_1 & \mathscr{D}_2 \\
P \Rightarrow Q & P \\
\hline
\multicolumn{2}{c}{Q}
\end{array}
$$

where the deductions of $P \Rightarrow Q$ and P have the *same* set of premises, Γ. Equivalently, the rule in sequent format that corresponds to the ⇒-elimination rule of Definition 1.1 is

$$\frac{\Gamma \to P \Rightarrow Q \quad \Delta \to P}{\Gamma, \Delta \to Q} \quad (\Rightarrow\text{-}elim'),$$

where Γ, Δ must be interpreted as the union of Γ and Δ.

A moment of reflection will reveal that the resulting proof systems are equivalent (i.e., every proof in one system can be converted to a proof in the other system and vice versa); if you are not sure, do Problem 1.15. The version of the ⇒-elimination rule in Definition 1.1 may be considered preferable because it gives us the ability to make the sets of premises labeling leaves smaller. On the other hand, after experimenting with the construction of proofs, one gets the feeling that every proof can be simplified to a "unique minimal" proof, if we define "minimal" in a suitable sense, namely, that a minimal proof never contains an elimination rule immediately following an introduction rule (for more on this, see Section 1.11). Then, it turns out that to define the notion of uniqueness of proofs, the second version is preferable. However, it is important to realize that in general, a proposition may possess distinct minimal proofs.

In principle, it does not matter which of the two systems $\mathcal{N}_m^{\Rightarrow}$ or $\mathcal{NG}_m^{\Rightarrow}$ we use to construct deductions; it is basically a matter of taste. My experience is that I make fewer mistakes with the Gentzen-sequent style system $\mathcal{NG}_m^{\Rightarrow}$.

We now describe the inference rules dealing with the connectives ∧, ∨ and ⊥.

1.3 Adding ∧, ∨, ⊥; The Proof Systems $\mathcal{N}_c^{\Rightarrow,\wedge,\vee,\perp}$ and $\mathcal{NG}_c^{\Rightarrow,\wedge,\vee,\perp}$

Recall that $\neg P$ is an abbreviation for $P \Rightarrow \perp$.

Definition 1.3. The axioms, inference rules, and deduction trees for *(propositional) classical logic* are defined as follows.

Axioms:

(i) Every one-node tree labeled with a single proposition P is a deduction tree for P with set of premises $\{P\}$.

(ii) The tree

$$\frac{\Gamma, P}{P}$$

is a deduction tree for P with multiset of premises $\Gamma \cup \{P\}$.

The $\Rightarrow$-*introduction rule*:

If $\mathscr{D}$ is a deduction of Q from the premises in $\Gamma \cup \{P\}$, then

$$\begin{array}{c} \Gamma, P^x \\ \mathscr{D} \\ Q \\ \hline P \Rightarrow Q \end{array} \quad x$$

is a deduction tree for $P \Rightarrow Q$ from Γ. All premises P labeled x are discharged.

The $\Rightarrow$-*elimination rule (or modus ponens)*:

If $\mathscr{D}_1$ is a deduction tree for $P \Rightarrow Q$ from the premises Γ, and $\mathscr{D}_2$ is a deduction for P from the premises Δ, then

$$\begin{array}{cc} \Gamma & \Delta \\ \mathscr{D}_1 & \mathscr{D}_2 \\ P \Rightarrow Q & P \\ \hline & Q \end{array}$$

is a deduction tree for Q from the premises in $\Gamma \cup \Delta$.

The $\wedge$-*introduction rule*:

If $\mathscr{D}_1$ is a deduction tree for P from the premises Γ, and $\mathscr{D}_2$ is a deduction for Q from the premises Δ, then

$$\begin{array}{cc} \Gamma & \Delta \\ \mathscr{D}_1 & \mathscr{D}_2 \\ P & Q \\ \hline & P \wedge Q \end{array}$$

is a deduction tree for $P \wedge Q$ from the premises in $\Gamma \cup \Delta$.

The $\wedge$-*elimination rule*:

If $\mathscr{D}$ is a deduction tree for $P \wedge Q$ from the premises Γ, then

$$\begin{array}{cc} \Gamma & \Gamma \\ \mathscr{D} & \mathscr{D} \\ P \wedge Q & P \wedge Q \\ \hline P & \hline Q \end{array}$$

are deduction trees for P and Q from the premises Γ.

The $\vee$-*introduction rule*:

If $\mathcal{D}$ is a deduction tree for P or for Q from the premises Γ, then

$$
\begin{array}{cc}
\Gamma & \Gamma \\
\mathcal{D} & \mathcal{D} \\
\underline{P} & \underline{Q} \\
P \vee Q & P \vee Q
\end{array}
$$

are deduction trees for $P \vee Q$ from the premises in Γ.

The $\vee$-*elimination rule*:

If $\mathcal{D}_1$ is a deduction tree for $P \vee Q$ from the premises Γ, $\mathcal{D}_2$ is a deduction for R from the premises in $\Delta \cup \{P\}$, and $\mathcal{D}_3$ is a deduction for R from the premises in $\Lambda \cup \{Q\}$, then

$$
\begin{array}{ccc}
\Gamma & \Delta, P^x & \Lambda, Q^y \\
\mathcal{D}_1 & \mathcal{D}_2 & \mathcal{D}_3 \\
P \vee Q & R & R \\
\hline
& R &
\end{array} \quad x,y
$$

is a deduction tree for R from the premises in $\Gamma \cup \Delta \cup \Lambda$. All premises P labeled x and all premises Q labeled y are discharged.

The $\perp$-*elimination rule*:

If $\mathcal{D}$ is a deduction tree for $\perp$ from the premises Γ, then

$$
\begin{array}{c}
\Gamma \\
\mathcal{D} \\
\underline{\perp} \\
P
\end{array}
$$

is a deduction tree for P from the premises Γ, for *any* proposition P.

The *proof-by-contradiction rule* (also known as *reductio ad absurdum rule*, for short *RAA*):

If $\mathcal{D}$ is a deduction tree for $\perp$ from the premises in $\Gamma \cup \{\neg P\}$, then

$$
\begin{array}{c}
\Gamma, \neg P^x \\
\mathcal{D} \\
\underline{\perp} \\
P
\end{array} \quad x
$$

is a deduction tree for P from the premises Γ. All premises $\neg P$ labeled x, are discharged.

Because $\neg P$ is an abbreviation for $P \Rightarrow \perp$, the $\neg$-introduction rule is a special case of the $\Rightarrow$-introduction rule (with $Q = \perp$). However, it is worth stating it explicitly.

The ¬-*introduction rule*:

If $\mathscr{D}$ is a deduction tree for $\bot$ from the premises in $\Gamma \cup \{P\}$, then

$$\Gamma, P^x$$

$$\mathscr{D}$$

$$\frac{\bot}{\neg P} \ x$$

is a deduction tree for $\neg P$ from the premises Γ. All premises P labeled x, are discharged.

The above rule can be viewed as a proof-by-contradiction principle applied to negated propositions.

Similarly, the ¬-elimination rule is a special case of ⇒-elimination applied to $\neg P \,(= P \Rightarrow \bot)$ and P.

The ¬-*elimination rule*:

If $\mathscr{D}_1$ is a deduction tree for $\neg P$ from the premises Γ, and $\mathscr{D}_2$ is a deduction for P from the premises Δ, then

$$\frac{\begin{array}{cc} \Gamma & \Delta \\ \mathscr{D}_1 & \mathscr{D}_2 \\ \neg P & P \end{array}}{\bot}$$

is a deduction tree for $\bot$ from the premises in $\Gamma \cup \Delta$.

In the above axioms and rules, Γ, Δ, or Λ may be empty; P, Q, R denote arbitrary propositions built up from the atoms in **PS**; $\mathscr{D}, \mathscr{D}_1, \mathscr{D}_2$ denote deductions, possibly a one-node tree; and all the premises labeled x or y are discharged.

A *deduction tree* is either a one-node tree labeled with a single proposition or a tree constructed using the above axioms and inference rules. A *proof tree* is a deduction tree such that *all its premises* are discharged. The above proof system is denoted $\mathscr{N}_c^{\Rightarrow,\wedge,\vee,\bot}$ (here, the subscript c stands for *classical*).

The system obtained by removing the proof-by-contradiction (RAA) rule is called *(propositional) intuitionistic logic* and is denoted $\mathscr{N}_i^{\Rightarrow,\wedge,\vee,\bot}$. The system obtained by deleting both the $\bot$-elimination rule and the proof-by-contradiction rule is called *(propositional) minimal logic* and is denoted $\mathscr{N}_m^{\Rightarrow,\wedge,\vee,\bot}$

The version of $\mathscr{N}_c^{\Rightarrow,\wedge,\vee,\bot}$ in terms of Gentzen sequents is the following.

Definition 1.4. The axioms and inference rules of the system $\mathscr{N}\mathscr{G}_c^{\Rightarrow,\wedge,\vee,\bot}$ (of *propositional classical logic, Gentzen-sequent style*) are listed below.

$$\Gamma, x : P \to P \quad \text{(Axioms)}$$

$$\frac{\Gamma, x : P \to Q}{\Gamma \to P \Rightarrow Q} \quad (\Rightarrow\text{-}intro)$$

$$\frac{\Gamma \to P \Rightarrow Q \quad \Gamma \to P}{\Gamma \to Q} \quad (\Rightarrow\text{-elim})$$

$$\frac{\Gamma \to P \quad \Gamma \to Q}{\Gamma \to P \wedge Q} \quad (\wedge\text{-intro})$$

$$\frac{\Gamma \to P \wedge Q}{\Gamma \to P} \quad (\wedge\text{-elim}) \qquad \frac{\Gamma \to P \wedge Q}{\Gamma \to Q} \quad (\wedge\text{-elim})$$

$$\frac{\Gamma \to P}{\Gamma \to P \vee Q} \quad (\vee\text{-intro}) \qquad \frac{\Gamma \to Q}{\Gamma \to P \vee Q} \quad (\vee\text{-intro})$$

$$\frac{\Gamma \to P \vee Q \quad \Gamma, x: P \to R \quad \Gamma, y: Q \to R}{\Gamma \to R} \quad (\vee\text{-elim})$$

$$\frac{\Gamma \to \perp}{\Gamma \to P} \quad (\perp\text{-elim})$$

$$\frac{\Gamma, x: \neg P \to \perp}{\Gamma \to P} \quad (\text{by-contra})$$

$$\frac{\Gamma, x: P \to \perp}{\Gamma \to \neg P} \quad (\neg\text{-introduction})$$

$$\frac{\Gamma \to \neg P \quad \Gamma \to P}{\Gamma \to \perp} \quad (\neg\text{-elimination})$$

A *deduction tree* is either a one-node tree labeled with an axiom or a tree constructed using the above inference rules. A *proof tree* is a deduction tree whose conclusion is a sequent with an empty set of premises (a sequent of the form $\emptyset \to P$).

The rule ($\perp$-*elim*) is trivial (does nothing) when $P = \perp$, therefore from now on we assume that $P \neq \perp$. *Propositional minimal logic*, denoted $\mathcal{NG}_m^{\Rightarrow,\wedge,\vee,\perp}$, is obtained by dropping the ($\perp$-*elim*) and (*by-contra*) rules. *Propositional intuitionistic logic*, denoted $\mathcal{NG}_i^{\Rightarrow,\wedge,\vee,\perp}$, is obtained by dropping the (*by-contra*) rule.

When we say that a proposition P is *provable from* Γ, we mean that we can construct a proof tree whose conclusion is P and whose set of premises is Γ, in one of the systems $\mathcal{N}_c^{\Rightarrow,\wedge,\vee,\perp}$ or $\mathcal{NG}_c^{\Rightarrow,\wedge,\vee,\perp}$. Therefore, when we use the word "provable" unqualified, we mean provable in *classical logic*. If P is provable from Γ in one of the intuitionistic systems $\mathcal{N}_i^{\Rightarrow,\wedge,\vee,\perp}$ or $\mathcal{NG}_i^{\Rightarrow,\wedge,\vee,\perp}$, then we say *intuitionistically provable* (and similarly, if P is provable from Γ in one of the systems $\mathcal{N}_m^{\Rightarrow,\wedge,\vee,\perp}$ or $\mathcal{NG}_m^{\Rightarrow,\wedge,\vee,\perp}$, then we say *provable in minimal logic*). When P is provable from Γ, most people write $\Gamma \vdash P$, or $\vdash \Gamma \to P$, sometimes with the name of the corresponding proof system tagged as a subscript on the sign $\vdash$ if necessary to avoid ambiguities. When Γ is empty, we just say P is provable (provable in intuitionistic logic, and so on) and write $\vdash P$.

We treat *logical equivalence* as a derived connective: that is, we view $P \equiv Q$ as an abbreviation for $(P \Rightarrow Q) \wedge (Q \Rightarrow P)$. In view of the inference rules for $\wedge$, we see that to prove a logical equivalence $P \equiv Q$, we just have to prove both implications $P \Rightarrow Q$ and $Q \Rightarrow P$.

The equivalence of the systems $\mathscr{N}_m^{\Rightarrow,\wedge,\vee,\perp}$ and $\mathscr{N}\mathscr{G}_m^{\Rightarrow,\wedge,\vee,\perp}$ (as well as the systems $\mathscr{N}_c^{\Rightarrow,\wedge,\vee,\perp}$ and $\mathscr{N}\mathscr{G}_c^{\Rightarrow,\wedge,\vee,\perp}$) is not hard to show but is a bit tedious, see Problem 1.15.

In view of the $\neg$-elimination rule, we may be tempted to interpret the provability of a negation $\neg P$ as "P is not provable." Indeed, if $\neg P$ and P were both provable, then $\perp$ would be provable. So, P should not be provable if $\neg P$ is. However, if P is not provable, then $\neg P$ is **not** provable in general. There are plenty of propositions such that neither P nor $\neg P$ is provable (for instance, P, with P an atomic proposition). Thus, the fact that P is not provable is not equivalent to the provability of $\neg P$ and we should not interpret $\neg P$ as "P is not provable."

Let us now make some (much-needed) comments about the above inference rules. There is no need to repeat our comments regarding the $\Rightarrow$-rules.

The $\wedge$-introduction rule says that in order to prove a conjunction $P \wedge Q$ from some premises Γ, all we have to do is to prove *both* that P is provable from Γ *and* that Q is provable from Γ. The $\wedge$-elimination rule says that once we have proved $P \wedge Q$ from Γ, then P (and Q) is also provable from Γ. This makes sense intuitively as $P \wedge Q$ is "stronger" than P and Q separately ($P \wedge Q$ is true iff both P and Q are true).

The $\vee$-introduction rule says that if P (or Q) has been proved from Γ, then $P \vee Q$ is also provable from Γ. Again, this makes sense intuitively as $P \vee Q$ is "weaker" than P and Q.

The $\vee$-elimination rule formalizes the *proof-by-cases* method. It is a more subtle rule. The idea is that if we know that in the case where P is already assumed to be provable and similarly in the case where Q is already assumed to be provable that we can prove R (also using premises in Γ), then if $P \vee Q$ is also provable from Γ, as we have "covered both cases," it should be possible to prove R from Γ only (i.e., the premises P and Q are discarded). For example, if remain$1(n)$ is the proposition that asserts n is a whole number of the form $4k + 1$ and remain$3(n)$ is the proposition that asserts n is a whole number of the form $4k + 3$ (for some whole number k), then we can prove the implication

$$(\text{remain}1(n) \vee \text{remain}3(n)) \Rightarrow \text{odd}(n),$$

where odd(n) asserts that n is odd, namely, that n is of the form $2h + 1$ for some h.

To prove the above implication we first assume the premise, remain$1(n) \vee$ remain$3(n)$. Next, we assume each of the alternatives in this proposition. When we assume remain$1(n)$, we have $n = 4k + 1 = 2(2k) + 1$ for some k, so n is odd. When we assume remain$3(n)$, we have $n = 4k + 3 = 2(2k + 1) + 1$, so again, n is odd. By $\vee$-elimination, we conclude that odd(n) follows from the premise remain$1(n) \vee$ remain$3(n)$, and by $\Rightarrow$-introduction, we obtain a proof of our implication.

The $\perp$-elimination rule formalizes the principle that once a false statement has been established, then anything should be provable.

The $\neg$-introduction rule is a proof-by-contradiction principle applied to negated propositions. In order to prove $\neg P$, we assume P and we derive a contradiction ($\perp$). It is a more restrictive principle than the classical proof-by-contradiction rule (RAA). Indeed, if the proposition P to be proven is not a negation (P is not of the form $\neg Q$), then the $\neg$-introduction rule cannot be applied. On the other hand, the classical proof-by-contradiction rule can be applied but we have to assume $\neg P$ as a premise. For further comments on the difference between the $\neg$-introduction rule and the classical proof-by-contradiction rule, see Section 1.4.

The proof-by-contradiction rule formalizes the method of proof by contradiction. That is, in order to prove that P can be deduced from some premises Γ, one may assume the negation $\neg P$ of P (intuitively, assume that P is false) and then derive a contradiction from Γ and $\neg P$ (i.e., derive falsity). Then, P actually follows from Γ *without using* $\neg P$ *as a premise*, that is, $\neg P$ is discharged. For example, let us prove by contradiction that if n^2 is odd, then n itself must be odd, where n is a natural number.

According to the proof-by-contradiction rule, let us assume that n is not odd, which means that n is even. (Actually, in this step we are using a property of the natural numbers that is proved by induction but let's not worry about that right now) But to say that n is even means that $n = 2k$ for some k and then $n^2 = 4k^2 = 2(2k^2)$, so n^2 is even, contradicting the assumption that n^2 is odd. By the proof-by-contradiction rule, we conclude that n must be odd.

Remark: If the proposition to be proved, P, is of the form $\neg Q$, then if we use the proof-by-contradiction rule, we have to assume the premise $\neg\neg Q$ and then derive a contradiction. Because we are using classical logic, we often make implicit use of the fact that $\neg\neg Q$ is equivalent to Q (see Proposition 1.2) and instead of assuming $\neg\neg Q$ as a premise, we assume Q as a premise. But then, observe that we are really using $\neg$-introduction.

In summary, when trying to prove a proposition P by contradiction, proceed as follows.

(1) If P is a negated formula (P is of the form $\neg Q$), then use the $\neg$-introduction rule; that is, assume Q as a premise and derive a contradiction.
(2) If P is *not* a negated formula, then use the the proof-by-contradiction rule; that is, assume $\neg P$ as a premise and derive a contradiction.

Most people, I believe, will be comfortable with the rules of minimal logic and will agree that they constitute a "reasonable" formalization of the rules of reasoning involving $\Rightarrow$, $\wedge$, and $\vee$. Indeed, these rules seem to express the intuitive meaning of the connectives $\Rightarrow$, $\wedge$, and $\vee$. However, some may question the two rules $\perp$-elimination and proof-by-contradiction. Indeed, their meaning is not as clear and, certainly, the proof-by-contradiction rule introduces a form of indirect reasoning that is somewhat worrisome.

The problem has to do with the meaning of disjunction and negation and more generally, with the notion of *constructivity* in mathematics. In fact, in the early

1900s, some mathematicians, especially L. Brouwer (1881–1966), questioned the validity of the proof-by-contradiction rule, among other principles.

Fig. 1.2 L. E. J. Brouwer, 1881–1966

Two specific cases illustrate the problem, namely, the propositions

$$P \vee \neg P \quad \text{and} \quad \neg\neg P \Rightarrow P.$$

As we show shortly, the above propositions are both provable in classical logic. Now, Brouwer and some mathematicians belonging to his school of thought (the so-called "intuitionists" or "constructivists") advocate that in order to prove a disjunction $P \vee Q$ (from some premises Γ) one has to either exhibit a proof of P or a proof or Q (from Γ). However, it can be shown that this fails for $P \vee \neg P$. The fact that $P \vee \neg P$ is provable (in classical logic) **does not** imply (in general) that either P is provable or that $\neg P$ is provable. That $P \vee \neg P$ is provable is sometimes called the *principle (or law) of the excluded middle*. In intuitionistic logic, $P \vee \neg P$ is **not** provable (in general). Of course, if one gives up the proof-by-contradiction rule, then fewer propositions become provable. On the other hand, one may claim that the propositions that remain provable have more constructive proofs and thus, feel on safer grounds.

A similar controversy arises with the proposition $\neg\neg P \Rightarrow P$ (*double-negation rule*) If we give up the proof-by-contradiction rule, then this formula is no longer provable (i.e., $\neg\neg P$ is no longer equivalent to P). Perhaps this relates to the fact that if one says

" I don't have no money,"

then this does not mean that this person has money. (Similarly with "I don't get no satisfaction."). However, note that one can still prove $P \Rightarrow \neg\neg P$ in minimal logic (try doing it). Even stranger, $\neg\neg\neg P \Rightarrow \neg P$ is provable in intuitionistic (and minimal) logic, so $\neg\neg\neg P$ and $\neg P$ are equivalent intuitionistically.

Remark: Suppose we have a deduction

$$\Gamma, \neg P$$

$$\mathscr{D}$$

$$\bot$$

as in the proof-by-contradiction rule. Then, by $\neg$-introduction, we get a deduction of $\neg\neg P$ from Γ:

$$\Gamma, \neg P^x$$
$$\mathscr{D}$$
$$\frac{\perp}{\neg\neg P} \quad x$$

So, if we knew that $\neg\neg P$ was equivalent to P (actually, if we knew that $\neg\neg P \Rightarrow P$ is provable) then the proof-by-contradiction rule would be justified as a valid rule (it follows from modus ponens). We can view the proof-by-contradiction rule as a sort of act of faith that consists in saying that if we can derive an inconsistency (i.e., chaos) by assuming the falsity of a statement P, then P has to hold in the first place. It not so clear that such an act of faith is justified and the intuitionists refuse to take it.

Constructivity in mathematics is a fascinating subject but it is a topic that is really outside the scope of this course. What we hope is that our brief and very incomplete discussion of constructivity issues made the reader aware that the rules of logic are not cast in stone and that, in particular, there isn't **only one** logic.

We feel safe in saying that most mathematicians work with classical logic and only a few of them have reservations about using the proof-by-contradiction rule. Nevertheless, intuitionistic logic has its advantages, especially when it comes to proving the correctess of programs (a branch of computer science). We come back to this point several times in this course.

In the rest of this section, we make further useful remarks about (classical) logic and give some explicit examples of proofs illustrating the inference rules of classical logic. We begin by proving that $P \vee \neg P$ is provable in classical logic.

Proposition 1.1. *The proposition $P \vee \neg P$ is provable in classical logic.*

Proof. We prove that $P \vee (P \Rightarrow \perp)$ is provable by using the proof-by-contradiction rule as shown below:

$$\cfrac{\cfrac{((P \vee (P \Rightarrow \perp)) \Rightarrow \perp)^y \quad \cfrac{\cfrac{P^x}{P \vee (P \Rightarrow \perp)}}{\cfrac{\perp}{P \Rightarrow \perp} \quad x}}{((P \vee (P \Rightarrow \perp)) \Rightarrow \perp)^y \qquad P \vee (P \Rightarrow \perp)}}{\cfrac{\perp}{P \vee (P \Rightarrow \perp)} \quad y \text{ (by-contra)}}$$

$\square$

Next, we consider the equivalence of P and $\neg\neg P$.

Proposition 1.2. *The proposition $P \Rightarrow \neg\neg P$ is provable in minimal logic. The proposition $\neg\neg P \Rightarrow P$ is provable in classical logic. Therefore, in classical logic, P is equivalent to $\neg\neg P$.*

Proof. We leave that $P \Rightarrow \neg\neg P$ is provable in minimal logic as an exercise. Below is a proof of $\neg\neg P \Rightarrow P$ using the proof-by-contradiction rule:

$$\cfrac{\cfrac{\cfrac{((P \Rightarrow \bot) \Rightarrow \bot)^y \qquad (P \Rightarrow \bot)^x}{\cfrac{\bot}{P} \; x \text{ (by-contra)}}}{((P \Rightarrow \bot) \Rightarrow \bot) \Rightarrow P}}{} \; y$$

$\square$

The next proposition shows why $\bot$ can be viewed as the "ultimate" contradiction.

Proposition 1.3. *In intuitionistic logic, the propositions $\bot$ and $P \wedge \neg P$ are equivalent for all P. Thus, $\bot$ and $P \wedge \neg P$ are also equivalent in classical propositional logic*

Proof. We need to show that both $\bot \Rightarrow (P \wedge \neg P)$ and $(P \wedge \neg P) \Rightarrow \bot$ are provable in intuitionistic logic. The provability of $\bot \Rightarrow (P \wedge \neg P)$ is an immediate consequence or $\bot$-elimination, with $\Gamma = \emptyset$. For $(P \wedge \neg P) \Rightarrow \bot$, we have the following proof.

$$\cfrac{\cfrac{\cfrac{(P \wedge \neg P)^x}{\neg P} \qquad \cfrac{(P \wedge \neg P)^x}{P}}{\bot}}{(P \wedge \neg P) \Rightarrow \bot} \; x$$

$\square$

So, in intuitionistic logic (and also in classical logic), $\bot$ is equivalent to $P \wedge \neg P$ for all P. This means that $\bot$ is the "ultimate" contradiction; it corresponds to total inconsistency. By the way, we could have the bad luck that the system $\mathcal{N}_c^{\Rightarrow,\wedge,\vee,\bot}$ (or $\mathcal{N}_i^{\Rightarrow,\wedge,\vee,\bot}$ or even $\mathcal{N}_m^{\Rightarrow,\wedge,\vee,\bot}$) is *inconsistent*, that is, that $\bot$ is provable. Fortunately, this is not the case, although this is hard to prove. (It is also the case that $P \vee \neg P$ and $\neg\neg P \Rightarrow P$ are **not** provable in intuitionistic logic, but this too is hard to prove.)

1.4 Clearing Up Differences Among ¬-Introduction, ⊥-Elimination, and RAA

The differences between the rules, ¬-introduction, ⊥-elimination, and the proof-by-contradiction rule (RAA) are often unclear to the uninitiated reader and this tends to cause confusion. In this section, we try to clear up some common misconceptions about these rules.

Confusion 1. Why is RAA not a special case of $\neg$-introduction?

$$\Gamma, P^x \qquad\qquad\qquad\qquad \Gamma, \neg P^x$$
$$\mathscr{D} \qquad\qquad\qquad\qquad\qquad \mathscr{D}$$
$$\frac{\perp}{\neg P} \; x(\neg\text{-intro}) \qquad\qquad \frac{\perp}{P} \; x(\text{RAA})$$

The only apparent difference between $\neg$-introduction (on the left) and RAA (on the right) is that in RAA, the premise P is negated but the conclusion is not, whereas in $\neg$-introduction the premise P is not negated but the conclusion is.

The important difference is that the conclusion of RAA is **not** negated. If we had applied $\neg$-introduction instead of RAA on the right, we would have obtained

$$\Gamma, \neg P^x$$
$$\mathscr{D}$$
$$\frac{\perp}{\neg\neg P} \; x(\neg\text{-intro})$$

where the conclusion would have been $\neg\neg P$ as opposed to P. However, as we already said earlier, $\neg\neg P \Rightarrow P$ is **not** provable intuitionistically. Consequently, RAA **is not** a special case of $\neg$-introduction. On the other hand, one may view $\neg$-introduction as a "constructive" version of RAA applying to negated propositions (propositions of the form $\neg P$).

Confusion 2. Is there any difference between $\perp$-elimination and RAA?

$$\Gamma \qquad\qquad\qquad\qquad\qquad \Gamma, \neg P^x$$
$$\mathscr{D} \qquad\qquad\qquad\qquad\qquad \mathscr{D}$$
$$\frac{\perp}{P} \; (\perp\text{-elim}) \qquad\qquad \frac{\perp}{P} \; x(\text{RAA})$$

The difference is that $\perp$-elimination does not discharge any of its premises. In fact, RAA is a stronger rule that implies $\perp$-elimination as we now demonstate.

RAA implies $\perp$-Elimination

Suppose we have a deduction

$$\Gamma$$
$$\mathscr{D}$$
$$\perp$$

Then, for any proposition P, we can add the premise $\neg P$ to every leaf of the above deduction tree and we get the deduction tree

$$\Gamma, \neg P$$
$$\mathscr{D}'$$
$$\perp$$

We can now apply RAA to get the following deduction tree of P from Γ (because $\neg P$ is discharged), and this is just the result of $\bot$-elimination:

$$\Gamma, \neg P^x$$
$$\mathscr{D}'$$
$$\frac{\bot}{P} \quad x(\text{RAA})$$

The above considerations also show that RAA is obtained from $\neg$-introduction by adding the new rule of $\neg\neg$-*elimination* (also called *double-negation elimination*):

$$\Gamma$$
$$\mathscr{D}$$
$$\frac{\neg\neg P}{P} \quad (\neg\neg\text{-elimination})$$

Some authors prefer adding the $\neg\neg$-elimination rule to intuitionistic logic instead of RAA in order to obtain classical logic. As we just demonstrated, the two additions are equivalent: by adding either RAA or $\neg\neg$-elimination to intuitionistic logic, we get classical logic.

There is another way to obtain RAA from the rules of intuitionistic logic, this time, using the propositions of the form $P \vee \neg P$. We saw in Proposition 1.1 that all formulae of the form $P \vee \neg P$ are provable in classical logic (using RAA).

Confusion 3. Are propositions of the form $P \vee \neg P$ provable in intuitionistic logic?

The answer is **no**, which may be disturbing to some readers. In fact, it is quite difficult to prove that propositions of the form $P \vee \neg P$ are not provable in intuitionistic logic. One method consists in using the fact that intuitionistic proofs can be normalized (see Section 1.11 for more on normalization of proofs). Another method uses Kripke models (see Section 1.8 and van Dalen [23]).

Part of the difficulty in understanding at some intuitive level why propositions of the form $P \vee \neg P$ are not provable in intuitionistic logic is that the notion of truth based on the truth values **true** and **false** is deeply rooted in all of us. In this frame of mind, it seems ridiculous to question the provability of $P \vee \neg P$, because its truth value is **true** whether P is assigned the value **true** or **false**. Classical two-valued truth values semantics is too crude for intuitionistic logic.

Another difficulty is that it is tempting to equate the notion of truth and the notion of provability. Unfortunately, because classical truth values semantics is too crude for intuitionistic logic, there are propositions that are universally true (i.e., they evaluate to **true** for all possible truth assignments of the atomic letters in them) and yet they are **not** provable intuitionistically. The propositions $P \vee \neg P$ and $\neg\neg P \Rightarrow P$ are such examples.

One of the major motivations for advocating intuitionistic logic is that it yields proofs that are more constructive than classical proofs. For example, in classical logic, when we prove a disjunction $P \vee Q$, we generally can't conclude that either P

or Q is provable, as exemplified by $P \vee \neg P$. A more interesting example involving a nonconstructive proof of a disjunction is given in Section 1.5. But, in intuitionistic logic, from a proof of $P \vee Q$, it is possible to extract either a proof of P or a proof of Q (and similarly for existential statements; see Section 1.9). This property is not easy to prove. It is a consequence of the normal form for intuitionistic proofs (see Section 1.11).

In brief, besides being a fun intellectual game, intuitionistic logic is only an interesting alternative to classical logic if we care about the constructive nature of our proofs. But then, we are forced to abandon the classical two-valued truth values semantics and adopt other semantics such as Kripke semantics. If we do not care about the constructive nature of our proofs and if we want to stick to two-valued truth values semantics, then we should stick to classical logic. Most people do that, so don't feel bad if you are not comfortable with intuitionistic logic.

One way to gauge how intuitionisic logic differs from classical logic is to ask what kind of propositions need to be added to intuitionisic logic in order to get classical logic. It turns out that if all the propositions of the form $P \vee \neg P$ are considered to be axioms, then RAA follows from some of the rules of intuitionistic logic.

RAA Holds in Intuitionistic Logic + All Axioms $P \vee \neg P$.
The proof involves a subtle use of the ⊥-elimination and ∨-elimination rules which may be a bit puzzling. Assume, as we do when we use the proof-by-contradiction rule (RAA) that we have a deduction

$$\Gamma, \neg P$$
$$\mathcal{D}$$
$$\bot$$

Here is the deduction tree demonstrating that RAA is a derived rule:

$$
\cfrac{P \vee \neg P \qquad \cfrac{P^x}{P} \qquad \cfrac{\cfrac{\Gamma, \neg P^y}{\mathcal{D}}\ \ \cfrac{\bot}{P}\ (\bot\text{-elim})}{} }{P} \ \ x,y\ (\vee\text{-elim})
$$

At first glance, the rightmost subtree

$$\Gamma, \neg P^y$$
$$\mathcal{D}$$
$$\cfrac{\bot}{P} \quad (\bot\text{-elim})$$

appears to use RAA and our argument looks circular. But this is not so because the premise $\neg P$ labeled y is *not* discharged in the step that yields P as conclusion; the step that yields P is a ⊥-elimination step. The premise $\neg P$ labeled y is actually discharged by the ∨-elimination rule (and so is the premise P labeled x). So, our argument establishing RAA is not circular after all.

In conclusion, intuitionistic logic is obtained from classical logic by *taking away the proof-by-contradiction rule (RAA)*. In this more restrictive proof system, we obtain more constructive proofs. In that sense, the situation is better than in classical logic. The major drawback is that we can't think in terms of classical truth values semantics anymore.

Conversely, classical logic is obtained from intuitionistic logic in at least three ways:

1. Add the proof-by-contradiction rule (RAA).
2. Add the $\neg\neg$-elimination rule.
3. Add all propositions of the form $P \vee \neg P$ as axioms.

1.5 De Morgan Laws and Other Rules of Classical Logic

In classical logic, we have the de Morgan laws.

Proposition 1.4. *The following equivalences (de Morgan laws) are provable in classical logic.*

$$\neg(P \wedge Q) \equiv \neg P \vee \neg Q$$
$$\neg(P \vee Q) \equiv \neg P \wedge \neg Q.$$

In fact, $\neg(P \vee Q) \equiv \neg P \wedge \neg Q$ and $(\neg P \vee \neg Q) \Rightarrow \neg(P \wedge Q)$ are provable in intuitionistic logic. The proposition $(P \wedge \neg Q) \Rightarrow \neg(P \Rightarrow Q)$ is provable in intuitionistic logic and $\neg(P \Rightarrow Q) \Rightarrow (P \wedge \neg Q)$ is provable in classical logic. Therefore, $\neg(P \Rightarrow Q)$ and $P \wedge \neg Q$ are equivalent in classical logic. Furthermore, $P \Rightarrow Q$ and $\neg P \vee Q$ are equivalent in classical logic and $(\neg P \vee Q) \Rightarrow (P \Rightarrow Q)$ is provable in intuitionistic logic.

Proof. We only prove the very last part of Proposition 1.4 leaving the other parts as a series of exercises. Here is an intuitionistic proof of $(\neg P \vee Q) \Rightarrow (P \Rightarrow Q)$:

$$
\cfrac{
(\neg P \vee Q)^w \qquad
\cfrac{\cfrac{\cfrac{\neg P^z \qquad P^x}{\bot}}{Q}}{P \Rightarrow Q} x \qquad
\cfrac{\cfrac{P^y \qquad Q^t}{Q}}{P \Rightarrow Q} y
}{
\cfrac{P \Rightarrow Q}{(\neg P \vee Q) \Rightarrow (P \Rightarrow Q)} w
} z,t
$$

Here is a classical proof of $(P \Rightarrow Q) \Rightarrow (\neg P \vee Q)$:

$$
\cfrac{
  (P \Rightarrow Q)^z \qquad
  \cfrac{
    (\neg(\neg P \vee Q))^y \qquad
    \cfrac{
      \cfrac{\neg P^x}{\neg P \vee Q}
    }{
      \cfrac{\dfrac{\bot}{P}}{}
    } \; x \text{ RAA}
  }{
    \cfrac{
      (\neg(\neg P \vee Q))^y \qquad
      \cfrac{Q}{\neg P \vee Q}
    }{
      \cfrac{\dfrac{\bot}{\neg P \vee Q}}{} \; y \text{ RAA}
    }
  }
}{
  (P \Rightarrow Q) \Rightarrow (\neg P \vee Q)
} \; z
$$

The other proofs are left as exercises. $\square$

Propositions 1.2 and 1.4 show a property that is very specific to classical logic, namely, that the logical connectives $\Rightarrow, \wedge, \vee, \neg$ are not independent. For example, we have $P \wedge Q \equiv \neg(\neg P \vee \neg Q)$, which shows that $\wedge$ can be expressed in terms of $\vee$ and $\neg$. In intuitionistic logic, $\wedge$ and $\vee$ cannot be expressed in terms of each other via negation.

The fact that the logical connectives $\Rightarrow, \wedge, \vee, \neg$ are not independent in classical logic suggests the following question. Are there propositions, written in terms of $\Rightarrow$ only, that are provable classically but not provable intuitionistically?

The answer is yes. For instance, the proposition $((P \Rightarrow Q) \Rightarrow P) \Rightarrow P$ (known as *Peirce's law*) is provable classically (do it) but it can be shown that it is not provable intuitionistically.

In addition to the proof-by-cases method and the proof-by-contradiction method, we also have the proof-by-contrapositive method valid in classical logic:

Proof-by-contrapositive rule:

$$
\cfrac{
  \begin{array}{c}
  \Gamma, \neg Q^x \\
  \mathscr{D} \\
  \neg P
  \end{array}
}{
  P \Rightarrow Q
} \; x
$$

This rule says that in order to prove an implication $P \Rightarrow Q$ (from Γ), one may assume $\neg Q$ as proved, and then deduce that $\neg P$ is provable from Γ and $\neg Q$. This inference rule is valid in classical logic because we can construct the following deduction.

$$\frac{\begin{array}{c} \Gamma, \neg Q^x \\ \mathcal{D} \\ \dfrac{\neg P \qquad\qquad P^y}{\dfrac{\perp}{Q}} \; x \text{ (by-contra)} \end{array}}{P \Rightarrow Q} \; y$$

1.6 Formal Versus Informal Proofs; Some Examples

In this section, we give some explicit examples of proofs illustrating the proof principles that we just discussed. But first, it should be said that *it is practically impossible to write formal proofs* (i.e., proofs written as proof trees using the rules of one of the systems presented earlier) of "real" statements that are not "toy propositions." This is because it would be extremely tedious and time-consuming to write such proofs and these proofs would be huge and thus very hard to read.

As we said before it is possible in principle to write formalized proofs however, most of us will never do so. So, what *do* we do?

Well, we construct "informal" proofs in which we still make use of the logical rules that we have presented but we take shortcuts and sometimes we even omit proof steps (some elimination rules, such as $\wedge$-elimination and some introduction rules, such as $\vee$-introduction) and we use a natural language (here, presumably, English) rather than formal symbols (we say "and" for $\wedge$, "or" for $\vee$, etc.). As an example of a shortcut, when using the $\vee$-elimination rule, in most cases, the disjunction $P \vee Q$ has an "obvious proof" because P and Q "exhaust all the cases," in the sense that Q subsumes $\neg P$ (or P subsumes $\neg Q$) and classically, $P \vee \neg P$ is an axiom. Also, we implicitly keep track of the open premises of a proof in our head rather than explicitly discharge premises when required. This may be the biggest source of mistakes and we should make sure that when we have finished a proof, there are no "dangling premises," that is, premises that were never used in constructing the proof. If we are "lucky," some of these premises are in fact unnecessary and we should discard them. Otherwise, this indicates that there is something wrong with our proof and we should make sure that every premise is indeed used somewhere in the proof or else look for a counterexample.

We urge our readers to read Chapter 3 of Gowers [10] which contains very illuminating remarks about the notion of proof in mathematics.

The next question is then, "How does one write good informal proofs?"

It is very hard to answer such a question because the notion of a "good" proof is quite subjective and partly a "social" concept. Nevertheless, people have been writing informal proofs for centuries so there are at least many examples of what to do (and what not to do). As for everything else, practicing a sport, playing a music intrument, knowing "good" wines, and so on, *the more you practice, the better you become*. Knowing the theory of swimming is fine but you have to get wet and do some actual swimming. Similarly, knowing the proof rules is important but you have to put them to use.

Write proofs as much as you can. Find good proof writers (like good swimmers, good tennis players, etc.), try to figure out why they write clear and easily readable proofs and try to emulate what they do. Don't follow bad examples (it will take you a little while to "smell" a bad proof style).

Another important point is that nonformalized proofs make heavy use of *modus ponens*. This is because, when we search for a proof, we rarely (if ever) go back to first principles. This would result in extremely long proofs that would be basically incomprehensible. Instead, we search in our "database" of facts for a proposition of the form $P \Rightarrow Q$ (an auxiliary lemma) that is already known to be proved, and if we are smart enough (lucky enough), we find that we can prove P and thus we deduce Q, the proposition that we really need to prove. Generally, we have to go through several steps involving auxiliary lemmas. This is why it is important to build up a database of proven facts as large as possible about a mathematical field: numbers, trees, graphs, surfaces, and so on. This way, we increase the chance that we will be able to prove some fact about some some field of mathematics. On the other hand, one might argue that it might be better to start fresh and not to know much about a problem in order to tackle it. Somehow, knowing too much may hinder one's creativity. There are indeed a few examples of this phenomenon where very smart people solve a difficult problem basically "out of the blue," having little if any knowledge about the problem area. However, I claim that these cases are few and that the average human being has a better chance of solving a problem if she or he possesses a larger rather than a smaller database of mathematical facts in a problem area. Like any sport, it is also crucial to keep practicing (constructing proofs).

Let us conclude our discussion with a concrete example illustrating the usefulnes of auxiliary lemmas.

Say we wish to prove the implication

$$\neg(P \wedge Q) \Rightarrow \big((\neg P \wedge \neg Q) \vee (\neg P \wedge Q) \vee (P \wedge \neg Q)\big). \tag{$*$}$$

It can be shown that the above proposition is not provable intuitionistically, so we have to use the proof-by-contradiction method in our proof. One quickly realizes that any proof ends up re-proving basic properties of $\wedge$ and $\vee$, such as associativity, commutativity, idempotence, distributivity, and so on, some of the de Morgan laws, and that the complete proof is very large. However, if we allow ourselves to use the de Morgan laws as well as various basic properties of $\wedge$ and $\vee$, such as distributivity,

$$(A \wedge B) \vee C \equiv (A \wedge C) \vee (B \wedge C),$$

commutativity of $\wedge$ and $\vee$ ($A \wedge B \equiv B \wedge A$, $A \vee B \equiv B \vee A$), associativity of $\wedge$ and $\vee$ ($A \wedge (B \wedge C) \equiv (A \wedge B) \wedge C$, $A \vee (B \vee C) \equiv (A \vee B) \vee C$), and the idempotence of $\wedge$ and $\vee$ ($A \wedge A \equiv A$, $A \vee A \equiv A$), then we get

$$(\neg P \wedge \neg Q) \vee (\neg P \wedge Q) \vee (P \wedge \neg Q) \equiv (\neg P \wedge \neg Q) \vee (\neg P \wedge \neg Q)$$
$$\vee (\neg P \wedge Q) \vee (P \wedge \neg Q)$$
$$\equiv (\neg P \wedge \neg Q) \vee (\neg P \wedge Q)$$
$$\vee (\neg P \wedge \neg Q) \vee (P \wedge \neg Q)$$
$$\equiv (\neg P \wedge (\neg Q \vee Q)) \vee (\neg P \wedge \neg Q) \vee (P \wedge \neg Q)$$
$$\equiv \neg P \vee (\neg P \wedge \neg Q) \vee (P \wedge \neg Q)$$
$$\equiv \neg P \vee ((\neg P \vee P) \wedge \neg Q)$$
$$\equiv \neg P \vee \neg Q,$$

where we make implicit uses of commutativity and associativity, and the fact that $R \wedge (P \vee \neg P) \equiv R$, and by de Morgan,

$$\neg (P \wedge Q) \equiv \neg P \vee \neg Q,$$

using auxiliary lemmas, we end up proving $(*)$ without too much pain.

And now, we return to some explicit examples of informal proofs.

Recall that the *set of integers* is the set

$$\mathbb{Z} = \{\ldots, -2, -1, 0, 1, 2, \ldots\}$$

and that the *set of natural numbers* is the set

$$\mathbb{N} = \{0, 1, 2, \ldots\}.$$

(Some authors exclude 0 from $\mathbb{N}$. We don't like this discrimination against zero.) An integer is *even* if it is divisible by 2, that is, if it can be written as $2k$, where $k \in \mathbb{Z}$. An integer is *odd* if it is not divisible by 2, that is, if it can be written as $2k + 1$, where $k \in \mathbb{Z}$. The following facts are essentially obvious.

(a) The sum of even integers is even.
(b) The sum of an even integer and of an odd integer is odd.
(c) The sum of two odd integers is even.
(d) The product of odd integers is odd.
(e) The product of an even integer with any integer is even.

Now, we prove the following fact using the proof-by-cases method.

Proposition 1.5. *Let a, b, c be odd integers. For any integers p and q, if p and q are not both even, then*
$$ap^2 + bpq + cq^2$$
is odd.

Proof. We consider the three cases:

1. p and q are odd. In this case as a, b, and c are odd, by (d) all the products ap^2, bpq, and cq^2 are odd. By (c), $ap^2 + bpq$ is even and by (b), $ap^2 + bpq + cq^2$ is odd.

2. p is even and q is odd. In this case, by (e), both ap^2 and bpq are even and by (d), cq^2 is odd. But then, by (a), $ap^2 + bpq$ is even and by (b), $ap^2 + bpq + cq^2$ is odd.

3. p is odd and q is even. This case is analogous to the previous case, except that p and q are interchanged. The reader should have no trouble filling in the details.

All three cases exhaust all possibilities for p and q not to be both even, thus the proof is complete by the $\vee$-elimination rule (applied twice). $\square$

The set of *rational numbers* $\mathbb{Q}$ consists of all fractions p/q, where $p, q \in \mathbb{Z}$, with $q \neq 0$. The set of real numbers is denoted by $\mathbb{R}$. A real number, $a \in \mathbb{R}$, is said to be *irrational* if it cannot be expressed as a number in $\mathbb{Q}$ (a fraction).

We now use Proposition 1.5 and the proof by contradiction method to prove the following.

Proposition 1.6. *Let a, b, c be odd integers. Then, the equation*

$$aX^2 + bX + c = 0$$

has no rational solution X. Equivalently, every zero of the above equation is irrational.

Proof. We proceed by contradiction (by this, we mean that we use the proof-by-contradiction rule). So, assume that there is a rational solution $X = p/q$. We may assume that p and q have no common divisor, which implies that p and q are not both even. As $q \neq 0$, if $aX^2 + bX + c = 0$, then by multiplying by q^2, we get

$$ap^2 + bpq + cq^2 = 0.$$

However, as p and q are not both even and a, b, c are odd, we know from Proposition 1.5 that $ap^2 + bpq + cq^2$ is odd. This contradicts the fact that $p^2 + bpq + cq^2 = 0$ and thus, finishes the proof. $\square$

Remark: A closer look at the proof of Proposition 1.6 shows that rather than using the proof-by-contradiction rule we really used $\neg$-introduction (a "constructive" version of RAA).

As as example of the proof-by-contrapositive method, we prove that if an integer n^2 is even, then n must be even.

Observe that if an integer is not even then it is odd (and vice versa). This fact may seem quite obvious but to prove it actually requires using *induction* (which we haven't officially met yet). A rigorous proof is given in Section 1.10.

Now, the contrapositive of our statement is: if n is odd, then n^2 is odd. But, to say that n is odd is to say that $n = 2k + 1$ and then, $n^2 = (2k + 1)^2 = 4k^2 + 4k + 1 = 2(2k^2 + 2k) + 1$, which shows that n^2 is odd.

As it is, because the above proof uses the proof-by-contrapositive method, it is not constructive. Thus, the question arises, is there a constructive proof of the above fact?

Indeed there is a constructive proof if we observe that every integer n is either even or odd but not both. Now, one might object that we just relied on the law of the excluded middle but there is a way to circumvent this problem by using *induction*; see Section 1.10 for a rigorous proof.

Now, because *an integer is odd iff it is not even*, we may proceed to prove that *if n^2 is even, then n is not odd*, by using our constructive version of the proof-by-contradiction principle, namely, $\neg$-introduction.

Therefore, assume that n^2 is even and that n is odd. Then, $n = 2k + 1$, which implies that $n^2 = 4k^2 + 4k + 1 = 2(2k^2 + 2k) + 1$, an odd number, contradicting the fact that n^2 is asssumed to be even. □

As another illustration of the proof methods that we have just presented, let us prove that $\sqrt{2}$ is irrational, which means that $\sqrt{2}$ is *not* rational. The reader may also want to look at the proof given by Gowers in Chapter 3 of his book [10]. Obviously, our proof is similar but we emphasize step (2) a little more.

Because we are trying to prove that $\sqrt{2}$ is not rational, let us use our constructive version of the proof-by-contradiction principle, namely, $\neg$-introduction. Thus, let us assume that $\sqrt{2}$ is rational and derive a contradiction. Here are the steps of the proof.

1. If $\sqrt{2}$ is rational, then there exist some integers $p, q \in \mathbb{Z}$, with $q \neq 0$, so that $\sqrt{2} = p/q$.
2. Any fraction p/q is equal to some fraction r/s, where r and s are not both even.
3. By (2), we may assume that
$$\sqrt{2} = \frac{p}{q},$$
   where $p, q \in \mathbb{Z}$ are *not both even* and with $q \neq 0$.
4. By (3), because $q \neq 0$, by multiplying both sides by q, we get
$$q\sqrt{2} = p.$$
5. By (4), by squaring both sides, we get
$$2q^2 = p^2.$$
6. Inasmuch as $p^2 = 2q^2$, the number p^2 must be even. By a fact previously established, *p itself is even*; that is, $p = 2s$, for some $s \in \mathbb{Z}$.
7. By (6), if we substitute $2s$ for p in the equation in (5) we get $2q^2 = 4s^2$. By dividing both sides by 2, we get
$$q^2 = 2s^2.$$
8. By (7), we see that q^2 is even, from which we deduce (as above) that q *itself is even*.
9. Now, assuming that $\sqrt{2} = p/q$ where p and q are *not both even* (and $q \neq 0$), we concluded that *both p and q are even* (as shown in (6) and(8)), reaching

a contradiction. Therefore, by negation introduction, we proved that $\sqrt{2}$ is *not* rational.

A closer examination of the steps of the above proof reveals that the only step that may require further justification is step (2): that any fraction p/q is equal to some fraction r/s where r and s are not both even.

This fact does require a proof and the proof uses the division algorithm, which itself requires induction (see Section 5.3, Theorem 5.7). Besides this point, all the other steps only require simple arithmetic properties of the integers and are constructive.

Remark: Actually, every fraction p/q is equal to some fraction r/s where r and s have no common divisor except 1. This follows from the fact that every pair of integers has a *greatest common divisor* (a *gcd*; see Section 5.4) and r and s are obtained by dividing p and q by their gcd. Using this fact and Euclid's proposition (Proposition 5.9), we can obtain a shorter proof of the irrationality of $\sqrt{2}$. First, we may assume that p and q have no common divisor besides 1 (we say that p and q are *relatively prime*). From (5), we have

$$2q^2 = p^2,$$

so q divides p^2. However, q and p are relatively prime and as q divides $p^2 = p \times p$, by Euclid's proposition, q divides p. But because 1 is the only common divisor of p and q, we must have $q = 1$. Now, we get $p^2 = 2$, which is impossible inasmuch as 2 is not a perfect square.

The above argument can be easily adapted to prove that if the positive integer n is not a perfect square, then $\sqrt{n}$ is not rational.

Let us return briefly to the issue of constructivity in classical logic, in particular when it comes to disjunctions. Consider the question: are there two irrational real numbers a and b such that a^b is rational? Here is a way to prove that this is indeed the case. Consider the number $\sqrt{2}^{\sqrt{2}}$. If this number is rational, then $a = \sqrt{2}$ and $b = \sqrt{2}$ is an answer to our question (because we already know that $\sqrt{2}$ is irrational). Now, observe that

$$\left(\sqrt{2}^{\sqrt{2}}\right)^{\sqrt{2}} = \sqrt{2}^{\sqrt{2}\times\sqrt{2}} = \sqrt{2}^2 = 2 \quad \text{is rational.}$$

Thus, if $\sqrt{2}^{\sqrt{2}}$ is irrational, then $a = \sqrt{2}^{\sqrt{2}}$ and $b = \sqrt{2}$ is an answer to our question. So, we proved that

($\sqrt{2}$ is irrational and $\sqrt{2}^{\sqrt{2}}$ is rational) or

($\sqrt{2}^{\sqrt{2}}$ and $\sqrt{2}$ are irrational and $\left(\sqrt{2}^{\sqrt{2}}\right)^{\sqrt{2}}$ is rational).

However, the above proof does not tell us whether $\sqrt{2}^{\sqrt{2}}$ is rational!

We see one of the shortcomings of classical reasoning: certain statements (in particular, disjunctive or existential) are provable but their proof does not provide an explicit answer. It is in that sense that classical logic is not constructive.

Many more examples of nonconstructive arguments in classical logic can be given.

Remark: Actually, it turns out that another irrational number b can be found so that $\sqrt{2}^b$ is rational and the proof that b is not rational is fairly simple. It also turns out that the exact nature of $\sqrt{2}^{\sqrt{2}}$ (rational or irrational) is known. The answers to these puzzles can be found in Section 1.9.

1.7 Truth Values Semantics for Classical Logic Soundness and Completeness

So far, even though we have deliberately focused on proof theory and ignored semantic issues, we feel that we can't postpone any longer a discussion of the truth values semantics for classical propositional logic.

We all learned early on that the logical connectives $\Rightarrow$, $\wedge$, $\vee$, and $\neg$ can be interpreted as Boolean functions, that is, functions whose arguments and whose values range over the set of *truth values*,

$$\mathbf{BOOL} = \{\mathbf{true}, \mathbf{false}\}.$$

These functions are given by the following *truth tables*.

P	Q	$P \Rightarrow Q$	$P \wedge Q$	$P \vee Q$	$\neg P$
true	**true**	**true**	**true**	**true**	**false**
true	**false**	**false**	**false**	**true**	**false**
false	**true**	**true**	**false**	**true**	**true**
false	**false**	**true**	**false**	**false**	**true**

Now, any proposition P built up over the set of atomic propositions **PS** (our propositional symbols) contains a finite set of propositional letters, say

$$\{P_1, \ldots, P_m\}.$$

If we assign some truth value (from **BOOL**) to each symbol P_i then we can "compute" the *truth value* of P under this assignment by using recursively using the truth tables above. For example, the proposition $\mathbf{P}_1 \Rightarrow (\mathbf{P}_1 \Rightarrow \mathbf{P}_2)$, under the truth assignment v given by

$$\mathbf{P}_1 = \mathbf{true}, \ \mathbf{P}_2 = \mathbf{false},$$

evaluates to **false**. Indeed, the truth value, $v(\mathbf{P}_1 \Rightarrow (\mathbf{P}_1 \Rightarrow \mathbf{P}_2))$, is computed recursively as

$$v(\mathbf{P}_1 \Rightarrow (\mathbf{P}_1 \Rightarrow \mathbf{P}_2)) = v(\mathbf{P}_1) \Rightarrow v(\mathbf{P}_1 \Rightarrow \mathbf{P}_2).$$

Now, $v(\mathbf{P}_1) = \mathbf{true}$ and $v(\mathbf{P}_1 \Rightarrow \mathbf{P}_2)$ is computed recursively as

$$v(\mathbf{P}_1 \Rightarrow \mathbf{P}_2) = v(\mathbf{P}_1) \Rightarrow v(\mathbf{P}_2).$$

Because $v(\mathbf{P}_1) = \textbf{true}$ and $v(\mathbf{P}_2) = \textbf{false}$, using our truth table, we get

$$v(\mathbf{P}_1 \Rightarrow \mathbf{P}_2) = \textbf{true} \Rightarrow \textbf{false} = \textbf{false}.$$

Plugging this into the right-hand side of $v(\mathbf{P}_1 \Rightarrow (\mathbf{P}_1 \Rightarrow \mathbf{P}_2))$, we finally get

$$v(\mathbf{P}_1 \Rightarrow (\mathbf{P}_1 \Rightarrow \mathbf{P}_2)) = \textbf{true} \Rightarrow \textbf{false} = \textbf{false}.$$

However, under the truth assignment v given by

$$\mathbf{P}_1 = \textbf{true}, \mathbf{P}_2 = \textbf{true},$$

we find that our proposition evaluates to **true**.

If we now consider the proposition

$$P = (\mathbf{P}_1 \Rightarrow (\mathbf{P}_2 \Rightarrow \mathbf{P}_1)),$$

then it is easy to see that P evaluates to **true** for all four possible truth assignments for $\mathbf{P}_1$ and $\mathbf{P}_2$.

Note that to be rigorous, we need to justify why it is legitimate to define a function by recursion. This is done in Section 2.5.

Definition 1.5. We say that a proposition P is *satisfiable* iff it evaluates to **true** for *some* truth assignment (taking values in **BOOL**) of the propositional symbols occurring in P and otherwise we say that it is *unsatisfiable*. A proposition P is *valid* (or a *tautology*) iff it evaluates to **true** for *all* truth assignments of the propositional symbols occurring in P.

The problem of deciding whether a proposition is satisfiable is called the *satisfiability problem* and is sometimes denoted by SAT. The problem of deciding whether a proposition is valid is called the *validity problem*.

For example, the proposition

$$P = (\mathbf{P}_1 \vee \neg\mathbf{P}_2 \vee \neg\mathbf{P}_3) \wedge (\neg\mathbf{P}_1 \vee \neg\mathbf{P}_3) \wedge (\mathbf{P}_1 \vee \mathbf{P}_2 \vee \mathbf{P}_4) \wedge (\neg\mathbf{P}_3 \vee \mathbf{P}_4) \wedge (\neg\mathbf{P}_1 \vee \mathbf{P}_4)$$

is satisfiable because it evaluates to **true** under the truth assignment $\mathbf{P}_1 = \textbf{true}$, $\mathbf{P}_2 = \textbf{false}$, $\mathbf{P}_3 = \textbf{false}$, and $\mathbf{P}_4 = \textbf{true}$. On the other hand, the proposition

$$Q = (\mathbf{P}_1 \vee \mathbf{P}_2 \vee \mathbf{P}_3) \wedge (\neg\mathbf{P}_1 \vee \mathbf{P}_2) \wedge (\neg\mathbf{P}_2 \vee \mathbf{P}_3) \wedge (\mathbf{P}_1 \vee \neg\mathbf{P}_3) \wedge (\neg\mathbf{P}_1 \vee \neg\mathbf{P}_2 \vee \neg\mathbf{P}_3)$$

is unsatisfiable as one can verify by trying all eight truth assignments for $\mathbf{P}_1, \mathbf{P}_2, \mathbf{P}_3$. The reader should also verify that the proposition

$$R = (\neg\mathbf{P}_1 \wedge \neg\mathbf{P}_2 \wedge \neg\mathbf{P}_3) \vee (\mathbf{P}_1 \wedge \neg\mathbf{P}_2) \vee (\mathbf{P}_2 \wedge \neg\mathbf{P}_3) \vee (\neg\mathbf{P}_1 \wedge \mathbf{P}_3) \vee (\mathbf{P}_1 \wedge \mathbf{P}_2 \wedge \mathbf{P}_3)$$

is valid (observe that the proposition R is the negation of the proposition Q).

The satisfiability problem is a famous problem in computer science because of its complexity. Try it; solving it is not as easy as you think. The difficulty is that if a proposition P contains n distinct propositional letters, then there are 2^n possible truth assignments and checking all of them is practically impossible when n is large.

In fact, the satisfiability problem turns out to be an *NP-complete* problem, a very important concept that you will learn about in a course on the theory of computation and complexity. Very good expositions of this kind of material are found in Hopcroft, Motwani, and Ullman [12] and Lewis and Papadimitriou [16]. The validity problem is also important and it is related to SAT. Indeed, it is easy to see that a proposition P is valid iff $\neg P$ is unsatisfiable.

What's the relationship between validity and provability in the system $\mathcal{N}_c^{\Rightarrow, \wedge, \vee, \perp}$ (or $\mathcal{N}\mathcal{G}_c^{\Rightarrow, \wedge, \vee, \perp}$)?

Remarkably, in classical logic, validity and provability are equivalent.

In order to prove the above claim, we need to do two things:

(1) Prove that if a proposition P is provable in the system $\mathcal{N}_c^{\Rightarrow, \wedge, \vee, \perp}$ (or the system $\mathcal{N}\mathcal{G}_c^{\Rightarrow, \wedge, \vee, \perp}$), then it is valid. This is known as *soundness* or *consistency* (of the proof system).

(2) Prove that if a proposition P is valid, then it has a proof in the system $\mathcal{N}_c^{\Rightarrow, \wedge, \vee, \perp}$ (or $\mathcal{N}\mathcal{G}_c^{\Rightarrow, \wedge, \vee, \perp}$). This is known as the *completeness* (of the proof system).

In general, it is relatively easy to prove (1) but proving (2) can be quite complicated. In fact, some proof systems are *not* complete with respect to certain semantics. For instance, the proof system for intuitionistic logic $\mathcal{N}_i^{\Rightarrow, \wedge, \vee, \perp}$ (or $\mathcal{N}\mathcal{G}_i^{\Rightarrow, \wedge, \vee, \perp}$) is *not complete* with respect to truth values semantics. As an example, $((P \Rightarrow Q) \Rightarrow P) \Rightarrow P$ (known as *Peirce's law*), is valid but it can be shown that it cannot be proved in intuitionistic logic.

In this book, we content ourselves with soundness.

Proposition 1.7. (*Soundness of* $\mathcal{N}_c^{\Rightarrow, \wedge, \vee, \perp}$ *and* $\mathcal{N}\mathcal{G}_c^{\Rightarrow, \wedge, \vee, \perp}$) *If a proposition P is provable in the system* $\mathcal{N}_c^{\Rightarrow, \wedge, \vee, \perp}$ *(or* $\mathcal{N}\mathcal{G}_c^{\Rightarrow, \wedge, \vee, \perp}$*), then it is valid (according to the truth values semantics).*

Sketch of Proof. It is enough to prove that if there is a deduction of a proposition P from a set of premises Γ then for every truth assignment for which all the propositions in Γ evaluate to **true**, then P evaluates to **true**. However, this is clear for the axioms and every inference rule preserves that property.

Now, if P is provable, a proof of P has an empty set of premises and so P evaluates to **true** for all truth assignments, which means that P is valid. $\square$

Theorem 1.1. (*Completeness of* $\mathcal{N}_c^{\Rightarrow, \wedge, \vee, \perp}$ *and* $\mathcal{N}\mathcal{G}_c^{\Rightarrow, \wedge, \vee, \perp}$) *If a proposition P is valid (according to the truth values semantics), then P is provable in the system* $\mathcal{N}_c^{\Rightarrow, \wedge, \vee, \perp}$ *(or* $\mathcal{N}\mathcal{G}_c^{\Rightarrow, \wedge, \vee, \perp}$*).*

Proofs of completeness for classical logic can be found in van Dalen [23] or Gallier [4] (but for a different proof system).

Soundness (Proposition 1.7) has a very useful consequence: in order to prove that a proposition P is *not provable*, it is enough to find a truth assignment for which P evaluates to **false**. We say that such a truth assignment is a *counterexample* for P (or that P can be *falsified*). For example, no propositional symbol $\mathbf{P}_i$ is provable because it is falsified by the truth assignment $\mathbf{P}_i = \mathbf{false}$.

The soundness of the proof system $\mathcal{N}_c^{\Rightarrow,\wedge,\vee,\perp}$ (or $\mathcal{N}\mathcal{G}_c^{\Rightarrow,\wedge,\vee,\perp}$) also has the extremely important consequence that $\perp$ *cannot be proved* in this system, which means that *contradictory statements* cannot be derived. This is by no means obvious at first sight, but reassuring. It is also possible to prove that the proof system $\mathcal{N}_c^{\Rightarrow,\wedge,\vee,\perp}$ is consistent (i.e., $\perp$ cannot be proved) by purely proof-theoretic means involving proof normalization (See Section 1.11), but this requires a lot more work.

Note that completeness amounts to the fact that every unprovable formula has a counterexample.

Remark: Truth values semantics is not the right kind of semantics for intuitionistic logic; it is too coarse. A more subtle kind of semantics is required. Among the various semantics for intuitionistic logic, one of the most natural is the notion of the *Kripke model*. Then, again, soundness and completeness hold for intuitionistic proof systems (see Section 1.8 and van Dalen [23]).

Fig. 1.3 Saul Kripke, 1940–

1.8 Kripke Models for Intuitionistic Logic
Soundness and Completeness

In this section, we briefly describe the semantics of intuitionistic propositional logic in terms of Kripke models. This section has been included to quench the thirst of those readers who can't wait to see what kind of decent semantics can be given for intuitionistic propositional logic and it can be safely omitted. We recommend reviewing the material of Section 5.1 before reading this section.

In classical truth values semantics based on **BOOL** = {**true**, **false**}, we might say that truth is absolute. The idea of Kripke semantics is that there is a set of worlds W together with a partial ordering $\leq$ on W, and that truth depends on in which world we are. Furthermore, as we "go up" from a world u to a world v with $u \leq v$, truth "can only increase," that is, whatever is true in world u remains true in world v. Also, the truth of some propositions, such as $P \Rightarrow Q$ or $\neg P$, depends on "future worlds." With this type of semantics, which is no longer absolute, we can capture exactly the essence of intuitionistic logic. We now make these ideas precise.

Definition 1.6. A *Kripke model* for intuitionistic propositional logic is a pair $\mathcal{K} = (W, \varphi)$ where W is a partially ordered (nonempty) set called a *set of worlds* and φ is a function $\varphi \colon W \to \mathbf{BOOL^{PS}}$ such that for every $u \in W$, the function $\varphi(u) \colon \mathbf{PS} \to \mathbf{BOOL}$ is an assignment of truth values to the propositional symbols in $\mathbf{PS}$ satisfying the following property. For all $u, v \in W$, for all $\mathbf{P}_i \in \mathbf{PS}$,

$$\text{if } u \leq v \text{ and } \varphi(u)(\mathbf{P}_i) = \mathbf{true}, \text{ then } \varphi(v)(\mathbf{P}_i) = \mathbf{true}.$$

As we said in our informal comments, truth can't decrease when we move from a world u to a world v with $u \leq v$ but truth can increase; it is possible that $\varphi(u)(\mathbf{P}_i) = \mathbf{false}$ and yet, $\varphi(v)(\mathbf{P}_i) = \mathbf{true}$. We use Kripke models to define the semantics of propositions as follows.

Definition 1.7. Given a Kripke model $\mathcal{K} = (W, \varphi)$, for every $u \in W$ and for every proposition P we say that P *is satisfied by* $\mathcal{K}$ *at* u and we write $\varphi(u)(P) = \mathbf{true}$ iff

(a) $\varphi(u)(\mathbf{P}_i) = \mathbf{true}$, if $P = \mathbf{P}_i \in \mathbf{PS}$.
(b) $\varphi(u)(Q) = \mathbf{true}$ and $\varphi(u)(R) = \mathbf{true}$, if $P = Q \wedge R$.
(c) $\varphi(u)(Q) = \mathbf{true}$ or $\varphi(u)(R) = \mathbf{true}$, if $P = Q \vee R$.
(d) For all v such that $u \leq v$, if $\varphi(v)(Q) = \mathbf{true}$, then $\varphi(v)(R) = \mathbf{true}$, if $P = Q \Rightarrow R$.
(e) For all v such that $u \leq v$, $\varphi(v)(Q) = \mathbf{false}$, if $P = \neg Q$.
(f) $\varphi(u)(\bot) = \mathbf{false}$; that is, $\bot$ is not satisfied by $\mathcal{K}$ at u (for any $\mathcal{K}$ and any u).

We say that P is *valid in* $\mathcal{K}$ (or that $\mathcal{K}$ is a *model* of P) iff P is satisfied by $\mathcal{K} = (W, \varphi)$ at u for all $u \in W$ and we say that P is *intuitionistically valid* iff P is valid in every Kripke model $\mathcal{K}$.

When P is satisfied by $\mathcal{K}$ at u we also say that P *is true at* u *in* $\mathcal{K}$. Note that the truth at $u \in W$ of a proposition of the form $Q \Rightarrow R$ or $\neg Q$ depends on the truth of Q and R at all "future worlds," $v \in W$, with $u \leq v$. Observe that classical truth values semantics corresponds to the special case where W consists of a single element (a single world).

If $W = \{0, 1\}$ ordered so that $0 \leq 1$ and if φ is given by

$$\varphi(0)(\mathbf{P}_i) = \mathbf{false}$$
$$\varphi(1)(\mathbf{P}_i) = \mathbf{true},$$

then $\mathcal{K}_{\mathrm{bad}} = (W, \varphi)$ is a Kripke structure. The reader should check that the proposition $P = (\mathbf{P}_i \vee \neg \mathbf{P}_i)$ has the value $\mathbf{false}$ at 0 because $\varphi(0)(\mathbf{P}_i) = \mathbf{false}$ but $\varphi(1)(\mathbf{P}_i) = \mathbf{true}$, so clause (e) fails for $\neg \mathbf{P}_i$ at $u = 0$. Therefore, $P = (\mathbf{P}_i \vee \neg \mathbf{P}_i)$ is not valid in $\mathcal{K}_{\mathrm{bad}}$ and thus, it is not intuitionistically valid. We escaped the classical truth values semantics by using a universe with two worlds. The reader should also check that

$$\varphi(u)(\neg\neg P) = \textbf{true} \quad \text{iff} \quad \text{for all } v \text{ such that } u \leq v$$
$$\text{there is some } w \text{ with } v \leq w \text{ so that } \varphi(w)(P) = \textbf{true}.$$

This shows that in Kripke semantics, $\neg\neg P$ is weaker than P, in the sense that $\varphi(u)(\neg\neg P) = \textbf{true}$ does not necessarily imply that $\varphi(u)(P) = \textbf{true}$. The reader should also check that the proposition $\neg\neg P_i \Rightarrow P_i$ is not valid in the Kripke structure $\mathcal{K}_{bad}$.

As we said in the previous section, Kripke semantics is a perfect fit to intuitionistic provability in the sense that soundness and completeness hold.

Proposition 1.8. (*Soundness of* $\mathcal{N}_i^{\Rightarrow,\wedge,\vee,\perp}$ *and* $\mathcal{NG}_i^{\Rightarrow,\wedge,\vee,\perp}$) *If a proposition P is provable in the system* $\mathcal{N}_i^{\Rightarrow,\wedge,\vee,\perp}$ *(or* $\mathcal{NG}_i^{\Rightarrow,\wedge,\vee,\perp}$*), then it is valid in every Kripke model, that is, it is intuitionistically valid.*

Proposition 1.8 is not hard to prove. We consider any deduction of a proposition P from a set of premises Γ and we prove that for every Kripke model $\mathcal{K} = (W, \varphi)$, for every $u \in W$, if every premise in Γ is satisfied by $\mathcal{K}$ at u, then P is also satisfied by $\mathcal{K}$ at u. This is obvious for the axioms and it is easy to see that the inference rules preserve this property.

Completeness also holds, but it is harder to prove (see van Dalen [23]).

Theorem 1.2. (*Completeness of* $\mathcal{N}_i^{\Rightarrow,\wedge,\vee,\perp}$ *and* $\mathcal{NG}_i^{\Rightarrow,\wedge,\vee,\perp}$) *If a proposition P is intuitionistically valid, then P is provable in the system* $\mathcal{N}_i^{\Rightarrow,\wedge,\vee,\perp}$ *(or* $\mathcal{NG}_i^{\Rightarrow,\wedge,\vee,\perp}$*).*

Another proof of completeness for a different proof system for propositional intuitionistic logic (a Gentzen-sequent calculus equivalent to $\mathcal{NG}_i^{\Rightarrow,\wedge,\vee,\perp}$) is given in Takeuti [21]. We find this proof more instructive than van Dalen's proof. This proof also shows that if a proposition P is not intuitionistically provable, then there is a Kripke model $\mathcal{K}$ where W is a *finite tree* in which P is not valid. Such a Kripke model is called a *counterexample* for P.

We now add quantifiers to our language and give the corresponding inference rules.

1.9 Adding Quantifiers; The Proof Systems $\mathcal{N}_c^{\Rightarrow,\wedge,\vee,\forall,\exists,\perp}$, $\mathcal{NG}_c^{\Rightarrow,\wedge,\vee,\forall,\exists,\perp}$

As we mentioned in Section 1.1, atomic propositions may contain variables. The intention is that such variables correspond to arbitrary objects. An example is

$$\text{human}(x) \Rightarrow \text{needs-to-drink}(x).$$

Now, in mathematics, we usually prove universal statements, that is statements that hold for all possible "objects," or existential statements, that is, statements asserting the existence of some object satisfying a given property. As we saw earlier, we assert that every human needs to drink by writing the proposition

$$\forall x(\text{human}(x) \Rightarrow \text{needs-to-drink}(x)).$$

Observe that once the quantifier $\forall$ (pronounced "for all" or "for every") is applied to the variable x, the variable x becomes a placeholder and replacing x by y or any other variable does not change anything. What matters is the locations to which the outer x points in the inner proposition. We say that x is a *bound variable* (sometimes a "dummy variable").

If we want to assert that some human needs to drink we write

$$\exists x(\text{human}(x) \Rightarrow \text{needs-to-drink}(x));$$

Again, once the quantifier $\exists$ (pronounced "there exists") is applied to the variable x, the variable x becomes a placeholder. However, the intended meaning of the second proposition is very different and weaker than the first. It only asserts the existence of some object satisfying the statement

$$\text{human}(x) \Rightarrow \text{needs-to-drink}(x).$$

Statements may contain variables that are not bound by quantifiers. For example, in

$$\forall y \, \text{parent}(x, y)$$

the variable y is bound but the variable x is not. Here, the intended meaning of $\text{parent}(x, y)$ is that x is a parent of y. Variables that are not bound are called *free*. The proposition

$$\forall y \exists x \, \text{parent}(x, y),$$

which contains only bound variables is meant to assert that every y has some parent x. Typically, in mathematics, we only prove statements without free variables. However, statements with free variables may occur during intermediate stages of a proof.

The intuitive meaning of the statement $\forall x P$ is that P holds for all possible objects x and the intuitive meaning of the statement $\exists x P$ is that P holds for some object x. Thus, we see that it would be useful to use symbols to denote various objects. For example, if we want to assert some facts about the "parent" predicate, we may want to introduce some *constant symbols* (for short, constants) such as "Jean," "Mia," and so on and write

$$\text{parent}(\text{Jean}, \text{Mia})$$

to assert that Jean is a parent of Mia. Often, we also have to use *function symbols* (or *operators*, *constructors*), for instance, to write a statement about numbers: $+$, $*$, and so on. Using constant symbols, function symbols, and variables, we can form *terms*, such as

$$(x * x + 1) * (3 * y + 2).$$

In addition to function symbols, we also use *predicate symbols*, which are names for atomic properties. We have already seen several examples of predicate symbols: "human," "parent." So, in general, when we try to prove properties of certain classes of objects (people, numbers, strings, graphs, and so on), we assume that we have a certain *alphabet* consisting of constant symbols, function symbols, and predicate symbols. Using these symbols and an infinite supply of variables (assumed distinct from the variables we use to label premises) we can form *terms* and *predicate terms*. We say that we have a *(logical) language*. Using this language, we can write compound statements.

Let us be a little more precise. In a *first-order language* **L** in addition to the logical connectives $\Rightarrow, \wedge, \vee, \neg, \perp, \forall$, and $\exists$, we have a set **L** of *nonlogical symbols* consisting of

(i) A set **CS** of *constant symbols*, $c_1, c_2, \ldots,$.
(ii) A set **FS** of *function symbols*, $f_1, f_2, \ldots,$. Each function symbol f has a *rank* $n_f \geq 1$, which is the number of arguments of f.
(iii) A set **PS** of *predicate symbols*, $P_1, P_2, \ldots,$. Each predicate symbol P has a *rank* $n_P \geq 0$, which is the number of arguments of P. Predicate symbols of rank 0 are *propositional symbols* as in earlier sections.
(iv) The *equality predicate* $=$ is added to our language when we want to deal with equations.
(v) First-order variables $t_1, t_2, \ldots$ used to form *quantified formulae*.

The difference between function symbols and predicate symbols is that function symbols are interpreted as functions defined on a structure (e.g., addition, $+$, on $\mathbb{N}$), whereas predicate symbols are interpreted as properties of objects, that is, they take the value **true** or **false**. An example is the language of *Peano arithmetic*, $\mathbf{L} = \{0, S, +, *, =\}$. Here, the intended structure is $\mathbb{N}$, 0 is of course zero, S is interpreted as the function $S(n) = n + 1$, the symbol $+$ is addition, $*$ is multiplication, and $=$ is equality.

Using a first-order language **L**, we can form terms, predicate terms, and formulae. The *terms over* **L** are the following expressions.

(i) Every variable t is a term.
(ii) Every constant symbol $c \in \mathbf{CS}$, is a term.
(iii) If $f \in \mathbf{FS}$ is a function symbol taking n arguments and $\tau_1, \ldots, \tau_n$ are terms already constructed, then $f(\tau_1, \ldots, \tau_n)$ is a term.

The *predicate terms over* **L** are the following expressions.

(i) If $P \in \mathbf{PS}$ is a predicate symbol taking n arguments and $\tau_1, \ldots, \tau_n$ are terms already constructed, then $P(\tau_1, \ldots, \tau_n)$ is a predicate term. When $n = 0$, the predicate symbol P is a predicate term called a propositional symbol.
(ii) When we allow the equality predicate, for any two terms τ_1 and τ_2, the expression $\tau_1 = \tau_2$ is a predicate term. It is usually called an *equation*.

The *(first-order) formulae over* **L** are the following expressions.

(i) Every predicate term $P(\tau_1, \ldots, \tau_n)$ is an atomic formula. i.   des all propositional letters. We also view $\perp$ (and sometimes $\top$) as an atomic formula.

(ii) When we allow the equality predicate, every equation $\tau_1 = \tau_2$ is an atomic formula.

(iii) If P and Q are formulae already constructed, then $P \Rightarrow Q, P \wedge Q, P \vee Q, \neg P$ are compound formulae. We treat $P \equiv Q$ as an abbreviation for $(P \Rightarrow Q) \wedge (Q \Rightarrow P)$, as before.

(iv) If P is a formula already constructed and t is any variable, then $\forall t P$ and $\exists t P$ are *quantified* compound formulae.

All this can be made very precise but this is quite tedious. Our primary goal is to explain the basic rules of logic and not to teach a full-fledged logic course. We hope that our intuitive explanations will suffice and we now come to the heart of the matter, the inference rules for the quantifiers. Once again, for a complete treatment, readers are referred to Gallier [4], van Dalen [23], or Huth and Ryan [14].

Unlike the rules for $\Rightarrow, \vee, \wedge$ and $\perp$, which are rather straightforward, the rules for quantifiers are more subtle due to the presence of variables (occurring in terms and predicates). We have to be careful to forbid inferences that would yield "wrong" results and for this we have to be very precise about the way we use free variables. More specifically, we have to exercise care when we make *substitutions* of terms for variables in propositions. For example, say we have the predicate "odd," intended to express that a number is odd. Now, we can substitute the term $(2y + 1)^2$ for x in $\text{odd}(x)$ and obtain

$$\text{odd}((2y + 1)^2).$$

More generally, if $P(t_1, t_2, \ldots, t_n)$ is a statement containing the free variables $t_1, \ldots, t_n$ and if $\tau_1, \ldots, \tau_n$ are terms, we can form the new statement

$$P[\tau_1/t_1, \ldots, \tau_n/t_n]$$

obtained by substituting the term τ_i for all free occurrences of the variable t_i, for $i = 1, \ldots, n$. By the way, we denote terms by the Greek letter τ because we use the letter t for a variable and using t for both variables and terms would be confusing; sorry.

However, if $P(t_1, t_2, \ldots, t_n)$ contains quantifiers, some bad things can happen; namely, some of the variables occurring in some term τ_i may become quantified when τ_i is substituted for t_i. For example, consider

$$\forall x \exists y P(x, y, z)$$

which contains the free variable z and substitute the term $x + y$ for z: we get

$$\forall x \exists y P(x, y, x + y).$$

We see that the variables x and y occurring in the term $x + y$ become bound variables after substitution. We say that there is a "capture of variables."

This is not what we intended to happen. To fix this problem, we recall that bound variables are really place holders, so they can be renamed without changing anything. Therefore, we can rename the bound variables x and y in $\forall x \exists y P(x, y, z)$ to u and v, getting the statement $\forall u \exists v P(u, v, z)$ and now, the result of the substitution is

$$\forall u \exists v P(u, v, x + y).$$

Again, all this needs to be explained very carefully but this can be done.

Finally, here are the inference rules for the quantifiers, first stated in a natural deduction style and then in sequent style. It is assumed that we use two disjoint sets of variables for labeling premises $(x, y, \ldots)$ and free variables $(t, u, v, \ldots)$. As we show, the $\forall$-introduction rule and the $\exists$-elimination rule involve a crucial restriction on the occurrences of certain variables. Remember, *variables are terms*.

Definition 1.8. The *inference rules for the quantifiers* are

$\forall$-*introduction*:

If $\mathscr{D}$ is a deduction tree for $P[u/t]$ from the premises Γ, then

$$
\begin{array}{c}
\Gamma \\
\mathscr{D} \\
P[u/t] \\
\hline
\forall t P
\end{array}
$$

is a deduction tree for $\forall t P$ from the premises Γ. Here, u must be a variable that does not occur free in any of the propositions in Γ or in $\forall t P$. The notation $P[u/t]$ stands for the result of substituting u for all free occurrences of t in P.

Recall that Γ denotes the set of premises of the deduction tree $\mathscr{D}$, so if $\mathscr{D}$ only has one node, then $\Gamma = \{P[u/t]\}$ and t should not occur in P.

$\forall$-*elimination*:

If $\mathscr{D}$ is a deduction tree for $\forall t P$ from the premises Γ, then

$$
\begin{array}{c}
\Gamma \\
\mathscr{D} \\
\forall t P \\
\hline
P[\tau/t]
\end{array}
$$

is a deduction tree for $P[\tau/t]$ from the premises Γ. Here τ is an arbitrary term and it is assumed that bound variables in P have been renamed so that none of the variables in τ are captured after substitution.

$\exists$-*introduction*:

If $\mathscr{D}$ is a deduction tree for $P[\tau/t]$ from the premises Γ, then

$$
\begin{array}{c}
\Gamma \\
\mathscr{D} \\
P[\tau/t] \\
\hline
\exists t P
\end{array}
$$

is a deduction tree for $\exists tP$ from the premises Γ. As in $\forall$-elimination, τ is an arbitrary term and the same proviso on bound variables in P applies.

$\exists$-*elimination*:

If $\mathscr{D}_1$ is a deduction tree for $\exists tP$ from the premises Γ, and if $\mathscr{D}_2$ is a deduction tree for C from the premises in $\Delta \cup \{P[u/t]\}$, then

$$
\begin{array}{cc}
\Gamma & \Delta, P[u/t]^x \\
\mathscr{D}_1 & \mathscr{D}_2 \\
\exists tP & C \\
\hline
& C
\end{array} \quad x
$$

is a deduction tree of C from the set of premises in $\Gamma \cup \Delta$. Here, u must be a variable that does not occur free in any of the propositions in Δ, $\exists tP$, or C, and all premises $P[u/t]$ labeled x are discharged.

In the $\forall$-introduction and the $\exists$-elimination rules, the variable u is called the *eigenvariable* of the inference.

In the above rules, Γ or Δ may be empty; P, C denote arbitrary propositions constructed from a first-order language $\mathbf{L}$; $\mathscr{D}, \mathscr{D}_1, \mathscr{D}_2$ are deductions, possibly a one-node tree; and t is *any* variable.

The system of *first-order classical logic* $\mathscr{N}_c^{\Rightarrow, \vee, \wedge, \perp, \forall, \exists}$ is obtained by adding the above rules to the system of propositional classical logic $\mathscr{N}_c^{\Rightarrow, \vee, \wedge, \perp}$. The system of *first-order intuitionistic logic* $\mathscr{N}_i^{\Rightarrow, \vee, \wedge, \perp, \forall, \exists}$ is obtained by adding the above rules to the system of propositional intuitionistic logic $\mathscr{N}_i^{\Rightarrow, \vee, \wedge, \perp}$. Deduction trees and proof trees are defined as in the propositional case except that the quantifier rules are also allowed.

Using sequents, the quantifier rules in first-order logic are expressed as follows:

Definition 1.9. The *inference rules for the quantifiers in Gentzen-sequent style* are

$$
\frac{\Gamma \to P[u/t]}{\Gamma \to \forall tP} \quad (\forall\text{-}intro) \qquad \frac{\Gamma \to \forall tP}{\Gamma \to P[\tau/t]} \quad (\forall\text{-}elim)
$$

where in ($\forall$-*intro*), u does not occur free in Γ or $\forall tP$;

$$
\frac{\Gamma \to P[\tau/t]}{\Gamma \to \exists tP} \quad (\exists\text{-}intro) \qquad \frac{\Gamma \to \exists tP \quad z: P[u/t], \Gamma \to C}{\Gamma \to C} \quad (\exists\text{-}elim),
$$

where in ($\exists$-*elim*), u does not occur free in Γ, $\exists tP$, or C. Again, t is *any* variable.

The variable u is called the *eigenvariable* of the inference. The systems $\mathscr{N}\mathscr{G}_c^{\Rightarrow, \vee, \wedge, \perp, \forall, \exists}$ and $\mathscr{N}\mathscr{G}_i^{\Rightarrow, \vee, \wedge, \perp, \forall, \exists}$ are defined from the systems $\mathscr{N}\mathscr{G}_c^{\Rightarrow, \vee, \wedge, \perp}$ and $\mathscr{N}\mathscr{G}_i^{\Rightarrow, \vee, \wedge, \perp}$, respectively, by adding the above rules. As usual, a *deduction tree* is a either a one-node tree or a tree constructed using the above rules and a *proof tree* is a deduction tree whose conclusion is a sequent with an empty set of premises (a sequent of the form $\emptyset \to P$).

When we say that a proposition P is *provable from* Γ we mean that we can construct a proof tree whose conclusion is P and whose set of premises is Γ in one of the systems $\mathcal{N}_c^{\Rightarrow,\wedge,\vee,\perp,\forall,\exists}$ or $\mathcal{NG}_c^{\Rightarrow,\wedge,\vee,\perp,\forall,\exists}$. Therefore, as in propositional logic, when we use the word "provable" unqualified, we mean provable in *classical logic*. Otherwise, we say *intuitionistically provable*.

In order to prove that the proof systems $\mathcal{N}_c^{\Rightarrow,\wedge,\vee,\perp,\forall,\exists}$ and $\mathcal{NG}_c^{\Rightarrow,\wedge,\vee,\perp,\forall,\exists}$ are equivalent (and similarly for $\mathcal{N}_i^{\Rightarrow,\wedge,\vee,\perp,\forall,\exists}$ and $\mathcal{NG}_i^{\Rightarrow,\wedge,\vee,\perp,\forall,\exists}$), we need to prove that every deduction in the system $\mathcal{N}_c^{\Rightarrow,\wedge,\vee,\perp,\forall,\exists}$ using the $\exists$-elimination rule

$$
\frac{\begin{array}{cc} \Gamma & \Delta, P[u/t]^x \\ \mathcal{D}_1 & \mathcal{D}_2 \\ \exists t P & C \end{array}}{C} \; x
$$

(with the usual restriction on u) can be converted to a deduction using the following version of the $\exists$-elimination rule.

$$
\frac{\begin{array}{cc} \Gamma & \Gamma, P[u/t]^x \\ \mathcal{D}_1 & \mathcal{D}_2 \\ \exists t P & C \end{array}}{C} \; x
$$

where u is a variable that does not occur free in any of the propositions in Γ, $\exists t P$, or C. We leave the details as Problem 1.17.

A first look at the above rules shows that universal formulae $\forall t P$ behave somewhat like infinite conjunctions and that existential formulae $\exists t P$ behave somewhat like infinite disjunctions.

The $\forall$-introduction rule looks a little strange but the idea behind it is actually very simple: Because u is totally unconstrained, if $P[u/t]$ is provable (from Γ), then intuitively $P[u/t]$ holds of any arbitrary object, and so, the statement $\forall t P$ should also be provable (from Γ). Note that the tree

$$
\frac{P[u/t]}{\forall t P}
$$

is generally an *illegal deduction* because the deduction tree above $\forall t P$ is a one-node tree consisting of the single premise $P[u/t]$, and u occurs in $P[u/t]$ unless t does not occur in P.

The meaning of the $\forall$-elimination is that if $\forall t P$ is provable (from Γ), then P holds for all objects and so, in particular for the object denoted by the term τ; that is, $P[\tau/t]$ should be provable (from Γ).

The $\exists$-introduction rule is dual to the $\forall$-elimination rule. If $P[\tau/t]$ is provable (from Γ), this means that the object denoted by τ satisfies P, so $\exists t P$ should be provable (this latter formula asserts the existence of some object satisfying P, and τ is such an object).

The $\exists$-elimination rule is reminiscent of the $\vee$-elimination rule and is a little more tricky. It goes as follows. Suppose that we proved $\exists t P$ (from Γ). Moreover, suppose that for every possible case $P[u/t]$ we were able to prove C (from Γ). Then, as we have "exhausted" all possible cases and as we know from the provability of $\exists t P$ that some case must hold, we can conclude that C is provable (from Γ) without using $P[u/t]$ as a premise.

Like the $\vee$-elimination rule, the $\exists$-elimination rule is not very constructive. It allows making a conclusion (C) by considering alternatives without knowing which one actually occurs.

Remark: Analogously to disjunction, in (first-order) intuitionistic logic, if an existential statement $\exists t P$ is provable, then from any proof of $\exists t P$, some term τ can be extracted so that $P[\tau/t]$ is provable. Such a term τ is called a *witness*. The witness property is not easy to prove. It follows from the fact that intuitionistic proofs have a normal form (see Section 1.11). However, no such property holds in classical logic (for instance, see the a^b rational with a, b irrational example revisited below).

Here is an example of a proof in the system $\mathcal{N}_c^{\Rightarrow,\vee,\wedge,\perp,\forall,\exists}$ (actually, in the system $\mathcal{N}_i^{\Rightarrow,\vee,\wedge,\perp,\forall,\exists}$) of the formula $\forall t (P \wedge Q) \Rightarrow \forall t P \wedge \forall t Q$.

$$
\cfrac{\cfrac{\cfrac{\cfrac{\forall t(P\wedge Q)^x}{P[u/t]\wedge Q[u/t]}}{\cfrac{P[u/t]}{\forall t P}}\qquad \cfrac{\cfrac{\forall t(P\wedge Q)^x}{P[u/t]\wedge Q[u/t]}}{\cfrac{Q[u/t]}{\forall t Q}}}{\forall t P \wedge \forall t Q}}{\forall t(P\wedge Q)\Rightarrow \forall t P \wedge \forall t Q} \; x
$$

In the above proof, u is a new variable, that is, a variable that does not occur free in P or Q. We also have used some basic properties of substitutions such as

$$
\begin{aligned}
(P\wedge Q)[\tau/t] &= P[\tau/t]\wedge Q[\tau/t]\\
(P\vee Q)[\tau/t] &= P[\tau/t]\vee Q[\tau/t]\\
(P\Rightarrow Q)[\tau/t] &= P[\tau/t]\Rightarrow Q[\tau/t]\\
(\neg P)[\tau/t] &= \neg P[\tau/t]\\
(\forall s P)[\tau/t] &= \forall s P[\tau/t]\\
(\exists s P)[\tau/t] &= \exists s P[\tau/t],
\end{aligned}
$$

for any term τ such that no variable in τ is captured during the substitution (in particular, in the last two cases, the variable s does not occur in τ).

The reader should show that $\forall t P \wedge \forall t Q \Rightarrow \forall t (P \wedge Q)$ is also provable in the system $\mathcal{N}_i^{\Rightarrow,\vee,\wedge,\perp,\forall,\exists}$. However, in general, one can't just replace $\forall$ by $\exists$ (or $\wedge$ by $\vee$) and still obtain provable statements. For example, $\exists t P \wedge \exists t Q \Rightarrow \exists t (P \wedge Q)$ is not provable at all.

Here are some useful equivalences involving quantifiers. The first two are analogous to the de Morgan laws for $\wedge$ and $\vee$.

Proposition 1.9. *The following equivalences are provable in classical first-order logic.*

$$\neg\forall t P \equiv \exists t \neg P$$
$$\neg\exists t P \equiv \forall t \neg P$$
$$\forall t (P \wedge Q) \equiv \forall t P \wedge \forall t Q$$
$$\exists t (P \vee Q) \equiv \exists t P \vee \exists t Q.$$

In fact, the last three and $\exists t \neg P \Rightarrow \neg\forall t P$ are provable intuitionistically. Moreover, the formulae

$$\exists t (P \wedge Q) \Rightarrow \exists t P \wedge \exists t Q \quad and \quad \forall t P \vee \forall t Q \Rightarrow \forall t (P \vee Q)$$

are provable in intuitionistic first-order logic (and thus, also in classical first-order logic).

Proof. Left as an exercise to the reader. $\square$

Remark: We can illustrate, again, the fact that classical logic allows for nonconstructive proofs by re-examining the example at the end of Section 1.3. There, we proved that if $\sqrt{2}^{\sqrt{2}}$ is rational, then $a = \sqrt{2}$ and $b = \sqrt{2}$ are both irrational numbers such that a^b is rational and if $\sqrt{2}^{\sqrt{2}}$ is irrational then $a = \sqrt{2}^{\sqrt{2}}$ and $b = \sqrt{2}$ are both irrational numbers such that a^b is rational. By $\exists$-introduction, we deduce that if $\sqrt{2}^{\sqrt{2}}$ is rational then there exist some irrational numbers a, b so that a^b is rational and if $\sqrt{2}^{\sqrt{2}}$ is irrational then there exist some irrational numbers a, b so that a^b is rational. In classical logic, as $P \vee \neg P$ is provable, by $\vee$-elimination, we just proved that there exist some irrational numbers a and b so that a^b is rational.

However, this argument does not give us explicitly numbers a and b with the required properties. It only tells us that such numbers must exist. Now, it turns out that $\sqrt{2}^{\sqrt{2}}$ is indeed irrational (this follows from the Gel'fond–Schneider theorem, a hard theorem in number theory). Furthermore, there are also simpler explicit solutions such as $a = \sqrt{2}$ and $b = \log_2 9$, as the reader should check.

We conclude this section by giving an example of a "wrong proof." Here is an example in which the $\forall$-introduction rule is applied illegally, and thus, yields a statement that is actually false (not provable). In the incorrect "proof" below, P is an atomic predicate symbol taking two arguments (e.g., "parent") and 0 is a constant denoting zero:

$$\frac{P(u,0)^x}{\forall t P(t,0)} \quad \text{illegal step!}$$

$$\frac{}{P(t,0) \Rightarrow \forall t P(t,0)} \quad x$$

$$\frac{}{\forall s(P(s,0) \Rightarrow \forall t P(t,0))}$$

$$\frac{}{P(0,0) \Rightarrow \forall t P(t,0)}$$

The problem is that the variable u occurs free in the premise $P[u/t,0] = P(u,0)$ and therefore, the application of the $\forall$-introduction rule in the first step is illegal. However, note that this premise is discharged in the second step and so, the application of the $\forall$-introduction rule in the third step is legal. The (false) conclusion of this faulty proof is that $P(0,0) \Rightarrow \forall t P(t,0)$ is provable. Indeed, there are plenty of properties such that the fact that the single instance $P(0,0)$ holds does not imply that $P(t,0)$ holds for all t.

Remark: The above example shows why it is desirable to have premises that are universally quantified. A premise of the form $\forall t P$ can be instantiated to $P[u/t]$, using $\forall$-elimination, where u is a brand new variable. Later on, it may be possible to use $\forall$-introduction without running into trouble with free occurrences of u in the premises. But we still have to be very careful when we use $\forall$-introduction or $\exists$-elimination.

Before concluding this section, let us give a few more examples of proofs using the rules for the quantifiers. First, let us prove that

$$\forall t P \equiv \forall u P[u/t],$$

where u is any variable not free in $\forall t P$ and such that u is not captured during the substitution. This rule allows us to rename bound variables (under very mild conditions). We have the proofs

$$\frac{(\forall t P)^\alpha}{\dfrac{P[u/t]}{\dfrac{\forall u P[u/t]}{\forall t P \Rightarrow \forall u P[u/t]}}} \quad \alpha$$

and

$$\frac{(\forall u P[u/t])^\alpha}{\dfrac{P[u/t]}{\dfrac{\forall t P}{\forall u P[u/t] \Rightarrow \forall t P}}} \quad \alpha$$

Here is now a proof (intuitionistic) of

$$\exists t(P \Rightarrow Q) \Rightarrow (\forall t P \Rightarrow Q),$$

where t does not occur (free or bound) in Q.

$$
\cfrac{(\exists t(P \Rightarrow Q))^z \quad \cfrac{\cfrac{(P[u/t] \Rightarrow Q)^x \quad \cfrac{(\forall t P)^y \quad}{P[u/t]}}{Q} \; x}{\cfrac{Q}{\forall t P \Rightarrow Q} \; y}}{\exists t(P \Rightarrow Q) \Rightarrow (\forall t P \Rightarrow Q)} \; z
$$

In the above proof, u is a new variable that does not occur in Q, $\forall t P$, or $\exists t(P \Rightarrow Q)$. Because t does not occur in Q, we have

$$(P \Rightarrow Q)[u/t] = P[u/t] \Rightarrow Q.$$

The converse requires (RAA) and is a bit more complicated. Here is a classical proof:

$$
\cfrac{\cfrac{\cfrac{(\neg \exists t(P \Rightarrow Q))^y \quad \cfrac{\cfrac{P[u/t]^\alpha, Q^\beta}{Q}}{\cfrac{Q}{P[u/t] \Rightarrow Q}\;\alpha}}{\exists t(P \Rightarrow Q)}}{\cfrac{\perp}{\neg Q}\;\beta} \quad \cfrac{(\forall t P \Rightarrow Q)^x \quad \cfrac{(\neg \exists t(P \Rightarrow Q))^y \quad \cfrac{P[u/t] \Rightarrow Q}{\exists t(P \Rightarrow Q)}}{\cfrac{\perp}{P[u/t]}\;\delta\,(RAA)}}{\forall t P}}{\cfrac{\cfrac{\perp}{\exists t(P \Rightarrow Q)}\;y\,(RAA)}{(\forall t P \Rightarrow Q) \Rightarrow \exists t(P \Rightarrow Q)}\;x}
$$

where the subproof for $P[u/t] \Rightarrow Q$ uses $\cfrac{\neg P[u/t]^\delta \quad P[u/t]^\gamma}{\cfrac{\perp}{Q}}\;\gamma$ and the bottom combines Q.

Next, we give intuitionistic proofs of

$$(\exists t P \wedge Q) \Rightarrow \exists t(P \wedge Q)$$

and

$$\exists t(P \wedge Q) \Rightarrow (\exists t P \wedge Q),$$

where t does not occur (free or bound) in Q.

Here is an intuitionistic proof of the first implication:

$$
\cfrac{
\cfrac{(\exists t P \wedge Q)^x}{\exists t P}
\qquad
\cfrac{P[u/t]^y \qquad \cfrac{(\exists t P \wedge Q)^x}{Q}}{\cfrac{P[u/t] \wedge Q}{\exists t(P \wedge Q)}}
}{
\cfrac{\exists t(P \wedge Q)}{(\exists t P \wedge Q) \Rightarrow \exists t(P \wedge Q)} \ x
} \ y
$$

In the above proof, u is a new variable that does not occur in $\exists t P$ or Q. Because t does not occur in Q, we have

$$(P \wedge Q)[u/t] = P[u/t] \wedge Q.$$

Here is an intuitionistic proof of the converse:

$$
\cfrac{
\cfrac{(\exists t(P \wedge Q))^x \qquad \cfrac{\cfrac{(P[u/t] \wedge Q)^y}{P[u/t]}}{\exists t P}}{\exists t P} \ y
\qquad
\cfrac{(\exists t(P \wedge Q))^x \qquad \cfrac{(P[u/t] \wedge Q)^z}{Q}}{Q} \ z
}{
\cfrac{\exists t P \wedge Q}{\exists t(P \wedge Q) \Rightarrow (\exists t P \wedge Q)} \ x
}
$$

Finally, we give a proof (intuitionistic) of

$$(\forall t P \vee Q) \Rightarrow \forall t(P \vee Q),$$

where t does not occur (free or bound) in Q.

$$
\cfrac{
(\forall t P \vee Q)^z \qquad
\cfrac{
\cfrac{\cfrac{(\forall t P)^x}{P[u/t]}}{\cfrac{P[u/t] \vee Q}{\forall t(P \vee Q)}}
\qquad
\cfrac{Q^y}{\cfrac{P[u/t] \vee Q}{\forall t(P \vee Q)}}
}{\forall t(P \vee Q)} \ x,y
}{
\cfrac{\forall t(P \vee Q)}{(\forall t P \vee Q) \Rightarrow \forall t(P \vee Q)} \ z
}
$$

In the above proof, u is a new variable that does not occur in $\forall t P$ or Q. Because t does not occur in Q, we have

$$(P \vee Q)[u/t] = P[u/t] \vee Q.$$

The converse requires (RAA).

The useful above equivalences (and more) are summarized in the following propositions.

Proposition 1.10. *(1) The following equivalences are provable in classical first-order logic, provided that t does not occur (free or bound) in Q.*

$$\forall t P \wedge Q \equiv \forall t (P \wedge Q)$$
$$\exists t P \vee Q \equiv \exists t (P \vee Q)$$
$$\exists t P \wedge Q \equiv \exists t (P \wedge Q)$$
$$\forall t P \vee Q \equiv \forall t (P \vee Q).$$

Furthermore, the first three are provable intuitionistically and so is $(\forall t P \vee Q) \Rightarrow \forall t (P \vee Q).$

(2) The following equivalences are provable in classical logic, provided that t does not occur (free or bound) in P.

$$\forall t (P \Rightarrow Q) \equiv (P \Rightarrow \forall t Q)$$
$$\exists t (P \Rightarrow Q) \equiv (P \Rightarrow \exists t Q).$$

Furthermore, the first one is provable intuitionistically and so is $\exists t (P \Rightarrow Q) \Rightarrow (P \Rightarrow \exists t Q).$

(3) The following equivalences are provable in classical logic, provided that t does not occur (free or bound) in Q.

$$\forall t (P \Rightarrow Q) \equiv (\exists t P \Rightarrow Q)$$
$$\exists t (P \Rightarrow Q) \equiv (\forall t P \Rightarrow Q).$$

Furthermore, the first one is provable intuitionistically and so is $\exists t (P \Rightarrow Q) \Rightarrow (\forall t P \Rightarrow Q).$

Proofs that have not been supplied are left as exercises.

Obviously, every first-order formula that is provable intuitionistically is also provable classically and we know that there are formulae that are provable classically but *not* provable intuitionistically. Therefore, it appears that classical logic is more general than intuitionistic logic. However, this not not quite so because there is a way of translating classical logic into intuitionistic logic. To be more precise, every classical formula A can be translated into a formula A^* where A^* is classically equivalent to A and A is provable classically iff A^* is provable intuitionistically. Various translations are known, all based on a "trick" involving double-negation (This is because $\neg\neg\neg A$ and $\neg A$ are intuitionistically equivalent). Translations were given by Kolmogorov (1925), Gödel (1933), and Gentzen (1933).

For example, Gödel used the following translation.

Fig. 1.4 Andrey N. Kolmogorov, 1903–1987 (left) and Kurt Gödel, 1906–1978 (right)

$$A^* = \neg\neg A, \quad \text{if } A \text{ is atomic,}$$
$$(\neg A)^* = \neg A^*,$$
$$(A \wedge B)^* = (A^* \wedge B^*),$$
$$(A \Rightarrow B)^* = \neg(A^* \wedge \neg B^*),$$
$$(A \vee B)^* = \neg(\neg A^* \wedge \neg B^*),$$
$$(\forall x A)^* = \forall x A^*,$$
$$(\exists x A)^* = \neg\forall x \neg A^*.$$

Actually, if we restrict our attention to propositions (i.e., formulae without quantifiers), a theorem of V. Glivenko (1929) states that if a proposition A is provable classically, then $\neg\neg A$ is provable intuitionistically. In view of these results, the proponents of intuitionistic logic claim that classical logic is really a special case of intuitionistic logic. However, the above translations have some undesirable properties, as noticed by Girard. For more details on all this, see Gallier [5].

1.10 First-Order Theories

The way we presented deduction trees and proof trees may have given our readers the impression that the set of premises Γ was just an auxiliary notion. Indeed, in all of our examples, Γ ends up being empty. However, nonempty Γs are crucially needed if we want to develop theories about various kinds of structures and objects, such as the natural numbers, groups, rings, fields, trees, graphs, sets, and the like. Indeed, we need to make definitions about the objects we want to study and we need to state some axioms asserting the main properties of these objects. We do this by putting these definitions and axioms in Γ. Actually, we have to allow Γ to be infinite but we still require that our deduction trees be finite; they can only use finitely many of the formulae in Γ. We are then interested in all formulae P such that $\Delta \rightarrow P$ is provable, where Δ is any finite subset of Γ; the set of all such Ps is called a *theory* (or *first-order theory*). Of course we have the usual problem of consistency: if we are not careful, our theory may be inconsistent, that is, it may consist of all formulae.

Let us give two examples of theories.

Our first example is the *theory of equality*. Indeed, our readers may have noticed that we have avoided dealing with the equality relation. In practice, we can't do that.

Given a language **L** with a given supply of constant, function, and predicate symbols, the theory of equality consists of the following formulae taken as axioms.

$$\forall x(x = x)$$
$$\forall x_1 \cdots \forall x_n \forall y_1 \cdots \forall y_n[(x_1 = y_1 \wedge \cdots \wedge x_n = y_n) \Rightarrow f(x_1, \ldots, x_n) = f(y_1, \ldots, y_n)]$$
$$\forall x_1 \cdots \forall x_n \forall y_1 \cdots \forall y_n[(x_1 = y_1 \wedge \cdots \wedge x_n = y_n) \wedge P(x_1, \ldots, x_n) \Rightarrow P(y_1, \ldots, y_n)],$$

for all function symbols (of *n* arguments) and all predicate symbols (of *n* arguments), including the equality predicate, =, itself.

It is not immediately clear from the above axioms that = is symmetric and transitive but this can be shown easily.

Our second example is the first-order theory of the natural numbers known as *Peano arithmetic* (for short, *PA*).

Fig. 1.5 Giuseppe Peano, 1858–932

Here, we have the constant 0 (zero), the unary function symbol S (for successor function; the intended meaning is $S(n) = n + 1$) and the binary function symbols $+$ (for addition) and $*$ (for multiplication). In addition to the axioms for the theory of equality we have the following axioms:

$$\forall x \neg (S(x) = 0)$$
$$\forall x \forall y (S(x) = S(y) \Rightarrow x = y)$$
$$\forall x (x + 0 = x)$$
$$\forall x \forall y (x + S(y) = S(x + y))$$
$$\forall x (x * 0 = 0)$$
$$\forall x \forall y (x * S(y) = x * y + x)$$
$$[A(0) \wedge \forall x (A(x) \Rightarrow A(S(x)))] \Rightarrow \forall n A(n),$$

where A is any first-order formula with one free variable.

This last axiom is the *induction axiom*. Observe how $+$ and $*$ are defined recursively in terms of 0 and S and that there are infinitely many induction axioms (countably many).

Many properties that hold for the natural numbers (i.e., are true when the symbols $0, S, +, *$ have their usual interpretation and all variables range over the natural numbers) can be proved in this theory (Peano arithmetic), but not all. This is another very famous result of Gödel known as *Gödel's incompleteness theorem* (1931). However, the topic of incompleteness is definitely outside the scope of this course, so we do not say any more about it.

Fig. 1.6 Kurt Gödel with Albert Einstein

However, we feel that it should be intructive for the reader to see how simple properties of the natural numbers can be derived (in principle) in Peano arithmetic.

First, it is convenient to introduce abbreviations for the terms of the form $S^n(0)$, which represent the natural numbers. Thus, we add a countable supply of constants, $0, 1, 2, 3, \ldots$, to denote the natural numbers and add the axioms

$$n = S^n(0),$$

for all natural numbers n. We also write $n + 1$ for $S(n)$.

Let us illustrate the use of the quantifier rules involving terms ($\forall$-elimination and $\exists$-introduction) by proving some simple properties of the natural numbers, namely, being even or odd. We also prove a property of the natural number that we used before (in the proof that $\sqrt{2}$ is irrational), namely, that every natural number is either even or odd. For this, we add the predicate symbols, "even" and "odd", to our language, and assume the following axioms defining these predicates:

$$\forall n(\text{even}(n) \equiv \exists k(n = 2 * k))$$
$$\forall n(\text{odd}(n) \equiv \exists k(n = 2 * k + 1)).$$

Consider the term, $2 * (m + 1) * (m + 2) + 1$, where m is any given natural number. We would like to prove that $\text{odd}(2 * (m + 1) * (m + 2) + 1)$ is provable in Peano arithmetic.

As an auxiliary lemma, we first prove that

$$\forall x \, \text{odd}(2 * x + 1),$$

is provable in Peano arithmetic. Let p be a variable not occurring in any of the axioms of Peano arithmetic (the variable p stands for an arbitrary natural number). From the axiom,

$$\forall n (\text{odd}(n) \equiv \exists k (n = 2 * k + 1)),$$

by $\forall$-elimination where the term $2 * p + 1$ is substituted for the variable n we get

$$\text{odd}(2 * p + 1) \equiv \exists k (2 * p + 1 = 2 * k + 1). \qquad (*)$$

Now, we can think of the provable equation $2 * p + 1 = 2 * p + 1$ as

$$(2 * p + 1 = 2 * k + 1)[p/k],$$

so, by $\exists$-introduction, we can conclude that

$$\exists k (2 * p + 1 = 2 * k + 1),$$

which, by $(*)$, implies that
$$\text{odd}(2 * p + 1).$$

But now, because p is a variable not occurring free in the axioms of Peano arithmetic, by $\forall$-introduction, we conclude that

$$\forall x \, \text{odd}(2 * x + 1).$$

Finally, if we use $\forall$-elimination where we substitute the term, $\tau = (m + 1) * (m + 2)$, for x, we get
$$\text{odd}(2 * (m + 1) * (m + 2) + 1),$$

as claimed.

Now, we wish to prove the formula:

$$\forall n (\text{even}(n) \lor \text{odd}(n)).$$

We use the induction principle of Peano arithmetic with

$$A(n) = \text{even}(n) \lor \text{odd}(n).$$

For the base case, $n = 0$, because $0 = 2 * 0$ (which can be proved from the Peano axioms), we see that $\text{even}(0)$ holds and so $\text{even}(0) \lor \text{odd}(0)$ is proved.

For $n = 1$, because $1 = 2 * 0 + 1$ (which can be proved from the Peano axioms), we see that $\text{odd}(1)$ holds and so $\text{even}(1) \lor \text{odd}(1)$ is proved.

For the induction step, we may assume that $A(n)$ has been proved and we need to prove that $A(n + 1)$ holds.

So, assume that $\text{even}(n) \lor \text{odd}(n)$ holds. We do a proof by cases.

(a) If even(n) holds, by definition this means that $n = 2k$ for some k and then, $n + 1 = 2k + 1$, which again, by definition means that odd$(n + 1)$ holds and thus, even$(n + 1) \vee$ odd$(n + 1)$ holds.

(b) If odd(n) holds, by definition this means that $n = 2k + 1$ for some k and then, $n + 1 = 2k + 2 = 2(k + 1)$, which again, by definition means that even$(n + 1)$ holds and thus, even$(n + 1) \vee$ odd$(n + 1)$ holds.

By $\vee$-elimination, we conclude that even$(n + 1) \vee$ odd$(n + 1)$ holds, establishing the induction step.

Therefore, using induction, we have proved that

$$\forall n(\text{even}(n) \vee \text{odd}(n)).$$

Actually, we know that even(n) and odd(n) are mutually exclusive, which means that

$$\forall n \neg(\text{even}(n) \wedge \text{odd}(n))$$

holds, but how do we prove it?

We can do this using induction. For $n = 0$, the statement odd(0) means that $0 = 2k + 1 = S(2k)$, for some k. However, the first axiom of Peano arithmetic states that $S(x) \neq 0$ for all x, so we get a contradiction.

For the induction step, assume that $\neg(\text{even}(n) \wedge \text{odd}(n))$ holds. We need to prove that $\neg(\text{even}(n + 1) \wedge \text{odd}(n + 1))$ holds and we can do this by using our constructive proof-by-contradiction rule. So, assume that even$(n + 1) \wedge$ odd$(n + 1)$ holds. At this stage, we realize that if we could prove that

$$\forall n(\text{even}(n + 1) \Rightarrow \text{odd}(n)) \tag{$*$}$$

and

$$\forall n(\text{odd}(n + 1) \Rightarrow \text{even}(n)) \tag{$**$}$$

then even$(n + 1) \wedge$ odd$(n + 1)$ would imply even$(n) \wedge$ odd(n), contradicting the assumption $\neg(\text{even}(n) \wedge \text{odd}(n))$. Therefore, the proof is complete if we can prove $(*)$ and $(**)$.

Let's consider the implication $(*)$ leaving the proof of $(**)$ as an exercise.

Assume that even$(n + 1)$ holds. Then, $n + 1 = 2k$, for some natural number k. We can't have $k = 0$ because otherwise we would have $n + 1 = 0$, contradicting one of the Peano axioms. But then, k is of the form $k = h + 1$, for some natural number, h, so

$$n + 1 = 2k = 2(h + 1) = 2h + 2 = (2h + 1) + 1.$$

By the second Peano axiom, we must have

$$n = 2h + 1,$$

which proves that n is odd, as desired.

In that last proof, we made implicit use of the fact that every natural number n different from zero is of the form $n = m + 1$, for some natural number m which is formalized as

$$\forall n((n \neq 0) \Rightarrow \exists m(n = m + 1)).$$

This is easily proved by induction.

Having done all this work, we have finally proved (∗) and after proving (∗∗), we will have proved that

$$\forall n \neg (\text{even}(n) \wedge \text{odd}(n)).$$

It is also easy to prove that

$$\forall n (\text{even}(n) \vee \text{odd}(n))$$

and

$$\forall n \neg (\text{even}(n) \wedge \text{odd}(n))$$

together imply that

$$\forall n (\text{even}(n) \equiv \neg \text{odd}(n)) \quad \text{and} \quad \forall n (\text{odd}(n) \equiv \neg \text{even}(n)),$$

are provable, facts that we used several times in Section 1.6. This is because, if

$$\forall x(P \vee Q) \quad \text{and} \quad \forall x \neg (P \wedge Q)$$

can be deduced intuitionistically from a set of premises, Γ, then

$$\forall x(P \equiv \neg Q) \quad \text{and} \quad \forall x(Q \equiv \neg P)$$

can also be deduced intuitionistically from Γ. It also follows that $\forall x(\neg \neg P \equiv P)$ and $\forall x(\neg \neg Q \equiv Q)$ can be deduced intuitionistically from Γ.

Remark: Even though we proved that every nonzero natural number n is of the form $n = m + 1$, for some natural number m, the expression $n - 1$ does not make sense because the predecessor function $n \mapsto n - 1$ has not been defined yet in our logical system. We need to define a function symbol "pred" satisfying the axioms:

$$\text{pred}(0) = 0$$
$$\forall n(\text{pred}(n + 1) = n).$$

For simplicity of notation, we write $n - 1$ instead of $\text{pred}(n)$. Then, we can prove that if $k \neq 0$, then $2k - 1 = 2(k - 1) + 1$ (which really should be written as $\text{pred}(2k) = 2\text{pred}(k) + 1$). This can indeed be done by induction; we leave the details as an exercise. We can also define substraction, $-$, as a function sastisfying the axioms

$$\forall n(n - 0 = n)$$
$$\forall n \forall m(n - (m + 1) = \text{pred}(n - m)).$$

It is then possible to prove the usual properties of subtraction (by induction).

These examples of proofs in the theory of Peano arithmetic illustrate the fact that constructing proofs in an axiomatized theory is a very laborious and tedious process.

Many small technical lemmas need to be established from the axioms, which renders these proofs very lengthy and often unintuitive. It is therefore important to build up a database of useful basic facts if we wish to prove, with a certain amount of comfort, properties of objects whose properties are defined by an axiomatic theory (such as the natural numbers). However, when in doubt, we can always go back to the formal theory and try to prove rigorously the facts that we are not sure about, even though this is usually a tedious and painful process. Human provers navigate in a "spectrum of formality," most of the time constructing informal proofs containing quite a few (harmless) shortcuts, sometimes making extra efforts to construct more formalized and rigorous arguments if the need arises.

Now, what if the theory of Peano arithmetic were inconsistent! How do know that Peano arithmetic does not imply any contradiction? This is an important and hard question that motivated a lot of the work of Gentzen. An easy answer is that the *standard model* $\mathbb{N}$ of the natural numbers under addition and multiplication validates all the axioms of Peano arithmetic. Therefore, if both P and $\neg P$ could be proved from the Peano axioms, then both P and $\neg P$ would be true in $\mathbb{N}$, which is absurd. To make all this rigorous, we need to define the notion of *truth in a structure*, a notion explained in every logic book. It should be noted that the constructivists will object to the above method for showing the consistency of Peano arithmetic, because it assumes that the infinite set $\mathbb{N}$ exists as a completed entity. Until further notice, we have faith in the consistency of Peano arithmetic (so far, no inconsistency has been found).

Another very interesting theory is *set theory*. There are a number of axiomatizations of set theory and we discuss one of them (ZF) very briefly in Section 1.12.

Several times in this chapter, we have claimed that certain formulae are not provable in some logical system. What kind of reasoning do we use to validate such claims? In the next section, we briefly address this question as well as related ones.

1.11 Decision Procedures, Proof Normalization, Counterexamples

In the previous sections, we saw how the rules of mathematical reasoning can be formalized in various natural deduction systems and we defined a precise notion of proof. We observed that finding a proof for a given proposition was not a simple matter, nor was it to acertain that a proposition is unprovable. Thus, it is natural to ask the following question.

The Decision Problem: Is there a general procedure that takes any arbitrary proposition P as input, always terminates in a finite number of steps, and tells us whether P is provable?

Clearly, it would be very nice if such a procedure existed, especially if it also produced a proof of P when P is provable.

Unfortunately, for rich enough languages, such as first-order logic, it is impossible to find such a procedure. This deep result known as the *undecidability of the decision problem* or *Church's theorem* was proved by A. Church in 1936 (actually, Church proved the undecidability of the validity problem but, by Gödel's completeness theorem, validity and provability are equivalent).

Fig. 1.7 Alonzo Church, 1903–1995 (left) and Alan Turing, 1912–1954 (right)

Proving Church's theorem is hard and a lot of work. One needs to develop a good deal of what is called the *theory of computation*. This involves defining models of computation such as *Turing machines* and proving other deep results such as the *undecidability of the halting problem* and the *undecidability of the Post correspondence problem*, among other things, see Hopcroft, Motwani, and Ullman [12] and Lewis and Papadimitriou [16].

So, our hopes to find a "universal theorem prover" are crushed. However, if we restrict ourselves to propositional logic, classical or intuitionistic, it turns out that procedures solving the decision problem do exist and they even produce a proof of the input proposition when that proposition is provable.

Unfortunately, proving that such procedures exist and are correct in the propositional case is rather difficult, especially for intuitionistic logic. The difficulties have a lot to do with our choice of a natural deduction system. Indeed, even for the system $\mathcal{N}_m^{\Rightarrow}$ (or $\mathcal{N}\mathcal{G}_m^{\Rightarrow}$), provable propositions may have infinitely many proofs. This makes the search process impossible; when do we know how to stop, especially if a proposition is not provable. The problem is that proofs may contain redundancies (Gentzen said "detours"). A typical example of redundancy is when an elimination immediately follows an introduction, as in the following example in which $\mathcal{D}_1$ denotes a deduction with conclusion $\Gamma, x: A \to B$ and $\mathcal{D}_2$ denotes a deduction with conclusion $\Gamma \to A$.

$$
\cfrac{\cfrac{\begin{array}{c} \mathcal{D}_1 \\ \Gamma, x: A \to B \end{array}}{\Gamma \to A \Rightarrow B} \qquad \begin{array}{c} \mathcal{D}_2 \\ \Gamma \to A \end{array}}{\Gamma \to B}
$$

Intuitively, it should be possible to construct a deduction for $\Gamma \to B$ from the two deductions $\mathcal{D}_1$ and $\mathcal{D}_2$ without using at all the hypothesis $x: A$. This is indeed the case. If we look closely at the deduction $\mathcal{D}_1$, from the shape of the inference rules, assumptions are never created, and the leaves must be labeled with expressions of the form $\Gamma', \Delta, x: A, y: C \to C$ or $\Gamma, \Delta, x: A \to A$, where $y \neq x$ and either $\Gamma = \Gamma'$ or $\Gamma = \Gamma', y: C$. We can form a new deduction for $\Gamma \to B$ as follows. In $\mathscr{D}_1$, wherever a leaf of the form $\Gamma, \Delta, x: A \to A$ occurs, replace it by the deduction obtained from $\mathscr{D}_2$ by adding Δ to the premise of each sequent in $\mathscr{D}_2$. Actually, one should be careful to first make a fresh copy of $\mathscr{D}_2$ by renaming all the variables so that clashes with variables in $\mathscr{D}_1$ are avoided. Finally, delete the assumption $x: A$ from the premise of every sequent in the resulting proof. The resulting deduction is obtained by a kind of substitution and may be denoted as $\mathscr{D}_1[\mathscr{D}_2/x]$, with some minor abuse of notation. Note that the assumptions $x: A$ occurring in the leaves of the form $\Gamma', \Delta, x: A, y: C \to C$ were never used anyway. The step that consists in transforming the above redundant proof figure into the deduction $\mathscr{D}_1[\mathscr{D}_2/x]$ is called a *reduction step* or *normalization step*.

The idea of *proof normalization* goes back to Gentzen ([7], 1935). Gentzen noted that (formal) proofs can contain redundancies, or "detours," and that most complications in the analysis of proofs are due to these redundancies. Thus, Gentzen had the idea that the analysis of proofs would be simplified if it were possible to show that every proof can be converted to an equivalent irredundant proof, a proof in normal form. Gentzen proved a technical result to that effect, the "cut-elimination theorem," for a sequent-calculus formulation of first-order logic [7]. Cut-free proofs are direct, in the sense that they never use auxiliary lemmas via the cut rule.

Remark: It is important to note that Gentzen's result gives a particular algorithm to produce a proof in normal form. Thus we know that every proof can be reduced to some normal form using a specific strategy, but there may be more than one normal form, and certain normalization strategies may not terminate.

About 30 years later, Prawitz ([17], 1965) reconsidered the issue of proof normalization, but in the framework of natural deduction rather than the framework of sequent calculi.[1] Prawitz explained very clearly what redundancies are in systems of natural deduction, and he proved that every proof can be reduced to a normal form. Furthermore, this normal form is unique. A few years later, Prawitz ([18], 1971) showed that in fact, every reduction sequence terminates, a property also called *strong normalization*.

A remarkable connection between proof normalization and the notion of computation must also be mentioned. Curry (1958) made the remarkably insightful observation that certain typed combinators can be viewed as representations of proofs (in a Hilbert system) of certain propositions. (See in Curry and Feys [2] (1958), Chapter 9E, pages 312–315.)

Building up on this observation, Howard ([13], 1969) described a general correspondence among propositions and types, proofs in natural deduction and certain typed λ-terms, and proof normalization and β-reduction. (The simply typed λ-calculus was invented by Church, 1940). This correspondence, usually referred

---

[1] This is somewhat ironical, inasmuch as Gentzen began his investigations using a natural deduction system, but decided to switch to sequent calculi (known as Gentzen systems) for technical reasons.

Fig. 1.8 Haskell B. Curry, 1900–1982

to as the *Curry–Howard isomorphism* or *formulae-as-types principle*, is fundamental and very fruitful.

The Curry/Howard isomorphism establishes a deep correspondence between the notion of proof and the notion of computation. Furthermore, and this is the deepest aspect of the Curry/Howard isomorphism, proof normalization corresponds to term reduction in the λ-calculus associated with the proof system. To make the story short, the correspondence between proofs in intuitionistic logic and typed λ-terms on one hand and between proof normalization and β-conversion, can be used to translate results about typed λ-terms into results about proofs in intuitionistic logic.

In summary, using some suitable intuitionistic sequent calculi and Gentzen's cut elimination theorem or some suitable typed λ-calculi and (strong) normalization results about them, it is possible to prove that there is a decision procedure for propositional intuitionistic logic. However, it can also be shown that the time-complexity of any such procedure is very high. As a matter of fact, it was shown by Statman (1979) that deciding whether a proposition is intuitionisticaly provable is P-space complete [19]. Here, we are alluding to *complexity theory*, another active area of computer science, Hopcroft, Motwani, and Ullman [12] and Lewis and Papadimitriou [16].

Readers who wish to learn more about these topics can read my two survey papers Gallier [6] (On the Correspondence Between Proofs and λ-Terms) and Gallier [5] (A Tutorial on Proof Systems and Typed λ-Calculi), both available on the website http://www.cis.upenn.edu/˜jean/gbooks/logic.html and the excellent introduction to proof theory by Troelstra and Schwichtenberg [22].

Anybody who really wants to understand logic should of course take a look at Kleene [15] (the famous "I.M."), but this is not recommended to beginners.

Let us return to the question of deciding whether a proposition is not provable. To simplify the discussion, let us restrict our attention to propositional classical logic. So far, we have presented a very *proof-theoretic* view of logic, that is, a view based on the notion of provability as opposed to a more *semantic* view of based on the notions of truth and models. A possible excuse for our bias is that, as Peter Andrews (from CMU) puts it, "truth is elusive." Therefore, it is simpler to understand what truth is in terms of the more "mechanical" notion of provability. (Peter Andrews even gave the subtitle

To Truth Through Proof

Fig. 1.9 Stephen C. Kleene, 1909–1994

to his logic book Andrews [1].)

Fig. 1.10 Peter Andrews, 1937–

However, mathematicians are not mechanical theorem provers (even if they prove lots of stuff). Indeed, mathematicians almost always think of the objects they deal with (functions, curves, surfaces, groups, rings, etc.) as rather concrete objects (even if they may not seem concrete to the uninitiated) and not as abstract entities solely characterized by arcane axioms.

It is indeed natural and fruitful to try to interpret formal statements semantically. For propositional classical logic, this can be done quite easily if we interpret atomic propositional letters using the truth values **true** and **false**, as explained in Section 1.7. Then, the crucial point that *every provable proposition (say in $\mathcal{N}\mathcal{G}_c^{\Rightarrow,\vee,\wedge,\perp}$) has the value* **true** *no matter how we assign truth values to the letters in our proposition.* In this case, we say that *P is valid.*

The fact that provability implies validity is called *soundness* or *consistency* of the proof system. The soundness of the proof system $\mathcal{N}\mathcal{G}_c^{\Rightarrow,\vee,\wedge,\perp}$ is easy to prove, as sketched in Section 1.7.

We now have a method to show that a proposition P is not provable: Find some truth assignment that makes P **false**.

Such an assignment falsifying P is called a *counterexample*. If P has a counterexample, then it can't be provable because if it were, then by soundness it would be **true** for all possible truth assignments.

But now, another question comes up. If a proposition is not provable, can we always find a counterexample for it? Equivalently, *is every valid proposition prov-*

able? If every valid proposition is provable, we say that our proof system is *complete* (this is the *completeness* of our system).

The system $\mathcal{N}\mathcal{G}_c^{\Rightarrow,\vee,\wedge,\perp}$ is indeed complete. In fact, *all* the classical systems that we have discussed are sound and complete. Completeness is usually a lot harder to prove than soundness. For first-order classical logic, this is known as *Gödel's completeness theorem* (1929). Again, we refer our readers to Gallier [4], van Dalen [23], or or Huth and Ryan [14] for a thorough discussion of these matters. In the first-order case, one has to define *first-order structures* (or *first-order models*).

What about intuitionistic logic?

Well, one has to come up with a richer notion of semantics because it is no longer true that if a proposition is valid (in the sense of our two-valued semantics using **true**, **false**), then it is provable. Several semantics have been given for intuitionistic logic. In our opinion, the most natural is the notion of the *Kripke model*, presented in Section 1.8. Then, again, soundness and completeness hold for intuitionistic proof systems, even in the first-order case (see Section 1.8 and van Dalen [23]).

In summary, semantic models can be used to provide *counterexamples* of unprovable propositions. This is a quick method to establish that a proposition is not provable.

We close this section by repeating something we said earlier: there isn't just one logic but instead, *many* logics. In addition to classical and intuitionistic logic (propositional and first-order), there are: modal logics, higher-order logics, and *linear logic*, a logic due to Jean-Yves Girard, attempting to unify classical and intuitionistic logic (among other goals).

Fig. 1.11 Jean-Yves Girard, 1947–

An excellent introduction to these logics can be found in Troelstra and Schwichtenberg [22]. We warn our readers that most presentations of linear logic are (very) difficult to follow. This is definitely true of Girard's seminal paper [9]. A more approachable version can be found in Girard, Lafont, and Taylor [8], but most readers will still wonder what hit them when they attempt to read it.

In computer science, there is also *dynamic logic*, used to prove properties of programs and *temporal logic* and its variants (originally invented by A. Pnueli), to prove properties of real-time systems. So, logic is alive and well.

1.12 Basics Concepts of Set Theory

Having learned some fundamental notions of logic, it is now a good place before proceeding to more interesting things, such as functions and relations, to go through a very quick review of some basic concepts of set theory. This section takes the very "naive" point of view that a set is a collection of objects, the collection being regarded as a single object. Having first-order logic at our disposal, we could formalize set theory very rigorously in terms of axioms. This was done by Zermelo first (1908) and in a more satisfactory form by Zermelo and Fraenkel in 1921, in a theory known as the "Zermelo–Fraenkel" (ZF) axioms. Another axiomatization was given by John von Neumann in 1925 and later improved by Bernays in 1937. A modification of Bernay's axioms was used by Kurt Gödel in 1940. This approach is now known as "von Neumann–Bernays" (VNB) or "Gödel–Bernays" (GB) set theory. There are many books that give an axiomatic presentation of set theory. Among them, we recommend Enderton [3], which we find remarkably clear and elegant, Suppes [20] (a little more advanced), and Halmos [11], a classic (at a more elementary level).

Fig. 1.12 Ernst F. Zermelo, 1871–1953 (left), Adolf A. Fraenkel, 1891–1965 (middle left), John von Neumann, 1903–1957 (middle right) and Paul I. Bernays, 1888–1977 (right)

However, it must be said that set theory was first created by Georg Cantor (1845–1918) between 1871 and 1879. However, Cantor's work was not unanimously well received by all mathematicians.

Fig. 1.13 Georg F. L. P. Cantor, 1845–1918

Cantor regarded infinite objects as objects to be treated in much the same way as finite sets, a point of view that was shocking to a number of very prominent mathematicians who bitterly attacked him (among them, the powerful Kronecker). Also, it turns out that some paradoxes in set theory popped up in the early 1900s, in particular, Russell's paradox.

Fig. 1.14 Bertrand A. W. Russell, 1872–1970

Russell's paradox (found by Russell in 1902) has to to with the
"set of all sets that are not members of themselves,"

which we denote by

$$R = \{x \mid x \notin x\}.$$

(In general, the notation $\{x \mid P\}$ stand for the set of all objects satisfying the property P.)

Now, classically, either $R \in R$ or $R \notin R$. However, if $R \in R$, then the definition of R says that $R \notin R$; if $R \notin R$, then again, the definition of R says that $R \in R$.

So, we have a contradiction and the existence of such a set is a paradox. The problem is that we are allowing a property (here, $P(x) = x \notin x$), which is "too wild" and circular in nature. As we show, the way out, as found by Zermelo, is to place a restriction on the property P and to also make sure that P picks out elements from some already given set (see the subset axioms below).

The apparition of these paradoxes prompted mathematicians, with Hilbert among its leaders, to put set theory on firmer ground. This was achieved by Zermelo, Fraenkel, von Neumann, Bernays, and Gödel, to name only the major players.

In what follows, we are assuming that we are working in classical logic. We introduce various operations on sets using definitions involving the logical connectives $\land$, $\lor$, $\neg$, $\forall$, and $\exists$. In order to ensure the existence of some of these sets requires some of the *axioms of set theory*, but we are rather casual about that.

Given a set A we write that some object a is an element of (belongs to) the set A as

$$a \in A$$

and that a is not an element of A (does not belong to A) as

$$a \notin A.$$

When are two sets A and B equal? This corresponds to the first axiom of set theory, called the

Extensionality Axiom

Two sets A and B are equal iff they have exactly the same elements; that is,

$$\forall x(x \in A \Rightarrow x \in B) \land \forall x(x \in B \Rightarrow x \in A).$$

The above says: every element of A is an element of B and conversely.

There is a special set having no elements at all, the *empty set*, denoted $\emptyset$. This is the following.

Empty Set Axiom There is a set having no members. This set is denoted $\emptyset$ and it is characterized by the property

$$\forall x(x \notin \emptyset).$$

Remark: Beginners often wonder whether there is more than one empty set. For example, is the empty set of professors distinct from the empty set of potatoes?

The answer is, by the extensionality axiom, there is only *one* empty set.

Given any two objects a and b, we can form the set $\{a,b\}$ containing exactly these two objects. Amazingly enough, this must also be an axiom:

Pairing Axiom

Given any two objects a and b (think sets), there is a set $\{a,b\}$ having as members just a and b.

Observe that if a and b are identical, then we have the set $\{a,a\}$, which is denoted by $\{a\}$ and is called a *singleton set* (this set has a as its only element).

To form bigger sets, we use the union operation. This too requires an axiom.

Union Axiom (Version 1)

For any two sets A and B, there is a set $A \cup B$ called the *union of A and B* defined by

$$x \in A \cup B \quad \text{iff} \quad (x \in A) \lor (x \in B).$$

This reads, x is a member of $A \cup B$ if either x belongs to A or x belongs to B (or both). We also write

$$A \cup B = \{x \mid x \in A \quad \text{or} \quad x \in B\}.$$

Using the union operation, we can form bigger sets by taking unions with singletons. For example, we can form

$$\{a,b,c\} = \{a,b\} \cup \{c\}.$$

Remark: We can systematically construct bigger and bigger sets by the following method: Given any set A let

$$A^+ = A \cup \{A\}.$$

If we start from the empty set, we obtain sets that can be used to define the natural numbers and the $+$ operation corresponds to the successor function on the natural numbers (i.e., $n \mapsto n+1$).

Another operation is the power set formation. It is indeed a "powerful" operation, in the sense that it allows us to form very big sets. For this, it is helpful to define the notion of inclusion between sets. Given any two sets, A and B, we say that A *is a subset of B* (or that A *is included in B*), denoted $A \subseteq B$, iff every element of A is also an element of B, that is,

$$\forall x (x \in A \Rightarrow x \in B).$$

We say that A *is a proper subset of B* iff $A \subseteq B$ and $A \neq B$. This implies that that there is some $b \in B$ with $b \notin A$. We usually write $A \subset B$.

Observe that the equality of two sets can be expressed by

$$A = B \quad \text{iff} \quad A \subseteq B \quad \text{and} \quad B \subseteq A.$$

Power Set Axiom

Given any set A, there is a set $\mathscr{P}(A)$ (also denoted 2^A) called the *power set of A* whose members are exactly the subsets of A; that is,

$$X \in \mathscr{P}(A) \quad \text{iff} \quad X \subseteq A.$$

For example, if $A = \{a, b, c\}$, then

$$\mathscr{P}(A) = \{\emptyset, \{a\}, \{b\}, \{c\}, \{a,b\}, \{a,c\}, \{b,c\}, \{a,b,c\}\},$$

a set containing eight elements. Note that the empty set and A itself are always members of $\mathscr{P}(A)$.

Remark: If A has n elements, it is not hard to show that $\mathscr{P}(A)$ has 2^n elements. For this reason, many people, including me, prefer the notation 2^A for the power set of A.

At this stage, we define intersection and complementation. For this, given any set A and given a property P (specified by a first-order formula) we need to be able to define the subset of A consisting of those elements satisfying P. This subset is denoted by

$$\{x \in A \mid P\}.$$

Unfortunately, there are problems with this construction. If the formula P is somehow a circular definition and refers to the subset that we are trying to define, then some paradoxes may arise.

The way out is to place a restriction on the formula used to define our subsets, and this leads to the subset axioms, first formulated by Zermelo. These axioms are also called *comprehension axioms* or *axioms of separation*.

Subset Axioms

For every first-order formula P we have the axiom:

$$\forall A \exists X \forall x (x \in X \quad \text{iff} \quad (x \in A) \wedge P),$$

where P does *not* contain X as a free variable. (However, P may contain x free.)

The subset axioms says that for every set A there is a set X consisting exactly of those elements of A so that P holds. For short, we usually write

$$X = \{x \in A \mid P\}.$$

As an example, consider the formula

$$P(B, x) = x \in B.$$

Then, the subset axiom says

$$\forall A \exists X \forall x (x \in A \wedge x \in B),$$

which means that X is the set of elements that belong both to A and B. This is called the *intersection of A and B*, denoted by $A \cap B$. Note that

$$A \cap B = \{x \mid x \in A \quad \text{and} \quad x \in B\}.$$

We can also define the *relative complement of B in A*, denoted $A - B$, given by the formula $P(B, x) = x \notin B$, so that

$$A - B = \{x \mid x \in A \quad \text{and} \quad x \notin B\}.$$

In particular, if A is any given set and B is any subset of A, the set $A - B$ is also denoted $\overline{B}$ and is called the *complement of B*. Because $\wedge, \vee$, and $\neg$ satisfy the de Morgan laws (remember, we are dealing with classical logic), for any set X, the operations of union, intersection, and complementation on subsets of X satisfy various identities, in particular the de Morgan laws

$$\overline{A \cap B} = \overline{A} \cup \overline{B}$$
$$\overline{A \cup B} = \overline{A} \cap \overline{B}$$
$$\overline{\overline{A}} = A,$$

and various associativity, commutativity, and distributivity laws.

So far, the union axiom only applies to two sets but later on we need to form infinite unions. Thus, it is necessary to generalize our union axiom as follows.

Union Axiom (Final Version)

Given any set X (think of X as a set of sets), there is a set $\bigcup X$ defined so that

$$x \in \bigcup X \quad \text{iff} \quad \exists B (B \in X \wedge x \in B).$$

This says that $\bigcup X$ consists of all elements that belong to some member of X.

If we take $X = \{A, B\}$, where A and B are two sets, we see that

$$\bigcup \{A, B\} = A \cup B,$$

and so, our final version of the union axiom subsumes our previous union axiom which we now discard in favor of the more general version.

Observe that

$$\bigcup \{A\} = A, \quad \bigcup \{A_1, \ldots, A_n\} = A_1 \cup \cdots \cup A_n.$$

and in particular, $\bigcup \emptyset = \emptyset$.

Using the subset axioms, we can also define infinite intersections. For every nonempty set X there is a set $\bigcap X$ defined by

$$x \in \bigcap X \quad \text{iff} \quad \forall B (B \in X \Rightarrow x \in B).$$

The existence of $\bigcap X$ is justified as follows: Because X is nonempty, it contains some set, A; let

$$P(X, x) = \forall B (B \in X \Rightarrow x \in B).$$

Then, the subset axioms asserts the existence of a set Y so that for every x,

$$x \in Y \quad \text{iff} \quad x \in A \quad \text{and} \quad P(X, x)$$

which is equivalent to

$$x \in Y \quad \text{iff} \quad P(X, x).$$

Therefore, the set Y is our desired set, $\bigcap X$.

Observe that

$$\bigcap \{A, B\} = A \cap B, \quad \bigcap \{A_1, \ldots, A_n\} = A_1 \cap \cdots \cap A_n.$$

Note that $\bigcap \emptyset$ is not defined. Intuitively, it would have to be the set of all sets, but such a set does not exist, as we now show. This is basically a version of Russell's paradox.

Theorem 1.3. *(Russell) There is no set of all sets, that is, there is no set to which every other set belongs.*

Proof. Let A be any set. We construct a set B that does not belong to A. If the set of all sets existed, then we could produce a set that does not belong to it, a contradiction. Let

$$B = \{a \in A \mid a \notin a\}.$$

We claim that $B \notin A$. We proceed by contradiction, so assume $B \in A$. However, by the definition of B, we have

$$B \in B \quad \text{iff} \quad B \in A \quad \text{and} \quad B \notin B.$$

Because $B \in A$, the above is equivalent to

$$B \in B \quad \text{iff} \quad B \notin B,$$

which is a contradiction. Therefore, $B \notin A$ and we deduce that there is no set of all sets. □

Remarks:

(1) We should justify why the equivalence $B \in B$ iff $B \notin B$ is a contradiction. What we mean by "a contradiction" is that if the above equivalence holds, then we can derive $\bot$ (falsity) and thus, all propositions become provable. This is because we can show that for any proposition P if $P \equiv \neg P$ is provable, then every proposition is provable. We leave the proof of this fact as an easy exercise for the reader. By the way, this holds classically as well as intuitionistically.

(2) We said that in the subset axioms, the variable X is not allowed to occur free in P. A slight modification of Russell's paradox shows that allowing X to be free in P leads to paradoxical sets. For example, pick A to be any nonempty set and set $P(X,x) = x \notin X$. Then, look at the (alleged) set

$$X = \{x \in A \mid x \notin X\}.$$

As an exercise, the reader should show that X is empty iff X is nonempty,

This is as far as we can go with the elementary notions of set theory that we have introduced so far. In order to proceed further, we need to define relations and functions, which is the object of the next chapter.

The reader may also wonder why we have not yet discussed infinite sets. This is because we don't know how to show that they exist. Again, perhaps surprisingly, this takes another axiom, the *axiom of infinity*. We also have to define when a set is infinite. However, we do not go into this right now. Instead, we accept that the set of natural numbers $\mathbb{N}$ exists and is infinite. Once we have the notion of a function, we will be able to show that other sets are infinite by comparing their "size" with that of $\mathbb{N}$ (This is the purpose of *cardinal numbers*, but this would lead us too far afield).

Remark: In an axiomatic presentation of set theory, the natural numbers can be defined from the empty set using the operation $A \mapsto A^+ = A \cup \{A\}$ introduced just after the union axiom. The idea due to von Neumann is that the natural numbers, $0, 1, 2, 3, \ldots$, can be viewed as concise notations for the following sets.

$$0 = \emptyset$$
$$1 = 0^+ = \{\emptyset\} = \{0\}$$
$$2 = 1^+ = \{\emptyset, \{\emptyset\}\} = \{0, 1\}$$
$$3 = 2^+ = \{\emptyset, \{\emptyset\}, \{\emptyset, \{\emptyset\}\}\} = \{0, 1, 2\}$$
$$\vdots$$
$$n + 1 = n^+ = \{0, 1, 2, \ldots, n\}$$
$$\vdots$$

Fig. 1.15 John von Neumann

However, the above subsumes induction. Thus, we have to proceed in a different way to avoid circularities.

Definition 1.10. We say that a set X is *inductive* iff

(1) $\emptyset \in X$.
(2) For every $A \in X$, we have $A^+ \in X$.

Axiom of Infinity

There is some inductive set.

Having done this, we make the following.

Definition 1.11. A *natural number* is a set that belongs to every inductive set.

Using the subset axioms, we can show that there is a set whose members are exactly the natural numbers. The argument is very similar to the one used to prove that arbitrary intersections exist. By the axiom of infinity, there is some inductive set, say A. Now consider the property $P(x)$ which asserts that x belongs to every inductive set. By the subset axioms applied to P, there is a set, $\mathbb{N}$, such that

$$x \in \mathbb{N} \quad \text{iff} \quad x \in A \quad \text{and} \quad P(x)$$

and because A is inductive and P says that x belongs to every inductive set, the above is equivalent to

$$x \in \mathbb{N} \quad \text{iff} \quad P(x);$$

that is, $x \in \mathbb{N}$ iff x belongs to every inductive set. Therefore, the set of all natural numbers $\mathbb{N}$ does exist. The set $\mathbb{N}$ is also denoted ω. We can now easily show the following.

Theorem 1.4. *The set $\mathbb{N}$ is inductive and it is a subset of every inductive set.*

Proof. Recall that $\emptyset$ belongs to every inductive set; so, $\emptyset$ is a natural number (0). As $\mathbb{N}$ is the set of natural numbers, $\emptyset (= 0)$ belongs to $\mathbb{N}$. Secondly, if $n \in \mathbb{N}$, this means that n belongs to every inductive set (n is a natural number), which implies that $n^+ = n + 1$ belongs to every inductive set, which means that $n + 1$ is a natural number, that is, $n + 1 \in \mathbb{N}$. Because $\mathbb{N}$ is the set of natural numbers and because every natural number belongs to every inductive set, we conclude that $\mathbb{N}$ is a subset of every inductive set. □

It would be tempting to view $\mathbb{N}$ as the intersection of the family of inductive sets, but unfortunately this family is not a set; it is too "big" to be a set.

As a consequence of the above fact, we obtain the following.

Induction Principle for $\mathbb{N}$: Any inductive subset of $\mathbb{N}$ is equal to $\mathbb{N}$ itself.

Now, in our setting, $0 = \emptyset$ and $n^+ = n + 1$, so the above principle can be restated as follows.

Induction Principle for $\mathbb{N}$ (Version 2): For any subset, $S \subseteq \mathbb{N}$, if $0 \in S$ and $n + 1 \in S$ whenever $n \in S$, then $S = \mathbb{N}$.

We show how to rephrase this induction principle a little more conveniently in terms of the notion of function in the next chapter.

Remarks:

1. We still don't know what an infinite set is or, for that matter, that $\mathbb{N}$ is infinite. This is shown in the next chapter (see Corollary 2.2).
2. Zermelo–Fraenkel set theory (+ Choice) has three more axioms that we did not discuss: The *axiom of choice*, the *replacement axioms* and the *regularity axiom*. For our purposes, only the axiom of choice is needed and we introduce it in Chapter 2. Let us just say that the replacement axioms are needed to deal with ordinals and cardinals and that the regularity axiom is needed to show that every set is grounded. For more about these axioms, see Enderton [3], Chapter 7. The regularity axiom also implies that no set can be a member of itself, an eventuality that is not ruled out by our current set of axioms.

As we said at the beginning of this section, set theory can be axiomatized in first-order logic. To illustrate the generality and expressiveness of first-order logic, we conclude this section by stating the axioms of *Zermelo–Fraenkel set theory* (for short, *ZF*) as first-order formulae. The language of Zermelo–Fraenkel set theory consists of the constant $\emptyset$ (for the empty set), the equality symbol, and of the binary predicate symbol $\in$ for set membership. It is convenient to abbreviate $\neg(x = y)$ as

$x \neq y$ and $\neg(x \in y)$ as $x \notin y$. The axioms are the equality axioms plus the following seven axioms.

$$\forall A \forall B (\forall x (x \in A \equiv x \in B) \Rightarrow A = B)$$
$$\forall x (x \notin \emptyset)$$
$$\forall a \forall b \exists Z \forall x (x \in Z \equiv (x = a \lor x = b))$$
$$\forall X \exists Y \forall x (x \in Y \equiv \exists B (B \in X \land x \in B))$$
$$\forall A \exists Y \forall X (X \in Y \equiv \forall z (z \in X \Rightarrow z \in A))$$
$$\forall A \exists X \forall x (x \in X \equiv (x \in A) \land P)$$
$$\exists X (\emptyset \in X \land \forall y (y \in X \Rightarrow y \cup \{y\} \in X)),$$

where P is any first-order formula that does not contain X free.

- Axiom (1) is the extensionality axiom.
- Axiom (2) is the empty set axiom.
- Axiom (3) asserts the existence of a set Y whose only members are a and b. By extensionality, this set is unique and it is denoted $\{a, b\}$. We also denote $\{a, a\}$ by $\{a\}$.
- Axiom (4) asserts the existence of set Y which is the union of all the sets that belong to X. By extensionality, this set is unique and it is denoted $\bigcup X$. When $X = \{A, B\}$, we write $\bigcup \{A, B\} = A \cup B$.
- Axiom (5) asserts the existence of set Y which is the set of all subsets of A (the power set of A). By extensionality, this set is unique and it is denoted $\mathscr{P}(A)$ or 2^A.
- Axioms (6) are the subset axioms (or axioms of separation).
- Axiom (7) is the infinity axiom, stated using the abbreviations introduced above.

For a comprehensive treatment of axiomatic theory (including the missing three axioms), see [3] and Suppes [20].

1.13 Summary

The main goal of this chapter is to describe precisely the logical rules used in mathematical reasoning and the notion of a mathematical proof. A brief introduction to set theory is also provided. We decided to describe the rules of reasoning in a formalism known as a natural deduction system because the logical rules of such a system mimic rather closely the informal rules that (nearly) everybody uses when constructing a proof in everyday life. Another advantage of natural deduction systems is that it is very easy to present various versions of the rules involving negation and thus, to explain why the "proof-by-contradiction" proof rule or the "law of the excluded middle" allow for the derivation of "nonconstructive" proofs. This is a subtle point often not even touched in traditional presentations of logic. However, inasmuch as most of our readers write computer programs and expect that their programs will not just promise to give an answer but will actually produce results, we feel that they will grasp rather easily the difference between constructive and nonconstructive proofs, and appreciate the latter, even if they are harder to find.

- We describe the syntax of *propositional logic.*
- The proof rules for *implication* are defined in a *natural deduction system* (Prawitz-style).
- *Deductions* proceed from *assumptions* (or *premises*) using *inference rules.*
- The process of *discharging* (or *closing*) a premise is explained. A *proof* is a deduction in which all the premises have been discharged.
- We explain how we can *search* for a proof using a combined bottom-up and top-down process.
- We propose another mechanism for decribing the process of discharging a premise and this leads to a formulation of the rules in terms of *sequents* and to a *Gentzen system.*
- We introduce falsity $\perp$ and negation $\neg P$ as an abbrevation for $P \Rightarrow \perp$. We describe the inference rules for conjunction, disjunction, and negation, in both Prawitz style and Gentzen-sequent style *natural deduction systems*
- One of the rules for negation is the *proof-by-contradiction* rule (also known as *RAA*).
- We define *intuitionistic* and *classical* logic.
- We introduce the notion of a *constructive* (or *intuitionistic*) proof and discuss the two nonconstructive culprits: $P \vee \neg P$ (the *law of the excluded middle*) and $\neg\neg P \Rightarrow P$ (*double-negation rule*).
- We show that $P \vee \neg P$ and $\neg\neg P \Rightarrow P$ are provable in classical logic
- We clear up some potential confusion involving the various versions of the rules regarding negation.

  1. RAA is not a special case of $\neg$-introduction.
  2. RAA is not equivalent to $\perp$-elimination; in fact, it implies it.
  3. Not all propositions of the form $P \vee \neg P$ are provable in intuitionistic logic. However, RAA holds in intuitionistic logic plus all propositions of the form $P \vee \neg P$.
  4. We define *double-negation elimination.*

- We present the *de Morgan laws* and prove their validity in classical logic.
- We present the *proof-by-contrapositive rule* and show that it is valid in classical logic.
- We give some examples of proofs of "real" statements.
- We give an example of a nonconstructive proof of the statement: there are two irrational numbers, a and b, so that a^b is rational.
- We explain the *truth-value semantics* of propositional logic.
- We define the *truth tables* for the propositional connectives
- We define the notions of *satisfiability, unsatisfiability, validity,* and *tautology.*
- We define the *satisfiability problem* and the *validity problem* (for classical propositional logic).

- We mention the *NP-completeness* of satisfiability.
- We discuss *soundness* (or *consistency*) and *completeness*.
- We state the *soundness and completeness theorems* for propositional classical logic formulated in natural deduction.
- We explain how to use *counterexamples* to prove that certain propositions are not provable.
- We give a brief introduction to *Kripke semantics* for propositional intuitionistic logic.
- We define *Kripke models* (based on a *set of worlds*).
- We define *validity* in a Kripke model.
- We state the the *soundness and completeness theorems* for propositional intuitionistic logic formulated in natural deduction.
- We add *first-order quantifiers* ("for all" $\forall$ and "there exists" $\exists$) to the language of propositional logic and define *first-order logic*.
- We describe *free* and *bound* variables.
- We give inference rules for the quantifiers in Prawitz-style and Gentzen sequent-style *natural deduction systems*.
- We explain the *eigenvariable restriction* in the $\forall$-introduction and $\exists$-elimination rules.
- We prove some "de Morgan"-type rules for the quantified formulae valid in classical logic.
- We discuss the nonconstructiveness of proofs of certain existential statements.
- We explain briefly how classical logic can be translated into intuitionistic logic (the Gödel translation).
- We define *first-order theories* and give the example of *Peano arithmetic*.
- We revisit the *decision problem* and mention the *undecidability of the decision problem* for first-order logic (*Church's theorem*).
- We discuss the notion of *detours* in proofs and the notion of *proof normalization*.
- We mention *strong normalization*.
- We mention the correspondence between propositions and types and proofs and typed λ-terms (the *Curry–Howard isomorphism*).
- We mention *Gödel's completeness theorem* for first-order logic.
- Again, we mention the use of *counterexamples*.
- We mention *Gödel's incompleteness theorem*.
- We present informally the axioms of *Zermelo–Fraenkel set theory* (ZF).
- We present *Russell's paradox*, a warning against "self-referential" definitions of sets.
- We define the *empty set* ($\emptyset$), the set $\{a,b\}$, whose elements are a and b, the *union* $A \cup B$, of two sets A and B, and the *power set* 2^A, of A.
- We state carefully Zermelo's *subset axioms* for defining the subset $\{x \in A \mid P\}$ of elements of a given set A satisfying a property P.
- Then, we define the *intersection* $A \cap B$, and the *relative complement* $A - B$, of two sets A and B.
- We also define the *union* $\bigcup A$ and the *intersection* $\bigcap A$, of a set of sets A.

- We show that one should avoid sets that are "too big;" in particular, we prove that there is no *set of all sets*.
- We define the *natural numbers* "a la Von Neumann."
- We define *inductive sets* and state the *axiom of infinity*.
- We show that the natural numbers form an inductive set $\mathbb{N}$, and thus, obtain an *induction principle for* $\mathbb{N}$.
- We summarize the axioms of Zermelo–Fraenkel set theory in first-order logic.

Problems

1.1. (a) Give a proof of the proposition $P \Rightarrow (Q \Rightarrow P)$ in the system $\mathcal{N}_m^{\Rightarrow}$.

(b) Prove that if there are deduction trees of $P \Rightarrow Q$ and $Q \Rightarrow R$ from the set of premises Γ in the system $\mathcal{N}_m^{\Rightarrow}$, then there is a deduction tree for $P \Rightarrow R$ from Γ in $\mathcal{N}_m^{\Rightarrow}$.

1.2. Give a proof of the proposition $(P \Rightarrow Q) \Rightarrow ((P \Rightarrow (Q \Rightarrow R)) \Rightarrow (P \Rightarrow R))$ in the system $\mathcal{N}_m^{\Rightarrow}$.

1.3. (a) Prove the "de Morgan" laws in classical logic:

$$\neg(P \wedge Q) \equiv \neg P \vee \neg Q$$
$$\neg(P \vee Q) \equiv \neg P \wedge \neg Q.$$

(b) Prove that $\neg(P \vee Q) \equiv \neg P \wedge \neg Q$ is also provable in intuitionistic logic.

(c) Prove that the proposition $(P \wedge \neg Q) \Rightarrow \neg(P \Rightarrow Q)$ is provable in intuitionistic logic and $\neg(P \Rightarrow Q) \Rightarrow (P \wedge \neg Q)$ is provable in classical logic.

1.4. (a) Show that $P \Rightarrow \neg\neg P$ is provable in intuitionistic logic.

(b) Show that $\neg\neg\neg P$ and $\neg P$ are equivalent in intuitionistic logic.

1.5. Show that if we assume that all propositions of the form $P \vee \neg P$ are provable, then the proof-by-contradiction rule can be established from the rules of intuitionistic logic.

1.6. Recall that an integer is *even* if it is divisible by 2, that is, if it can be written as $2k$, where $k \in \mathbb{Z}$. An integer is *odd* if it is not divisible by 2, that is, if it can be written as $2k + 1$, where $k \in \mathbb{Z}$. Prove the following facts.

(a) The sum of even integers is even.
(b) The sum of an even integer and of an odd integer is odd.
(c) The sum of two odd integers is even.
(d) The product of odd integers is odd.
(e) The product of an even integer with any integer is even.

1.7. (a) Show that if we assume that all propositions of the form

$$P \Rightarrow (Q \Rightarrow R)$$

are axioms (where P, Q, R are arbitrary propositions), then *every proposition* is provable.

(b) Show that if P is provable (intuitionistically or classically), then $Q \Rightarrow P$ is also provable for *every* proposition Q.

1.8. (a) Give intuitionistic proofs for the equivalences

$$P \vee P \equiv P$$
$$P \wedge P \equiv P$$
$$P \vee Q \equiv Q \vee P$$
$$P \wedge Q \equiv Q \wedge P.$$

(b) Give intuitionistic proofs for the equivalences

$$P \wedge (P \vee Q) \equiv P$$
$$P \vee (P \wedge Q) \equiv P.$$

1.9. Give intuitionistic proofs for the propositions

$$P \Rightarrow (Q \Rightarrow (P \wedge Q))$$
$$(P \Rightarrow Q) \Rightarrow ((P \Rightarrow \neg Q) \Rightarrow \neg P)$$
$$(P \Rightarrow R) \Rightarrow ((Q \Rightarrow R) \Rightarrow ((P \vee Q) \Rightarrow R)).$$

1.10. Prove that the following equivalences are provable intuitionistically:

$$P \wedge (P \Rightarrow Q) \equiv P \wedge Q$$
$$Q \wedge (P \Rightarrow Q) \equiv Q$$
$$(P \Rightarrow (Q \wedge R)) \equiv ((P \Rightarrow Q) \wedge (P \Rightarrow R)).$$

1.11. Give intuitionistic proofs for

$$(P \Rightarrow Q) \Rightarrow \neg\neg(\neg P \vee Q)$$
$$\neg\neg(\neg\neg P \Rightarrow P).$$

1.12. Give an intuitionistic proof for $\neg\neg(P \vee \neg P)$.

1.13. Give intuitionistic proofs for the propositions

$$(P \vee \neg P) \Rightarrow (\neg\neg P \Rightarrow P) \quad \text{and} \quad (\neg\neg P \Rightarrow P) \Rightarrow (P \vee \neg P).$$

Hint. For the second implication, you may want to use Problem 1.12.

1.14. Give intuitionistic proofs for the propositions

$$(P \Rightarrow Q) \Rightarrow \neg\neg(\neg P \vee Q) \quad \text{and} \quad (\neg P \Rightarrow Q) \Rightarrow \neg\neg(P \vee Q).$$

1.15. (1) Prove that every deduction (in $\mathscr{N}_c^{\Rightarrow,\wedge,\vee,\perp}$ or $\mathscr{N}_i^{\Rightarrow,\wedge,\vee,\perp}$) of a proposition P from any set of premises Γ,

$$\Gamma$$
$$\mathscr{D}$$
$$P$$

can be converted to a deduction of P from the set of premises $\Gamma \cup \Delta$, where Δ is any set of propositions (not necessarily disjoint from Γ):

$$\Gamma \cup \Delta$$
$$\mathscr{D}'$$
$$P$$

(2) Consider the proof system obtained by changing the following rules of the proof system $\mathscr{N}_c^{\Rightarrow,\wedge,\vee,\perp}$ given in Definition 1.3:

$\Rightarrow$-*elimination rule*:

$$
\begin{array}{cc}
\Gamma & \Delta \\
\mathscr{D}_1 & \mathscr{D}_2 \\
P \Rightarrow Q & P \\
\hline
& Q
\end{array}
$$

$\wedge$-*introduction rule*:

$$
\begin{array}{cc}
\Gamma & \Delta \\
\mathscr{D}_1 & \mathscr{D}_2 \\
P & Q \\
\hline
& P \wedge Q
\end{array}
$$

$\vee$-*elimination rule*:

$$
\begin{array}{ccc}
\Gamma & \Delta, P^x & \Lambda, Q^y \\
\mathscr{D}_1 & \mathscr{D}_2 & \mathscr{D}_3 \\
P \vee Q & R & R \\
\hline
& R &
\end{array} \quad x,y
$$

$\neg$-*elimination rule*:

$$
\begin{array}{cc}
\Gamma & \Delta \\
\mathscr{D}_1 & \mathscr{D}_2 \\
\neg P & P \\
\hline
& \perp
\end{array}
$$

to the following rules using the same set of premises Γ:

$\Rightarrow$-*elimination rule'*:

$$
\begin{array}{cc}
\Gamma & \Gamma \\
\mathscr{D}_1 & \mathscr{D}_2 \\
P \Rightarrow Q & P \\
\hline
\multicolumn{2}{c}{Q}
\end{array}
$$

$\wedge$-*introduction rule'*:

$$
\begin{array}{cc}
\Gamma & \Gamma \\
\mathscr{D}_1 & \mathscr{D}_2 \\
P & Q \\
\hline
\multicolumn{2}{c}{P \wedge Q}
\end{array}
$$

$\vee$-*elimination rule'*:

$$
\begin{array}{ccc}
\Gamma & \Gamma, P^x & \Gamma, Q^y \\
\mathscr{D}_1 & \mathscr{D}_2 & \mathscr{D}_3 \\
P \vee Q & R & R \\
\hline
\multicolumn{3}{c}{R}
\end{array} \quad x,y
$$

$\neg$-*elimination rule'*:

$$
\begin{array}{cc}
\Gamma & \Gamma \\
\mathscr{D}_1 & \mathscr{D}_2 \\
\neg P & P \\
\hline
\multicolumn{2}{c}{\bot}
\end{array}
$$

and call the resulting proof system $\mathscr{N}_c'^{\Rightarrow,\wedge,\vee,\bot}$.

Prove that every deduction in $\mathscr{N}_c'^{\Rightarrow,\wedge,\vee,\bot}$ of a proposition P from a set of premises Γ can be converted to a deduction of P from Γ in $\mathscr{N}_c^{\Rightarrow,\wedge,\vee,\bot}$.

Hint. Use induction on deduction trees and part (1) of the problem.

Conclude from the above that the same set of propositions is provable in the systems $\mathscr{N}_c^{\Rightarrow,\wedge,\vee,\bot}$ and $\mathscr{N}_c'^{\Rightarrow,\wedge,\vee,\bot}$ and similarly for $\mathscr{N}_i^{\Rightarrow,\wedge,\vee,\bot}$ and $\mathscr{N}_i'^{\Rightarrow,\wedge,\vee,\bot}$, the systems obtained by dropping the proof-by-contradiction rule.

1.16. Prove that the following version of the $\vee$-elimination rule formulated in Gentzen-sequent style is a consequence of the rules of intuitionistic logic:

$$
\frac{\Gamma, x: P \to R \quad \Gamma, y: Q \to R}{\Gamma, z: P \vee Q \to R}
$$

Conversely, if we assume that the above rule holds, then prove that the $\vee$-elimination rule

$$
\frac{\Gamma \to P \vee Q \quad \Gamma, x: P \to R \quad \Gamma, y: Q \to R}{\Gamma \to R} \quad (\vee\text{-}elim)
$$

follows from the rules of intuitionistic logic (of course, excluding the $\vee$-elimination rule).

1.17. (1) Prove that every deduction (in the proof system $\mathcal{N}_c^{\Rightarrow,\wedge,\vee,\perp,\forall,\exists}$ or the proof system $\mathcal{N}_i^{\Rightarrow,\wedge,\vee,\perp,\forall,\exists}$) of a formula P from any set of premises Γ,

$$\Gamma$$
$$\mathcal{D}$$
$$P$$

can be converted to a deduction of P from the set of premises $\Gamma \cup \Delta$, where Δ is any set of formulae (not necessarily disjoint from Γ) such that all the variable free in Γ are distinct from all the eigenvariables occurring in the original deduction:

$$\Gamma \cup \Delta$$
$$\mathcal{D}'$$
$$P$$

(2) Let $\Gamma = \{P_1, \dots, P_n\}$ be a set of first-order formulae. If $\{t_1, \dots, t_m\}$ is the set of all variables occurring free in the formulae in Γ, we denote by $\Gamma[\tau_1/t_1, \dots, \tau_m/t_m]$ the set of substituted formulae

$$\{P_1[\tau_1/t_1, \dots, \tau_m/t_m], \dots, P_m[\tau_1/t_1, \dots, \tau_m/t_m]\},$$

where $\tau_1, \dots, \tau_m$ are any terms such that no capture takes place when the substitutions are made.

Given any deduction of a formula P from a set of premises Γ (in the proof system $\mathcal{N}_c^{\Rightarrow,\wedge,\vee,\perp,\forall,\exists}$ or the proof system $\mathcal{N}_i^{\Rightarrow,\wedge,\vee,\perp,\forall,\exists}$), if $\{t_1, \dots, t_m\}$ is the set of all the variables occurring free in the formulae in Γ, then prove that there is a deduction of $P[u_1/t_1, \dots, u_m/t_m]$ from $\Gamma[u_1/t_1, \dots, u_m/t_m]$, where $\{u_1, \dots, u_m\}$ is a set of variables not occurring free or bound in any of the propositions in Γ or in P (e.g., a set of "fresh" variables).

Hint. Use induction on deduction trees.

(3) Prove that every deduction (in the proof system $\mathcal{N}_c^{\Rightarrow,\wedge,\vee,\perp,\forall,\exists}$ or the proof system $\mathcal{N}_i^{\Rightarrow,\wedge,\vee,\perp,\forall,\exists}$) using the $\exists$-elimination rule

$$
\begin{array}{cc}
\Gamma & \Delta, P[u/t]^x \\
\mathcal{D}_1 & \mathcal{D}_2 \\
\exists t\, P & C \\
\hline
\multicolumn{2}{c}{C}
\end{array} \quad x
$$

(with the usual restriction on u) can be converted to a deduction using the following version of the $\exists$-elimination rule.

$$
\begin{array}{cc}
\Gamma & \Gamma, P[u/t]^x \\
\mathcal{D}_1 & \mathcal{D}_2 \\
\exists t\, P & C \\
\hline
\multicolumn{2}{c}{C}
\end{array} \quad x
$$

where u is a variable that does not occur free in any of the propositions in Γ, $\exists t\, P$, or C.

Hint. Use induction on deduction trees and parts (1) and (2) of this problem.

1.18. (a) Give intuitionistic proofs for the distributivity of $\wedge$ over $\vee$ and of $\vee$ over $\wedge$:

$$P \wedge (Q \vee R) \equiv (P \wedge Q) \vee (P \wedge R)$$
$$P \vee (Q \wedge R) \equiv (P \vee Q) \wedge (P \vee R).$$

(b) Give intuitionistic proofs for the associativity of $\wedge$ and $\vee$:

$$P \wedge (Q \wedge R) \equiv (P \wedge Q) \wedge R$$
$$P \vee (Q \vee R) \equiv (P \vee Q) \vee R.$$

1.19. Recall that in Problem 1.1 we proved that if $P \Rightarrow Q$ and $Q \Rightarrow R$ are provable, then $P \Rightarrow R$ is provable. Deduce from this fact that if $P \equiv Q$ and $Q \equiv R$ hold, then $P \equiv R$ holds (intuitionistically or classically).

Prove that if $P \equiv Q$ holds then $Q \equiv P$ holds (intuitionistically or classically). Finally, check that $P \equiv P$ holds (intuitionistically or classically).

1.20. Prove (intuitionistically or classically) that if $P_1 \Rightarrow Q_1$ and $P_2 \Rightarrow Q_2$ then

1. $(P_1 \wedge P_2) \Rightarrow (Q_1 \wedge Q_2)$
2. $(P_1 \vee P_2) \Rightarrow (Q_1 \vee Q_2)$.

(b) Prove (intuitionistically or classically) that if $Q_1 \Rightarrow P_1$ and $P_2 \Rightarrow Q_2$ then

1. $(P_1 \Rightarrow P_2) \Rightarrow (Q_1 \Rightarrow Q_2)$
2. $\neg P_1 \Rightarrow \neg Q_1$.

(c) Prove (intuitionistically or classically) that if $P \Rightarrow Q$, then

1. $\forall t P \Rightarrow \forall t Q$
2. $\exists t P \Rightarrow \exists t Q$.

(d) Prove (intuitionistically or classically) that if $P_1 \equiv Q_1$ and $P_2 \equiv Q_2$ then

1. $(P_1 \wedge P_2) \equiv (Q_1 \wedge Q_2)$
2. $(P_1 \vee P_2) \equiv (Q_1 \vee Q_2)$
3. $(P_1 \Rightarrow P_2) \equiv (Q_1 \Rightarrow Q_2)$
4. $\neg P_1 \equiv \neg Q_1$
5. $\forall t P_1 \equiv \forall t Q_1$
6. $\exists t P_1 \equiv \exists t Q_1$.

1.21. Show that the following are provable in classical first-order logic:

$$\neg \forall t P \equiv \exists t \neg P$$
$$\neg \exists t P \equiv \forall t \neg P$$
$$\forall t (P \wedge Q) \equiv \forall t P \wedge \forall t Q$$
$$\exists t (P \vee Q) \equiv \exists t P \vee \exists t Q.$$

(b) Moreover, show that the propositions $\exists t(P \wedge Q) \Rightarrow \exists t P \wedge \exists t Q$ and $\forall t P \vee \forall t Q \Rightarrow \forall t (P \vee Q)$ are provable in intuitionistic first-order logic (and thus, also in classical first-order logic).

(c) Prove intuitionistically that

$$\exists x \forall y P \Rightarrow \forall y \exists x P.$$

Give an informal argument to the effect that the converse, $\forall y \exists x P \Rightarrow \exists x \forall y P$, is not provable, even classically.

1.22. (a) Assume that Q is a formula that does **not** contain the variable t (free or bound). Give a classical proof of

$$\forall t(P \vee Q) \Rightarrow (\forall t P \vee Q).$$

(b) If P is a proposition, write $P(x)$ for $P[x/t]$ and $P(y)$ for $P[y/t]$, where x and y are distinct variables that do not occur in the orginal proposition P. Give an intuitionistic proof for

$$\neg \forall x \exists y (\neg P(x) \wedge P(y)).$$

(c) Give a classical proof for

$$\exists x \forall y (P(x) \vee \neg P(y)).$$

Hint. Negate the above, then use some identities we've shown (such as de Morgan) and reduce the problem to part (b).

1.23. (a) Let $X = \{X_i \mid 1 \le i \le n\}$ be a finite family of sets. Prove that if $X_{i+1} \subseteq X_i$ for all i, with $1 \le i \le n - 1$, then

$$\bigcap X = X_n.$$

Prove that if $X_i \subseteq X_{i+1}$ for all i, with $1 \le i \le n - 1$, then

$$\bigcup X = X_n.$$

(b) Recall that $\mathbb{N}_+ = \mathbb{N} - \{0\} = \{1, 2, 3, \ldots, n, \ldots\}$. Give an example of an infinite family of sets, $X = \{X_i \mid i \in \mathbb{N}_+\}$, such that

1. $X_{i+1} \subseteq X_i$ for all $i \ge 1$.
2. X_i is infinite for every $i \ge 1$.
3. $\bigcap X$ has a single element.

(c) Give an example of an infinite family of sets, $X = \{X_i \mid i \in \mathbb{N}_+\}$, such that

1. $X_{i+1} \subseteq X_i$ for all $i \ge 1$.
2. X_i is infinite for every $i \ge 1$.
3. $\bigcap X = \emptyset$.

1.24. Prove that the following propositions are provable intuitionistically:

$$(P \Rightarrow \neg P) \equiv \neg P, \qquad (\neg P \Rightarrow P) \equiv \neg\neg P.$$

Use these to conlude that if the equivalence $P \equiv \neg P$ is provable intuitionistically, then *every* proposition is provable (intuitionistically).

1.25. (1) Prove that if we assume that all propositions of the form,

$$((P \Rightarrow Q) \Rightarrow P) \Rightarrow P,$$

are axioms (Peirce's law), then $\neg\neg P \Rightarrow P$ becomes provable in intuitionistic logic. Thus, another way to get classical logic from intuitionistic logic is to add Peirce's law to intuitionistic logic.

Hint. Pick Q in a suitable way and use Problem 1.24.

(2) Prove $((P \Rightarrow Q) \Rightarrow P) \Rightarrow P$ in classical logic.

Hint. Use the de Morgan laws.

1.26. Let A be any nonempty set. Prove that the definition

$$X = \{a \in A \mid a \notin X\}$$

yields a "set," X, such that X is empty iff X is nonempty and therefore does not define a set, after all.

1.27. Prove the following fact: if

$$\begin{array}{ccc} \Gamma & & \Gamma, R \\ \mathcal{D}_1 & \text{and} & \mathcal{D}_2 \\ P \vee Q & & Q \end{array}$$

are deduction trees provable intuitionistically, then there is a deduction tree

$$\begin{array}{c} \Gamma, P \Rightarrow R \\ \mathcal{D} \\ Q \end{array}$$

for Q from the premises in $\Gamma \cup \{P \Rightarrow S\}$.

1.28. Recall that the constant $\top$ stands for **true**. So, we add to our proof systems (intuitionistic and classical) all axioms of the form

$$\frac{\overbrace{P_1, \ldots, P_1}^{k_1}, \ldots, \overbrace{P_i, \ldots, P_i}^{k_i}, \ldots, \overbrace{P_n, \ldots, P_n}^{k_n}}{\top}$$

where $k_i \geq 1$ and $n \geq 0$; note that $n = 0$ is allowed, which amounts to the one-node tree, $\top$.

(a) Prove that the following equivalences hold intuitionistically.

$$P \vee \top \equiv \top$$
$$P \wedge \top \equiv P.$$

Prove that if P is intuitionistically (or classically) provable, then $P \equiv \top$ is also provable intuitionistically (or classically). In particular, in classical logic, $P \vee \neg P \equiv \top$. Also prove that

$$P \vee \bot \equiv P$$
$$P \wedge \bot \equiv \bot$$

hold intuitionistically.

(b) In the rest of this problem, we are dealing only with classical logic. The connective *exclusive or*, denoted $\oplus$, is defined by

$$P \oplus Q \equiv (P \wedge \neg Q) \vee (\neg P \wedge Q).$$

In solving the following questions, you will find that constructing proofs using the rules of classical logic is very tedious because these proofs are very long. Instead, use some identities from previous problems.

Prove the equivalence

$$\neg P \equiv P \oplus \top.$$

(c) Prove that

$$P \oplus P \equiv \bot$$
$$P \oplus Q \equiv Q \oplus P$$
$$(P \oplus Q) \oplus R \equiv P \oplus (Q \oplus R).$$

(d) Prove the equivalence

$$P \vee Q \equiv (P \wedge Q) \oplus (P \oplus Q).$$

1.29. Give a classical proof of

$$\neg(P \Rightarrow \neg Q) \Rightarrow (P \wedge Q).$$

1.30. (a) Prove that the rule

$$
\begin{array}{cc}
\Gamma & \Delta \\
\mathscr{D}_1 & \mathscr{D}_2 \\
P \Rightarrow Q & \neg Q \\
\hline
\multicolumn{2}{c}{\neg P}
\end{array}
$$

can be derived from the other rules of intuitionistic logic.

(b) Give an intuitionistic proof of $\neg P$ from $\Gamma = \{\neg(\neg P \vee Q), P \Rightarrow Q\}$ or equivalently, an intuitionistic proof of

$$\left(\neg(\neg P \vee Q) \wedge (P \Rightarrow Q)\right) \Rightarrow \neg P.$$

1.31. (a) Give intuitionistic proofs for the equivalences

$$\exists x \exists y P \equiv \exists y \exists x P \quad \text{and} \quad \forall x \forall y P \equiv \forall y \forall x P.$$

(b) Give intuitionistic proofs for

$$(\forall t P \wedge Q) \Rightarrow \forall t (P \wedge Q) \quad \text{and} \quad \forall t (P \wedge Q) \Rightarrow (\forall t P \wedge Q),$$

where t does not occur (free or bound) in Q.

(c) Give intuitionistic proofs for

$$(\exists t P \vee Q) \Rightarrow \exists t (P \vee Q) \quad \text{and} \quad \exists t (P \vee Q) \Rightarrow (\exists t P \vee Q),$$

where t does not occur (free or bound) in Q.

1.32. An integer, $n \in \mathbb{Z}$, is divisible by 3 iff $n = 3k$, for some $k \in \mathbb{Z}$. Thus (by the division theorem), an integer, $n \in \mathbb{Z}$, is not divisible by 3 iff it is of the form $n = 3k + 1, 3k + 2$, for some $k \in \mathbb{Z}$ (you don't have to prove this).

Prove that for any integer, $n \in \mathbb{Z}$, if n^2 is divisible by 3, then n is divisible by 3.

Hint. Prove the contrapositive. If n of the form $n = 3k + 1, 3k + 2$, then so is n^2 (for a different k).

1.33. Use Problem 1.32 to prove that $\sqrt{3}$ is irrational, that is, $\sqrt{3}$ can't be written as $\sqrt{3} = p/q$, with $p, q \in \mathbb{Z}$ and $q \neq 0$.

1.34. Give an intuitionistic proof of the proposition

$$\left((P \Rightarrow R) \wedge (Q \Rightarrow R)\right) \equiv \left((P \vee Q) \Rightarrow R\right).$$

1.35. Give an intuitionistic proof of the proposition

$$\left((P \wedge Q) \Rightarrow R\right) \equiv \left(P \Rightarrow (Q \Rightarrow R)\right).$$

1.36. (a) Give an intuitionistic proof of the proposition $(P \wedge Q) \Rightarrow (P \vee Q)$.

(b) Prove that the proposition $(P \vee Q) \Rightarrow (P \wedge Q)$ is not valid, where P, Q, are propositional symbols.

(c) Prove that the proposition $(P \vee Q) \Rightarrow (P \wedge Q)$ is not provable in general and that if we assume that *all* propositions of the form $(P \vee Q) \Rightarrow (P \wedge Q)$ are axioms, then *every* proposition becomes provable intuitionistically.

1.37. Give the details of the proof of Proposition 1.7; namely, if a proposition P is provable in the system $\mathcal{N}_c^{\Rightarrow, \wedge, \vee, \perp}$ (or $\mathcal{N}\mathcal{G}_c^{\Rightarrow, \wedge, \vee, \perp}$), then it is valid (according to the truth values semantics).

1.38. Give the details of the proof of Theorem 1.8; namely, if a proposition P is provable in the system $\mathcal{N}_i^{\Rightarrow,\wedge,\vee,\perp}$ (or $\mathcal{N}\mathcal{G}_i^{\Rightarrow,\wedge,\vee,\perp}$), then it is valid in every Kripke model; that is, it is intuitionistically valid.

1.39. Prove that $b = \log_2 9$ is irrational. Then, prove that $a = \sqrt{2}$ and $b = \log_2 9$ are two irrational numbers such that a^b is rational.

1.40. (1) Prove that if $\forall x \neg (P \wedge Q)$ can be deduced intuitionistically from a set of premises Γ, then $\forall x(P \Rightarrow \neg Q)$ and $\forall x(Q \Rightarrow \neg P)$ can also be deduced intuitionistically from Γ.

(2) Prove that if $\forall x(P \vee Q)$ can be deduced intuitionistically from a set of premises Γ, then $\forall x(\neg P \Rightarrow Q)$ and $\forall x(\neg Q \Rightarrow P)$ can also be deduced intuitionistically from Γ.

Conclude that if

$$\forall x(P \vee Q) \quad \text{and} \quad \forall x \neg (P \wedge Q)$$

can be deduced intuitionistically from a set of premises Γ, then

$$\forall x(P \equiv \neg Q) \quad \text{and} \quad \forall x(Q \equiv \neg P)$$

can also be deduced intuitionistically from Γ.

(3) Prove that if $\forall x(P \Rightarrow Q)$ can be deduced intuitionistically from a set of premises Γ, then $\forall x(\neg Q \Rightarrow \neg P)$ can also be deduced intuitionistically from Γ. Use this to prove that if

$$\forall x(P \equiv \neg Q) \quad \text{and} \quad \forall x(Q \equiv \neg P)$$

can be deduced intuitionistically from a set of premises Γ, then $\forall x(\neg\neg P \equiv P)$ and $\forall x(\neg\neg Q \equiv Q)$ can be deduced intuitionistically from Γ.

1.41. Prove that the formula,

$$\forall x \operatorname{even}(2 * x),$$

is provable in Peano arithmetic. Prove that

$$\operatorname{even}(2 * (n + 1) * (n + 3)),$$

is provable in Peano arithmetic for any natural number n.

1.42. A first-order formula A is said to be in *prenex-form* if either

(1) A is a quantifier-free formula.
(2) $A = \forall t B$ or $A = \exists t B$, where B is in prenex-form.

In other words, a formula is in prenex form iff it is of the form

$$Q_1 t_1 Q_2 t_2 \cdots Q_m t_m P,$$

where P is quantifier-free and where $Q_1 Q_2 \cdots Q_m$ is a string of quantifiers, $Q_i \in \{\forall, \exists\}$.

Prove that every first-order formula A is classically equivalent to a formula B in prenex form.

1.43. Even though natural deduction proof systems for classical propositional logic are complete (with respect to the truth values semantics), they are not adequate for designing algorithms searching for proofs (because of the amount of nondeterminism involved).

Gentzen designed a different kind of proof system using *sequents* (later refined by Kleene, Smullyan, and others) that is far better suited for the design of automated theorem provers. Using such a proof system (a *sequent calculus*), it is relatively easy to design a procedure that terminates for all input propositions P and either certifies that P is (classically) valid or else returns some (or all) falsifying truth assignment(s) for P. In fact, if P is valid, the tree returned by the algorithm can be viewed as a proof of P in this proof system.

For this miniproject, we describe a *Gentzen sequent-calculus G'* for propositional logic that lends itself well to the implementation of algorithms searching for proofs or falsifying truth assignments of propositions.

Such algorithms build trees whose nodes are labeled with pairs of sets called sequents. A *sequent* is a pair of sets of propositions denoted by

$$P_1, \ldots, P_m \to Q_1, \ldots, Q_n,$$

with $m, n \geq 0$. Symbolically, a sequent is usally denoted $\Gamma \to \Delta$, where Γ and Δ are two finite sets of propositions (not necessarily disjoint).

For example,

$$\to P \Rightarrow (Q \Rightarrow P), \ P \vee Q \to, \ P, Q \to P \wedge Q$$

are sequents. The sequent $\to$, where both $\Gamma = \Delta = \emptyset$ corresponds to falsity.

The choice of the symbol $\to$ to separate the two sets of propositions Γ and Δ is commonly used and was introduced by Gentzen but there is nothing special about it. If you don't like it, you may replace it by any symbol of your choice as long as that symbol does not clash with the logical connectives ($\Rightarrow, \wedge, \vee, \neg$). For example, you could denote a sequent

$$P_1, \ldots, P_m; Q_1, \ldots, Q_n,$$

using the semicolon as a separator.

Given a truth assignment v to the propositional letters in the propositions P_i and Q_j, we say that v *satisfies the sequent*, $P_1, \ldots, P_m \to Q_1, \ldots, Q_n$, iff

$$v((P_1 \wedge \cdots \wedge P_m) \Rightarrow (Q_1 \vee \cdots \vee Q_n)) = \textbf{true},$$

or equivalently, v *falsifies the sequent*, $P_1, \ldots, P_m \to Q_1, \ldots, Q_n$, iff

$$v(P_1 \wedge \cdots \wedge P_m \wedge \neg Q_1 \wedge \cdots \wedge \neg Q_n) = \textbf{true},$$

iff

$$v(P_i) = \textbf{true}, \ 1 \leq i \leq m \quad \text{and} \quad v(Q_j) = \textbf{false}, \ 1 \leq j \leq n.$$

A sequent is *valid* iff it is satisfied by all truth assignments iff it cannot be falsified.

Note that a sequent $P_1, \ldots, P_m \to Q_1, \ldots, Q_n$ can be falsified iff some truth assignment satisfies all of $P_1, \ldots, P_m$ and falsifies all of $Q_1, \ldots, Q_n$. In particular, if $\{P_1, \ldots, P_m\}$ and $\{Q_1, \ldots, Q_n\}$ have some common proposition (they have a nonempty intersection), then the sequent, $P_1, \ldots, P_m \to Q_1, \ldots, Q_n$, is valid. On the other hand if all the P_is and Q_js are propositional letters and $\{P_1, \ldots, P_m\}$ and $\{Q_1, \ldots, Q_n\}$ are disjoint (they have no symbol in common), then the sequent, $P_1, \ldots, P_m \to Q_1, \ldots, Q_n$, is falsified by the truth assignment v where $v(P_i) = $ **true**, for $i = 1, \ldots m$ and $v(Q_j) = $ **false**, for $j = 1, \ldots, n$.

The main idea behind the design of the proof system G' is to systematically *try to falsify a sequent*. If such an attempt fails, the sequent is valid and a proof tree is found. Otherwise, all falsifying truth assignments are returned. In some sense

failure to falsify is success (in finding a proof).

The rules of G' are designed so that the conclusion of a rule is falsified by a truth assignment v iff its single premise of one of its two premises is falsified by v. Thus, these rules can be viewed as *two-way* rules that can either be read bottom-up or top-down.

Here are the axioms and the rules of the *sequent calculus G'*:

Axioms: $\Gamma, P \to P, \Delta$

Inference rules:

$$\frac{\Gamma, P, Q, \Delta \to \Lambda}{\Gamma, P \wedge Q, \Delta \to \Lambda} \quad \wedge\text{: left} \qquad\qquad \frac{\Gamma \to \Delta, P, \Lambda \quad \Gamma \to \Delta, Q, \Lambda}{\Gamma \to \Delta, P \wedge Q, \Lambda} \quad \wedge\text{: right}$$

$$\frac{\Gamma, P, \Delta \to \Lambda \quad \Gamma, Q, \Delta \to \Lambda}{\Gamma, P \vee Q, \Delta \to \Lambda} \quad \vee\text{: left} \qquad\qquad \frac{\Gamma \to \Delta, P, Q, \Lambda}{\Gamma \to \Delta, P \vee Q, \Lambda} \quad \vee\text{: right}$$

$$\frac{\Gamma, \Delta \to P, \Lambda \quad Q, \Gamma, \Delta \to \Lambda}{\Gamma, P \Rightarrow Q, \Delta \to \Lambda} \quad \Rightarrow\text{: left} \qquad\qquad \frac{P, \Gamma \to Q, \Delta, \Lambda}{\Gamma \to \Delta, P \Rightarrow Q, \Lambda} \quad \Rightarrow\text{: right}$$

$$\frac{\Gamma, \Delta \to P, \Lambda}{\Gamma, \neg P, \Delta \to \Lambda} \quad \neg\text{: left} \qquad\qquad \frac{P, \Gamma \to \Delta, \Lambda}{\Gamma \to \Delta, \neg P, \Lambda} \quad \neg\text{: right}$$

where Γ, Δ, Λ are any finite sets of propositions, possibly the empty set.

A *deduction tree* is either a one-node tree labeled with a sequent or a tree constructed according to the rules of system G'. A *proof tree* (or *proof*) is a deduction tree whose leaves are *all* axioms. A proof tree for a proposition P is a proof tree for the sequent $\to P$ (with an empty left-hand side).

For example,

$$P, Q \to P$$

is a proof tree.

Here is a proof tree for $(P \Rightarrow Q) \Rightarrow (\neg Q \Rightarrow \neg P)$:

$$\cfrac{\cfrac{\cfrac{P, \neg Q \to P}{\neg Q \to \neg P, P}}{\to P, (\neg Q \Rightarrow \neg P)} \qquad \cfrac{\cfrac{Q \to Q, \neg P}{\neg Q, Q \to \neg P}}{Q \to (\neg Q \Rightarrow \neg P)}}{\cfrac{(P \Rightarrow Q) \to (\neg Q \Rightarrow \neg P)}{\to (P \Rightarrow Q) \Rightarrow (\neg Q \Rightarrow \neg P)}}$$

The following is a deduction tree but not a proof tree,

$$\cfrac{\cfrac{\cfrac{P, R \to P}{R \to \neg P, P}}{\to P, (R \Rightarrow \neg P)} \qquad \cfrac{\cfrac{R, Q, P \to}{R, Q \to \neg P}}{Q \to (R \Rightarrow \neg P)}}{\cfrac{(P \Rightarrow Q) \to (R \Rightarrow \neg P)}{\to (P \Rightarrow Q) \Rightarrow (R \Rightarrow \neg P)}}$$

because its rightmost leaf, $R, Q, P \to$, is falsified by the truth assignment $v(P) = v(Q) = v(R) = \textbf{true}$, which also falsifies $(P \Rightarrow Q) \Rightarrow (R \Rightarrow \neg P)$.

Let us call a sequent $P_1, \ldots, P_m \to Q_1, \ldots, Q_n$ *finished* if either it is an axiom ($P_i = Q_j$ for some i and some j) or all the propositions P_i and Q_j are atomic and $\{P_1, \ldots, P_m\} \cap \{Q_1, \ldots, Q_n\} = \emptyset$. We also say that a deduction tree is finished if all its leaves are finished sequents.

The beauty of the system G' is that for every sequent, $P_1, \ldots, P_m \to Q_1, \ldots, Q_n$, the process of building a deduction tree from this sequent *always terminates with a tree where all leaves are finished independently of the order in which the rules are applied*. Therefore, we can apply any strategy we want when we build a deduction tree and we are sure that we will get a deduction tree with all its leaves finished. If all the leaves are axioms, then we have a proof tree and the sequent is valid, or else all the leaves that are not axioms yield a falsifying assignment, and all falsifying assignments for the root sequent are found this way.

If we only want to know whether a proposition (or a sequent) is valid, we can stop as soon as we find a finished sequent that is not an axiom because in this case, the input sequent is falsifiable.

(1) Prove that for every sequent $P_1, \ldots, P_m \to Q_1, \ldots, Q_n$ any sequence of applications of the rules of G' terminates with a deduction tree whose leaves are all finished sequents (a finished deduction tree).

Hint. Define the number of connectives $c(P)$ in a proposition P as follows.

(1) If P is a propositional symbol, then

$$c(P) = 0.$$

(2) If $P = \neg Q$, then

$$c(\neg Q) = c(Q) + 1.$$

(3) If $P = Q * R$, where $* \in \{\Rightarrow, \vee, \wedge\}$, then

$$c(Q * R) = c(Q) + c(R) + 1.$$

Given a sequent,

$$\Gamma \rightarrow \Delta = P_1, \ldots, P_m \rightarrow Q_1, \ldots, Q_n,$$

define the number of connectives, $c(\Gamma \rightarrow \Delta)$, in $\Gamma \rightarrow \Delta$ by

$$c(\Gamma \rightarrow \Delta) = c(P_1) + \cdots + c(P_m) + c(Q_1) + \cdots + c(Q_n).$$

Prove that the application of every rule decreases the number of connectives in the premise(s) of the rule.

(2) Prove that for every sequent $P_1, \ldots, P_m \rightarrow Q_1, \ldots, Q_n$ for every finished deduction tree T constructed from $P_1, \ldots, P_m \rightarrow Q_1, \ldots, Q_n$ using the rules of G', every truth assignment v satisfies $P_1, \ldots, P_m \rightarrow Q_1, \ldots, Q_n$ iff v satisfies every leaf of T. Equivalently, a truth assignment v falsifies $P_1, \ldots, P_m \rightarrow Q_1, \ldots, Q_n$ iff v falsifies some leaf of T.

Deduce from the above that a sequent is valid iff all leaves of every finished deduction tree T are axioms. Furthermore, if a sequent is not valid, then for every finished deduction tree T, for that sequent, every falsifying assignment for that sequent is a falsifying assignment of some leaf of the tree, T.

(3) **Programming Project**:
Design an algorithm taking any sequent as input and constructing a finished deduction tree. If the deduction tree is a proof tree, output this proof tree in some fashion (such a tree can be quite big so you may have to find ways of "flattening" these trees). If the sequent is falsifiable, stop when the algorithm encounters the first leaf that is not an axiom and output the corresponding falsifying truth assignment.

I suggest using a *depth-first expansion strategy* for constructing a deduction tree. What this means is that when building a deduction tree, the algorithm will proceed recursively as follows. Given a nonfinished sequent

$$A_1, \ldots, A_p \rightarrow B_1, \ldots, B_q,$$

if A_i is the *leftmost* nonatomic proposition if such proposition occurs on the left or if B_j is the leftmost nonatomic proposition if all the A_is are atomic, then

(1) The sequent is of the form

$$\Gamma, A_i, \Delta \rightarrow \Lambda,$$

with A_i the leftmost nonatomic proposition. Then either

(a) $A_i = C_i \wedge D_i$ or $A_i = \neg C_i$, in which case either we recursively construct a (finished) deduction tree

$$\mathscr{D}_1$$
$$\Gamma, C_i, D_i, \Delta \rightarrow \Lambda$$

to get the deduction tree

$$\mathcal{D}_1$$
$$\Gamma, C_i, D_i, \Delta \to \Lambda$$
$$\overline{\Gamma, C_i \wedge D_i, \Delta \to \Lambda}$$

or we recursively construct a (finished) deduction tree

$$\mathcal{D}_1$$
$$\Gamma, \Delta \to C_i, \Lambda$$

to get the deduction tree

$$\mathcal{D}_1$$
$$\Gamma, \Delta \to C_i, \Lambda$$
$$\overline{\Gamma, \neg C_i, \Delta \to \Lambda}$$

or

(b) $A_i = C_i \vee D_i$ or $A_i = C_i \Rightarrow D_i$, in which case either we recursively construct two (finished) deduction trees

$$\mathcal{D}_1 \qquad\qquad \mathcal{D}_2$$
$$\Gamma, C_i, \Delta \to \Lambda \quad \text{and} \quad \Gamma, D_i, \Delta \to \Lambda$$

to get the deduction tree

$$\mathcal{D}_1 \qquad\qquad\qquad \mathcal{D}_2$$
$$\Gamma, C_i, \Delta \to \Lambda \qquad\qquad \Gamma, D_i, \Delta \to \Lambda$$
$$\overline{\qquad\qquad \Gamma, C_i \vee D_i, \Delta \to \Lambda \qquad\qquad}$$

or we recursively construct two (finished) deduction trees

$$\mathcal{D}_1 \qquad\qquad \mathcal{D}_2$$
$$\Gamma, \Delta \to C_i, \Lambda \quad \text{and} \quad D_i, \Gamma, \Delta \to \Lambda$$

to get the deduction tree

$$\mathcal{D}_1 \qquad\qquad\qquad \mathcal{D}_2$$
$$\Gamma, \Delta \to C_i, \Lambda \qquad\qquad D_i, \Gamma, \Delta \to \Lambda$$
$$\overline{\qquad\qquad \Gamma, C_i \Rightarrow D_i, \Delta \to \Lambda \qquad\qquad}$$

(2) The nonfinished sequent is of the form

$$\Gamma \to \Delta, B_j, \Lambda,$$

with B_j the leftmost nonatomic proposition. Then either

(a) $B_j = C_j \vee D_j$ or $B_j = C_j \Rightarrow D_j$, or $B_j = \neg C_j$, in which case either we recursively construct a (finished) deduction tree

$$\mathcal{D}_1$$
$$\Gamma \to \Delta, C_j, D_j, \Lambda$$

to get the deduction tree

$$\frac{\begin{array}{c}\mathscr{D}_1\\ \Gamma \rightarrow \Delta, C_j, D_j, \Lambda\end{array}}{\Gamma \rightarrow \Delta, C_j \vee D_j, \Lambda}$$

or we recursively construct a (finished) deduction tree

$$\begin{array}{c}\mathscr{D}_1\\ C_j, \Gamma \rightarrow D_j, \Delta, \Lambda\end{array}$$

to get the deduction tree

$$\frac{\begin{array}{c}\mathscr{D}_1\\ C_j, \Gamma \rightarrow D_j, \Delta, \Lambda\end{array}}{\Gamma \rightarrow \Delta, C_j \Rightarrow D_j, \Lambda}$$

or we recursively construct a (finished) deduction tree

$$\begin{array}{c}\mathscr{D}_1\\ C_j, \Gamma \rightarrow \Delta, \Lambda\end{array}$$

to get the deduction tree

$$\frac{\begin{array}{c}\mathscr{D}_1\\ C_j, \Gamma \rightarrow \Delta, \Lambda\end{array}}{\Gamma \rightarrow \Delta, \neg C_j, \Lambda}$$

or

(b) $B_j = C_j \wedge D_j$, in which case we recursively construct two (finished) deduction trees

$$\begin{array}{ccc}\mathscr{D}_1 & & \mathscr{D}_2\\ \Gamma \rightarrow \Delta, C_j, \Lambda & \text{and} & \Gamma \rightarrow \Delta, D_j, \Lambda\end{array}$$

to get the deduction tree

$$\frac{\begin{array}{ccc}\mathscr{D}_1 & & \mathscr{D}_2\\ \Gamma \rightarrow \Delta, C_j, \Lambda & & \Gamma \rightarrow \Delta, D_j, \Lambda\end{array}}{\Gamma \rightarrow \Delta, C_j \wedge D_j, \Lambda}$$

If you prefer, you can apply a *breadth-first expansion strategy* for constructing a deduction tree.

1.44. Let A and be B be any two sets of sets.

(1) Prove that

$$\left(\bigcup A\right) \cup \left(\bigcup B\right) = \bigcup (A \cup B).$$

(2) Assume that A and B are nonempty. Prove that

$$\left(\bigcap A\right) \cap \left(\bigcap B\right) = \bigcap (A \cup B).$$

(3) Assume that A and B are nonempty. Prove that

$$\bigcup (A \cap B) \subseteq \left(\bigcup A\right) \cap \left(\bigcup B\right)$$

and give a counterexample of the inclusion

$$\left(\bigcup A\right) \cap \left(\bigcup B\right) \subseteq \bigcup (A \cap B).$$

Hint. Reduce the above questions to the provability of certain formulae that you have already proved in a previous assignment (you need **not** re-prove these formulae).

1.45. A set A is said to be *transitive* iff for all $a \in A$ and all $x \in a$, then $x \in A$, or equivalently, for all $a \in A$,

$$a \in A \Rightarrow a \subseteq A.$$

(1) Check that a set A is transitive iff

$$\bigcup A \subseteq A$$

iff

$$A \subseteq 2^A.$$

(2) Recall the definition of the von Neumann successor of a set A given by

$$A^+ = A \cup \{A\}.$$

Prove that if A is a transitive set, then

$$\bigcup (A^+) = A.$$

(3) Recall the von Neumann definition of the natural numbers. Check that for every natural number m

$$m \in m^+ \text{ and } m \subseteq m^+.$$

Prove that every natural number is a transitive set.
Hint. Use induction.

(4) Prove that for any two von Neumann natural numbers m and n, if $m^+ = n^+$, then $m = n$.

(5) Prove that the set, $\mathbb{N}$, of natural numbers is a transitive set.
Hint. Use induction.

References

1. Peter B. Andrews. *An Introduction to Mathematical Logic and Type Theory: To truth Through Proof.* New York: Academic Press, 1986.
2. H.B. Curry and R. Feys. *Combinatory Logic, Vol. I.* Studies in Logic. Amsterdam: North-Holland, third edition, 1974.
3. Herbert B. Enderton. *Elements of Set Theory.* New York: Academic Press, first edition, 1977.
4. Jean H. Gallier. *Logic for Computer Science.* New York: Harper and Row, 1986.
5. Jean Gallier. Constructive logics. Part I: A tutorial on proof systems and typed λ-calculi. *Theoretical Computer Science,* 110(2):249–339, 1993.
6. Jean Gallier. On the Correspondence Between Proofs and λ-Terms. In Philippe de Groote, editor, *Cahiers Du Centre de Logique, Vol. 8,* pages 55–138. Louvain-La-Neuve: Academia, 1995.
7. G. Gentzen. Investigations into logical deduction. In M.E. Szabo, editor, *The Collected Papers of Gerhard Gentzen.* Amsterdam: North-Holland, 1969.
8. J.-Y. Girard, Y. Lafont, and P. Taylor. *Proofs and Types,* volume 7 of *Cambridge Tracts in Theoretical Computer Science.* Cambridge, UK: Cambridge University Press, 1989.
9. Jean-Yves Girard. Linear logic. *Theoretical Computer Science,* 50:1–102, 1987.
10. Timothy Gowers. *Mathematics: A Very Short Introduction.* Oxford, UK: Oxford University Press, first edition, 2002.
11. Paul R. Halmos. *Naive Set Theory.* Undergraduate Text in Mathematics. New York: Springer Verlag, first edition, 1974.
12. John E. Hopcroft, Rajeev Motwani, and Jeffrey D. Ullman. *Introduction to Automata, Languages and Computation.* Reading, MA: Addison Wesley, third edition, 2006.
13. W. A. Howard. The formulae-as-types notion of construction. In J. P. Seldin and J. R. Hindley, editors, *To H. B. Curry: Essays on Combinatory Logic, Lambda Calculus and Formalism,* pages 479–490. London: Academic Press, 1980. Reprint of manuscript first published in 1969.
14. Michael Huth and Mark Ryan. *Logic in Computer Science. Modelling and reasonning about systems.* Cambridge, UK: Cambridge University Press, 2000.
15. S. Kleene. *Introduction to Metamathematics.* Amsterdam: North-Holland, seventh edition, 1952.
16. Harry Lewis and Christos H. Papadimitriou. *Elements of the Theory of Computation.* Englewood Cliffs, NJ: Prentice-Hall, second edition, 1997.
17. D. Prawitz. *Natural Deduction, A Proof-Theoretical Study.* Stockholm: Almquist & Wiksell, 1965.
18. D. Prawitz. Ideas and results in proof theory. In J.E. Fenstad, editor, *Proc. 2nd Scand. Log. Symp.,* pages 235–307. New York: North-Holland, 1971.
19. R. Statman. Intuitionistic propositional logic is polynomial-space complete. *Theoretical Computer Science,* 9(1):67–72, 1979.
20. Patrick Suppes. *Axiomatic Set Theory.* New York: Dover, first edition, 1972.
21. G. Takeuti. *Proof Theory,* volume 81 of *Studies in Logic.* Amsterdam: North-Holland, 1975.
22. A.S. Troelstra and H. Schwichtenberg. *Basic Proof Theory,* volume 43 of *Cambridge Tracts in Theoretical Computer Science.* Cambridge, UK: Cambridge University Press, 1996.
23. D. van Dalen. *Logic and Structure.* New York: Universitext. Springer Verlag, second edition, 1980.

Chapter 2
Relations, Functions, Partial Functions

2.1 What is a Function?

We use functions all the time in mathematics and in computer science. But, what exactly is a function?

Roughly speaking, a function f is a rule or mechanism that takes input values in some *input domain*, say X, and produces output values in some *output domain*, say Y, in such a way that to each input $x \in X$ corresponds a *unique* output value $y \in Y$, denoted $f(x)$. We usually write $y = f(x)$, or better, $x \mapsto f(x)$.

Often, functions are defined by some sort of closed expression (a formula), but not always. For example, the formula

$$y = 2x$$

defines a function. Here, we can take both the input and output domain to be $\mathbb{R}$, the set of real numbers. Instead, we could have taken $\mathbb{N}$, the set of natural numbers; this gives us a different function. In the above example, $2x$ makes sense for all input x, whether the input domain is $\mathbb{N}$ or $\mathbb{R}$, so our formula yields a function defined for all of its input values.

Now, look at the function defined by the formula

$$y = \frac{x}{2}.$$

If the input and output domains are both $\mathbb{R}$, again this function is well defined. However, what if we assume that the input and output domains are both $\mathbb{N}$? This time, we have a problem when x is odd. For example, $3/2$ is not an integer, so our function is not defined for all of its input values. It is actually a *partial function*, a concept that subsumes the notion of a function but is more general. Observe that this partial function is defined for the set of even natural numbers (sometimes denoted $2\mathbb{N}$) and this set is called the *domain* (of definition) of f. If we enlarge the output domain to be $\mathbb{Q}$, the set of rational numbers, then our partial function is defined for all inputs.

J. Gallier, *Discrete Mathematics*, Universitext,
DOI 10.1007/978-1-4419-8047-2_2, © Springer Science+Business Media, LLC 2011

Another example of a partial function is given by

$$y = \frac{x+1}{x^2 - 3x + 2},$$

assuming that both the input and output domains are $\mathbb{R}$. Observe that for $x = 1$ and $x = 2$, the denominator vanishes, so we get the undefined fractions $2/0$ and $3/0$. This partial function "blows up" for $x = 1$ and $x = 2$, its value is "infinity" $(= \infty)$, which is not an element of $\mathbb{R}$. So, the domain of f is $\mathbb{R} - \{1, 2\}$.

In summary, partial functions need not be defined for all of their input values and we need to pay close attention to both the input and the output domain of our partial functions.

The following example illustrates another difficulty: consider the partial function given by

$$y = \sqrt{x}.$$

If we assume that the input domain is $\mathbb{R}$ and that the output domain is $\mathbb{R}^+ = \{x \in \mathbb{R} \mid x \geq 0\}$, then this partial function is not defined for negative values of x. To fix this problem, we can extend the output domain to be $\mathbb{C}$, the complex numbers. Then we can make sense of $\sqrt{x}$ when $x < 0$. However, a new problem comes up: every negative number x has two complex square roots, $-i\sqrt{-x}$ and $+i\sqrt{-x}$ (where i is "the" square root of -1). Which of the two should we pick?

In this case, we could systematically pick $+i\sqrt{-x}$ but what if we extend the input domain to be $\mathbb{C}$? Then, it is not clear which of the two complex roots should be picked, as there is no obvious total order on $\mathbb{C}$. We can treat f as a *multivalued function*, that is, a function that may return several possible outputs for a given input value.

Experience shows that it is awkward to deal with multivalued functions and that it is best to treat them as relations (or to change the output domain to be a power set, which is equivalent to viewing the function as a relation).

Let us give one more example showing that it is not always easy to make sure that a formula is a proper definition of a function. Consider the function from $\mathbb{R}$ to $\mathbb{R}$ given by

$$f(x) = 1 + \sum_{n=1}^{\infty} \frac{x^n}{n!}.$$

Here, $n!$ is the function *factorial*, defined by

$$n! = n \cdot (n-1) \cdots 2 \cdot 1.$$

How do we make sense of this infinite expression? Well, that's where analysis comes in, with the notion of limit of a series, and so on. It turns out that $f(x)$ is the exponential function $f(x) = e^x$. Actually, e^x is even defined when x is a complex number or even a square matrix (with real or complex entries). Don't panic, we do not use such functions in this course.

Another issue comes up, that is, the notion of *computability*. In all of our examples, and for most (partial) functions we will ever need to compute, it is clear that it is possible to give a mechanical procedure, that is, a computer program that computes our functions (even if it hard to write such a program or if such a program takes a very long time to compute the output from the input).

Unfortunately, there are functions that, although well defined mathematically, are not computable.[1] For an example, let us go back to first-order logic and the notion of provable proposition. Given a finite (or countably infinite) alphabet of function, predicate, constant symbols, and a countable supply of variables, it is quite clear that the set $\mathscr{F}$ of all propositions built up from these symbols and variables can be enumerated systematically. We can define the function Prov with input domain $\mathscr{F}$ and output domain $\{0, 1\}$, so that, for every proposition $P \in \mathscr{F}$,

$$\mathrm{Prov}(P) = \begin{cases} 1 & \text{if } P \text{ is provable (classically)} \\ 0 & \text{if } P \text{ is not provable (classically).} \end{cases}$$

Mathematically, for every proposition, $P \in \mathscr{F}$, either P is provable or it is not, so this function makes sense. However, by Church's theorem (see Section 1.11), we know that there is **no** computer program that will terminate for all input propositions and give an answer in a finite number of steps. So, although the function Prov makes sense as an abstract function, it is not computable.

Is this a paradox? No, if we are careful when defining a function not to incorporate in the definition any notion of computability and instead to take a more abstract and, in some some sense, naive view of a function as some kind of input/output process given by pairs ⟨input value, output value⟩ (without worrying about the way the output is "computed" from the input).

A rigorous way to proceed is to use the notion of ordered pair and of graph of a function. Before we do so, let us point out some facts about "functions" that were revealed by our examples:

1. In order to define a "function," in addition to defining its input/output behavior, it is also important to specify what is its *input domain* and its *output domain.*
2. Some "functions" may not be defined for all of their input values; a function can be a *partial function.*
3. The input/output behavior of a "function" can be defined by a set of ordered pairs. As we show next, this is the *graph* of the function.

We are now going to formalize the notion of function (possibly partial) using the concept of ordered pair.

---

[1] This can be proved quickly using the notion of *countable set* defined later in this chapter. The set of functions from $\mathbb{N}$ to itself is not countable but computer programs are finite strings over a finite alphabet, so the set of computer programs is countable.

2.2 Ordered Pairs, Cartesian Products, Relations, Functions, Partial Functions

Given two sets A and B, one of the basic constructions of set theory is the formation of an *ordered pair*, $\langle a, b \rangle$, where $a \in A$ and $b \in B$. Sometimes, we also write (a, b) for an ordered pair. The main property of ordered pairs is that if $\langle a_1, b_1 \rangle$ and $\langle a_2, b_2 \rangle$ are ordered pairs, where $a_1, a_2 \in A$ and $b_1, b_2 \in B$, then

$$\langle a_1, b_1 \rangle = \langle a_2, b_2 \rangle \text{ iff } a_1 = a_2 \text{ and } b_1 = b_2.$$

Observe that this property implies that

$$\langle a, b \rangle \neq \langle b, a \rangle,$$

unless $a = b$. Thus, the ordered pair $\langle a, b \rangle$ is not a notational variant for the set $\{a, b\}$; implicit to the notion of ordered pair is the fact that there is an order (even though we have not yet defined this notion) among the elements of the pair. Indeed, in $\langle a, b \rangle$, the element a comes first and b comes second. Accordingly, given an ordered pair $p = \langle a, b \rangle$, we denote a by $pr_1(p)$ and b by $pr_2(p)$ (*first and second projection* or *first and second coordinate*).

Remark: Readers who like set theory will be happy to hear that an ordered pair $\langle a, b \rangle$ can be defined as the set $\{\{a\}, \{a, b\}\}$. This definition is due to K. Kuratowski, 1921. An earlier (more complicated) definition given by N. Wiener in 1914 is $\{\{\{a\}, \emptyset\}, \{\{b\}\}\}$.

Fig. 2.1 Kazimierz Kuratowski, 1896–1980

Now, from set theory, it can be shown that given two sets A and B, the set of all ordered pairs $\langle a, b \rangle$, with $a \in A$ and $b \in B$, is a set denoted $A \times B$ and called the *Cartesian product of A and B* (in that order). The set $A \times B$ is also called the *cross-product* of A and B.

By convention, we agree that $\emptyset \times B = A \times \emptyset = \emptyset$. To simplify the terminology, we often say *pair* for *ordered pair*, with the understanding that pairs are always ordered (otherwise, we should say *set*).

Of course, given three sets, A, B, C, we can form $(A \times B) \times C$ and we call its elements (ordered) *triples* (or *triplets*). To simplify the notation, we write $\langle a, b, c \rangle$ instead of $\langle \langle a, b \rangle, c \rangle$ and $A \times B \times C$ instead of $(A \times B) \times C$.

More generally, given n sets $A_1, \ldots, A_n$ ($n \geq 2$), we define the set of *n-tuples*, $A_1 \times A_2 \times \cdots \times A_n$, as $(\cdots ((A_1 \times A_2) \times A_3) \times \cdots) \times A_n$. An element of $A_1 \times A_2 \times \cdots \times A_n$ is denoted by $\langle a_1, \ldots, a_n \rangle$ (an *n-tuple*). We agree that when $n = 1$, we just have A_1 and a 1-tuple is just an element of A_1.

We now have all we need to define relations.

Definition 2.1. Given two sets A and B, a (binary) *relation between A and B* is any triple $\langle A, R, B \rangle$, where $R \subseteq A \times B$ is any set of ordered pairs from $A \times B$. When $\langle a, b \rangle \in R$, we also write aRb and we say that *a and b are related by R*. The set

$$dom(R) = \{a \in A \mid \exists b \in B, \langle a, b \rangle \in R\}$$

is called the *domain of R* and the set

$$range(R) = \{b \in B \mid \exists a \in A, \langle a, b \rangle \in R\}$$

is called the *range of R*. Note that $dom(R) \subseteq A$ and $range(R) \subseteq B$. When $A = B$, we often say that R *is a (binary) relation over A*.

Sometimes, the term *correspondence between A and B* is used instead of the term relation between A and B and the word *relation* is reserved for the case where $A = B$.

It is worth emphasizing that two relations $\langle A, R, B \rangle$ and $\langle A', R', B' \rangle$ are equal iff $A = A'$, $B = B'$, and $R = R'$. In particular, if $R = R'$ but either $A \neq A'$ or $B \neq B'$, then the relations $\langle A, R, B \rangle$ and $\langle A', R', B' \rangle$ *are considered to be different*. For simplicity, we usually refer to a relation $\langle A, R, B \rangle$ as a relation $R \subseteq A \times B$.

Among all relations between A and B, we mention three relations that play a special role:

1. $R = \emptyset$, the *empty relation*. Note that $dom(\emptyset) = range(\emptyset) = \emptyset$. This is not a very exciting relation.
2. When $A = B$, we have the *identity relation*,

$$id_A = \{\langle a, a \rangle \mid a \in A\}.$$

The identity relation relates every element to itself, and that's it. Note that $dom(id_A) = range(id_A) = A$.
3. The relation $A \times B$ itself. This relation relates every element of A to every element of B. Note that $dom(A \times B) = A$ and $range(A \times B) = B$.

Relations can be represented graphically by pictures often called graphs. (Beware, the term "graph" is very much overloaded. Later on, we define what a graph is.) We depict the elements of both sets A and B as points (perhaps with different colors) and we indicate that $a \in A$ and $b \in B$ are related (i.e., $\langle a, b \rangle \in R$) by drawing an oriented edge (an arrow) starting from a (its source) and ending in b (its target). Here is an example:

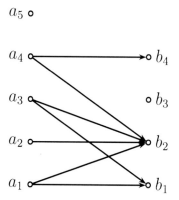

Fig. 2.2 A binary relation, R

In Figure 2.2, $A = \{a_1, a_2, a_3, a_4, a_5\}$ and $B = \{b_1, b_2, b_3, b_4\}$. Observe that a_5 is not related to any element of B, b_3 is not related to any element of A, and that some elements of A, namely, a_1, a_3, a_4, are related to several elements of B.

Now, given a relation $R \subseteq A \times B$, some element $a \in A$ may be related to several distinct elements $b \in B$. If so, R does not correspond to our notion of a function, because we want our functions to be single-valued. So, we impose a natural condition on relations to get relations that correspond to functions.

Definition 2.2. We say that a relation R between two sets A and B is *functional* if for every $a \in A$, there is *at most one* $b \in B$ so that $\langle a, b \rangle \in R$. Equivalently, R is functional if for all $a \in A$ and all $b_1, b_2 \in B$, if $\langle a, b_1 \rangle \in R$ and $\langle a, b_2 \rangle \in R$, then $b_1 = b_2$.

The picture in Figure 2.3 shows an example of a functional relation.

Using Definition 2.2, we can give a rigorous definition of a function (partial or not).

Definition 2.3. A *partial function* f is a triple $f = \langle A, G, B \rangle$, where A is a set called the *input domain of* f, B is a set called the *output domain of* f (sometimes *codomain of* f), and $G \subseteq A \times B$ is a functional relation called the *graph of* f (see Figure 2.4); we let $graph(f) = G$. We write $f: A \to B$ to indicate that A is the input domain of f and that B is the codomain of f and we let $dom(f) = dom(G)$ and $range(f) = range(G)$. For every $a \in dom(f)$, the unique element $b \in B$, so that $\langle a, b \rangle \in graph(f)$ is denoted by $f(a)$ (so, $b = f(a)$). Often we say that $b = f(a)$ is the *image of a by* f. The range of f is also called the *image of* f and is denoted $\text{Im}(f)$. If $dom(f) = A$, we say that f is a *total function*, for short, a *function with domain A*.

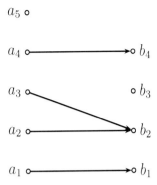

Fig. 2.3 A functional relation G

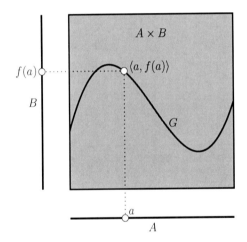

Fig. 2.4 A (partial) function $\langle A, G, B \rangle$

As in the case of relations, it is worth emphasizing that two functions (partial or total) $f = \langle A, G, B \rangle$ and $f' = \langle A', G', B' \rangle$ are equal iff $A = A'$, $B = B'$, and $G = G'$. In particular, if $G = G'$ but either $A \neq A'$ or $B \neq B'$, then the functions (partial or total) f and f' *are considered to be different*.

Observe that most computer programs are not defined for all inputs. For example, programs designed to run on numerical inputs will typically crash when given strings as input. Thus, most computer programs compute partial functions that are not total and it may be very hard to figure out what is the domain of these functions.

This is a strong motivation for considering the notion of a partial function and not just the notion of a (total) function.

Remarks:

1. If $f = \langle A, G, B \rangle$ is a partial function and $b = f(a)$ for some $a \in dom(f)$, we say that f *maps a to b*; we may write $f : a \mapsto b$. For any $b \in B$, the set

$$\{a \in A \mid f(a) = b\}$$

   is denoted $f^{-1}(b)$ and called the *inverse image* or *preimage of b by f*. (It is also called the *fibre of f above b*. We explain this peculiar language later on.) Note that $f^{-1}(b) \neq \emptyset$ iff b is in the image (range) of f. Often, a function, partial or not, is called a *map*.

2. Note that Definition 2.3 allows $A = \emptyset$. In this case, we must have $G = \emptyset$ and, technically, $\langle \emptyset, \emptyset, B \rangle$ is a total function. It is the *empty function from $\emptyset$ to B*.

3. When a partial function is a total function, we don't call it a "partial total function," but simply a "function." The usual practice is that the term "function" refers to a total function. However, sometimes we say "total function" to stress that a function is indeed defined on all of its input domain.

4. Note that if a partial function $f = \langle A, G, B \rangle$ is not a total function, then $dom(f) \neq A$ and for all $a \in A - dom(f)$, there is **no** $b \in B$ so that $\langle a, b \rangle \in graph(f)$. This corresponds to the intuitive fact that f does not produce any output for any value not in its domain of definition. We can imagine that f "blows up" for this input (as in the situation where the denominator of a fraction is 0) or that the program computing f loops indefinitely for that input.

5. If $f = \langle A, G, B \rangle$ is a total function and $A \neq \emptyset$, then $B \neq \emptyset$.

6. For any set A, the identity relation id_A, is actually a function $id_A : A \to A$.

7. Given any two sets A and B, the rules $\langle a, b \rangle \mapsto a = pr_1(\langle a, b \rangle)$ and $\langle a, b \rangle \mapsto b = pr_2(\langle a, b \rangle)$ make pr_1 and pr_2 into functions $pr_1 : A \times B \to A$ and $pr_2 : A \times B \to B$ called the *first and second projections*.

8. A function $f : A \to B$ is sometimes denoted $A \xrightarrow{f} B$. Some authors use a different kind of arrow to indicate that f is partial, for example, a dotted or dashed arrow. We do not go that far.

9. The set of all functions, $f : A \to B$, is denoted by B^A. If A and B are finite, A has m elements and B has n elements, it is easy to prove that B^A has n^m elements.

The reader might wonder why, in the definition of a (total) function, $f : A \to B$, we do not require $B = \text{Im} f$, inasmuch as we require that $dom(f) = A$.

The reason has to do with experience and convenience. It turns out that in most cases, we know what the domain of a function is, but it may be very hard to determine exactly what its image is. Thus, it is more convenient to be flexible about the codomain. As long as we know that f maps into B, we are satisfied.

For example, consider functions $f : \mathbb{R} \to \mathbb{R}^2$ from the real line into the plane. The image of such a function is a *curve* in the plane $\mathbb{R}^2$. Actually, to really get "decent" curves we need to impose some reasonable conditions on f, for example, to be differentiable. Even continuity may yield very strange curves (see Section 2.10).

But even for a very well-behaved function, f, it may be very hard to figure out what the image of f is. Consider the function $t \mapsto (x(t), y(t))$ given by

$$x(t) = \frac{t(1+t^2)}{1+t^4}$$
$$y(t) = \frac{t(1-t^2)}{1+t^4}.$$

The curve that is the image of this function, shown in Figure 2.5, is called the "lemniscate of Bernoulli."

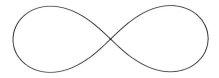

Fig. 2.5 Lemniscate of Bernoulli

Observe that this curve has a self-intersection at the origin, which is not so obvious at first glance.

2.3 Induction Principles on $\mathbb{N}$

Now that we have the notion of function, we can restate the induction principle (Version 2) stated at the end of Section 1.12 to make it more flexible. We define a *property of the natural numbers* as any function, $P: \mathbb{N} \to \{\textbf{true}, \textbf{false}\}$. The idea is that $P(n)$ holds iff $P(n) = \textbf{true}$, else $P(n) = \textbf{false}$. Then, we have the following principle.

Principle of Induction for $\mathbb{N}$ (Version 3).
Let P be any property of the natural numbers. In order to prove that $P(n)$ holds for all $n \in \mathbb{N}$, it is enough to prove that

(1) $P(0)$ holds.
(2) For every $n \in \mathbb{N}$, the implication $P(n) \Rightarrow P(n+1)$ holds.

As a formula, (1) and (2) can be written

$$[P(0) \wedge (\forall n \in \mathbb{N})(P(n) \Rightarrow P(n+1))] \Rightarrow (\forall n \in \mathbb{N})P(n).$$

Step (1) is usually called the *basis* or *base step* of the induction and step (2) is called the *induction step*. In step (2), $P(n)$ is called the *induction hypothesis*. That the above induction principle is valid is given by the following.

Proposition 2.1. *The principle of induction stated above is valid.*

Proof. Let
$$S = \{n \in \mathbb{N} \mid P(n) = \textbf{true}\}.$$

By the induction principle (Version 2) stated at the end of Section 1.12, it is enough to prove that S is inductive, because then $S = \mathbb{N}$ and we are done.

Because $P(0)$ hold, we have $0 \in S$. Now, if $n \in S$ (i.e., if $P(n)$ holds), because $P(n) \Rightarrow P(n+1)$ holds for every n we deduce that $P(n+1)$ holds; that is, $n+1 \in S$. Therefore, S is inductive as claimed and this finishes the proof. $\square$

Induction is a very valuable tool for proving properties of the natural numbers and we make extensive use of it. We also show other more powerful induction principles. Let us give some examples illustrating how it is used.

We begin by finding a formula for the sum

$$1 + 2 + 3 + \cdots + n,$$

where $n \in \mathbb{N}$. If we compute this sum for small values of n, say $n = 0, 1, 2, 3, 4, 5, 6$ we get

$$0 = 0$$
$$1 = 1$$
$$1 + 2 = 3$$
$$1 + 2 + 3 = 6$$
$$1 + 2 + 3 + 4 = 10$$
$$1 + 2 + 3 + 4 + 5 = 15$$
$$1 + 2 + 3 + 4 + 5 + 6 = 21.$$

What is the pattern?

After a moment of reflection, we see that

$$0 = (0 \times 1)/2$$
$$1 = (1 \times 2)/2$$
$$3 = (2 \times 3)/2$$
$$6 = (3 \times 4)/2$$
$$10 = (4 \times 5)/2$$
$$15 = (5 \times 6)/2$$
$$21 = (6 \times 7)/2,$$

so we conjecture

Claim 1:
$$1+2+3+\cdots+n = \frac{n(n+1)}{2},$$

where $n \in \mathbb{N}$.

For the basis of the induction, where $n = 0$, we get $0 = 0$, so the base step holds.
For the induction step, for any $n \in \mathbb{N}$, assume that

$$1+2+3+\cdots+n = \frac{n(n+1)}{2}.$$

Consider $1+2+3+\cdots+n+(n+1)$. Then, using the induction hypothesis, we have

$$
\begin{aligned}
1+2+3+\cdots+n+(n+1) &= \frac{n(n+1)}{2}+n+1 \\
&= \frac{n(n+1)+2(n+1)}{2} \\
&= \frac{(n+1)(n+2)}{2},
\end{aligned}
$$

establishing the induction hypothesis and therefore proving our formula.   □
Next, let us find a formula for the sum of the first $n+1$ odd numbers:

$$1+3+5+\cdots+2n+1,$$

where $n \in \mathbb{N}$. If we compute this sum for small values of n, say $n = 0,1,2,3,4,5,6$
we get

$$
\begin{aligned}
1 &= 1 \\
1+3 &= 4 \\
1+3+5 &= 9 \\
1+3+5+7 &= 16 \\
1+3+5+7+9 &= 25 \\
1+3+5+7+9+11 &= 36 \\
1+3+5+7+9+11+13 &= 49.
\end{aligned}
$$

This time, it is clear what the pattern is: we get perfect squares. Thus, we conjecture
Claim 2:
$$1+3+5+\cdots+2n+1 = (n+1)^2,$$

where $n \in \mathbb{N}$.

For the basis of the induction, where $n = 0$, we get $1 = 1^2$, so the base step holds.
For the induction step, for any $n \in \mathbb{N}$, assume that

$$1+3+5+\cdots+2n+1 = (n+1)^2.$$

Consider $1 + 3 + 5 + \cdots + 2n + 1 + 2(n + 1) + 1 = 1 + 3 + 5 + \cdots + 2n + 1 + 2n + 3$. Then, using the induction hypothesis, we have

$$
\begin{aligned}
1 + 3 + 5 + \cdots + 2n + 1 + 2n + 3 &= (n + 1)^2 + 2n + 3 \\
&= n^2 + 2n + 1 + 2n + 3 = n^2 + 4n + 4 \\
&= (n + 2)^2.
\end{aligned}
$$

Therefore, the induction step holds and this completes the proof by induction. □

The two formulae that we just discussed are subject to a nice geometric interpetation that suggests a closed-form expression for each sum and this is often the case for sums of special kinds of numbers. For the first formula, if we represent n as a sequence of n "bullets," then we can form a rectangular array with n rows and $n + 1$ columns showing that the desired sum is half of the number of bullets in the array, which is indeed $n(n + 1)/2$, as shown below for $n = 5$:

$$
\begin{array}{ccccccc}
\bullet & \circ & \circ & \circ & \circ & \circ \\
\bullet & \bullet & \circ & \circ & \circ & \circ \\
\bullet & \bullet & \bullet & \circ & \circ & \circ \\
\bullet & \bullet & \bullet & \bullet & \circ & \circ \\
\bullet & \bullet & \bullet & \bullet & \bullet & \circ \\
\end{array}
$$

Thus, we see that the numbers

$$
\Delta_n = \frac{n(n + 1)}{2},
$$

have a simple geometric interpretation in terms of triangles of bullets; for example, $\Delta_4 = 10$ is represented by the triangle

For this reason, the numbers Δ_n are often called *triangular numbers*. A natural question then arises; what is the sum

$$
\Delta_1 + \Delta_2 + \Delta_3 + \cdots + \Delta_n?
$$

The reader should compute these sums for small values of n and try to guess a formula that should then be proved correct by induction. It is not too hard to find a nice formula for these sums. The reader may also want to find a geometric interpretation for the above sums (stacks of cannon balls).

In order to get a geometric interpretation for the sum

$$
1 + 3 + 5 + \cdots + 2n + 1,
$$

we represent $2n + 1$ using $2n + 1$ bullets displayed in a V-shape; for example, $7 = 2 \times 3 + 1$ is represented by

Then, the sum $1 + 3 + 5 + \cdots + 2n + 1$ corresponds to the square

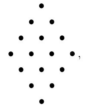

which clearly reveals that

$$1 + 3 + 5 + \cdots + 2n + 1 = (n+1)^2.$$

A natural question is then; what is the sum

$$1^2 + 2^2 + 3^2 + \cdots + n^2?$$

Again, the reader should compute these sums for small values of n, then guess a formula and check its correctness by induction. It is not too difficult to find such a formula. For a fascinating discussion of all sorts of numbers and their geometric interpretations (including the numbers we just introduced), the reader is urged to read Chapter 2 of Conway and Guy [1].

Sometimes, it is necessary to prove a property $P(n)$ for all natural numbers $n \geq m$, where $m > 0$. Our induction principle does not seem to apply because the base case is not $n = 0$. However, we can define the property $Q(n)$ given by

$$Q(n) = P(m+n), \ n \in \mathbb{N},$$

and because $Q(n)$ holds for all $n \in \mathbb{N}$ iff $P(k)$ holds for all $k \geq m$, we can apply our induction principle to prove $Q(n)$ for all $n \in \mathbb{N}$ and thus, $P(k)$, for all $k \geq m$ (note, $k = m + n$). Of course, this amounts to considering that the base case is $n = m$ and this is what we always do without any further justification. Here is an example.

Let us prove that

$$(3n)^2 \leq 2^n, \ \text{for all } n \geq 10.$$

The base case is $n = 10$. For $n = 10$, we get

$$(3 \times 10)^2 = 30^2 = 900 \leq 1024 = 2^{10},$$

which is indeed true. Let us now prove the induction step. Assuming that $(3n)^2 \leq 2^n$ holds for all $n \geq 10$, we want to prove that $(3(n+1))^2 \leq 2^{n+1}$. As

$$(3(n+1))^2 = (3n+3)^2 = (3n)^2 + 18n + 9,$$

if we can prove that $18n + 9 \leq (3n)^2$ when $n \geq 10$, using the induction hypothesis, $(3n)^2 \leq 2^n$, we have

$$(3(n+1))^2 = (3n)^2 + 18n + 9 \leq (3n)^2 + (3n)^2 \leq 2^n + 2^n = 2^{n+1},$$

establishing the induction step. However,

$$(3n)^2 - (18n + 9) = (3n - 3)^2 - 18$$

and $(3n - 3)^2 \geq 18$ as soon as $n \geq 3$, so $18n + 9 \leq (3n)^2$ when $n \geq 10$, as required.

Observe that the formula $(3n)^2 \leq 2^n$ fails for $n = 9$, because $(3 \times 9)^2 = 27^2 = 729$ and $2^9 = 512$, but $729 > 512$. Thus, the base has to be $n = 10$.

There is another induction principle which is often more flexible than our original induction principle. This principle, called *complete induction* (or sometimes *strong induction*), is stated below.

Complete Induction Principle for $\mathbb{N}$.

In order to prove that a predicate $P(n)$ holds for all $n \in \mathbb{N}$ it is enough to prove that

(1) $P(0)$ holds (the base case).
(2) For every $m \in \mathbb{N}$, if $(\forall k \in \mathbb{N})(k \leq m \Rightarrow P(k))$ then $P(m+1)$.

The difference between ordinary induction and complete induction is that in complete induction, the induction hypothesis $(\forall k \in \mathbb{N})(k \leq m \Rightarrow P(k))$ assumes that $P(k)$ holds for all $k \leq m$ and not just for m (as in ordinary induction), in order to deduce $P(m+1)$. This gives us more proving power as we have more knowledge in order to prove $P(m+1)$. Complete induction is discussed more extensively in Section 5.3 and its validity is proved as a consequence of the fact that every nonempty subset of $\mathbb{N}$ has a smallest element but we can also justify its validity as follows. Define $Q(m)$ by

$$Q(m) = (\forall k \in \mathbb{N})(k \leq m \Rightarrow P(k)).$$

Then, it is an easy exercise to show that if we apply our (ordinary) induction principle to $Q(m)$ (induction principle, Version 3), then we get the principle of complete induction. Here is an example of a proof using complete induction.

Define the sequence of natural numbers F_n (*Fibonacci sequence*) by

$$F_0 = 1, \ F_1 = 1, \ F_{n+2} = F_{n+1} + F_n, \ n \geq 0.$$

We claim that

$$F_n \geq \frac{3^{n-2}}{2^{n-3}}, \ n \geq 3.$$

Fig. 2.6 Leonardo P. Fibonacci, 1170–1250

The base case corresponds to $n = 3$, where

$$F_3 = 3 \geq \frac{3^1}{2^0} = 3,$$

which is true. Note that we also need to consider the case $n = 4$ by itself before we do the induction step because even though $F_4 = F_3 + F_2$, the induction hypothesis only applies to F_3 ($n \geq 3$ in the inequality above). We have

$$F_4 = 5 \geq \frac{3^2}{2^1} = \frac{9}{2},$$

which is true because $10 > 9$. Now for the induction step where $n \geq 3$, we have

$$
\begin{aligned}
F_{n+2} &= F_{n+1} + F_n \\
&\geq \frac{3^{n-1}}{2^{n-2}} + \frac{3^{n-2}}{2^{n-3}} \\
&\geq \frac{3^{n-2}}{2^{n-3}}\left(1 + \frac{3}{2}\right) = \frac{3^{n-2}}{2^{n-3}}\frac{5}{2} \geq \frac{3^{n-2}}{2^{n-3}}\frac{9}{4} = \frac{3^n}{2^{n-1}},
\end{aligned}
$$

since $5/2 > 9/4$, which concludes the proof of the induction step. Observe that we used the induction hypothesis for both F_{n+1} and F_n in order to deduce that it holds for F_{n+2}. This is where we needed the extra power of complete induction.

Remark: The Fibonacci sequence F_n is really a function from $\mathbb{N}$ to $\mathbb{N}$ defined recursively but we haven't proved yet that recursive definitions are legitimate methods for defining functions. In fact, certain restrictions are needed on the kind of recursion used to define functions. This topic is explored further in Section 2.5. Using results from Section 2.5, it can be shown that the Fibonacci sequence is a well-defined function (but this does not follow immediately from Theorem 2.1).

Induction proofs can be subtle and it might be instructive to see some examples of *faulty* induction proofs.

Assertion 1: For every natural numbers $n \geq 1$, the number $n^2 - n + 11$ is an odd prime (recall that a prime number is a natural number $p \geq 2$, which is only divisible by 1 and itself).

Proof. We use induction on $n \geq 1$. For the base case $n = 1$, we have $1^2 - 1 + 11 = 11$, which is an odd prime, so the induction step holds.

Assume inductively that $n^2 - n + 11$ is prime. Then, as

$$(n+1)^2 - (n+1) + 11 = n^2 + 2n + 1 - n - 1 + 11 = n^2 + n + 11,$$

we see that

$$(n+1)^2 - (n+1) + 11 = n^2 - n + 11 + 2n.$$

By the induction hypothesis, $n^2 - n + 11$ is an odd prime p, and because $2n$ is even, $p + 2n$ is odd and therefore prime, establishing the induction hypothesis. $\square$

If we compute $n^2 - n + 11$ for $n = 1, 2, \ldots, 10$, we find that these numbers are indeed all prime, but for $n = 11$, we get

$$121 = 11^2 - 11 + 11 = 11 \times 11,$$

which is not prime.

Where is the mistake?

What is wrong is the induction step: the fact that $n^2 - n + 11$ is prime does not imply that $(n + 1)^2 - (n + 1) + 11 = n^2 + n + 11$ is prime, as illustrated by $n = 10$. Our "proof" of the induction step is nonsense.

The lesson is: the fact that a statement holds for many values of $n \in \mathbb{N}$ does not imply that it holds for all $n \in \mathbb{N}$ (or all $n \geq k$, for some fixed $k \in \mathbb{N}$).

Interestingly, the prime numbers k, so that $n^2 - n + k$ is prime for $n = 1, 2, \ldots, k - 1$, are all known (there are only six of them). It can be shown that these are the prime numbers k such that $1 - 4k$ is a *Heegner number*, where the Heegner numbers are the nine integers:

$$-1, -2, -3, -7, -11, -19, -43, -67, -163.$$

The above results are hard to prove and require some deep theorems of number theory. What can also be shown (and you should prove it) is that no nonconstant polynomial takes prime numbers as values for all natural numbers.

Assertion 2: Every Fibonacci number F_n is even.

Proof. For the base case, $F_2 = 2$, which is even, so the base case holds.

Assume inductively that F_n is even for all $n \geq 2$. Then, as

$$F_{n+2} = F_{n+1} + F_n$$

and as both F_n and F_{n+1} are even by the induction hypothesis, we conclude that F_{n+2} is even. $\square$

However, Assertion 2 is clearly false, because the Fibonacci sequence begins with

$$1, 1, 2, 3, 5, 8, 13, 21, 34, \ldots.$$

This time, the mistake is that we did not check the two base cases, $F_0 = 1$ and $F_1 = 1$.

Our experience is that if an induction proof is wrong, then, in many cases, the base step is faulty. So, pay attention to the base step(s).

A useful way to produce new relations or functions is to compose them.

2.4 Composition of Relations and Functions

We begin with the definition of the composition of relations.

Definition 2.4. Given two relations $R \subseteq A \times B$ and $S \subseteq B \times C$, the *composition of R and S*, denoted $R \circ S$, is the relation between A and C defined by

$$R \circ S = \{\langle a, c \rangle \in A \times C \mid \exists b \in B, \langle a, b \rangle \in R \text{ and } \langle b, c \rangle \in S\}.$$

One should check that for any relation $R \subseteq A \times B$, we have $id_A \circ R = R$ and $R \circ id_B = R$. If R and S are the graphs of functions, possibly partial, is $R \circ S$ the graph of some function? The answer is yes, as shown in the following.

Proposition 2.2. Let $R \subseteq A \times B$ and $S \subseteq B \times C$ be two relations.

(a) If R and S are both functional relations, then $R \circ S$ is also a functional relation. Consequently, $R \circ S$ is the graph of some partial function.
(b) If $dom(R) = A$ and $dom(S) = B$, then $dom(R \circ S) = A$.
(c) If R is the graph of a (total) function from A to B and S is the graph of a (total) function from B to C, then $R \circ S$ is the graph of a (total) function from A to C.

Proof. (a) Assume that $\langle a, c_1 \rangle \in R \circ S$ and $\langle a, c_2 \rangle \in R \circ S$. By definition of $R \circ S$, there exist $b_1, b_2 \in B$ so that

$$\langle a, b_1 \rangle \in R, \ \langle b_1, c_1 \rangle \in S,$$
$$\langle a, b_2 \rangle \in R, \ \langle b_2, c_2 \rangle \in S.$$

As R is functional, $\langle a, b_1 \rangle \in R$ and $\langle a, b_2 \rangle \in R$ implies $b_1 = b_2$. Let $b = b_1 = b_2$, so that $\langle b_1, c_1 \rangle = \langle b, c_1 \rangle$ and $\langle b_2, c_2 \rangle = \langle b, c_2 \rangle$. But, S is also functional, so $\langle b, c_1 \rangle \in S$ and $\langle b, c_2 \rangle \in S$ implies that $c_1 = c_2$, which proves that $R \circ S$ is functional.

(b) If $A = \emptyset$ then $R = \emptyset$ and so $R \circ S = \emptyset$, which implies that $dom(R \circ S) = \emptyset = A$. If $A \neq \emptyset$, pick any $a \in A$. The fact that $dom(R) = A \neq \emptyset$ means that there is some $b \in B$ so that $\langle a, b \rangle \in R$ and so, $B \neq \emptyset$. As $dom(S) = B \neq \emptyset$, there is some $c \in C$ so that $\langle b, c \rangle \in S$. Then, by the definition of $R \circ S$, we see that $\langle a, c \rangle \in R \circ S$. The argument holds for any $a \in A$, therefore we deduce that $dom(R \circ S) = A$.

(c) If R and S are the graphs of partial functions, then this means that they are functional and (a) implies that $R \circ S$ is also functional. This shows that $R \circ S$ is the graph of the partial function $\langle A, R \circ S, C \rangle$. If R and S are the graphs of total functions, then $dom(R) = A$ and $dom(S) = B$. By (b), we deduce that $dom(R \circ S) = A$. By the first part of (c), $R \circ S$ is the graph of the partial function $\langle A, R \circ S, C \rangle$, which is a total function, inasmuch as $dom(R \circ S) = A$. $\square$

Proposition 2.2 shows that it is legitimate to define the composition of functions, possibly partial. Thus, we make the following definition.

Definition 2.5. Given two functions $f: A \to B$ and $g: B \to C$, possibly partial, the *composition of f and g*, denoted $g \circ f$, is the function (possibly partial)

$$g \circ f = \langle A, graph(f) \circ graph(g), C \rangle.$$

The reader must have noticed that the composition of two functions $f: A \to B$ and $g: B \to C$ is denoted $g \circ f$, whereas the graph of $g \circ f$ is denoted $graph(f) \circ graph(g)$. This "reversal" of the order in which function composition and relation composition are written is unfortunate and somewhat confusing.

Once again, we are the victims of tradition. The main reason for writing function composition as $g \circ f$ is that traditionally the result of applying a function f to an argument x is written $f(x)$. Then, $(g \circ f)(x) = g(f(x))$, because $z = (g \circ f)(x)$ iff there is some y so that $y = f(x)$ and $z = g(y)$; that is, $z = g(f(x))$. Some people, in particular algebraists, write function composition as $f \circ g$, but then, they write the result of applying a function f to an argument x as xf. With this convention, $x(f \circ g) = (xf)g$, which also makes sense.

We prefer to stick to the convention where we write $f(x)$ for the result of applying a function f to an argument x and, consequently, we use the notation $g \circ f$ for the composition of f with g, even though it is the opposite of the convention for writing the composition of relations.

Given any three relations, $R \subseteq A \times B$, $S \subseteq B \times C$, and $T \subseteq C \times D$, the reader should verify that

$$(R \circ S) \circ T = R \circ (S \circ T).$$

We say that composition is *associative*. Similarly, for any three functions (possibly partial), $f: A \to B$, $g: B \to C$, and $h: C \to D$, we have (associativity of function composition)

$$(h \circ g) \circ f = h \circ (g \circ f).$$

2.5 Recursion on $\mathbb{N}$

The following situation often occurs. We have some set A, some fixed element $a \in A$, some function $g: A \to A$, and we wish to define a new function $h: \mathbb{N} \to A$, so that

$$h(0) = a,$$
$$h(n+1) = g(h(n)) \text{ for all } n \in \mathbb{N}.$$

This way of defining h is called a *recursive definition* (or a definition by *primitive recursion*). I would be surprised if any computer scientist had any trouble with this "definition" of h but how can we justify rigorously that such a function exists and is unique?

Indeed, the existence (and uniqueness) of h requires proof. The proof, although not really hard, is surprisingly involved and in fact quite subtle. For those reasons, we do not give a proof of the following theorem but instead the main idea of the proof. The reader will find a complete proof in Enderton [2] (Chapter 4).

Theorem 2.1. *(Recursion theorem on $\mathbb{N}$) Given any set A, any fixed element $a \in A$, and any function $g \colon A \to A$, there is a unique function $h \colon \mathbb{N} \to A$, so that*

$$h(0) = a,$$
$$h(n+1) = g(h(n)) \text{ for all } n \in \mathbb{N}.$$

Proof. The idea is to approximate h. To do this, define a function f to be *acceptable* iff

1. $dom(f) \subseteq \mathbb{N}$ and $range(f) \subseteq A$.
2. If $0 \in dom(f)$, then $f(0) = a$.
3. If $n + 1 \in dom(f)$, then $n \in dom(f)$ and $f(n+1) = g(f(n))$.

Let $\mathscr{F}$ be the collection of all acceptable functions and set

$$h = \bigcup \mathscr{F}.$$

All we can say, so far, is that h is a relation. We claim that h is the desired function. For this, four things need to be proved:

1. The relation h is a function.
2. The function h is acceptable.
3. The function h has domain $\mathbb{N}$.
4. The function h is unique.

As expected, we make heavy use of induction in proving (1)–(4). For complete details, see Enderton [2] (Chapter 4). □

Theorem 2.1 is very important. Indeed, experience shows that it is used almost as much as induction. As an example, we show how to define addition on $\mathbb{N}$. Indeed, at the moment, we know what the natural numbers are but we don't know what are the arithmetic operations such as $+$ or $*$ (at least, not in our axiomatic treatment; of course, nobody needs an axiomatic treatment to know how to add or multiply).

How do we define $m + n$, where $m, n \in \mathbb{N}$?

If we try to use Theorem 2.1 directly, we seem to have a problem, because addition is a function of two arguments, but h and g in the theorem only take one argument. We can overcome this problem in two ways:

(1) We prove a generalization of Theorem 2.1 involving functions of several arguments, but with recursion only in a *single* argument. This can be done quite easily but we have to be a little careful.

(2) For any fixed m, we define $add_m(n)$ as $add_m(n) = m + n$; that is, we define addition of a *fixed* m to any n. Then, we let $m + n = add_m(n)$.

Solution (2) involves much less work, thus we follow it. Let S denote the successor function on $\mathbb{N}$, that is, the function given by

$$S(n) = n^+ = n + 1.$$

Then, using Theorem 2.1 with $a = m$ and $g = S$, we get a function, add_m, such that

$$add_m(0) = m,$$
$$add_m(n + 1) = S(add_m(n)) = add_m(n) + 1 \qquad \text{for all} \quad n \in \mathbb{N}.$$

Finally, for all $m, n \in \mathbb{N}$, we define $m + n$ by

$$m + n = add_m(n).$$

Now, we have our addition function on $\mathbb{N}$. But this is not the end of the story because we don't know yet that the above definition yields a function having the usual properties of addition, such as

$$m + 0 = m$$
$$m + n = n + m$$
$$(m + n) + p = m + (n + p).$$

To prove these properties, of course, we use induction.

We can also define multiplication. Mimicking what we did for addition, define $mult_m(n)$ by recursion as follows.

$$mult_m(0) = 0,$$
$$mult_m(n + 1) = mult_m(n) + m \text{ for all } n \in \mathbb{N}.$$

Then, we set

$$m \cdot n = mult_m(n).$$

Note how the recursive definition of $mult_m$ uses the adddition function $+$, previously defined. Again, to prove the usual properties of multiplication as well as the distributivity of $\cdot$ over $+$, we use induction. Using recursion, we can define many more arithmetic functions. For example, the reader should try defining exponentiation m^n.

We still haven't defined the usual ordering on the natural numbers but we do so later. Of course, we all know what it is and we do not refrain from using it. Still, it is interesting to give such a definition in our axiomatic framework.

2.6 Inverses of Functions and Relations

Given a function $f: A \to B$ (possibly partial), with $A \neq \emptyset$, suppose there is some function $g: B \to A$ (possibly partial), called a *left inverse of f*, such that

$$g \circ f = \mathrm{id}_A.$$

If such a g exists, we see that f must be total but more is true. Indeed, assume that $f(a) = f(b)$. Then, by applying g, we get

$$(g \circ f)(a) = g(f(a)) = g(f(b)) = (g \circ f)(b).$$

However, because $g \circ f = \mathrm{id}_A$, we have $(g \circ f)(a) = \mathrm{id}_A(a) = a$ and $(g \circ f)(b) = \mathrm{id}_A(b) = b$, so we deduce that
$$a = b.$$

Therefore, we showed that if a function f with nonempty domain has a left inverse, then f is total and has the property that for all $a, b \in A$, $f(a) = f(b)$ implies that $a = b$, or equivalently $a \neq b$ implies that $f(a) \neq f(b)$. We say that f is *injective*. As we show later, injectivity is a very desirable property of functions.

Remark: If $A = \emptyset$, then f is still considered to be injective. In this case, g is the empty partial function (and when $B = \emptyset$, both f and g are the empty function from $\emptyset$ to itself).

Now, suppose there is some function $h: B \to A$ (possibly partial) with $B \neq \emptyset$ called a *right inverse of f*, but this time, we have

$$f \circ h = \mathrm{id}_B.$$

If such an h exists, we see that it must be total but more is true. Indeed, for any $b \in B$, as $f \circ h = \mathrm{id}_B$, we have

$$f(h(b)) = (f \circ h)(b) = \mathrm{id}_B(b) = b.$$

Therefore, we showed that if a function f with nonempty codomain has a right inverse h then h is total and f has the property that for all $b \in B$, there is some $a \in A$, namely, $a = h(b)$, so that $f(a) = b$. In other words, $\mathrm{Im}(f) = B$ or equivalently, every element in B is the image by f of some element of A. We say that f is *surjective*. Again, surjectivity is a very desirable property of functions.

Remark: If $B = \emptyset$, then f is still considered to be surjective but h is not total unless $A = \emptyset$, in which case f is the empty function from $\emptyset$ to itself.

If a function has a left inverse (respectively, a right inverse), then it may have more than one left inverse (respectively, right inverse).

If a function (possibly partial) $f: A \to B$ with $A, B \neq \emptyset$ happens to have both a left inverse $g: B \to A$ and a right inverse $h: B \to A$, then we know that f and h are total. We claim that $g = h$, so that g is total and moreover g is uniquely determined by f.

Lemma 2.1. *Let $f: A \to B$ be any function and suppose that f has a left inverse $g: B \to A$ and a right inverse $h: B \to A$. Then, $g = h$ and, moreover, g is unique, which means that if $g': B \to A$ is any function that is both a left and a right inverse of f, then $g' = g$.*

Proof. Assume that

$$g \circ f = \mathrm{id}_A \text{ and } f \circ h = \mathrm{id}_B.$$

Then, we have

$$g = g \circ \mathrm{id}_B = g \circ (f \circ h) = (g \circ f) \circ h = \mathrm{id}_A \circ h = h.$$

Therefore, $g = h$. Now, if g' is any other left inverse of f and h' is any other right inverse of f, the above reasoning applied to g and h' shows that $g = h'$ and the same reasoning applied to g' and h' shows that $g' = h'$. Therefore, $g' = h' = g = h$, that is, g is uniquely determined by f. □

This leads to the following definition.

Definition 2.6. A function $f: A \to B$ is said to be *invertible* iff there is a function $g: B \to A$ which is both a left inverse and a right inverse; that is,

$$g \circ f = \mathrm{id}_A \text{ and } f \circ g = \mathrm{id}_B.$$

In this case, we know that g is unique and it is denoted f^{-1}.

From the above discussion, if a function is invertible, then it is both injective and surjective. This shows that a function *generally does not have an inverse*. In order to have an inverse a function needs to be injective and surjective, but this fails to be true for many functions. It turns out that if a function is injective and surjective then it has an inverse. We prove this in the next section.

The notion of inverse can also be defined for relations, but it is a somewhat weaker notion.

Definition 2.7. Given any relation $R \subseteq A \times B$, the *converse* or *inverse* of R is the relation $R^{-1} \subseteq B \times A$, defined by

$$R^{-1} = \{\langle b, a \rangle \in B \times A \mid \langle a, b \rangle \in R\}.$$

In other words, R^{-1} is obtained by swapping A and B and reversing the orientation of the arrows. Figure 2.7 below shows the inverse of the relation of Figure 2.2:

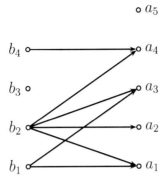

Fig. 2.7 The inverse of the relation R from Figure 2.2

Now, if R is the graph of a (partial) function f, beware that R^{-1} is generally *not* the graph of a function at all, because R^{-1} may not be functional. For example, the inverse of the graph G in Figure 2.3 is *not* functional; see below.

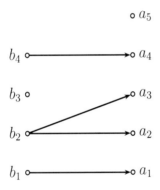

Fig. 2.8 The inverse, G^{-1}, of the graph of Figure 2.3

The above example shows that one has to be careful not to view a function as a relation in order to take its inverse. In general, this process does not produce a function. This only works if the function is invertible.

Given any two relations, $R \subseteq A \times B$ and $S \subseteq B \times C$, the reader should prove that

$$(R \circ S)^{-1} = S^{-1} \circ R^{-1}.$$

(Note the switch in the order of composition on the right-hand side.) Similarly, if $f : A \to B$ and $g : B \to C$ are any two invertible functions, then $g \circ f$ is invertible and

$$(g \circ f)^{-1} = f^{-1} \circ g^{-1}.$$

2.7 Injections, Surjections, Bijections, Permutations

We encountered injectivity and surjectivity in Section 2.6. For the record, let us give the following.

Definition 2.8. Given any function $f : A \to B$, we say that f *is injective* (or *one-to-one*) iff for all $a, b \in A$, if $f(a) = f(b)$, then $a = b$, or equivalently, if $a \neq b$, then $f(a) \neq f(b)$. We say that f *is surjective* (or *onto*) iff for every $b \in B$, there is some $a \in A$ so that $b = f(a)$, or equivalently if $\mathrm{Im}(f) = B$. The function f is *bijective* iff it is both injective and surjective. When $A = B$, a bijection $f : A \to A$ is called a *permutation of A*.

Remarks:

1. If $A = \emptyset$, then any function, $f : \emptyset \to B$ is (trivially) injective.
2. If $B = \emptyset$, then f is the empty function from $\emptyset$ to itself and it is (trivially) surjective.
3. A function, $f : A \to B$, is **not injective** iff **there exist** $a, b \in A$ with $a \neq b$ and **yet** $f(a) = f(b)$; see Figure 2.9.
4. A function, $f : A \to B$, is **not surjective** iff **for some** $b \in B$, **there is no** $a \in A$ with $b = f(a)$; see Figure 2.10.
5. We have $\mathrm{Im} f = \{b \in B \mid (\exists a \in A)(b = f(a))\}$, thus a function $f : A \to B$ is always surjective onto its image.
6. The notation $f : A \hookrightarrow B$ is often used to indicate that a function $f : A \to B$ is an injection.
7. If $A \neq \emptyset$, a function $f : A \to B$ is injective iff for every $b \in B$, there *at most one* $a \in A$ such that $b = f(a)$.
8. If $A \neq \emptyset$, a function $f : A \to B$ is surjective iff for every $b \in B$, there *at least one* $a \in A$ such that $b = f(a)$ iff $f^{-1}(b) \neq \emptyset$ for all $b \in B$.
9. If $A \neq \emptyset$, a function $f : A \to B$ is bijective iff for every $b \in B$, there is *a unique* $a \in A$ such that $b = f(a)$.
10. When A is the finite set $A = \{1, \ldots, n\}$, also denoted $[n]$, it is not hard to show that there are $n!$ permutations of $[n]$.

The function $f_1 : \mathbb{Z} \to \mathbb{Z}$ given by $f_1(x) = x + 1$ is injective and surjective. However, the function $f_2 : \mathbb{Z} \to \mathbb{Z}$ given by $f_2(x) = x^2$ is neither injective nor surjective (why?). The function $f_3 : \mathbb{Z} \to \mathbb{Z}$ given by $f_3(x) = 2x$ is injective but not surjective. The function $f_4 : \mathbb{Z} \to \mathbb{Z}$ given by

$$f_4(x) = \begin{cases} k & \text{if } x = 2k \\ k & \text{if } x = 2k + 1 \end{cases}$$

is surjective but not injective.

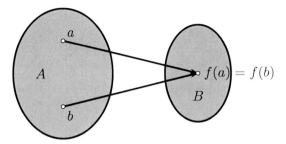

Fig. 2.9 A noninjective function

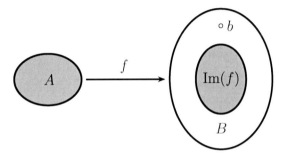

Fig. 2.10 A nonsurjective function

Remark: The reader should prove that if A and B are finite sets, A has m elements and B has n elements ($m \leq n$) then the set of injections from A to B has

$$\frac{n!}{(n-m)!}$$

elements. The following theorem relates the notions of injectivity and surjectivity to the existence of left and right inverses.

Theorem 2.2. *Let $f : A \to B$ be any function and assume $A \neq \emptyset$.*

(a) The function f is injective iff it has a left inverse g (i.e., a function $g : B \to A$ so that $g \circ f = \mathrm{id}_A$).
(b) The function f is surjective iff it has a right inverse h (i.e., a function $h : B \to A$ so that $f \circ h = \mathrm{id}_B$).
(c) The function f is invertible iff it is injective and surjective.

Proof. (a) We already proved in Section 2.6 that the existence of a left inverse implies injectivity. Now, assume f is injective. Then, for every $b \in range(f)$, there is a unique $a_b \in A$ so that $f(a_b) = b$. Because $A \neq \emptyset$, we may pick some a_0 in A. We define $g \colon B \to A$ by

$$g(b) = \begin{cases} a_b & \text{if } b \in range(f) \\ a_0 & \text{if } b \in B - range(f). \end{cases}$$

Then, $g(f(a)) = a$ for all $a \in A$, because $f(a) \in range(f)$ and a is the only element of A so that $f(a) = f(a)$. This shows that $g \circ f = id_A$, as required.

(b) We already proved in Section 2.6 that the existence of a right inverse implies surjectivity. For the converse, assume that f is surjective. As $A \neq \emptyset$ and f is a function (i.e., f is total), $B \neq \emptyset$. So, for every $b \in B$, the preimage $f^{-1}(b) = \{a \in A \mid f(a) = b\}$ is nonempty. We make a function $h \colon B \to A$ as follows. For each $b \in B$, pick some element $a_b \in f^{-1}(b)$ (which is nonempty) and let $h(b) = a_b$. By definition of $f^{-1}(b)$, we have $f(a_b) = b$ and so,

$$f(h(b)) = f(a_b) = b, \quad \text{for all } b \in B.$$

This shows that $f \circ h = id_B$, as required.

(c) If f is invertible, we proved in Section 2.6 that f is injective and surjective. Conversely, if f is both injective and surjective, by (a) the function f has a left inverse g and by (b) it has a right inverse h. However, by Lemma 2.1, $g = h$, which shows that f is invertible. $\square$

The alert reader may have noticed a "fast turn" in the proof of the converse in (b). Indeed, we constructed the function h by choosing, for each $b \in B$, some element in $f^{-1}(b)$. How do we justify this procedure from the axioms of set theory?

Well, we can't. For this we need another (historically somewhat controversial) axiom, the *axiom of choice*. This axiom has many equivalent forms. We state the following form which is intuitively quite plausible.

Axiom of Choice (Graph Version).
For every relation $R \subseteq A \times B$, there is a partial function $f \colon A \to B$, with $graph(f) \subseteq R$ and $dom(f) = dom(R)$.

We see immediately that the axiom of choice justifies the existence of the function h in part (b) of Theorem 2.2.

Remarks:

1. Let $f \colon A \to B$ and $g \colon B \to A$ be any two functions and assume that

$$g \circ f = id_A.$$

Thus, f is a right inverse of g and g is a left inverse of f. So, by Theorem 2.2 (a) and (b), we deduce that f is injective and g is surjective. In particular, this shows that any left inverse of an injection is a surjection and that any right inverse of a surjection is an injection.

2. Any right inverse h of a surjection $f: A \to B$ is called a *section* of f (which is an abbreviation for *cross-section*). This terminology can be better understood as follows: Because f is surjective, the preimage, $f^{-1}(b) = \{a \in A \mid f(a) = b\}$ of any element $b \in B$ is nonempty. Moreover, $f^{-1}(b_1) \cap f^{-1}(b_2) = \emptyset$ whenever $b_1 \neq b_2$. Therefore, the pairwise disjoint and nonempty subsets $f^{-1}(b)$, where $b \in B$, partition A. We can think of A as a big "blob" consisting of the union of the sets $f^{-1}(b)$ (called fibres) and lying over B. The function f maps each fibre, $f^{-1}(b)$ onto the element, $b \in B$. Then, any right inverse $h: B \to A$ of f picks out some element in each fibre, $f^{-1}(b)$, forming a sort of horizontal section of A shown as a curve in Figure 2.11.

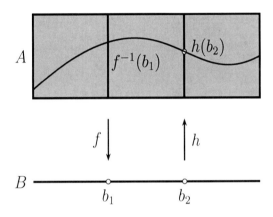

Fig. 2.11 A section h of a surjective function f.

3. Any left inverse g of an injection $f: A \to B$ is called a *retraction* of f. The terminology reflects the fact that intuitively, as f is injective (thus, g is surjective), B is bigger than A and because $g \circ f = \text{id}_A$, the function g "squeezes" B onto A in such a way that each point $b = f(a)$ in $\text{Im} f$ is mapped back to its ancestor $a \in A$. So, B is "retracted" onto A by g.

Before discussing direct and inverse images, we define the notion of restriction and extension of functions.

Definition 2.9. Given two functions, $f: A \to C$ and $g: B \to C$, with $A \subseteq B$, we say that f *is the restriction of* g *to* A if $graph(f) \subseteq graph(g)$; we write $f = g \upharpoonright A$. In this case, we also say that g *is an extension of* f *to* B.

2.8 Direct Image and Inverse Image

A function $f: X \to Y$ induces a function from 2^X to 2^Y also denoted f and a function from 2^Y to 2^X, as shown in the following definition.

Definition 2.10. Given any function $f: X \to Y$, we define the function $f: 2^X \to 2^Y$ so that, for every subset A of X,

$$f(A) = \{y \in Y \mid \exists x \in A, y = f(x)\}.$$

The subset $f(A)$ of Y is called the *direct image of A under f*, for short, the *image of A under f*. We also define the function $f^{-1}: 2^Y \to 2^X$ so that, for every subset B of Y,

$$f^{-1}(B) = \{x \in X \mid \exists y \in B, y = f(x)\}.$$

The subset $f^{-1}(B)$ of X is called the *inverse image of B under f* or the *preimage of B under f*.

Remarks:

1. The overloading of notation where f is used both for denoting the original function $f: X \to Y$ and the new function $f: 2^X \to 2^Y$ may be slightly confusing. If we observe that $f(\{x\}) = \{f(x)\}$, for all $x \in X$, we see that the new f is a natural extension of the old f to the subsets of X and so, using the same symbol f for both functions is quite natural after all. To avoid any confusion, some authors (including Enderton) use a different notation for $f(A)$, for example, $f[\![A]\!]$. We prefer not to introduce more notation and we hope that which f we are dealing with is made clear by the context.
2. The use of the notation f^{-1} for the function $f^{-1}: 2^Y \to 2^X$ may even be more confusing, because we know that f^{-1} is generally not a function from Y to X. However, it *is* a function from 2^Y to 2^X. Again, some authors use a different notation for $f^{-1}(B)$, for example, $f^{-1}[\![A]\!]$. We stick to $f^{-1}(B)$.
3. The set $f(A)$ is sometimes called the *push-forward of A along f* and $f^{-1}(B)$ is sometimes called the *pullback of B along f*.
4. Observe that $f^{-1}(y) = f^{-1}(\{y\})$, where $f^{-1}(y)$ is the preimage defined just after Definition 2.3.
5. Although this may seem counterintuitive, the function f^{-1} has a better behavior than f with respect to union, intersection, and complementation.

Some useful properties of $f: 2^X \to 2^Y$ and $f^{-1}: 2^Y \to 2^X$ are now stated without proof. The proofs are easy and left as exercises.

Proposition 2.3. *Given any function $f: X \to Y$, the following properties hold.*

(1) For any $B \subseteq Y$, we have
$$f(f^{-1}(B)) \subseteq B.$$

(2) If $f: X \to Y$ is surjective, then

$$f(f^{-1}(B)) = B.$$

(3) For any $A \subseteq X$, we have

$$A \subseteq f^{-1}(f(A)).$$

(4) If $f: X \to Y$ is injective, then

$$A = f^{-1}(f(A)).$$

The next proposition deals with the behavior of $f: 2^X \to 2^Y$ and $f^{-1}: 2^Y \to 2^X$ with respect to union, intersection, and complementation.

Proposition 2.4. *Given any function $f: X \to Y$ the following properties hold.*

(1) For all $A, B \subseteq X$, we have

$$f(A \cup B) = f(A) \cup f(B).$$

(2)

$$f(A \cap B) \subseteq f(A) \cap f(B).$$

Equality holds if $f: X \to Y$ is injective.

(3)

$$f(A) - f(B) \subseteq f(A - B).$$

Equality holds if $f: X \to Y$ is injective.

(4) For all $C, D \subseteq Y$, we have

$$f^{-1}(C \cup D) = f^{-1}(C) \cup f^{-1}(D).$$

(5)

$$f^{-1}(C \cap D) = f^{-1}(C) \cap f^{-1}(D).$$

(6)

$$f^{-1}(C - D) = f^{-1}(C) - f^{-1}(D).$$

As we can see from Proposition 2.4, the function $f^{-1}: 2^Y \to 2^X$ has better behavior than $f: 2^X \to 2^Y$ with respect to union, intersection, and complementation.

2.9 Equinumerosity; The Pigeonhole Principle and the Schröder–Bernstein Theorem

The notion of size of a set is fairly intuitive for finite sets but what does it mean for infinite sets? How do we give a precise meaning to the questions:

(a) Do X and Y have the same size?

(b) Does X have more elements than Y?

For finite sets, we can rely on the natural numbers. We count the elements in the two sets and compare the resulting numbers. If one of the two sets is finite and the other is infinite, it seems fair to say that the infinite set has more elements than the finite one.

But what if both sets are infinite?

Remark: A critical reader should object that we have not yet defined what a finite set is (or what an infinite set is). Indeed, we have not. This can be done in terms of the natural numbers but, for the time being, we rely on intuition. We should also point out that when it comes to infinite sets, experience shows that our intuition fails us miserably. So, we should be very careful.

Let us return to the case where we have two infinite sets. For example, consider $\mathbb{N}$ and the set of even natural numbers, $2\mathbb{N} = \{0, 2, 4, 6, \ldots\}$. Clearly, the second set is properly contained in the first. Does that make $\mathbb{N}$ bigger? On the other hand, the function $n \mapsto 2n$ is a bijection between the two sets, which seems to indicate that they have the same number of elements. Similarly, the set of squares of natural numbers, Squares $= \{0, 1, 4, 9, 16, 25, \ldots\}$ is properly contained in $\mathbb{N}$ and many natural numbers are missing from Squares. But, the map $n \mapsto n^2$ is a bijection between $\mathbb{N}$ and Squares, which seems to indicate that they have the same number of elements.

A more extreme example is provided by $\mathbb{N} \times \mathbb{N}$ and $\mathbb{N}$. Intuitively, $\mathbb{N} \times \mathbb{N}$ is two-dimensional and $\mathbb{N}$ is one-dimensional, so $\mathbb{N}$ seems much smaller than $\mathbb{N} \times \mathbb{N}$. However, it is possible to construct bijections between $\mathbb{N} \times \mathbb{N}$ and $\mathbb{N}$ (try to find one). In fact, such a function J has the graph partially shown below:

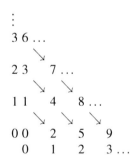

The function J corresponds to a certain way of enumerating pairs of integers. Note that the value of $m + n$ is constant along each diagonal, and consequently, we have

$$J(m,n) = 1 + 2 + \cdots + (m+n) + m,$$
$$= ((m+n)(m+n+1) + 2m)/2,$$
$$= ((m+n)^2 + 3m + n)/2.$$

For example, $J(2,1) = ((2+1)^2 + 3 \cdot 2 + 1)/2 = (9 + 6 + 1)/2 = 16/2 = 8$. The function

$$J(m,n) = \frac{1}{2}((m+n)^2 + 3m + n)$$

is a bijection but that's not so easy to prove.

Perhaps even more surprising, there are bijections between $\mathbb{N}$ and $\mathbb{Q}$. What about between $\mathbb{R} \times \mathbb{R}$ and $\mathbb{R}$? Again, the answer is yes, but that's harder to prove.

These examples suggest that the notion of bijection can be used to define rigorously when two sets have the same size. This leads to the concept of equinumerosity.

Definition 2.11. A set A is *equinumerous* to a set B, written $A \approx B$, iff there is a bijection $f \colon A \to B$. We say that A is *dominated* by B, written $A \preceq B$, iff there is an injection from A to B. Finally, we say that A is *strictly dominated* by B, written $A \prec B$, iff $A \preceq B$ and $A \not\approx B$.

Using the above concepts, we can give a precise definition of finiteness. First, recall that for any $n \in \mathbb{N}$, we defined $[n]$ as the set $[n] = \{1, 2, \ldots, n\}$, with $[0] = \emptyset$.

Definition 2.12. A set A is *finite* if it is equinumerous to a set of the form $[n]$, for some $n \in \mathbb{N}$. A set A is *infinite* iff it is not finite. We say that A is *countable* (or *denumerable*) iff A is dominated by $\mathbb{N}$.

Two pretty results due to Cantor (1873) are given in the next theorem. These are among the earliest results of set theory. We assume that the reader is familiar with the fact that every number, $x \in \mathbb{R}$, can be expressed in decimal expansion (possibly infinite). For example,

$$\pi = 3.14159265358979\cdots$$

Theorem 2.3. *(Cantor's Theorem) (a) The set $\mathbb{N}$ is not equinumerous to the set $\mathbb{R}$ of real numbers.*

(b) For every set A there is no surjection from A onto 2^A. Consequently, no set A is equinumerous to its power set 2^A.

Proof. (a) We use a famous proof method due to Cantor and known as a *diagonal argument*. We prove that if we assume there is a bijection $f \colon \mathbb{N} \to \mathbb{R}$, then there is a real number z not belonging to the image of f, contradicting the surjectivity of f. Now, if f exists, we can form a bi-infinite array

$$f(0) = k_0.d_{01}d_{02}d_{03}d_{04}\cdots,$$
$$f(1) = k_1.d_{11}d_{12}d_{13}d_{14}\cdots,$$
$$f(2) = k_2.d_{21}d_{22}d_{23}d_{24}\cdots,$$

$$\vdots$$

$$f(n) = k_n.d_{n1}d_{n2}\cdots d_{nn+1}\cdots,$$

$$\vdots$$

where k_n is the integer part of $f(n)$ and the d_{ni} are the decimals of $f(n)$, with $i \geq 1$.

The number

$$z = 0.d_1d_2d_3 \cdots d_{n+1} \cdots$$

is defined so that $d_{n+1} = 1$ if $d_{nn+1} \neq 1$, else $d_{n+1} = 2$ if $d_{nn+1} = 1$, for every $n \geq 0$, The definition of z shows that

$$d_{n+1} \neq d_{nn+1}, \text{ for all } n \geq 0,$$

which implies that z is not in the above array; that is, $z \notin \text{Im } f$.

(b) The proof is a variant of Russell's paradox. Assume that there is a surjection, $g \colon A \to 2^A$; we construct a set $B \subseteq A$ that is not in the image of g, a contradiction. Consider the set

$$B = \{a \in A \mid a \notin g(a)\}.$$

Obviously, $B \subseteq A$. However, for every $a \in A$,

$$a \in B \text{ iff } a \notin g(a),$$

which shows that $B \neq g(a)$ for all $a \in A$; that is, B is not in the image of g. $\quad\square$

As there is an obvious injection of $\mathbb{N}$ into $\mathbb{R}$, Theorem 2.3 shows that $\mathbb{N}$ is strictly dominated by $\mathbb{R}$. Also, as we have the injection $a \mapsto \{a\}$ from A into 2^A, we see that every set is strictly dominated by its power set. So, we can form sets as big as we want by repeatedly using the power set operation.

Remark: In fact, $\mathbb{R}$ is equinumerous to $2^{\mathbb{N}}$; see Problem 2.39

The following proposition shows an interesting connection between the notion of power set and certain sets of functions. To state this proposition, we need the concept of characteristic function of a subset.

Given any set X for any subset A of X, define the *characteristic function of A*, denoted χ_A, as the function $\chi_A \colon X \to \{0,1\}$ given by

$$\chi_A(x) = \begin{cases} 1 & \text{if } x \in A \\ 0 & \text{if } x \notin A. \end{cases}$$

In other words, χ_A tests membership in A. For any $x \in X$, $\chi_A(x) = 1$ iff $x \in A$. Observe that we obtain a function $\chi \colon 2^X \to \{0,1\}^X$ from the power set of X to the set of characteristic functions from X to $\{0,1\}$, given by

$$\chi(A) = \chi_A.$$

We also have the function, $\mathscr{S} \colon \{0,1\}^X \to 2^X$, mapping any characteristic function to the set that it defines and given by

$$\mathscr{S}(f) = \{x \in X \mid f(x) = 1\},$$

for every characteristic function, $f \in \{0,1\}^X$.

Proposition 2.5. *For any set X the function $\chi \colon 2^X \to \{0,1\}^X$ from the power set of X to the set of characteristic functions on X is a bijection whose inverse is $\mathscr{S} \colon \{0,1\}^X \to 2^X$.*

Proof. Simply check that $\chi \circ \mathscr{S} = \mathrm{id}$ and $\mathscr{S} \circ \chi = \mathrm{id}$, which is straightforward. $\square$

In view of Proposition 2.5, there is a bijection between the power set 2^X and the set of functions in $\{0,1\}^X$. If we write $2 = \{0,1\}$, then we see that the two sets look the same. This is the reason why the notation 2^X is often used for the power set (but others prefer $\mathscr{P}(X)$).

There are many other interesting results about equinumerosity. We only mention four more, all very important.

Theorem 2.4. *(Pigeonhole Principle) No set of the form $[n]$ is equinumerous to a proper subset of itself, where $n \in \mathbb{N}$,*

Proof. Although the pigeonhole principle seems obvious, the proof is not. In fact, the proof requires induction. We advise the reader to skip this proof and come back to it later after we have given more examples of proof by induction.

Suppose we can prove the following claim.

Claim. Whenever a function $f \colon [n] \to [n]$ is an injection, then it is a surjection onto $[n]$ (and thus, a bijection).

Observe that the above claim implies the pigeonhole principle. This is proved by contradiction. So, assume there is a function $f \colon [n] \to [n]$, such that f is injective and $\mathrm{Im}\, f = A \subseteq [n]$ with $A \neq [n]$; that is, f is a bijection between $[n]$ and A, a proper subset of $[n]$. Because $f \colon [n] \to [n]$ is injective, by the claim, we deduce that $f \colon [n] \to [n]$ is surjective, that is, $\mathrm{Im}\, f = [n]$, contradicting the fact that $\mathrm{Im}\, f = A \neq [n]$.

It remains to prove by induction on $n \in \mathbb{N}$ that if $f \colon [n] \to [n]$ is an injection, then it is a surjection (and thus, a bijection). For $n = 0$, f must be the empty function, which is a bijection.

Assume that the induction hypothesis holds for any $n \geq 0$ and consider any injection, $f \colon [n+1] \to [n+1]$. Observe that the restriction of f to $[n]$ is injective.

Case 1. The subset $[n]$ is closed under f; that is, $f([n]) \subseteq [n]$. Then, we know that $f \restriction [n]$ is injective and by the induction hypothesis, $f([n]) = [n]$. Because f is injective, we must have $f(n+1) = n+1$. Hence, f is surjective, as claimed.

Case 2. The subset $[n]$ is not closed under f; that is, there is some $p \leq n$ such that $f(p) = n+1$. We can create a new injection $\widehat{f}$ from $[n+1]$ to itself with the same image as f by interchanging two values of f so that $[n]$ closed under $\widehat{f}$. Define $\widehat{f}$ by

$$\widehat{f}(p) = f(n+1)$$
$$\widehat{f}(n+1) = f(p) = n+1$$
$$\widehat{f}(i) = f(i), \qquad 1 \leq i \leq n, i \neq p.$$

Then, $\widehat{f}$ is an injection from $[n+1]$ to itself and $[n]$ is closed under $\widehat{f}$. By Case 1, $\widehat{f}$ is surjective, and as $\mathrm{Im}\, f = \mathrm{Im}\, \widehat{f}$, we conclude that f is also surjective. $\square$

Corollary 2.1. *(Pigeonhole Principle for Finite Sets) No finite set is equinumerous to a proper subset of itself.*

Proof. To say that a set A is finite is to say that there is a bijection $g: A \to [n]$ for some $n \in \mathbb{N}$. Assume that there is a bijection f between A and some proper subset of A. Then, consider the function $g \circ f \circ g^{-1}$, from $[n]$ to itself, as shown in the diagram below:

$$
\begin{array}{ccc}
A & \xleftarrow{\ g^{-1}\ } & [n] \\
f \downarrow & & \downarrow g \circ f \circ g^{-1} \\
A & \xrightarrow[g]{} & [n]
\end{array}
$$

The rest of the proof consists in showing that $[n]$ would be equinumerous to a proper subset of itself, contradicting Theorem 2.4. We leave the details as an exercise. □

The pigeonhole principle is often used in the following way. If we have m distinct slots and $n > m$ distinct objects (the pigeons), then when we put all n objects into the m slots, two objects must end up in the same slot. This fact was apparently first stated explicitly by Dirichlet in 1834. As such, it is also known as *Dirichlet's box principle*.

Fig. 2.12 Johan Peter Gutav Lejeune Dirichlet, 1805–1859

Let A be a finite set. Then, by definition, there is a bijection $f: A \to [n]$ for some $n \in \mathbb{N}$. We claim that such an n is unique. Otherwise, there would be another bijection $g: A \to [p]$ for some $p \in \mathbb{N}$ with $n \neq p$. But now, we would have a bijection $g \circ f^{-1}$ between $[n]$ and $[p]$ with $n \neq p$. This would imply that there is either an injection from $[n]$ to a proper subset of itself or an injection from $[p]$ to a proper subset of itself,[2] contradicting the pigeonhole principle.

---

[2] Recall that $n + 1 = \{0, 1, \ldots, n\} = [n] \cup \{0\}$. Here in our argument, we are using the fact that for any two natural numbers n, p, either $n \subseteq p$ or $p \subseteq n$. This fact is indeed true but requires a proof. The proof uses induction and some special properties of the natural numbers implied by the definition of a natural number as a set that belongs to every inductive set. For details, see Enderton [2], Chapter 4.

If A is a finite set, the unique natural number, $n \in \mathbb{N}$, such that $A \approx [n]$ is called the *cardinality of* A and we write $|A| = n$ (or sometimes, card$(A) = n$).

Remark: The notion of cardinality also makes sense for infinite sets. What happens is that every set is equinumerous to a special kind of set (an initial ordinal) called a *cardinal* (or *cardinal number*). Let us simply mention that the cardinal number of $\mathbb{N}$ is denoted $\aleph_0$ (say "aleph" 0). A naive way to define the cardinality of a set X would be to define it as the equivalence class $\{Y \mid Y \approx X\}$ of all sets equinumerous to X. However, this does not work because the collection of sets Y such that $Y \approx X$, is not a set! In order to avoid this logical difficulty, one has to define the notion of a cardinal in a more subtle manner. One way to proceed is to first define *ordinals*, certain kinds of well-ordered sets. Then, assuming the axiom of choice, every set X is equinumerous to some ordinal and the cardinal $|X|$ of the set X is defined as the least ordinal equinumerous to X (an initial ordinal). The theory of ordinals and cardinals is thoroughly developed in Enderton [2] and Suppes [3] but it is beyond the scope of this book.

Corollary 2.2. *(a) Any set equinumerous to a proper subset of itself is infinite.*
(b) The set $\mathbb{N}$ is infinite.

Proof. Left as an exercise to the reader. $\square$

The image of a finite set by a function is also a finite set. In order to prove this important property we need the next two propositions. The first of these two propositions may appear trivial but again, a rigorous proof requires induction.

Proposition 2.6. *Let n be any positive natural number, let A be any nonempty set, and pick any element $a_0 \in A$. Then there exists a bijection $f: A \to [n+1]$ iff there exists a bijection $g: (A - \{a_0\}) \to [n]$.*

Proof. We proceed by induction on $n \geq 1$. The proof of the induction step is very similar to the proof of the induction step in Proposition 2.4. The details of the proof are left as an exercise to the reader. $\square$

Proposition 2.7. *For any function $f: A \to B$ if f is surjective and if A is a finite nonempty set, then B is also a finite set and there is an injection $h: B \to A$ such that $f \circ h = \mathrm{id}_B$. Moreover, $|B| \leq |A|$.*

Proof. The existence of an injection $h: B \to A$, such that $f \circ h = \mathrm{id}_B$, follows immediately from Theorem 2.2 (b), but the proof uses the axiom of choice, which seems a bit of an overkill. However, we can give an alternate proof avoiding the use of the axiom of choice by proceeding by induction on the cardinality of A.

If A has a single element, say a, because f is surjective, B is the one-element set (obviously finite), $B = \{f(a)\}$, and the function, $h: B \to A$, given by $g(f(a)) = a$ is obviously a bijection such that $f \circ h = \mathrm{id}_B$.

For the induction step, assume that A has $n+1$ elements. If f is a bijection, then $h = f^{-1}$ does the job and B is a finite set with $n+1$ elements.

If f is surjective but not injective, then there exist two distinct elements, $a', a'' \in A$, such that $f(a') = f(a'')$. If we let $A' = A - \{a''\}$ then, by Proposition 2.6, the set A' has n elements and the restriction f' of f to A' is surjective because for every $b \in B$, if $b \neq f(a')$, then by the surjectivity of f there is some $a \in A - \{a', a''\}$ such that $f'(a) = f(a) = b$ and if $b = f(a')$, then $f'(a') = f(a')$. By the induction hypothesis, B is a finite set and there is an injection $h': B \to A'$ such that $f' \circ h' = \mathrm{id}_B$. However, our injection $h': B \to A'$ can be viewed as an injection $h: B \to A$, which satisfies the identity $f \circ h = \mathrm{id}_B$, and this concludes the induction step.

Inasmuch as we have an injection $h: B \to A$ and A and B are finite sets, as every finite set has a uniquely defined cardinality, we deduce that $|B| \leq |A|$. $\square$

Corollary 2.3. *For any function $f: A \to B$, if A is a finite set, then the image $f(A)$ of f is also finite and $|f(A)| \leq |A|$.*

Proof. Any function $f: A \to B$ is surjective on its image $f(A)$, so the result is an immediate consequence of Proposition 2.7. $\square$

Corollary 2.4. *For any two sets A and B, if B is a finite set of cardinality n and is A is a proper subset of B, then A is also finite and A has cardinality $m < n$.*

Proof. Corollary 2.4 can be proved by induction on n using Proposition 2.6. Another proof goes as follows: Because $A \subseteq B$, the inclusion function $j: A \to B$ given by $j(a) = a$ for all $a \in A$, is obviously an injection. By Theorem 2.2(a), there is a surjection, $g: B \to A$. Because B is finite, by Proposition 2.7, the set A is also finite and because there is an injection $j: A \to B$, we have $m = |A| \leq |B| = n$. However, inasmuch as B is a proper subset of A, by the pigeonhole principle, we must have $m \neq n$, that is, $m < n$. $\square$

If A is an infinite set, then the image $f(A)$ is not finite in general but we still have the following fact.

Proposition 2.8. *For any function $f: A \to B$ we have $f(A) \preceq A$; that is, there is an injection from the image of f to A.*

Proof. Any function $f: A \to B$ is surjective on its image $f(A)$. By Theorem 2.2(b), there is an injection $h: f(B) \to A$, such that $f \circ h = \mathrm{id}_B$, which means that $f(A) \preceq A$. $\square$

Here are two more important facts that follow from the pigeonhole principle for finite sets and Proposition 2.7.

Proposition 2.9. *Let A be any finite set. For any function $f: A \to A$ the following properties hold.*

(a) If f is injective, then f is a bijection.
(b) If f is surjective, then f is a bijection.

The proof of Proposition 2.9 is left as an exercise (use Corollary 2.1 and Proposition 2.7).

Proposition 2.9 *only holds for finite sets*. Indeed, just after the remarks following Definition 2.8 we gave examples of functions defined on an infinite set for which Proposition 2.9 fails.

A convenient characterization of countable sets is stated below.

Proposition 2.10. *A nonempty set A is countable iff there is a surjection $g: \mathbb{N} \to A$ from $\mathbb{N}$ onto A.*

Proof. Recall that by definition, A is countable iff there is an injection $f: A \to \mathbb{N}$. The existence of a surjection $g: \mathbb{N} \to A$ follows from Theorem 2.2(a). Conversely, if there is a surjection $g: \mathbb{N} \to A$, then by Theorem 2.2(b), there is an injection $f: A \to \mathbb{N}$. However, the proof of Theorem 2.2(b) requires the axiom of choice. It is possible to avoid the axiom of choice by using the fact that every nonempty subset of $\mathbb{N}$ has a smallest element (see Theorem 5.3). $\square$

The following fact about infinite sets is also useful to know.

Theorem 2.5. *For every infinite set A, there is an injection from $\mathbb{N}$ into A.*

Proof. The proof of Theorem 2.5 is actually quite tricky. It requires a version of the axiom of choice and a subtle use of the recursion theorem (Theorem 2.1). Let us give a sketch of the proof.

The version of the axiom of choice that we need says that for every nonempty set A there is a function F (a *choice function*) such that the domain of F is $2^A - \{\emptyset\}$ (all nonempty subsets of A) and such that $F(B) \in B$ for every nonempty subset B of A.

We use the recursion theorem to define a function h from $\mathbb{N}$ to the set of finite subsets of A. The function h is defined by

$$h(0) = \emptyset$$
$$h(n+1) = h(n) \cup \{F(A - h(n))\}.$$

Because A is infinite and $h(n)$ is finite, $A - h(n)$ is nonempty and we use F to pick some element in $A - h(n)$, which we then add to the set $h(n)$, creating a new finite set $h(n+1)$. Now, we define $g: \mathbb{N} \to A$ by

$$g(n) = F(A - h(n))$$

for all $n \in \mathbb{N}$. Because $h(n)$ is finite and A is infinite, g is well defined. It remains to check that g is an injection. For this, we observe that $g(n) \notin h(n)$ because $F(A - h(n)) \in A - h(n)$; the details are left as an exercise. $\square$

The intuitive content of Theorem 2.5 is that $\mathbb{N}$ is the "smallest" infinite set.

An immediate consequence of Theorem 2.5 is that every infinite subset of $\mathbb{N}$ is equinumerous to $\mathbb{N}$.

Here is a characterization of infinite sets originally proposed by Dedekind in 1888.

Proposition 2.11. *A set A is infinite iff it is equinumerous to a proper subset of itself.*

Proof. If A is equinumerous to a proper subset of itself, then it must be infinite because otherwise the pigeonhole principle would be contradicted.

Conversely, assume A is infinite. By Theorem 2.5, there is an injection $f\colon \mathbb{N} \to A$. Define the function $g\colon A \to A$ as follows.

$$g(f(n)) = f(n+1) \quad \text{if} \quad n \in \mathbb{N}$$
$$g(a) = a \qquad\qquad \text{if} \quad a \notin \mathrm{Im}(f).$$

It is easy to check that g is a bijection of A onto $A - \{f(0)\}$, a proper subset of A. □

Let us give another application of the pigeonhole principle involving sequences of integers. Given a finite sequence S of integers $a_1, \ldots, a_n$, a *subsequence of S* is a sequence $b_1, \ldots, b_m$, obtained by deleting elements from the original sequence and keeping the remaining elements in the same order as they originally appeared. More precisely, $b_1, \ldots, b_m$ is a subsequence of $a_1, \ldots, a_n$ if there is an injection $g\colon \{1, \ldots, m\} \to \{1, \ldots, n\}$ such that $b_i = a_{g(i)}$ for all $i \in \{1, \ldots, m\}$ and $i \leq j$ implies $g(i) \leq g(j)$ for all $i, j \in \{1, \ldots, m\}$. For example, the sequence

$$1 \quad \mathbf{9} \quad 10 \quad \mathbf{8} \quad 3 \quad 7 \quad 5 \quad 2 \quad \mathbf{6} \quad \mathbf{4}$$

contains the subsequence

$$9 \quad 8 \quad 6 \quad 4.$$

An *increasing subsequence* is a subsequence whose elements are in strictly increasing order and a *decreasing subsequence* is a subsequence whose elements are in strictly decreasing order. For example, $9\,8\,6\,4$ is a decreasing subsequence of our original sequence. We now prove the following beautiful result due to Erdös and Szekeres.

Theorem 2.6. *(Erdös and Szekeres) Let n be any nonzero natural number. Every sequence of $n^2 + 1$ pairwise distinct natural numbers must contain either an increasing subsequence or a decreasing subsequence of length $n + 1$.*

Proof. The proof proceeds by contradiction. So, assume there is a sequence S of $n^2 + 1$ pairwise distinct natural numbers so that all increasing or decreasing subsequences of S have length at most n. We assign to every element s of the sequence S a pair of natural numbers (u_s, d_s), called a *label*, where u_s, is the length of a longest increasing subsequence of S that starts at s and where d_s is the length of a longest decreasing subsequence of S that starts at s.

There are no increasing or descreasing subsequences of length $n + 1$ in S, thus observe that $1 \leq u_s, d_s \leq n$ for all $s \in S$. Therefore,

Claim 1: There are at most n^2 distinct labels (u_s, d_s), where $s \in S$.

We also assert the following.

Claim 2: If s and t are any two distinct elements of S, then $(u_s, d_s) \neq (u_t, d_t)$.

We may assume that s precedes t in S because otherwise we interchange s and t in the following argument. Inasmuch as $s \neq t$, there are two cases:

(a) $s < t$. In this case, we know that there is an increasing subsequence of length u_t starting with t. If we insert s in front of this subsequence, we get an increasing subsequence of $u_t + 1$ elements starting at s. Then, as u_s is the maximal length of all increasing subsequences starting with s, we must have $u_t + 1 \leq u_s$; that is,

$$u_s > u_t,$$

which implies $(u_s, d_s) \neq (u_t, d_t)$.

(b) $s > t$. This case is similar to case (a), except that we consider a decreasing subsequence of length d_t starting with t. We conclude that

$$d_s > d_t,$$

which implies $(u_s, d_s) \neq (u_t, d_t)$.

Therefore, in all cases, we proved that s and t have distinct labels.

Now, by Claim 1, there are only n^2 distinct labels and S has $n^2 + 1$ elements so, by the pigeonhole principle, two elements of S must have the same label. But, this contradicts Claim 2, which says that distinct elements of S have distinct labels. Therefore, S must have either an increasing subsequence or a decreasing subsequence of length $n + 1$, as originally claimed. □

Remark: Note that this proof is not constructive in the sense that it does not produce the desired subsequence; it merely asserts that such a sequence exists.

Our next theorem is the historically famous Schröder–Bernstein theorem, sometimes called the "Cantor–Bernstein theorem." Cantor proved the theorem in 1897 but his proof used a principle equivalent to the axiom of choice. Schröder announced the theorem in an 1896 abstract. His proof, published in 1898, had problems and he published a correction in 1911. The first fully satisfactory proof was given by Felix Bernstein and was published in 1898 in a book by Emile Borel. A shorter proof was given later by Tarski (1955) as a consequence of his fixed point theorem. We postpone giving this proof until the section on lattices (see Section 5.2).

Theorem 2.7. *(Schröder–Bernstein Theorem)* *Given any two sets A and B, if there is an injection from A to B and an injection from B to A, then there is a bijection between A and B. Equivalently, if $A \preceq B$ and $B \preceq A$, then $A \approx B$.*

The Schröder–Bernstein theorem is quite a remarkable result and it is a main tool to develop cardinal arithmetic, a subject beyond the scope of this course.

Our third theorem is perhaps the one that is the more surprising from an intuitive point of view. If nothing else, it shows that our intuition about infinity is rather poor.

Theorem 2.8. *If A is any infinite set, then $A \times A$ is equinumerous to A.*

Fig. 2.13 Georg Cantor, 1845–1918 (left), Ernst Schröder, 1841–1902 (middle left), Felix Bernstein, 1878–1956 (middle right) and Emile Borel, 1871–1956 (right)

Proof. The proof is more involved than any of the proofs given so far and it makes use of the axiom of choice in the form known as *Zorn's lemma* (see Theorem 5.1). For these reasons, we omit the proof and instead refer the reader to Enderton [2] (Chapter 6). ☐

Fig. 2.14 Max August Zorn, 1906–1993

In particular, Theorem 2.8 implies that $\mathbb{R} \times \mathbb{R}$ is in bijection with $\mathbb{R}$. But, geometrically, $\mathbb{R} \times \mathbb{R}$ is a plane and $\mathbb{R}$ is a line and, intuitively, it is surprising that a plane and a line would have "the same number of points." Nevertheless, that's what mathematics tells us.

Remark: It is possible to give a bijection between $\mathbb{R} \times \mathbb{R}$ and $\mathbb{R}$ without using Theorem 2.8; see Problem 2.40.

Our fourth theorem also plays an important role in the theory of cardinal numbers.

Theorem 2.9. *(Cardinal Comparability) Given any two sets, A and B, either there is an injection from A to B or there is an injection from B to A (i.e., either $A \preceq B$ or $B \preceq A$).*

Proof. The proof requires the axiom of choice in a form known as the *well-ordering theorem*, which is also equivalent to Zorn's lemma. For details, see Enderton [2] (Chapters 6 and 7). ☐

Theorem 2.8 implies that there is a bijection between the closed line segment

$$[0,1] = \{x \in \mathbb{R} \mid 0 \le x \le 1\}$$

and the closed unit square

$$[0,1] \times [0,1] = \{(x,y) \in \mathbb{R}^2 \mid 0 \le x, y \le 1\}.$$

As an interlude, in the next section, we describe a famous space-filling function due to Hilbert. Such a function is obtained as the limit of a sequence of curves that can be defined recursively.

2.10 An Amazing Surjection: Hilbert's Space-Filling Curve

In the years 1890–1891, Giuseppe Peano and David Hilbert discovered examples of *space-filling functions* (also called *space-filling curves*). These are surjective functions from the line segment $[0,1]$ onto the unit square and thus their image is the whole unit square. Such functions defy intuition because they seem to contradict our intuition about the notion of dimension; a line segment is one-dimensional, yet the unit square is two-dimensional. They also seem to contradict our intuitive notion of area. Nevertheless, such functions do exist, even continuous ones, although to justify their existence rigorously requires some tools from mathematical analysis. Similar curves were found by others, among whom we mention Sierpinski, Moore, and Gosper.

Fig. 2.15 David Hilbert 1862–1943 and Waclaw Sierpinski, 1882–1969

We describe Hilbert's scheme for constructing such a square-filling curve. We define a sequence (h_n) of polygonal lines $h_n\colon [0,1] \to [0,1] \times [0,1]$, starting from the simple pattern h_0 (a "square cap" ⊓) shown on the left in Figure 2.16.

The curve h_{n+1} is obtained by scaling down h_n by a factor of $\frac{1}{2}$, and connecting the four copies of this scaled-down version of h_n obtained by rotating by $\pi/2$ (left lower part), rotating by $-\pi/2$, and translating right (right lower part), translating up (left upper part), and translating diagonally (right upper part), as illustrated in Figure 2.16.

Fig. 2.16 A sequence of Hilbert curves h_0, h_1, h_2

Fig. 2.17 The Hilbert curve h_5

It can be shown that the sequence (h_n) converges (uniformly) to a continuous curve $h: [0,1] \to [0,1] \times [0,1]$ whose trace is the entire square $[0,1] \times [0,1]$. The Hilbert curve h is surjective, continuous, and nowhere differentiable. It also has infinite length.

The curve h_5 is shown in Figure 2.17. You should try writing a computer program to plot these curves. By the way, it can be shown that no continuous square-filling function can be injective. It is also possible to define cube-filling curves and even higher-dimensional cube-filling curves.

Before we close this chapter and move on to special kinds of relations, namely, partial orders and equivalence relations, we illustrate how the notion of function can be used to define strings, multisets, and indexed families rigorously.

2.11 Strings, Multisets, Indexed Families

Strings play an important role in computer science and linguistics because they are the basic tokens of which languages are made. In fact, formal language theory takes the (somewhat crude) view that a language is a set of strings. A string is a finite sequence of letters, for example, "Jean", "Val", "Mia", "math", "gaga", "abab". Usually, we have some alphabet in mind and we form strings using letters from this alphabet. Strings are not sets; the order of the letters matters: "abab" and "baba" are different strings. What matters is the position of every letter. In the string "aba", the leftmost "a" is in position 1, "b" is in position 2, and the rightmost "b" is in position 3. All this suggests defining strings as certain kinds of functions whose domains are the sets $[n] = \{1, 2, \ldots, n\}$ (with $[0] = \emptyset$) encountered earlier. Here is the very beginning of the theory of formal languages.

Definition 2.13. An *alphabet* Σ is any **finite** set.

We often write $\Sigma = \{a_1, \ldots, a_k\}$. The a_i are called the *symbols* of the alphabet.

Remark: There are a few occasions where we allow infinite alphabets but normally an alphabet is assumed to be finite.
Examples:
$\Sigma = \{a\}$
$\Sigma = \{a, b, c\}$
$\Sigma = \{0, 1\}$
A string is a finite sequence of symbols. Technically, it is convenient to define strings as functions.

Definition 2.14. Given an alphabet Σ a *string over Σ (or simply a string) of length n* is any function

$$u \colon [n] \to \Sigma.$$

The integer n is the *length* of the string u, and it is denoted by $|u|$. When $n = 0$, the special string $u \colon [0] \to \Sigma$, of length 0 is called the *empty string, or null string*, and is denoted by ε.

Given a string $u \colon [n] \to \Sigma$ of length $n \geq 1$, $u(i)$ is the ith letter in the string u. For simplicity of notation, we denote the string u as

$$u = u_1 u_2 \dots u_n,$$

with each $u_i \in \Sigma$.

For example, if $\Sigma = \{a, b\}$ and $u\colon [3] \to \Sigma$ is defined such that $u(1) = a, u(2) = b$, and $u(3) = a$, we write

$$u = aba.$$

Strings of length 1 are functions $u\colon [1] \to \Sigma$ simply picking some element $u(1) = a_i$ in Σ. Thus, we identify every symbol $a_i \in \Sigma$ with the corresponding string of length 1.

The set of all strings over an alphabet Σ, including the empty string, is denoted as Σ^*. Observe that when $\Sigma = \emptyset$, then

$$\emptyset^* = \{\varepsilon\}.$$

When $\Sigma \neq \emptyset$, the set Σ^* is countably infinite. Later on, we show ways of ordering and enumerating strings.

Strings can be juxtaposed, or concatenated.

Definition 2.15. Given an alphabet Σ, given two strings $u\colon [m] \to \Sigma$ and $v\colon [n] \to \Sigma$, the *concatenation, $u \cdot v$, (also written uv) of u and v* is the string $uv\colon [m+n] \to \Sigma$, defined such that

$$uv(i) = \begin{cases} u(i) & \text{if } 1 \leq i \leq m, \\ v(i-m) & \text{if } m+1 \leq i \leq m+n. \end{cases}$$

In particular, $u\varepsilon = \varepsilon u = u$.

It is immediately verified that

$$u(vw) = (uv)w.$$

Thus, concatenation is a binary operation on Σ^* that is associative and has ε as an identity. Note that generally, $uv \neq vu$, for example, for $u = a$ and $v = b$.

Definition 2.16. Given an alphabet Σ, given any two strings $u, v \in \Sigma^*$, we define the following notions as follows.

u is a prefix of v iff there is some $y \in \Sigma^*$ such that

$$v = uy.$$

u is a suffix of v iff there is some $x \in \Sigma^*$ such that

$$v = xu.$$

u is a substring of v iff there are some $x, y \in \Sigma^*$ such that

$$v = xuy.$$

We say that u *is a proper prefix (suffix, substring) of* v iff u is a prefix (suffix, substring) of v and $u \neq v$.

For example, *ga* is a prefix of *gallier*, the string *lier* is a suffix of *gallier*, and *all* is a substring of *gallier*.

Finally, languages are defined as follows.

Definition 2.17. Given an alphabet Σ, a *language over* Σ *(or simply a language)* is any subset L of Σ^*.

The next step would be to introduce various formalisms to define languages, such as automata or grammars but you'll have to take another course to learn about these things.

We now consider multisets. We already encountered multisets in Section 1.2 when we defined the axioms of propositional logic. As for sets, in a multiset, the order of elements does not matter, but as in strings, multiple occurrences of elements matter. For example,

$$\{a,a,b,c,c,c\}$$

is a multiset with two occurrences of a, one occurrence of b, and three occurrences of c. This suggests defining a multiset as a function with range $\mathbb{N}$, to specify the multiplicity of each element.

Definition 2.18. Given any set S a *multiset M over S* is any function $M \colon S \to \mathbb{N}$. A *finite multiset M over S* is any function $M \colon S \to \mathbb{N}$ such that $M(a) \neq 0$ only for finitely many $a \in S$. If $M(a) = k > 0$, we say that *a appears with mutiplicity k in M*.

For example, if $S = \{a,b,c\}$, we may use the notation $\{a,a,a,b,c,c\}$ for the multiset where a has multiplicity 3, b has multiplicity 1, and c has multiplicity 2.

The empty multiset is the function having the constant value 0. The *cardinality* $|M|$ of a (finite) multiset is the number

$$|M| = \sum_{a \in S} M(a).$$

Note that this is well defined because $M(a) = 0$ for all but finitely many $a \in S$. For example,

$$|\{a,a,a,b,c,c\}| = 6.$$

We can define the *union* of multisets as follows. If M_1 and M_2 are two multisets, then $M_1 \cup M_2$ is the multiset given by

$$(M_1 \cup M_2)(a) = M_1(a) + M_2(a), \text{ for all } a \in S.$$

A multiset M_1 is a *submultiset* of a multiset M_2 if $M_1(a) \leq M_2(a)$ for all $a \in S$. The *difference of M_1 and M_2* is the multiset $M_1 - M_2$ given by

$$(M_1 - M_2)(a) = \begin{cases} M_1(a) - M_2(a) & \text{if } M_1(a) \geq M_2(a) \\ 0 & \text{if } M_1(a) < M_2(a). \end{cases}$$

Intersection of multisets can also be defined but we leave this as an exercise.

Let us now discuss indexed families. The Cartesian product construct, $A_1 \times A_2 \times \cdots \times A_n$, allows us to form finite indexed sequences, $\langle a_1, \ldots, a_n \rangle$, but there are situations where we need to have infinite indexed sequences. Typically, we want to be able to consider families of elements indexed by some index set of our choice, say I. We can do this as follows.

Definition 2.19. Given any X and any other set I, called the *index set*, the set of I-*indexed families (or sequences) of elements from X* is the set of all functions $A: I \to X$; such functions are usually denoted $A = (A_i)_{i \in I}$. When X is a set of sets, each A_i is some set in X and we call $(A_i)_{i \in I}$ a *family of sets (indexed by I)*.

Observe that if $I = [n] = \{1, \ldots, n\}$, then an I-indexed family is just a string over X. When $I = \mathbb{N}$, an $\mathbb{N}$-indexed family is called an *infinite sequence* or often just a *sequence*. In this case, we usually write (x_n) for such a sequence $((x_n)_{n \in \mathbb{N}}$, if we want to be more precise). Also, note that although the notion of indexed family may seem less general than the notion of arbitrary collection of sets, this is an illusion. Indeed, given any collection of sets X, we may choose the index set I to be X itself, in which case X appears as the range of the identity function, id: $X \to X$.

The point of indexed families is that the operations of union and intersection can be generalized in an interesting way. We can also form infinite Cartesian products, which are very useful in algebra and geometry.

Given any indexed family of sets $(A_i)_{i \in I}$, the *union of the family* $(A_i)_{i \in I}$, denoted $\bigcup_{i \in I} A_i$, is simply the union of the range of A; that is,

$$\bigcup_{i \in I} A_i = \bigcup range(A) = \{a \mid (\exists i \in I), a \in A_i\}.$$

Observe that when $I = \emptyset$, the union of the family is the empty set. When $I \neq \emptyset$, we say that we have a *nonempty family* (even though some of the A_i may be empty).

Similarly, if $I \neq \emptyset$, then the *intersection of the family* $(A_i)_{i \in I}$, denoted $\bigcap_{i \in I} A_i$, is simply the intersection of the range of A; that is,

$$\bigcap_{i \in I} A_i = \bigcap range(A) = \{a \mid (\forall i \in I), a \in A_i\}.$$

Unlike the situation for union, when $I = \emptyset$, the intersection of the family does not exist. It would be the set of all sets, which does not exist.

It is easy to see that the laws for union, intersection, and complementation generalize to families but we leave this to the exercises.

An important construct generalizing the notion of finite Cartesian product is the product of families.

Definition 2.20. Given any family of sets $(A_i)_{i \in I}$, the *product of the family* $(A_i)_{i \in I}$, denoted $\prod_{i \in I} A_i$, is the set

$$\prod_{i \in I} A_i = \{a: I \to \bigcup_{i \in I} A_i \mid (\forall i \in I), a(i) \in A_i\}.$$

Definition 2.20 says that the elements of the product $\prod_{i \in I} A_i$ are the functions $a: I \to \bigcup_{i \in I} A_i$, such that $a(i) \in A_i$ for every $i \in I$. We denote the members of $\prod_{i \in I} A_i$ by $(a_i)_{i \in I}$ and we usually call them *I-tuples*. When $I = \{1, \ldots, n\} = [n]$, the members of $\prod_{i \in [n]} A_i$ are the functions whose graph consists of the sets of pairs

$$\{\langle 1, a_1 \rangle, \langle 2, a_2 \rangle, \ldots, \langle n, a_n \rangle\}, \ a_i \in A_i, \ 1 \leq i \leq n,$$

and we see that the function

$$\{\langle 1, a_1 \rangle, \langle 2, a_2 \rangle, \ldots, \langle n, a_n \rangle\} \mapsto \langle a_1, \ldots, a_n \rangle$$

yields a bijection between $\prod_{i \in [n]} A_i$ and the Cartesian product $A_1 \times \cdots \times A_n$. Thus, if each A_i is nonempty, the product $\prod_{i \in [n]} A_i$ is nonempty. But what if I is infinite?

If I is infinite, we smell choice functions. That is, an element of $\prod_{i \in I} A_i$ is obtained by choosing for every $i \in I$ some $a_i \in A_i$. Indeed, the axiom of choice is needed to ensure that $\prod_{i \in I} A_i \neq \emptyset$ if $A_i \neq \emptyset$ for all $i \in I$. For the record, we state this version (among many) of the axiom of choice.

Axiom of Choice (Product Version)
For any family of sets, $(A_i)_{i \in I}$, if $I \neq \emptyset$ and $A_i \neq \emptyset$ for all $i \in I$, then $\prod_{i \in I} A_i \neq \emptyset$.

Given the product of a family of sets, $\prod_{i \in I} A_i$, for each $i \in I$, we have the function $pr_i \colon \prod_{i \in I} A_i \to A_i$, called the *ith projection function*, defined by

$$pr_i((a_i)_{i \in I}) = a_i.$$

2.12 Summary

This chapter deals with the notions of relations, partial functions and functions, and their basic properties. The notion of a function is used to define the concept of a finite set and to compare the "size" of infinite sets. In particular, we prove that the power set 2^A of any set A is always "strictly bigger" than A itself (Cantor's theorem).

- We give some examples of functions, emphasizing that a function has a set of input values and a set of output values but that a function may not be defined for all of its input values (it may be a *partial function*). A function is given by a set of ⟨ input, output ⟩ pairs.
- We define *ordered pairs* and the *Cartesian product* $A \times B$ of two sets A and B.
- We define the *first and second projection* of a pair.
- We define *binary relations* and their *domain* and *range*.
- We define the *identity relation*.
- We define *functional* relations.
- We define *partial functions*, *total functions*, the *graph* of a partial or total function, the *domain*, and the *range* of a (partial) function.
- We define the *preimage* or *inverse image* $f^{-1}(a)$ of an element a by a (partial) function f.

- The set of all functions from A to B is denoted B^A.
- We revisit the *induction principle for* $\mathbb{N}$ stated in terms of properties and give several examples of proofs by induction.
- We state the *complete induction principle for* $\mathbb{N}$ and prove its validity; we prove a property of the *Fibonacci numbers* by complete induction.
- We define the *composition* $R \circ S$ of two relations R and S.
- We prove some basic properties of the composition of functional relations.
- We define the *composition* $g \circ f$ of two (partial or total) functions, f and g.
- We describe the process of defining functions on $\mathbb{N}$ by *recursion* and state a basic result about the validity of such a process (The *recursion theorem on* $\mathbb{N}$).
- We define the *left inverse* and the *right inverse* of a function.
- We define *invertible* functions and prove the uniqueness of the inverse f^{-1} of a function f when it exists.
- We define the *inverse* or *converse* of a relation .
- We define, *injective, surjective*, and *bijective* functions.
- We characterize injectivity, surjectivity, and bijectivity in terms of left and right inverses.
- We observe that to prove that a surjective function has a right inverse, we need the *axiom of choice* (AC).
- We define *sections, retractions*, and the *restriction* of a function to a subset of its domain.
- We define *direct* and *inverse* images of a set under a function ($f(A)$, respectively, $f^{-1}(B)$).
- We prove some basic properties of direct and inverse images with respect to union, intersection, and relative complement.
- We define when two sets are *equinumerous* or when a set A *dominates* a set B.
- We give a bijection between $\mathbb{N} \times \mathbb{N}$ and $\mathbb{N}$.
- We define when a set if *finite* or *infinite*.
- We prove that $\mathbb{N}$ is not equinumerous to $\mathbb{R}$ (the real numbers), a result due to Cantor, and that there is no surjection from A to 2^A.
- We define the *characteristic function* χ_A of a subset A.
- We state and prove the *pigeonhole principle*.
- The set of natural numbers $\mathbb{N}$ is infinite.
- Every finite set A is equinumerous with a unique set $[n] = \{1, \ldots, n\}$ and the integer n is called the *cardinality of* A and is denoted $|A|$.
- If A is a finite set, then for every function $f \colon A \to B$ the image $f(A)$ of f is finite and $|f(A)| \leq |A|$.
- Any subset A of a finite set B is also finite and $|A| \leq |B|$.
- If A is a finite set, then every injection $f \colon A \to A$ is a bijection and every surjection $f \colon A \to A$ is a bijection.
- A set A is countable iff there is a surjection from $\mathbb{N}$ onto A.
- For every infinite set A there is an injection from $\mathbb{N}$ into A.
- A set A is infinite iff it is equinumerous to a proper subset of itself.
- We state the *Schröder–Bernstein theorem*.
- We state that every infinite set A is equinumerous to $A \times A$.

- We state the *cardinal comparability theorem*.
- We mention *Zorn's lemma*, one of the many versions of the axiom of choice.
- We describe *Hilbert's space-filling curve*.
- We define *strings* and *multisets*.
- We define the *product of a family of sets* and explain how the non-emptyness of such a product is equivalent to the axiom of choice.

Problems

2.1. Given any two sets A, B, prove that for all $a_1, a_2 \in A$ and all $b_1, b_2 \in B$,

$$\{\{a_1\}, \{a_1, b_1\}\} = \{\{a_2\}, \{a_2, b_2\}\}$$

iff

$$a_1 = a_2 \quad \text{and} \quad b_1 = b_2.$$

2.2. (a) Prove that the composition of two injective functions is injective. Prove that the composition of two surjective functions is surjective.

(b) Prove that a function $f: A \to B$ is injective iff for all functions $g, h: C \to A$,

$$\text{if } f \circ g = f \circ h, \text{ then } g = h.$$

(c) Prove that a function $f: A \to B$ is surjective iff for all functions $g, h: B \to C$,

$$\text{if } g \circ f = h \circ f, \text{ then } g = h.$$

2.3. (a) Prove that

$$\sum_{k=1}^{n} k^2 = \frac{n(n+1)(2n+1)}{6}.$$

(b) Prove that

$$\sum_{k=1}^{n} k^3 = \left(\sum_{k=1}^{n} k \right)^2.$$

2.4. Given any finite set A, let $|A|$ denote the number of elements in A.

(a) If A and B are finite sets, prove that

$$|A \cup B| = |A| + |B| - |A \cap B|.$$

(b) If A, B, and C are finite sets, prove that

$$|A \cup B \cup C| = |A| + |B| + |C| - |A \cap B| - |A \cap C| - |B \cap C| + |A \cap B \cap C|.$$

2.5. Prove that there is no set X such that

$$2^X \subseteq X.$$

Hint. Given any two sets A, B, if there is an injection from A to B, then there is a surjection from B to A.

2.6. Let $f: X \rightarrow Y$ be any function. (a) Prove that for any two subsets $A, B \subseteq X$ we have

$$f(A \cup B) = f(A) \cup f(B)$$
$$f(A \cap B) \subseteq f(A) \cap f(B).$$

Give an example of a function f and of two subsets A, B such that

$$f(A \cap B) \neq f(A) \cap f(B).$$

Prove that if $f: X \rightarrow Y$ is injective, then

$$f(A \cap B) = f(A) \cap f(B).$$

(b) For any two subsets $C, D \subseteq Y$, prove that

$$f^{-1}(C \cup D) = f^{-1}(C) \cup f^{-1}(D)$$
$$f^{-1}(C \cap D) = f^{-1}(C) \cap f^{-1}(D).$$

(c) Prove that for any two subsets $A \subseteq X$ and $C \subseteq Y$, we have

$$f(A) \subseteq C \quad \text{iff} \quad A \subseteq f^{-1}(C).$$

2.7. Prove that the set of natural numbers $\mathbb{N}$ is infinite. (Recall, a set X is finite iff there is a bijection from X to $[n] = \{1, \ldots, n\}$, where $n \in \mathbb{N}$ is a natural number with $[0] = \emptyset$. Thus, a set X is infinite iff there is no bijection from X to any $[n]$, with $n \in \mathbb{N}$.)

2.8. Let $R \subseteq A \times A$ be a relation. Prove that if $R \circ R = \mathrm{id}_A$, then R is the graph of a bijection whose inverse is equal to itself.

2.9. Given any three relations $R \subseteq A \times B$, $S \subseteq B \times C$, and $T \subseteq C \times D$, prove the associativity of composition:

$$(R \circ S) \circ T = R \circ (S \circ T).$$

2.10. Let $f: A \rightarrow A'$ and $g: B \rightarrow B'$ be two functions and define $h: A \times B \rightarrow A' \times B'$ by

$$h(\langle a, b \rangle) = \langle f(a), g(b) \rangle,$$

for all $a \in A$ and $b \in B$.

(a) Prove that if f and g are injective, then so is h.

Hint. Use the definition of injectivity, not the existence of a left inverse and do not proceed by contradiction.

(b) Prove that if f and g are surjective, then so is h.

Hint. Use the definition of surjectivity, not the existence of a right inverse and do not proceed by contradiction.

2.11. Let $f: A \to A'$ and $g: B \to B'$ be two injections. Prove that if $\operatorname{Im} f \cap \operatorname{Im} g = \emptyset$, then there is an injection from $A \cup B$ to $A' \cup B'$.
   Is the above still correct if $\operatorname{Im} f \cap \operatorname{Im} g \neq \emptyset$?

2.12. Let $[0, 1]$ and $(0, 1)$ denote the set of real numbers

$$[0, 1] = \{x \in \mathbb{R} \mid 0 \le x \le 1\}$$
$$(0, 1) = \{x \in \mathbb{R} \mid 0 < x < 1\}.$$

(a) Give a bijection $f: [0, 1] \to (0, 1)$.

Hint. There are such functions that are the identity almost everywhere but for a countably infinite set of points in $[0, 1]$.

(b) Consider the open square $(0, 1) \times (0, 1)$ and the closed square $[0, 1] \times [0, 1]$. Give a bijection $f: [0, 1] \times [0, 1] \to (0, 1) \times (0, 1)$.

2.13. Consider the function, $J: \mathbb{N} \times \mathbb{N} \to \mathbb{N}$, given by

$$J(m, n) = \frac{1}{2}[(m+n)^2 + 3m + n].$$

(a) Prove that for any $z \in \mathbb{N}$, if $J(m, n) = z$, then

$$8z + 1 = (2m + 2n + 1)^2 + 8m.$$

Deduce from the above that

$$2m + 2n + 1 \le \sqrt{8z + 1} < 2m + 2n + 3.$$

(b) If $x \mapsto \lfloor x \rfloor$ is the function from $\mathbb{R}$ to $\mathbb{N}$ (the *floor function*), where $\lfloor x \rfloor$ is the largest integer $\le x$ (e.g., $\lfloor 2.3 \rfloor = 2$, $\lfloor \sqrt{2} \rfloor = 1$), prove that

$$\lfloor \sqrt{8z + 1} \rfloor + 1 = 2m + 2n + 2 \text{ or } \lfloor \sqrt{8z + 1} \rfloor + 1 = 2m + 2n + 3,$$

so that

$$\lfloor (\lfloor \sqrt{8z + 1} \rfloor + 1)/2 \rfloor = m + n + 1.$$

(c) Because $J(m, n) = z$ means that

$$2z = (m + n)^2 + 3m + n,$$

prove that m and n are solutions of the system

$$m + n = \lfloor (\lfloor \sqrt{8z+1} \rfloor + 1)/2 \rfloor - 1$$
$$3m + n = 2z - (\lfloor (\lfloor \sqrt{8z+1} \rfloor + 1)/2 \rfloor - 1)^2.$$

If we let

$$Q_1(z) = \lfloor (\lfloor \sqrt{8z+1} \rfloor + 1)/2 \rfloor - 1$$
$$Q_2(z) = 2z - (\lfloor (\lfloor \sqrt{8z+1} \rfloor + 1)/2 \rfloor - 1)^2 = 2z - (Q_1(z))^2,$$

prove that $Q_2(z) - Q_1(z)$ is even and that

$$m = \frac{1}{2}(Q_2(z) - Q_1(z)) = K(z)$$

$$n = Q_1(z) - \frac{1}{2}(Q_2(z) - Q_1(z)) = L(z).$$

Conclude that J is a bijection between $\mathbb{N} \times \mathbb{N}$ and $\mathbb{N}$, with

$$m = K(J(m,n))$$
$$n = L(J(m,n)).$$

Remark: It can also be shown that $J(K(z), L(z)) = z$.

2.14. (i) In 3-dimensional space $\mathbb{R}^3$ the sphere S^2 is the set of points of coordinates (x,y,z) such that $x^2 + y^2 + z^2 = 1$. The point $N = (0,0,1)$ is called the *north pole*, and the point $S = (0,0,-1)$ is called the *south pole*. The *stereographic projection* map $\sigma_N : (S^2 - \{N\}) \to \mathbb{R}^2$ is defined as follows. For every point $M \neq N$ on S^2, the point $\sigma_N(M)$ is the intersection of the line through N and M and the equatorial plane of equation $z = 0$.

Prove that if M has coordinates (x,y,z) (with $x^2 + y^2 + z^2 = 1$), then

$$\sigma_N(M) = \left(\frac{x}{1-z}, \frac{y}{1-z} \right).$$

Hint. Recall that if $A = (a_1, a_2, a_3)$ and $B = (b_1, b_2, b_3)$ are any two distinct points in $\mathbb{R}^3$, then the unique line (AB) passing through A and B has parametric equations

$$x = (1-t)a_1 + tb_1$$
$$y = (1-t)a_2 + tb_2$$
$$z = (1-t)a_3 + tb_3,$$

which means that every point (x,y,z) on the line (AB) is of the above form, with $t \in \mathbb{R}$. Find the intersection of a line passing through the North pole and a point $M \neq N$ on the sphere S^2.

Prove that σ_N is bijective and that its inverse is given by the map $\tau_N : \mathbb{R}^2 \to (S^2 - \{N\})$ with

$$(x,y) \mapsto \left(\frac{2x}{x^2+y^2+1}, \frac{2y}{x^2+y^2+1}, \frac{x^2+y^2-1}{x^2+y^2+1} \right).$$

Hint. Find the intersection of a line passing through the North pole and some point P of the equatorial plane $z = 0$ with the sphere of equation

$$x^2 + y^2 + z^2 = 1.$$

Similarly, $\sigma_S \colon (S^2 - \{S\}) \to \mathbb{R}^2$ is defined as follows. For every point $M \neq S$ on S^2, the point $\sigma_S(M)$ is the intersection of the line through S and M and the plane of equation $z = 0$.
Prove that

$$\sigma_S(M) = \left(\frac{x}{1+z}, \frac{y}{1+z} \right).$$

Prove that σ_S is bijective and that its inverse is given by the map, $\tau_S \colon \mathbb{R}^2 \to (S^2 - \{S\})$, with

$$(x,y) \mapsto \left(\frac{2x}{x^2+y^2+1}, \frac{2y}{x^2+y^2+1}, \frac{1-x^2-y^2}{x^2+y^2+1} \right).$$

(ii) Give a bijection between the sphere S^2 and the equatorial plane of equation $z = 0$.

Hint. Use the stereographic projection and the method used in Problem 2.12, to define a bijection between $[0,1]$ and $(0,1)$.

2.15. (a) Give an example of a function $f \colon A \to A$ such that $f^2 = f \circ f = f$ and f is not the identity function.

(b) Prove that if a function $f \colon A \to A$ is not the identity function and $f^2 = f$, then f is not invertible.

(c) Give an example of an invertible function $f \colon A \to A$, such that $f^3 = f \circ f \circ f = f$, yet $f \circ f \neq f$.

(d) Give an example of a noninvertible function $f \colon A \to A$, such that $f^3 = f \circ f \circ f = f$, yet $f \circ f \neq f$.

2.16. Let X be any finite set.
(1) Prove that every injection $f \colon X \to X$ is actually a bijection.
(2) Prove that every surjection $f \colon X \to X$ is actually a bijection.
(3) Give counterexamples to both (1) and (2) when X is infinite.

2.17. (1) Let $(-1,1)$ be the set of real numbers

$$(-1,1) = \{ x \in \mathbb{R} \mid -1 < x < 1 \}.$$

Let $f \colon \mathbb{R} \to (-1,1)$ be the function given by

$$f(x) = \frac{x}{\sqrt{1+x^2}}.$$

Prove that f is a bijection. Find the inverse of f.

(2) Let $(0,1)$ be the set of real numbers

$$(0,1) = \{x \in \mathbb{R} \mid 0 < x < 1\}.$$

Give a bijection between $(-1,1)$ and $(0,1)$. Use (1) and (2) to give a bijection between $\mathbb{R}$ and $(0,1)$.

2.18. Let $D \subseteq \mathbb{R}^2$ be the subset of the real plane given by

$$D = \{(x,y) \in \mathbb{R}^2 \mid x^2 + y^2 < 1\},$$

that is, all points strictly inside of the unit circle $x^2 + y^2 = 1$. The set D is often called the *open unit disc*. Let $f \colon \mathbb{R}^2 \to D$ be the function given by

$$f(x,y) = \left(\frac{x}{\sqrt{1+x^2+y^2}}, \frac{y}{\sqrt{1+x^2+y^2}} \right).$$

(1) Prove that f is a bijection and find its inverse.

(2) Give a bijection between the sphere S^2 and the open unit disk D in the equatorial plane.

2.19. Prove by induction on n that

$$n^2 \leq 2^n \text{ for all } n \geq 4.$$

Hint. You need to show that $2n + 1 \leq n^2$ for all $n \geq 3$.

2.20. Let $f \colon A \to A$ be a function.

(a) Prove that if

$$f \circ f \circ f = f \circ f \text{ and } f \neq \mathrm{id}_A, \tag{$*$}$$

then f is neither injective nor surjective.

Hint. Proceed by contradiction and use the characterization of injections and surjections in terms of left and right inverses.

(b) Give a simple example of a function $f \colon \{a,b,c\} \to \{a,b,c\}$, satisfying the conditions of $(*)$.

2.21. Recall that a set A is infinite iff there is no bijection from $\{1,\dots,n\}$ onto A, for any natural number $n \in \mathbb{N}$. Prove that the set of odd natural numbers is infinite.

2.22. Consider the sum

$$\frac{3}{1 \cdot 4} + \frac{5}{4 \cdot 9} + \dots + \frac{2n+1}{n^2 \cdot (n+1)^2},$$

with $n \geq 1$.

Which of the following expressions is the sum of the above:

$$(1) \ \frac{n+2}{(n+1)^2} \qquad (2) \ \frac{n(n+2)}{(n+1)^2}.$$

Justify your answer.

Hint. Note that

$$n^4 + 6n^3 + 12n^2 + 10n + 3 = (n^3 + 3n^2 + 3n + 1)(n+3).$$

2.23. Consider the following version of the Fibonacci sequence starting from $F_0 = 0$ and defined by:

$$F_0 = 0$$
$$F_1 = 1$$
$$F_{n+2} = F_{n+1} + F_n, \ n \geq 0.$$

Prove the following identity, for any fixed $k \geq 1$ and all $n \geq 0$,

$$F_{n+k} = F_k F_{n+1} + F_{k-1} F_n.$$

2.24. Recall that the triangular numbers Δ_n are given by the formula

$$\Delta_n = \frac{n(n+1)}{2},$$

with $n \in \mathbb{N}$.
   (a) Prove that

$$\Delta_n + \Delta_{n+1} = (n+1)^2$$

and

$$\Delta_1 + \Delta_2 + \Delta_3 + \cdots + \Delta_n = \frac{n(n+1)(n+2)}{6}.$$

   (b) The numbers

$$T_n = \frac{n(n+1)(n+2)}{6}$$

are called *tetrahedral numbers*, due to their geometric interpretation as 3-D stacks of triangular numbers. Prove that

$$T_1 + T_2 + \cdots + T_n = \frac{n(n+1)(n+2)(n+3)}{24}.$$

Prove that

$$T_n + T_{n+1} = 1^2 + 2^2 + \cdots + (n+1)^2,$$

and from this, derive the formula

$$1^2 + 2^2 + \cdots + n^2 = \frac{n(n+1)(2n+1)}{6}.$$

   (c) The numbers

$$P_n = \frac{n(n+1)(n+2)(n+3)}{24}$$

are called *pentatope numbers*. The above numbers have a geometric interpretation in four dimensions as stacks of tetrahedral numbers. Prove that

$$P_1 + P_2 + \cdots + P_n = \frac{n(n+1)(n+2)(n+3)(n+4)}{120}.$$

Do you see a pattern? Can you formulate a conjecture and perhaps even prove it?

2.25. Consider the following table containing 11 copies of the triangular number, $\Delta_5 = 1 + 2 + 3 + 4 + 5$:

```
                    1                    1²
                  1 2 1                  2²
                1 2 3 2 1                3²
              1 2 3 4 3 2 1              4²
            1 2 3 4 5 4 3 2 1            5²
          1 2 3 4 5   5 4 3 2 1
      1²  2 3 4 5         5 4 3 2  1²
      2²  3 4 5             5 4 3  2²
      3²  4 5                 5 4  3²
      4²  5                     5  4²
      5²                          5²
```

Note that the above array splits into three triangles, one above the solid line and two below the solid line. Observe that the upward diagonals of the left lower triangle add up to $1^2, 2^2, 3^2, 4^2, 5^2$; similarly the downward diagonals of the right lower triangle add up to $1^2, 2^2, 3^2, 4^2, 5^2$, and the rows of the triangle above the solid line add up to $1^2, 2^2, 3^2, 4^2, 5^2$. Therefore,

$$3 \times (1^2 + 2^2 + 3^2 + 4^2 + 5^2) = 11 \times \Delta_5.$$

In general, use a generalization of the above array to prove that

$$3 \times (1^2 + 2^2 + 3^2 + \cdots + n^2) = (2n+1)\Delta_n,$$

which yields the familiar formula:

$$1^2 + 2^2 + 3^2 \cdots + n^2 = \frac{n(n+1)(2n+1)}{6}.$$

2.26. Consider the following table:

$$1 = 1^3$$
$$3 + 5 = 2^3$$
$$7 + 9 + 11 = 3^3$$
$$13 + 15 + 17 + 19 = 4^3$$
$$21 + 23 + 25 + 27 + 29 = 5^3$$
$$\cdots\cdots\cdots\cdots$$

(a) If we number the rows starting from $n = 1$, prove that the leftmost number on row n is $1 + (n - 1)n$. Then, prove that the sum of the numbers on row n (the n consecutive odd numbers beginning with $1 + (n - 1)n)$) is n^3.

(b) Use the triangular array in (a) to give a geometric proof of the identity

$$\sum_{k=1}^{n} k^3 = \left(\sum_{k=1}^{n} k \right)^2.$$

Hint. Recall that

$$1 + 3 + \cdots + 2n - 1 = n^2.$$

2.27. Let $f: A \to B$ be a function and define the function $g: B \to 2^A$ by

$$g(b) = f^{-1}(b) = \{a \in A \mid f(a) = b\},$$

for all $b \in B$. (a) Prove that if f is surjective, then g is injective.

(b) If g is injective, can we conclue that f is surjective?

2.28. Let X, Y, Z be any three nonempty sets and let $f: X \to Y$ be any function. Define the function $R_f: Z^Y \to Z^X$ (R_f, as a reminder that we compose with f on the right), by

$$R_f(h) = h \circ f,$$

for every function $h: Y \to Z$.

Let T be another nonempty set and let $g: Y \to T$ be any function.

(a) Prove that

$$R_{g \circ f} = R_f \circ R_g$$

and if $X = Y$ and $f = \mathrm{id}_X$, then

$$R_{\mathrm{id}_X}(h) = h,$$

for every function $h: X \to Z$.

(b) Use (a) to prove that if f is surjective, then R_f is injective and if f is injective, then R_f is surjective.

2.29. Let X, Y, Z be any three nonempty sets and let $g: Y \to Z$ be any function. Define the function $L_g: Y^X \to Z^X$ (L_g, as a reminder that we compose with g on the left), by

$$L_g(f) = g \circ f,$$

for every function $f: X \to Y$.

(a) Prove that if $Y = Z$ and $g = \mathrm{id}_Y$, then

$$L_{\mathrm{id}_Y}(f) = f,$$

for all $f: X \to Y$.

Let T be another nonempty set and let $h: Z \to T$ be any function. Prove that

$$L_{h \circ g} = L_h \circ L_g.$$

(b) Use (a) to prove that if g is injective, then $L_g: Y^X \to Z^X$ is also injective and if g is surjective, then $L_g: Y^X \to Z^X$ is also surjective.

2.30. Recall that given any two sets X, Y, every function $f: X \to Y$ induces a function $f: 2^X \to 2^Y$ such that for every subset $A \subseteq X$,

$$f(A) = \{f(a) \in Y \mid a \in A\}$$

and a function $f^{-1}: 2^Y \to 2^X$, such that, for every subset $B \subseteq Y$,

$$f^{-1}(B) = \{x \in X \mid f(x) \in B\}.$$

(a) Prove that if $f: X \to Y$ is injective, then so is $f: 2^X \to 2^Y$.

(b) Prove that if f is bijective then $f^{-1}(f(A)) = A$ and $f(f^{-1}(B)) = B$, for all $A \subseteq X$ and all $B \subseteq Y$. Deduce from this that $f: 2^X \to 2^Y$ is bijective.

(c) Prove that for any set A there is an injection from the set A^A of all functions from A to A to $2^{A \times A}$, the power set of $A \times A$. If A is infinite, prove that there is an injection from A^A to 2^A.

2.31. Recall that given any two sets X, Y, every function $f: X \to Y$ induces a function $f: 2^X \to 2^Y$ such that for every subset $A \subseteq X$,

$$f(A) = \{f(a) \in Y \mid a \in A\}$$

and a function $f^{-1}: 2^Y \to 2^X$, such that, for every subset $B \subseteq Y$,

$$f^{-1}(B) = \{x \in X \mid f(x) \in B\}.$$

(a) Prove that if $f: X \to Y$ is surjective, then so is $f: 2^X \to 2^Y$.

(b) If A is infinite, prove that there is a bijection from A^A to 2^A.

Hint. Prove that there is an injection from A^A to 2^A and an injection from 2^A to A^A.

2.32. (a) Finish the proof of Theorem 2.5, which states that for any infinite set X there is an injection from $\mathbb{N}$ into X. Use this to prove that there is a bijection between X and $X \times \mathbb{N}$.

(b) Prove that if a subset $A \subseteq \mathbb{N}$ of $\mathbb{N}$ is not finite, then there is a bijection between A and $\mathbb{N}$.

(c) Prove that every infinite set X can be written as a disjoint union $X = \bigcup_{i \in I} X_i$, where every X_i is in bijection with $\mathbb{N}$.

(d) If X is any set, finite or infinite, prove that if X has at least two elements then there is a bijection f of X leaving no element fixed (i.e., so that $f(x) \neq x$ for all $x \in X$).

2.33. Prove that if $(X_i)_{i \in I}$ is a family of sets and if I and all the X_i are countable, then $(X_i)_{i \in I}$ is also countable.

Hint. Define a surjection from $\mathbb{N} \times \mathbb{N}$ onto $(X_i)_{i \in I}$.

2.34. Consider the alphabet, $\Sigma = \{a, b\}$. We can enumerate all strings in $\{a, b\}^*$ as follows. Say that u precedes v if $|u| < |v|$ and if $|u| = |v|$, use the lexicographic (dictionary) order. The enumeration begins with

$$\varepsilon$$

$$a, b$$

$$aa, ab, ba, bb$$

$$aaa, aab, aba, abb, baa, bab, bba, bbb$$

We would like to define a function, $f: \{a, b\}^* \to \mathbb{N}$, such that $f(u)$ is the position of the string u in the above list, starting with $f(\varepsilon) = 0$. For example,

$$f(baa) = 11.$$

(a) Prove that if $u = u_1 \cdots u_n$ (with $u_j \in \{a, b\}$ and $n \geq 1$), then

$$f(u) = i_1 2^{n-1} + i_2 2^{n-2} + \cdots + i_{n-1} 2^1 + i_n$$
$$= 2^n - 1 + (i_1 - 1)2^{n-1} + (i_2 - 1)2^{n-2} + \cdots + (i_{n-1} - 1)2^1 + i_n - 1,$$

with $i_j = 1$ if $u_j = a$, else $i_j = 2$ if $u_j = b$.

(b) Prove that the above function is a bijection $f: \{a, b\}^* \to \mathbb{N}$.

(c) Consider any alphabet $\Sigma = \{a_1, \dots, a_m\}$, with $m \geq 2$. We can also list all strings in Σ^* as in (a). Prove that the listing function $f: \Sigma^* \to \mathbb{N}$ is given by $f(\varepsilon) = 0$ and if $u = a_{i_1} \cdots a_{i_n}$ (with $a_{i_j} \in \Sigma$ and $n \geq 1$) by

$$f(u) = i_1 m^{n-1} + i_2 m^{n-2} + \cdots + i_{n-1} m^1 + i_n$$
$$= \frac{m^n - 1}{m - 1} + (i_1 - 1)m^{n-1} + (i_2 - 1)m^{n-2} + \cdots + (i_{n-1} - 1)m^1 + i_n - 1,$$

Prove that the above function $f: \Sigma^* \to \mathbb{N}$ is a bijection.

(d) Consider any infinite set A and pick two distinct elements, a_1, a_2, in A. We would like to define a surjection from A^A to 2^A by mapping any function $f: A \to A$ to its image,

$$\text{Im} f = \{f(a) \mid a \in A\}.$$

The problem with the above definition is that the empty set is missed. To fix this problem, let f_0 be the function defined so that $f(a_0) = a_1$ and $f(a) = a_0$ for all

$a \in A - \{a_0\}$. Then, we define $S \colon A^A \to 2^A$ by

$$S(f) = \begin{cases} \emptyset & \text{if } f = f_0 \\ \operatorname{Im}(f) & \text{if } f \neq f_0. \end{cases}$$

Prove that the function $S \colon A^A \to 2^A$ is indeed a surjection.

(e) Assume that Σ is an infinite set and consider the set of all *finite* strings Σ^*. If Σ^n denotes the set of all strings of length n, observe that

$$\Sigma^* = \bigcup_{n \geq 0} \Sigma^n.$$

Prove that there is a bijection between Σ^* and Σ.

2.35. Let $\operatorname{Aut}(A)$ denote the set of all bijections from A to itself.

(a) Prove that there is a bijection between $\operatorname{Aut}(\mathbb{N})$ and $2^{\mathbb{N}}$.
Hint. Consider the map, $S \colon \operatorname{Aut}(\mathbb{N}) \to 2^{\mathbb{N} - \{0\}}$, given by

$$S(f) = \{n \in \mathbb{N} - \{0\} \mid f(n) = n\}$$

and prove that it is surjective. Also, there is a bijection between $\mathbb{N}$ and $\mathbb{N} - \{0\}$

(b) Prove that for any infinite set A there is a bijection between $\operatorname{Aut}(A)$ and 2^A.
Hint. Use results from Problem 2.32 and adapt the method of Part (a).

2.36. Recall that a set A is infinite iff there is no bijection from $\{1, \ldots, n\}$ onto A, for any natural number $n \in \mathbb{N}$. Prove that the set of even natural numbers is infinite.

2.37. Consider the sum

$$\frac{1}{1 \cdot 2} + \frac{1}{2 \cdot 3} + \cdots + \frac{1}{n \cdot (n+1)},$$

with $n \geq 1$.

Which of the following expressions is the sum of the above:

$$(1) \ \frac{1}{n+1} \quad (2) \ \frac{n}{n+1}.$$

Justify your answer.

2.38. Consider the triangular region T_1, defined by $0 \leq x \leq 1$ and $|y| \leq x$ a⟨e⟩ subset D_1, of this triangular region inside the closed unit disk, that is, for whi⟨⟩ also have $x^2 + y^2 \leq 1$. See Figure 2.18 where D_1 is shown shaded in gray.

(a) Prove that the map $f_1 \colon T_1 \to D_1$ defined so that

$$f_1(x,y) = \left(\frac{x^2}{\sqrt{x^2 + y^2}}, \frac{xy}{\sqrt{x^2 + y^2}} \right), \ x \neq 0$$

$$f_1(0,0) = (0,0),$$

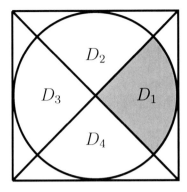

Fig. 2.18 The regions D_i

is bijective and that its inverse is given by

$$g_1(x,y) = \left(\sqrt{x^2+y^2}, \frac{y}{x}\sqrt{x^2+y^2}\right), \; x \neq 0$$
$$g_1(0,0) = (0,0).$$

If T_3 and D_3 are the regions obtained from T_3 and D_1 by the reflection about the y axis, $x \mapsto -x$, show that the map, $f_3 : T_3 \to D_3$, defined so that

$$f_3(x,y) = \left(-\frac{x^2}{\sqrt{x^2+y^2}}, -\frac{xy}{\sqrt{x^2+y^2}}\right), \; x \neq 0$$
$$f_3(0,0) = (0,0),$$

is bijective and that its inverse is given by

$$g_3(x,y) = \left(-\sqrt{x^2+y^2}, \frac{y}{x}\sqrt{x^2+y^2}\right), \; x \neq 0$$
$$g_3(0,0) = (0,0).$$

(b) Now consider the triangular region T_2 defined by $0 \leq y \leq 1$ and $|x| \leq y$ and the subset D_2, of this triangular region inside the closed unit disk, that is, for which we also have $x^2 + y^2 \leq 1$. The regions T_2 and D_2 are obtained from T_1 and D_1 by a counterclockwise rotation by the angle $\pi/2$.

Prove that the map $f_2 : T_2 \to D_2$ defined so that

$$f_2(x,y) = \left(\frac{xy}{\sqrt{x^2+y^2}}, \frac{y^2}{\sqrt{x^2+y^2}}\right), \; y \neq 0$$
$$f_2(0,0) = (0,0),$$

is bijective and that its inverse is given by

$$g_2(x,y) = \left(\frac{x}{y}\sqrt{x^2+y^2}, \sqrt{x^2+y^2}\right), \ y \neq 0$$
$$g_2(0,0) = (0,0).$$

If T_4 and D_4 are the regions obtained from T_2 and D_2 by the reflection about the x axis $y \mapsto -y$, show that the map $f_4 \colon T_4 \to D_4$, defined so that

$$f_4(x,y) = \left(-\frac{xy}{\sqrt{x^2+y^2}}, -\frac{y^2}{\sqrt{x^2+y^2}}\right), \ y \neq 0$$
$$f_4(0,0) = (0,0),$$

is bijective and that its inverse is given by

$$g_4(x,y) = \left(\frac{x}{y}\sqrt{x^2+y^2}, -\sqrt{x^2+y^2}\right), \ y \neq 0$$
$$g_4(0,0) = (0,0).$$

(c) Use the maps, f_1, f_2, f_3, f_4 to define a bijection between the closed square $[-1,1] \times [-1,1]$ and the closed unit disk $\overline{D} = \{(x,y) \in \mathbb{R}^2 \mid x^2+y^2 \leq 1\}$, which maps the boundary square to the boundary circle. Check that this bijection is continuous. Use this bijection to define a bijection between the closed unit disk $\overline{D}$ and the open unit disk $D = \{(x,y) \in \mathbb{R}^2 \mid x^2+y^2 < 1\}$.

2.39. The purpose of this problem is to prove that there is a bijection between $\mathbb{R}$ and $2^{\mathbb{N}}$. Using the results of Problem 2.17, it is sufficient to prove that there is a bijection betwen $(0,1)$ and $2^{\mathbb{N}}$. To do so, we represent the real numbers $r \in (0,1)$ in terms of their decimal expansions,

$$r = 0.r_1 r_2 \cdots r_n \cdots,$$

where $r_i \in \{0, 1, \ldots, 9\}$. However, some care must be exercised because this representation is ambiguous due to the possibility of having sequences containing the infinite suffix $9999 \cdots$. For example,

$$0.1200000000 \cdots = 0.1199999999 \cdots$$

Therefore, we only use representations not containing the infinite suffix $9999 \cdots$. Also recall that by Proposition 2.5, the power set $2^{\mathbb{N}}$ is in bijection with the set $\{0,1\}^{\mathbb{N}}$ of countably infinite binary sequences

$$b_0 b_1 \cdots b_n \cdots,$$

with $b_i \in \{0,1\}$.

(1) Prove that the function $f \colon \{0,1\}^{\mathbb{N}} \to (0,1)$ given by

$$f(b_0 b_1 \cdots b_n \cdots) = 0.1 b_0 b_1 \cdots b_n \cdots,$$

where $0.1b_0b_1\cdots b_n\cdots$ (with $b_n \in \{0,1\}$) is interpreted as a *decimal* (not binary) expansion, is an injection.

(2) Show that the image of the function f defined in (1) is the closed interval $[\frac{1}{10}, \frac{1}{9}]$ and thus, that f is not surjective.

(3) Every number, $k \in \{0,1,2,\ldots,9\}$ has a binary representation, $\mathrm{bin}(k)$, as a string of four bits; for example,

$$\mathrm{bin}(1) = 0001,\ \mathrm{bin}(2) = 0010,\ \mathrm{bin}(5) = 0101,\ \mathrm{bin}(6) = 0110,\ \mathrm{bin}(9) = 1001.$$

Prove that the function $g\colon (0,1) \to \{0,1\}^{\mathbb{N}}$ defined so that

$$g(0.r_1r_2\cdots r_n\cdots) = .\mathrm{bin}(r_1)\mathrm{bin}(r_2)\mathrm{bin}(r_1)\cdots\mathrm{bin}(r_n)\cdots$$

is an injection (Recall that we are assuming that the sequence $r_1r_2\cdots r_n\cdots$ does not contain the infinite suffix $99999\cdots$). Prove that g is not surjective.

(4) Use (1) and (3) to prove that there is a bijection between $\mathbb{R}$ and $2^{\mathbb{N}}$.

2.40. The purpose of this problem is to show that there is a bijection between $\mathbb{R} \times \mathbb{R}$ and $\mathbb{R}$. In view of the bijection between $\{0,1\}^{\mathbb{N}}$ and $\mathbb{R}$ given by Problem 2.39, it is enough to prove that there is a bijection between $\{0,1\}^{\mathbb{N}} \times \{0,1\}^{\mathbb{N}}$ and $\{0,1\}^{\mathbb{N}}$, where $\{0,1\}^{\mathbb{N}}$ is the set of countably infinite sequences of 0 and 1.

(1) Prove that the function $f\colon \{0,1\}^{\mathbb{N}} \times \{0,1\}^{\mathbb{N}} \to \{0,1\}^{\mathbb{N}}$ given by

$$f(a_0a_1\cdots a_n\cdots, b_0b_1\cdots b_n\cdots) = a_0b_0a_1b_1\cdots a_nb_n\cdots$$

is a bijection (here, $a_i, b_i \in \{0,1\}$).

(2) Suppose, as in Problem 2.39, that we represent the reals in $(0,1)$ by their decimal expansions not containing the infinite suffix $99999\cdots$. Define the function $h\colon (0,1) \times (0,1) \to (0,1)$ by

$$h(0.r_0r_1\cdots r_n\cdots, 0.s_0s_1\cdots s_n\cdots) = 0.r_0s_0r_1s_1\cdots r_ns_n\cdots$$

with $r_i, s_i \in \{0,1,2,\ldots,9\}$. Prove that h is injective but not surjective.

If we pick the decimal representations ending with the infinite suffix $99999\cdots$ rather that an infinite string of 0s, prove that h is also injective but still not surjective.

(3) Prove that for every positive natural number $n \in \mathbb{N}$, there is a bijection between $\mathbb{R}^n$ and $\mathbb{R}$.

2.41. Let E, F, G, be any arbitrary sets.

(1) Prove that there is a bijection

$$E^G \times F^G \longrightarrow (E \times F)^G.$$

(2) Prove that there is a bijection

$$(E^F)^G \longrightarrow E^{F \times G}.$$

(3) If F and G are disjoint, then prove that there is a bijection

$$E^F \times E^G \longrightarrow E^{F \cup G}.$$

2.42. Let E, F, G, be any arbitrary sets.

(1) Prove that if G is disjoint from both E and F and if $E \preceq F$, then $E \cup G \preceq F \cup G$.

(2) Prove that if $E \preceq F$, then $E \times G \preceq F \times G$.

(3) Prove that if $E \preceq F$, then $E^G \preceq F^G$.

(4) Prove that if E and G are not both empty and if $E \preceq F$, then $G^E \preceq G^F$.

References

1. John H. Conway and K. Guy, Richard. *The Book of Numbers*. Copernicus. New York: Springer-Verlag, first edition, 1996.
2. Herbert B. Enderton. *Elements of Set Theory*. New York: Academic Press, first edition, 1977.
3. Patrick Suppes. *Axiomatic Set Theory*. New York: Dover, first edition, 1972.

Chapter 3
Graphs, Part I: Basic Notions

3.1 Why Graphs? Some Motivations

Graphs are mathematical structures that have many applications in computer science, electrical engineering, and more widely in engineering as a whole, but also in sciences such as biology, linguistics, and sociology, among others. For example, relations among objects can usually be encoded by graphs. Whenever a system has a notion of state and a state transition function, graph methods may be applicable. Certain problems are naturally modeled by undirected graphs whereas others require directed graphs. Let us give a concrete example.

Suppose a city decides to create a public transportation system. It would be desirable if this system allowed transportation between certain locations considered important. Now, if this system consists of buses, the traffic will probably get worse so the city engineers decide that the traffic will be improved by making certain streets one-way streets. The problem then is, given a map of the city consisting of the important locations and of the two-way streets linking them, finding an orientation of the streets so that it is still possible to travel between any two locations. The problem requires finding a directed graph, given an undirected graph. Figure 3.1 shows the undirected graph corresponding to the city map and Figure 3.2 shows a proposed choice of one-way streets. Did the engineers do a good job or are there locations such that it is impossible to travel from one to the other while respecting the one-way signs?

The answer to this puzzle is revealed in Section 3.3.

There is a peculiar aspect of graph theory having to do with its terminology. Indeed, unlike most branches of mathematics, it appears that the terminology of graph theory is not standardized yet. This can be quite confusing to the beginner who has to struggle with many different and often inconsistent terms denoting the same concept, one of the worse being the notion of a *path*.

Our attitude has been to use terms that we feel are as simple as possible. As a result, we have not followed a single book. Among the many books on graph theory, we have been inspired by the classic texts, Harary [4], Berge [1], and Bollobas

J. Gallier, *Discrete Mathematics*, Universitext,
DOI 10.1007/978-1-4419-8047-2_3, © Springer Science+Business Media, LLC 2011

[2]. This chapter on graphs is heavily inspired by Sakarovitch [5], because we find

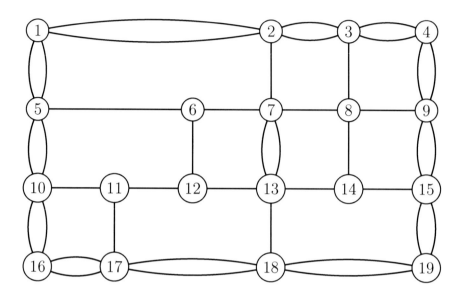

Fig. 3.1 An undirected graph modeling a city map

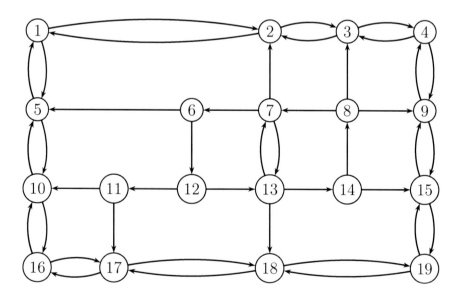

Fig. 3.2 A choice of one-way streets

Sakarovitch's book extremely clear and because it has more emphasis on applications than the previous two. Another more recent (and more advanced) text which is also excellent is Diestel [3].

Fig. 3.3 Claude Berge, 1926–2002 (left) and Frank Harary, 1921–2005 (right)

Many books begin by discussing undirected graphs and introduce directed graphs only later on. We disagree with this approach. Indeed, we feel that the notion of a directed graph is more fundamental than the notion of an undirected graph. For one thing, a unique undirected graph is obtained from a directed graph by forgetting the direction of the arcs, whereas there are many ways of orienting an undirected graph. Also, in general, we believe that most definitions about directed graphs are cleaner than the corresponding ones for undirected graphs (for instance, we claim that the definition of a directed graph is simpler than the definition of an undirected graph, and similarly for paths). Thus, we begin with directed graphs.

3.2 Directed Graphs

Informally, a directed graph consists of a set of nodes together with a set of oriented arcs (also called edges) between these nodes. Every arc has a single source (or initial point) and a single target (or endpoint), both of which are nodes. There are various ways of formalizing what a directed graph is and some decisions must be made. Two issues must be confronted:

1. Do we allow "loops," that is, arcs whose source and target are identical?
2. Do we allow "parallel arcs," that is, distinct arcs having the same source and target?

Every binary relation on a set can be represented as a directed graph with loops, thus our definition allows loops. The directed graphs used in automata theory must accomodate parallel arcs (usually labeled with different symbols), therefore our definition also allows parallel arcs. Thus we choose a more inclusive definition in order to accomodate as many applications as possible, even though some authors place restrictions on the definition of a graph, for example, forbidding loops and parallel

arcs (we call graphs without pa   el arcs, simple graphs). Before giving a formal
definition, let us say that graphs   e usually depicted by drawings (graphs!) where
the nodes are represented by c   les containing the node name and oriented line
segments labeled with their arc   me (see Figure 3.4).

Definition 3.1. A *directed grap*   *r digraph*) is a quadruple $G = (V, E, s, t)$, where
V is a set of *nodes or vertices*, E a set of *arcs or edges*, and $s, t \colon E \to V$ are two
functions, s being called the *sour   nction* and t the *target function*. Given an edge
$e \in E$, we also call $s(e)$ the *origi   source* of e, and $t(e)$ the *endpoint* or *target* of
e.

   If the context makes it clear   we are dealing only with directed graphs,
we usually say simply "graph" in   ead of "directed graph." A directed graph,
$G = (V, E, s, t)$, is *finite* iff both V and E are finite. In this case, $|V|$, the number
of nodes of G, is called the *order* of G.
   Example: Let G_1 be the directed graph defined such that
$E = \{e_1, e_2, e_3, e_4, e_5, e_6, e_7, e_8, e_9\}$,
$V = \{v_1, v_2, v_3, v_4, v_5, v_6\}$, and

$$s(e_1) = v_1, s(e_2) = v_2, s(e_3) = v_3, s(e_4) = v_4,$$
$$s(e_5) = v_2, s(e_6) = v_5, s(e_7) = v_5, s(e_8) = v_5, s(e_9) = v_6$$
$$t(e_1) = v_2, t(e_2) = v_3, t(e_3) = v_4, t(e_4) = v_2,$$
$$t(e_5) = v_5, t(e_6) = v_5, t(e_7) = v_6, t(e_8) = v_6, t(e_9) = v_4.$$

   The graph G_1 is represented by the diagram shown in Figure 3.4.
   It should be noted that there are many different ways of "drawing" a graph. Obviously, we would like as much as possible to avoid having too many intersecting
arrows but this is not always possible if we insist on drawing a graph on a sheet of
paper (on the plane).

Definition 3.2. Given a directed graph G, an edge $e \in E$, such that $s(e) = t(e)$ is
called a *loop* (or *self-loop*). Two edges $e, e' \in E$ are said to be *parallel edges* iff
$s(e) = s(e')$ and $t(e) = t(e')$. A directed graph is *simple* iff it has no parallel edges.

Remarks:

1. The functions s, t need not be injective or surjective. Thus, we allow "isolated
   vertices," that is, vertices that are not the source or the target of any edge.
2. When G is simple, every edge $e \in E$, is uniquely determined by the ordered pair
   of vertices (u, v), such that $u = s(e)$ and $v = t(e)$. In this case, we may denote
   the edge e by (uv) (some books also use the notation uv). Also, a graph without
   parallel edges can be defined as a pair (V, E), with $E \subseteq V \times V$. In other words,
   a simple graph is equivalent to a binary relation on a set ($E \subseteq V \times V$). This
   definition is often the one used to define directed graphs.
3. Given any edge $e \in E$, the nodes $s(e)$ and $t(e)$ are often called the *boundaries*
   of e and the expression $t(e) - s(e)$ is called the *boundary of e*.

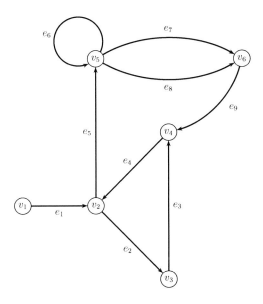

Fig. 3.4 A directed graph G_1

4. Given a graph $G = (V, E, s, t)$, we may also write $V(G)$ for V and $E(G)$ for E. Sometimes, we even drop s and t and simply write $G = (V, E)$ instead of $G = (V, E, s, t)$.
5. Some authors define a simple graph to be a graph without loops and without parallel edges.

Observe that the graph G_1 has the loop e_6 and the two parallel edges e_7 and e_8. When we draw pictures of graphs, we often omit the edge names (sometimes even the node names) as illustrated in Figure 3.5.

Definition 3.3. Given a directed graph G, for any edge $e \in E$, if $u = s(e)$ and $v = t(e)$, we say that

(i) The nodes u and v are *adjacent*.
(ii) The nodes u and v are *incident to the arc e*.
(iii) The arc e is *incident to the nodes u and v*.
(iv) Two edges $e, e' \in E$ are *adjacent* if they are incident to some common node (that is, either $s(e) = s(e')$ or $t(e) = t(e')$ or $t(e) = s(e')$ or $s(e) = t(e')$).

For any node $u \in V$, set

(a) $d_G^+(u) = |\{e \in E \mid s(e) = u\}|$, the *outer half-degree or outdegree of u*.
(b) $d_G^-(u) = |\{e \in E \mid t(e) = u\}|$, the *inner half-degree or indegree of u*.
(c) $d_G(u) = d_G^+(u) + d_G^-(u)$, the *degree of u*.

A graph is *regular* iff every node has the same degree.

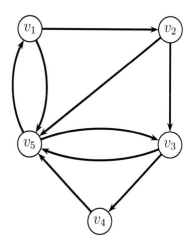

Fig. 3.5 A directed graph G_2

Note that d_G^+ (respectively, $d_G^-(u)$) counts the number of arcs "coming out from u," that is, whose source is u (respectively, counts the number of arcs "coming into u," i.e., whose target is u). For example, in the graph of Figure 3.5, $d_{G_2}^+(v_1) = 2$, $d_{G_2}^-(v_1) = 1$, $d_{G_2}^+(v_5) = 2$, $d_{G_2}^-(v_5) = 4$, $d_{G_2}^+(v_3) = 2$, $d_{G_2}^-(v_3) = 2$. Neither G_1 nor G_2 are regular graphs.

The first result of graph theory is the following simple but very useful proposition.

Proposition 3.1. *For any finite graph* $G = (V, E, s, t)$ *we have*

$$\sum_{u \in V} d_G^+(u) = \sum_{u \in V} d_G^-(u).$$

Proof. Every arc $e \in E$ has a single source and a single target and each side of the above equations simply counts the number of edges in the graph. □

Corollary 3.1. *For any finite graph* $G = (V, E, s, t)$ *we have*

$$\sum_{u \in V} d_G(u) = 2|E|;$$

that is, the sum of the degrees of all the nodes is equal to twice the number of edges.

Corollary 3.2. *For any finite graph* $G = (V, E, s, t)$ *there is an even number of nodes with an odd degree.*

The notion of homomorphism and isomorphism of graphs is fundamental.

Definition 3.4. Given two directed graphs $G_1 = (V_1, E_1, s_1, t_1)$ and $G_2 = (V_2, E_2, s_2, t_2)$, a *homomorphism* (or *morphism*) $f: G_1 \rightarrow G_2$ *from* G_1 *to* G_2 is a pair $f = (f^v, f^e)$ with $f^v: V_1 \rightarrow V_2$ and $f^e: E_1 \rightarrow E_2$ preserving incidence; that is, for every edge, $e \in E_1$, we have

$$s_2(f^e(e)) = f^v(s_1(e)) \text{ and } t_2(f^e(e)) = f^v(t_1(e)).$$

These conditions can also be expressed by saying that the following two diagrams commute:

$$
\begin{array}{ccc}
E_1 & \xrightarrow{f^e} & E_2 \\
{\scriptstyle s_1}\downarrow & & \downarrow{\scriptstyle s_2} \\
V_1 & \xrightarrow{f^v} & V_2
\end{array}
\qquad\qquad
\begin{array}{ccc}
E_1 & \xrightarrow{f^e} & E_2 \\
{\scriptstyle t_1}\downarrow & & \downarrow{\scriptstyle t_2} \\
V_1 & \xrightarrow{f^v} & V_2.
\end{array}
$$

Given three graphs G_1, G_2, G_3 and two homomorphisms $f: G_1 \rightarrow G_2$, $g: G_2 \rightarrow G_3$, with $f = (f^v, f^e)$ and $g = (g^v, g^e)$, it is easily checked that $(g^v \circ f^v, g^e \circ f^e)$ is a homomorphism from G_1 to G_3. The homomorphism $(g^v \circ f^v, g^e \circ f^e)$ is denoted $g \circ f$. Also, for any graph G, the map $\mathrm{id}_G = (\mathrm{id}_V, \mathrm{id}_E)$ is a homomorphism called the *identity homomorphism*. Then, a homomorphism $f: G_1 \rightarrow G_2$ is an *isomorphism* iff there is a homomorphism, $g: G_2 \rightarrow G_1$, such that

$$g \circ f = \mathrm{id}_{G_1} \text{ and } f \circ g = \mathrm{id}_{G_2}.$$

In this case, g is unique and it is called the *inverse* of f and denoted f^{-1}. If $f = (f^v, f^e)$ is an isomorphism, we see immediately that f^v and f^e are bijections. Checking whether two finite graphs are isomorphic is not as easy as it looks. In fact, no general efficient algorithm for checking graph isomorphism is known at this time and determining the exact complexity of this problem is a major open question in computer science. For example, the graphs G_3 and G_4 shown in Figure 3.6 are isomorphic. The bijection f^v is given by $f^v(v_i) = w_i$, for $i = 1, \ldots, 6$ and the reader will easily figure out the bijection on arcs. As we can see, isomorphic graphs can look quite different.

3.3 Paths in Digraphs; Strongly Connected Components

Many problems about graphs can be formulated as path existence problems. Given a directed graph G, intuitively, a path from a node u to a node v is a way to travel from u in v by following edges of the graph that "link up correctly." Unfortunately, if we look up the definition of a path in two different graph theory books, we are almost guaranteed to find different and usually clashing definitions. This has to do with the fact that for some authors, a path may not use the same edge more than once and for others, a path may not pass through the same node more than once. Moreover, when

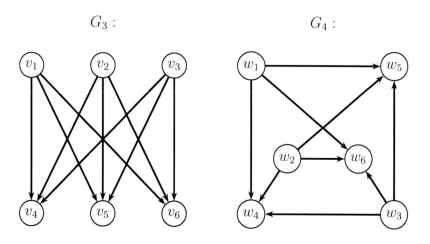

Fig. 3.6 Two isomorphic graphs, G_3 and G_4

parallel edges are present (i.e., when a graph is not simple), a sequence of nodes does not define a path unambiguously.

The terminology that we have chosen may not be standard, but it is used by a number of authors (some very distinguished, e.g., Fields medalists) and we believe that it is less taxing on one's memory (however, this point is probably the most debatable).

Definition 3.5. Given any digraph $G = (V, E, s, t)$, and any two nodes $u, v \in V$, a *path from u to v* is a triple, $\pi = (u, e_1 \cdots e_n, v)$, where $n \geq 1$ and $e_1 \cdots e_n$ is a sequence of edges, $e_i \in E$ (i.e., a nonempty string in E^*), such that

$$s(e_1) = u; t(e_n) = v; t(e_i) = s(e_{i+1}), \ 1 \leq i \leq n-1.$$

We call n the *length of the path* π and we write $|\pi| = n$. When $n = 0$, we have the *null path* (u, ε, u), from u to u (recall, ε denotes the empty string); the null path has length 0. If $u = v$, then π is called a *closed path*, else an *open path*. The path $\pi = (u, e_1 \cdots e_n, v)$ determines the sequence of nodes, $\text{nodes}(\pi) = \langle u_0, \ldots, u_n \rangle$, where $u_0 = u$, $u_n = v$ and $u_i = t(e_i)$, for $1 \leq i \leq n$. We also set $\text{nodes}((u, \varepsilon, u)) = \langle u, u \rangle$. A path $\pi = (u, e_1 \cdots e_n, v)$, is *edge-simple*, for short, *e-simple* iff $e_i \neq e_j$ for all $i \neq j$ (i.e., no edge in the path is used twice). A path π from u to v is *simple* iff no vertex in $\text{nodes}(\pi)$ occurs twice, except possibly for u if π is closed. Equivalently, if $\text{nodes}(\pi) = \langle u_0, \ldots, u_n \rangle$, then π is simple iff either

1. $u_i \neq u_j$ for all i, j with $i \neq j$ and $0 \leq i, j \leq n$, or π is closed (i.e., $u_0 = u_n$), in which case
2. $u_i \neq u_0 (= u_n)$ for all i with $1 \leq i \leq n-1$, and $u_i \neq u_j$ for all i, j with $i \neq j$ and $1 \leq i, j \leq n-1$.

The null path (u, ε, u), is considered e-simple and simple.

Remarks:

1. Other authors (such as Harary [4]) use the term *walk* for what we call a path. The term *trail* is also used for what we call an e-simple path and the term *path* for what we call a simple path. We decided to adopt the term "simple path" because it is prevalent in the computer science literature. However, note that Berge [1] and Sakarovitch [5] use the locution *elementary path* instead of simple path.
2. If a path π from u to v is simple, then every every node in the path occurs once except possibly u if $u = v$, so every edge in π occurs exactly once. Therefore, every simple path is an e-simple path.
3. If a digraph is not simple, then even if a sequence of nodes is of the form nodes(π) for some path, that sequence of nodes does not uniquely determine a path. For example, in the graph of Figure 3.7, the sequence $\langle v_2, v_5, v_6 \rangle$ corresponds to the two distinct paths $(v_2, e_5 e_7, v_6)$ and $(v_2, e_5 e_8, v_6)$.

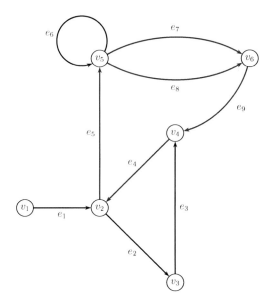

Fig. 3.7 A path in a directed graph G_1

In the graph G_1 from Figure 3.7,

$$(v_2, e_5 e_7 e_9 e_4 e_5 e_8, v_6)$$

is a path from v_2 to v_6 that is neither e-simple nor simple. The path

$$(v_2, e_2 e_3 e_4 e_5, v_5)$$

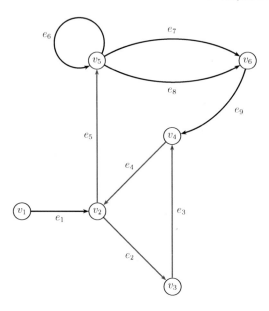

Fig. 3.8 An e-simple path in a directed graph G_1

is an e-simple path from v_2 to v_5 that is not simple and

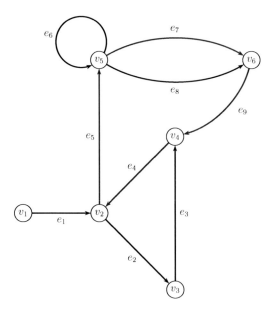

Fig. 3.9 Simple paths in a directed graph G_1

$$(v_2, e_5 e_7 e_9, v_4), \qquad (v_2, e_5 e_7 e_9 e_4, v_2)$$

are simple paths, the first one open and the second one closed.

Recall the notion of subsequence of a sequence defined just before stating Theorem 2.6. Then, if $\pi = (u, e_1 \cdots e_n, v)$ is any path from u to v in a digraph G a *subpath* of π is any path $\pi' = (u, e'_1 \cdots e'_m, v)$ such that $e'_1, \ldots, e'_m$ is a subsequence of $e_1, \ldots, e_n$. The following simple proposition is actually very important.

Proposition 3.2. *Let G be any digraph. (a) For any two nodes u, v in G, every nonnull path π from u to v contains a simple nonnull subpath.*

(b) If $|V| = n$, then every open simple path has length at most $n - 1$ and every closed simple path has length at most n.

Proof. (a) Let π be any nonnull path from u to v in G and let

$$S = \{k \in \mathbb{N} \mid k = |\pi'|, \quad \pi' \text{ is a nonnull subpath of } \pi\}.$$

The set $S \subseteq \mathbb{N}$ is nonempty because $|\pi| \in S$ and as $\mathbb{N}$ is well ordered (see Section 5.3 and Theorem 5.3), S has a least element, say $m \geq 1$. We claim that any subpath of π of length m is simple. Consider any such path, say $\pi' = (u, e'_1 \cdots e'_m, v)$; let

$$\text{nodes}(\pi') = \langle v_0, \ldots, v_m \rangle,$$

with $v_0 = u$ and $v_m = v$, and assume that π' is not simple. There are two cases:

(1) $u \neq v$. Then some node occurs twice in nodes(π'), say $v_i = v_j$, with $i < j$. Then, we can delete the path $(v_i, e'_{i+1}, \ldots, e'_j, v_j)$ from π' to obtain a nonnull (because $u \neq v$) subpath π'' of π' from u to v with $|\pi''| = |\pi'| - (j - i)$ and because $i < j$, we see that $|\pi''| < |\pi'|$, contradicting the minimality of m. Therefore, π' is a nonnull simple subpath of π.

(2) $u = v$. In this case, some node occurs twice in the sequence $\langle v_0, \ldots, v_{m-1} \rangle$. Then, as in (1), we can strictly shorten the path from v_0 to v_{m-1}. Even though the resulting path may be the null path, as the edge e'_m remains from the original path π', we get a nonnull path from u to u strictly shorter than π', contradicting the minimality of π'.

(b) As in (a), let π' be an open simple path from u to v and let

$$\text{nodes}(\pi') = \langle v_0, \ldots, v_m \rangle.$$

If $m \geq n = |V|$, as the above sequence has $m + 1 > n$ nodes, by the pigeonhole principle, some node must occur twice, contradicting the fact that π' is an open simple path. If π' is a nonnull closed path and $m \geq n + 1$, then the sequence $\langle v_0, \ldots, v_{m-1} \rangle$ has $m \geq n + 1$ nodes and by the pigeonhole principle, some node must occur twice, contradicting the fact that π' is a nonnull simple path. $\square$

Like strings, paths can be concatenated.

Definition 3.6. Two paths, $\pi = (u, e_1 \cdots e_m, v)$ and $\pi' = (u', e'_1 \cdots e'_n, v')$, in a digraph G can be *concatenated* iff $v = u'$ in which case their *concatenation* $\pi\pi'$ is the path

$$\pi\pi' = (u, e_1 \cdots e_m e'_1 \cdots e'_n, v').$$

We also let

$$(u, \varepsilon, u)\pi = \pi = \pi(v, \varepsilon, v).$$

Concatenation of paths is obviously associative and observe that $|\pi\pi'| = |\pi| + |\pi'|$.

Definition 3.7. Let $G = (V, E, s, t)$ be a digraph. We define the binary relation $\widehat{C}_G$ on V as follows. For all $u, v \in V$,

$u\widehat{C}_G v$ iff there is a path from u to v and there is a path from v to u.

When $u\widehat{C}_G v$, we say that *u and v are strongly connected.*

The relation $\widehat{C}_G$ is what is called an equivalence relation. The notion of an equivalence relation is discussed extensively in Chapter 5 (Section 5.6) but because it is a very important concept, we explain briefly what it is right now.

Repeating Definition 5.10, a binary relation R on a set X is an *equivalence relation* iff it is *reflexive*, *transitive*, and *symmetric*; that is:

(1) (*Reflexivity*): aRa, for all $a \in X$
(2) (*transitivity*): If aRb and bRc, then aRc, for all $a, b, c \in X$
(3) (*Symmetry*): If aRb, then bRa, for all $a, b \in X$

The main property of equivalence relations is that they partition the set X into nonempty, pairwise disjoint subsets called equivalence classes: For any $x \in X$, the set

$$[x]_R = \{y \in X \mid xRy\}$$

is the *equivalence class of x.* Each equivalence class $[x]_R$ is also denoted $\overline{x}_R$ and the subscript R is often omitted when no confusion arises.

For the reader's convenience, we repeat Proposition 5.11.

Let R be an equivalence relation on a set X. For any two elements $x, y \in X$, we have

$$xRy \text{ iff } [x] = [y].$$

Moreover, the equivalence classes of R satisfy the following properties.

(1) $[x] \neq \emptyset$, for all $x \in X$.
(2) If $[x] \neq [y]$ then $[x] \cap [y] = \emptyset$.
(3) $X = \bigcup_{x \in X} [x]$.

The relation $\widehat{C}_G$ is reflexive because we have the null path from u to u, symmetric by definition, and transitive because paths can be concatenated. The equivalence

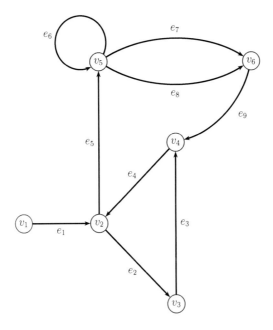

Fig. 3.10 A directed graph G_1 with two SCCs

classes of the relation $\widehat{C}_G$ are called the *strongly connected components of G (SCCs).*
A graph is *strongly connected* iff it has a single strongly connected component.

For example, we see that the graph G_1 of Figure 3.10 has two strongly connected
components

$$\{v_1\}, \quad \{v_2, v_3, v_4, v_5, v_6\},$$

inasmuch as there is a closed path

$$(v_4, e_4 e_2 e_3 e_4 e_5 e_7 e_9, v_4).$$

The graph G_2 of Figure 3.5 is strongly connected.

Let us give a simple algorithm for computing the strongly connected components
of a graph because this is often the key to solving many problems. The algorithm
works as follows. Given some vertex $u \in V$, the algorithm computes the two sets
$X^+(u)$ and $X^-(u)$, where

$$X^+(u) = \{v \in V \mid \text{there exists a path from } u \text{ to } v\}$$
$$X^-(u) = \{v \in V \mid \text{there exists a path from } v \text{ to } u\}.$$

Then, it is clear that the connected component $C(u)$ of u, is given by
$C(u) = X^+(u) \cap X^-(u).$

For simplicity, we assume that $X^+(u), X^-(u)$ and $C(u)$ are represented by linear
arrays. In order to make sure that the algorithm makes progress, we used a simple

marking scheme. We use the variable *total* to count how many nodes are in $X^+(u)$ (or in $X^-(u)$) and the variable *marked* to keep track of how many nodes in $X^+(u)$ (or in $X^-(u)$) have been processed so far. Whenever the algorithm considers some unprocessed node, the first thing it does is to increment *marked* by 1. Here is the algorithm in high-level form.

function *strcomp*(G: graph; u: node): set
   begin
      $X^+(u)[1] := u$; $X^-(u)[1] := u$; *total* := 1; *marked* := 0;
      while *marked* < *total* **do**
         marked := *marked* + 1; $v := X^+(u)[marked]$;
         for each $e \in E$
            if $(s(e) = v) \wedge (t(e) \notin X^+(u))$ **then**
               total := *total* + 1; $X^+(u)[total] := t(e)$ **endif**
         endfor
      endwhile;
      total := 1; *marked* := 0;
      while *marked* < *total* **do**
         marked := *marked* + 1; $v := X^-(u)[marked]$;
         for each $e \in E$
             if $(t(e) = v) \wedge (s(e) \notin X^-(u))$ **then**
               total := *total* + 1; $X^-(u)[total] := s(e)$ **endif**
         endfor
      endwhile;
      $C(u) = X^+(u) \cap X^-(u)$; *strcomp* := $C(u)$
   end

If we want to obtain all the strongly connected components (SCCs) of a finite graph G, we proceed as follows. Set $V_1 = V$, pick any node v_1 in V_1, and use the above algorithm to compute the strongly connected component C_1 of v_1. If $V_1 = C_1$, stop. Otherwise, let $V_2 = V_1 - C_1$. Again, pick any node v_2 in V_2 and determine the strongly connected component C_2 of v_2. If $V_2 = C_2$, stop. Otherwise, let $V_3 = V_2 - C_2$, pick v_3 in V_3, and continue in the same manner as before. Ultimately, this process will stop and produce all the strongly connected components $C_1, \ldots, C_k$ of G.

It should be noted that the function *strcomp* and the simple algorithm that we just described are "naive" algorithms that are not particularly efficient. Their main advantage is their simplicity. There are more efficient algorithms, in particular, there is a beautiful algorithm for computing the SCCs due to Robert Tarjan.

Going back to our city traffic problem from Section 3.1, if we compute the strongly connected components for the proposed solution shown in Figure 3.2, we find three SCCs

$$A = \{6, 7, 8, 12, 13, 14\}, \quad B = \{11\}, \quad C = \{1, 2, 3, 4, 5, 9, 10, 15, 16, 17, 18, 19\}.$$

shown in Figure 3.11.

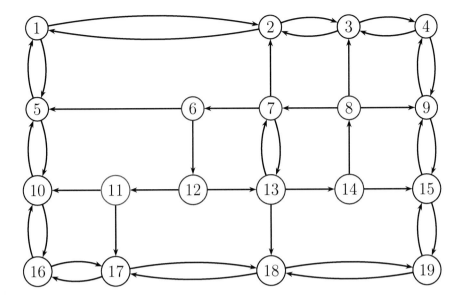

Fig. 3.11 The strongly connected components of the graph in Figure 3.2

Therefore, the city engineers did not do a good job. We show after proving Proposition 3.4 how to "fix" this faulty solution.

Note that the problem is that all the edges between the strongly connected components A and C go in the wrong direction.

Closed e-simple paths also play an important role.

Definition 3.8. Let $G = (V, E, s, t)$ be a digraph. A *circuit* is a closed e-simple path (i.e., no edge occurs twice) and a *simple circuit* is a simple closed path. The null path (u, ε, u) is a simple circuit.

Remark: A closed path is sometimes called a *pseudo-circuit*. In a pseudo-circuit, some edge may occur more than once.

The significance of simple circuits is revealed by the next proposition.

Proposition 3.3. *Let G be any digraph. (a) Every circuit π in G is the concatenation of pairwise edge-disjoint simple circuits.*

(b) A circuit is simple iff it is a minimal circuit, that is, iff it does not contain any proper circuit.

Proof. We proceed by induction on the length of π. The proposition is trivially true if π is the null path. Next, let $\pi = (u, e_1 \cdots e_m, u)$ be any nonnull circuit and let

$$\text{nodes}(\pi) = \langle v_0, \ldots, v_m \rangle,$$

with $v_0 = v_m = u$. If π is a simple circuit, we are done. Otherwise, some node occurs twice in the sequence $\langle v_0, \ldots, v_{m-1} \rangle$. Pick two occurrences of the same node, say $v_i = v_j$, with $i < j$, such that $j - i$ is minimal. Then, due to the minimality of $j - i$, no node occurs twice in $\langle v_i, \ldots, v_{j-1} \rangle$, which shows that $\pi_1 = (v_i, e_{i+1} \cdots e_j, v_i)$ is a simple circuit. Now we can write $\pi = \pi' \pi_1 \pi''$, with $|\pi'| < |\pi|$ and $|\pi''| < |\pi|$. Thus, we can apply the induction hypothesis to both π' and π'', which shows that π' and π'' are concatenations of simple circuits. Then π itself is the concatenation of simple circuits. All these simple circuits are pairwise edge-disjoint because π has no repeated edges.

(b) This is clear by definition of a simple circuit. $\square$

Remarks:

1. If u and v are two nodes that belong to a circuit π in G, (i.e., both u and v are incident to some edge in π), then u and v are strongly connected. Indeed, u and v are connected by a portion of the circuit π, and v and u are connected by the complementary portion of the circuit.

2. If π is a pseudo-circuit, the above proof shows that it is still possible to decompose π into simple circuits, but it may not be possible to write π as the concatenation of pairwise edge-disjoint simple circuits.

Given a graph G we can form a new and simpler graph from G by connecting the strongly connected components of G as shown below.

Definition 3.9. Let $G = (V, E, s, t)$ be a digraph. The *reduced graph* $\widehat{G}$ is the simple digraph whose set of nodes $\widehat{V} = V/\widehat{C}_G$ is the set of strongly connected components of V and whose set of edges $\widehat{E}$ is defined as follows.

$$(\widehat{u}, \widehat{v}) \in \widehat{E} \text{ iff } (\exists e \in E)(s(e) \in \widehat{u} \text{ and } t(e) \in \widehat{v}),$$

where we denote the strongly connected component of u by $\widehat{u}$.

That $\widehat{G}$ is "simpler" than G is the object of the next proposition.

Proposition 3.4. *Let G be any digraph. The reduced graph $\widehat{G}$ contains no circuits.*

Proof. Suppose that u and v are nodes of G and that u and v belong to two disjoint strongly connected components that belong to a circuit $\widehat{\pi}$ in $\widehat{G}$. Then the circuit $\widehat{\pi}$ yields a closed sequence of edges $e_1, \ldots, e_n$ between strongly connected components and we can arrange the numbering so that these components are $C_0, \ldots, C_n$, with $C_n = C_0$, with e_i an edge between $s(e_i) \in C_{i-1}$ and $t(e_i) \in C_i$ for $1 \leq i \leq n-1$, e_n an edge between between $s(e_n) \in C_{n-1}$ and $t(e_n) \in C_0$, $\widehat{u} = C_p$ and $\widehat{v} = C_q$, for some $p < q$. Now, we have $t(e_i) \in C_i$ and $s(e_{i+1}) \in C_i$ for $1 \leq i \leq n-1$ and $t(e_n) \in C_0$ and $s(e_1) \in C_0$ and as each C_i is strongly connected, we have simple paths from $t(e_i)$ to $s(e_{i+1})$ and from $t(e_n)$ to $s(e_1)$. Also, as $\widehat{u} = C_p$ and $\widehat{v} = C_q$ for some $p < q$, we have some simple paths from u to $s(e_{p+1})$ and from $t(e_q)$ to v. By concatenating the appropriate paths, we get a circuit in G containing u and v, showing that u and

v are strongly connected, contradicting that u and v belong to two disjoint strongly connected components. ☐

Remark: Digraphs without circuits are called *DAGs*. Such graphs have many nice properties. In particular, it is easy to see that any finite DAG has nodes with no incoming edges. Then, it is easy to see that finite DAGs are basically collections of trees with shared nodes.

The reduced graph of the graph shown in Figure 3.11 is shown in Figure 3.12, where its SCCs are labeled A, B, and C as shown below:

$$A = \{6, 7, 8, 12, 13, 14\}, \quad B = \{11\}, \quad C = \{1, 2, 3, 4, 5, 9, 10, 15, 16, 17, 18, 19\}.$$

The locations in the component A are inaccessible. Observe that changing the di-

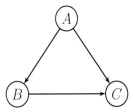

Fig. 3.12 The reduced graph of the graph in Figure 3.11

rection of *any* street between the strongly connected components A and C yields a solution, that is, a strongly connected graph. So, the engineers were not too far off after all.

A solution to our traffic problem obtained by changing the direction of the street between 13 and 18 is shown in Figure 3.13. Before discussing undirected graphs, let us collect various definitions having to do with the notion of subgraph.

Definition 3.10. Given any two digraphs $G = (V, E, s, t)$ and $G' = (V', E', s', t')$, we say that G' *is a subgraph of* G iff $V' \subseteq V$, $E' \subseteq E$, s' is the restriction of s to E' and t' is the restriction of t to E'. If G' is a subgraph of G and $V' = V$, we say that G' is a *spanning subgraph of* G. Given any subset V' of V, the *induced subgraph* $G\langle V' \rangle$ *of* G is the graph $(V', E_{V'}, s', t')$ whose set of edges is

$$E_{V'} = \{e \in E \mid s(e) \in V'; t(e) \in V'\}.$$

(Clearly, s' and t' are the restrictions of s and t to $E_{V'}$, respectively.) Given any subset, $E' \subseteq E$, the graph $G' = (V, E', s', t')$, where s' and t' are the restrictions of s and t to E', respectively, is called the *partial graph of* G *generated by* E'. The graph $(V', E' \cap E_{V'}, s', t')$ is a *partial subgraph of* G (here, s' and t' are the restrictions of s and t to $E' \cap E_{V'}$, respectively).

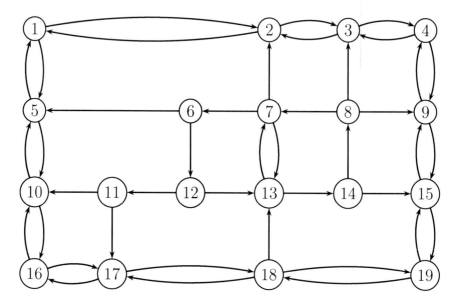

Fig. 3.13 A good choice of one-way streets

3.4 Undirected Graphs, Chains, Cycles, Connectivity

The edges of a graph express relationships among its nodes. Sometimes, these relationships are not symmetric, in which case it is desirable to use directed arcs as we have in the previous sections. However, there is a class of problems where these relationships are naturally symmetric or where there is no a priori preferred orientation of the arcs. For example, if V is the population of individuals that were students at Penn between 1900 until now and if we are interested in the relation where two people A and B are related iff they had the same professor in some course, then this relation is clearly symmetric. As a consequence, if we want to find the set of individuals who are related to a given individual A, it seems unnatural and, in fact, counterproductive, to model this relation using a directed graph.

As another example suppose we want to investigate the vulnerabilty of an Internet network under two kinds of attacks: (1) disabling a node and (2) cutting a link. Again, whether a link between two sites is oriented is irrelevant. What is important is that the two sites are either connected or disconnected.

These examples suggest that we should consider an "unoriented" version of a graph. How should we proceed?

One way to proceed is to still assume that we have a directed graph but to modify certain notions such as paths and circuits to account for the fact that such graphs are really "unoriented." In particular, we should redefine paths to allow edges to be traversed in the "wrong direction." Such an approach is possible but slightly awkward and ultimately it is really better to define undirected graphs. However, to show that this approach is feasible, let us give a new definition of a path that corresponds to the notion of path in an undirected graph.

Definition 3.11. Given any digraph $G = (V, E, s, t)$ and any two nodes $u, v \in V$, a *chain* (or *walk*) *from u to v* is a sequence $\pi = (u_0, e_1, u_1, e_2, u_2, \ldots, u_{n-1}, e_n, u_n)$, where $n \geq 1$; $u_i \in V$; $e_j \in E$ and

$$u_0 = u; \ u_n = v \text{ and } \{s(e_i), t(e_i)\} = \{u_{i-1}, u_i\}, \ 1 \leq i \leq n.$$

We call n the *length of the chain* π and we write $|\pi| = n$. When $n = 0$, we have the *null chain* (u, ε, u), from u to u, a chain of length 0. If $u = v$, then π is called a *closed chain*, else an *open chain*. The chain π determines the sequence of nodes: $\text{nodes}(\pi) = \langle u_0, \ldots, u_n \rangle$, with $\text{nodes}((u, \varepsilon, u)) = \langle u, u \rangle$. A chain π is *edge-simple*, for short, *e-simple* iff $e_i \neq e_j$ for all $i \neq j$ (i.e., no edge in the chain is used twice). A chain π from u to v is *simple* iff no vertex in $\text{nodes}(\pi)$ occurs twice, except possibly for u if π is closed. The null chain (u, ε, u) is considered *e*-simple and simple.

The main difference between Definition 3.11 and Definition 3.5 is that Definition 3.11 ignores the orientation: in a chain, an edge may be traversed backwards, from its endpoint back to its source. This implies that the reverse of a chain

$$\pi^R = (u_n, e_n, u_{n-1}, , \ldots, u_2, e_2, u_1, e_1, u_0)$$

is a chain from $v = u_n$ to $u = u_0$. In general, this fails for paths. Note, as before, that if G is a simple graph, then a chain is more simply defined by a sequence of nodes

$$(u_0, u_1, \ldots, u_n).$$

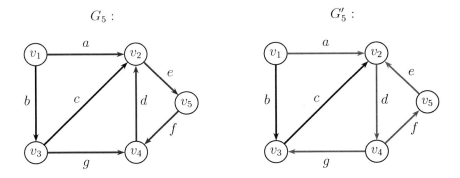

Fig. 3.14 The graphs G_5 and G_5'

For example, in the graph G_5 shown in Figure 3.14, we have the chains

$$(v_1,a,v_2,d,v_4,f,v_5,e,v_2,d,v_4,g,v_3),\ (v_1,a,v_2,d,v_4,f,v_5,e,v_2,c,v_3)$$

and

$$(v_1,a,v_2,d,v_4,g,v_3)$$

from v_1 to v_3.

Note that none of these chains are paths. The graph G_5' is obtained from the graph G_5 by reversing the direction of the edges d, f, e, and g, so that the above chains are actually paths in G_5'. The second chain is e-simple and the third is simple.

Chains are concatenated the same way as paths and the notion of subchain is analogous to the notion of subpath. The undirected version of Proposition 3.2 also holds. The proof is obtained by changing the word "path" to "chain".

Proposition 3.5. *Let G be any digraph. (a) For any two nodes u,v in G, every non-null chain π from u to v contains a simple nonnull subchain.*

(b) If $|V| = n$, then every open simple chain has length at most $n - 1$ and every closed simple chain has length at most n.

The undirected version of strong connectivity is the following:

Definition 3.12. Let $G = (V,E,s,t)$ be a digraph. We define the binary relation $\widetilde{C}_G$ on V as follows. For all $u,v \in V$,

$$u\widetilde{C}_G v \quad \text{iff} \quad \text{there is a chain from } u \text{ to } v.$$

When $u\widetilde{C}_G v$, we say that u *and v are connected.*

Oberve that the relation $\widetilde{C}_G$ is an equivalence relation. It is reflexive because we have the null chain from u to u, symmetric because the reverse of a chain is also a chain, and transitive because chains can be concatenated. The equivalence classes of the relation $\widetilde{C}_G$ are called the *connected components of G (CCs)*. A graph is *connected* iff it has a single connected component.

Observe that strong connectivity implies connectivity but the converse is false. For example, the graph G_1 of Figure 3.4 is connected but it is not strongly connected. The function *strcomp* and the method for computing the strongly connected components of a graph can easily be adapted to compute the connected components of a graph.

The undirected version of a circuit is the following.

Definition 3.13. Let $G = (V,E,s,t)$ be a digraph. A *cycle* is a closed e-simple chain (i.e., no edge occurs twice) and a *simple cycle* is a simple closed chain. The null chain (u,ε,u) is a simple cycle.

Remark: A closed chain is sometimes called a *pseudo-cycle*. The undirected version of Proposition 3.3 also holds. Again, the proof consists in changing the word "circuit" to "cycle".

Proposition 3.6. *Let G be any digraph. (a) Every cycle π in G is the concatenation of pairwise edge-disjoint simple cycles.*

    (b) A cycle is simple iff it is a minimal cycle, that is, iff it does not contain any proper cycle.

The reader should now be convinced that it is actually possible to use the notion of a directed graph to model a large class of problems where the notion of orientation is irrelevant. However, this is somewhat unnatural and often inconvenient, so it is desirable to introduce the notion of an undirected graph as a "first-class" object. How should we do that?

We could redefine the set of edges of an undirected graph to be of the form $E^+ \cup E^-$, where $E^+ = E$ is the original set of edges of a digraph and with

$$E^- = \{e^- \mid e^+ \in E^+, s(e^-) = t(e^+), t(e^-) = s(e^+)\},$$

each edge e^- being the "anti-edge" (opposite edge) of e^+. Such an approach is workable but experience shows that it not very satisfactory.

The solution adopted by most people is to relax the condition that every edge $e \in E$ be assigned an *ordered pair* $\langle u, v \rangle$ of nodes (with $u = s(e)$ and $v = t(e)$) to the condition that every edge $e \in E$ be assigned a *set* $\{u, v\}$ of nodes (with $u = v$ allowed). To this effect, let $[V]^2$ denote the subset of the power set consisting of all two-element subsets of V (the notation $\binom{V}{2}$ is sometimes used instead of $[V]^2$):

$$[V]^2 = \{\{u, v\} \in 2^V \mid u \neq v\}.$$

Definition 3.14. A *graph* is a triple $G = (V, E, st)$ where V is a set of *nodes or vertices*, E is a set of *arcs or edges*, and $st: E \to V \cup [V]^2$ is a function that assigns a set of *endpoints* (or *endnodes*) to every edge.

When we want to stress that we are dealing with an undirected graph as opposed to a digraph, we use the locution *undirected graph*. When we draw an undirected graph we suppress the tip on the extremity of an arc. For example, the undirected graph G_6 corresponding to the directed graph G_5, is shown in Figure 3.15.

Definition 3.15. Given a graph G, an edge $e \in E$ such that $st(e) \in V$ is called a *loop* (or *self-loop*). Two edges $e, e' \in E$ are said to be *parallel edges* iff $st(e) = st(e')$. A graph is *simple* iff it has no loops and no parallel edges.

Remarks:

1. The functions st need not be injective or surjective.
2. When G is simple, every edge $e \in E$ is uniquely determined by the set of vertices $\{u, v\}$ such that $\{u, v\} = st(e)$. In this case, we may denote the edge e by $\{u, v\}$ (some books also use the notation (uv) or even uv).

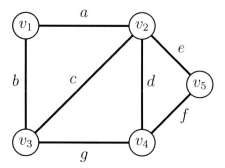

Fig. 3.15 The undirected graph G_6

3. Some authors call a graph with no loops but possibly parallel edges a *multigraph* and a graph with loops and parallel edges a *pseudograph*. We prefer to use the term graph for the most general concept.
4. Given an undirected graph $G = (V, E, st)$, we can form directed graphs from G by assigning an arbitrary orientation to the edges of G. This means that we assign to every set $st(e) = \{u, v\}$, where $u \neq v$, one of the two pairs (u, v) or (v, u) and define s and t such that $s(e) = u$ and $t(e) = v$ in the first case or such that $s(e) = v$ and $t(e) = u$ in the second case (when $u = v$, we have $s(e) = t(e) = u$)).
5. When a graph is simple, the function st is often omitted and we simply write (V, E), with the understanding that E is a set of two-element subsets of V.
6. The concepts or adjacency and incidence transfer immediately to (undirected) graphs.

It is clear that the definitions of chain, connectivity, and cycle (Definitions 3.11, 3.12, and 3.13) immediately apply to (undirected) graphs. For example, the notion of a chain in an undirected graph is defined as follows.

Definition 3.16. Given any graph $G = (V, E, st)$ and any two nodes $u, v \in V$, a *chain* (or *walk*) *from u to v* is a sequence $\pi = (u_0, e_1, u_1, e_2, u_2, \ldots, u_{n-1}, e_n, u_n)$, where $n \geq 1$; $u_i \in V$; $e_i \in E$ and

$$u_0 = u; \ u_n = v \text{ and } st(e_i) = \{u_{i-1}, u_i\}, \ 1 \leq i \leq n.$$

We call n the *length of the chain* π and we write $|\pi| = n$. When $n = 0$, we have the *null chain* (u, ε, u), from u to u, a chain of length 0. If $u = v$, then π is called a *closed chain*, else an *open chain*. The chain, π, determines the sequence of nodes, $\text{nodes}(\pi) = \langle u_0, \ldots, u_n \rangle$, with $\text{nodes}((u, \varepsilon, u)) = \langle u, u \rangle$. A chain π is *edge-simple*, for short, *e-simple* iff $e_i \neq e_j$ for all $i \neq j$ (i.e., no edge in the chain is used twice). A chain π from u to v is *simple* iff no vertex in $\text{nodes}(\pi)$ occurs twice, except possibly for u if π is closed. The null chain (u, ε, u) is considered *e*-simple and simple.

An *e*-simple chain is also called a *trail* (as in the case of directed graphs). Definitions 3.12 and 3.13 are adapted to undirected graphs in a similar fashion.

However, only the notion of *degree* (or *valency*) of a node applies to undirected graphs where it is given by

$$d_G(u) = |\{e \in E \mid u \in st(e)\}|.$$

We can check immediately that Corollary 3.1 and Corollary 3.2 apply to undirected graphs. For the reader's convenience, we restate these results.

Corollary 3.3. *For any finite undirected graph* $G = (V, E, st)$ *we have*

$$\sum_{u \in V} d_G(u) = 2|E|;$$

that is, the sum of the degrees of all the nodes is equal to twice the number of edges.

Corollary 3.4. *For any finite undirected graph* $G = (V, E, st)$, *there is an even number of nodes with an odd degree.*

Remark: When it is clear that we are dealing with undirected graphs, we sometimes allow ourselves some abuse of language. For example, we occasionally use the term path instead of chain.

An important class of graphs is the class of complete graphs. We define the *complete graph* K_n with n vertices ($n \geq 2$) as the simple undirected graph whose edges are all two-element subsets $\{i, j\}$, with $i, j \in \{1, 2, \ldots, n\}$ and $i \neq j$.

Even though the structure of complete graphs is quite simple, there are some very hard combinatorial problems involving them. For example, an amusing but very difficult problem involving edge colorings is the determination of Ramsey numbers.

A version of *Ramsey's theorem* says that: *for every pair,* (r, s), *of positive natural numbers, there is a least positive natural number,* $R(r, s)$, *such that for every coloring of the edges of the complete (undirected) graph on* $R(r, s)$ *vertices using the colors blue and red, either there is a complete subgraph with* r *vertices whose edges are all blue or there is a complete subgraph with* s *vertices whose edges are all red.*

So, $R(r, r)$ is the smallest number of vertices of a complete graph whose edges are colored either *blue* or *red* that must contain a complete subgraph with r vertices whose edges are all of the same color. It is called a *Ramsey number*. For details on Ramsey's theorems and Ramsey numbers, see Diestel [3], Chapter 9.

The graph shown in Figure 3.16 (left) is a complete graph on five vertices with a coloring of its edges so that there is no complete subgraph on three vertices whose edges are all of the same color. Thus, $R(3, 3) > 5$.

There are

$$2^{15} = 32768$$

2-colored complete graphs on 6 vertices. One of these graphs is shown in Figure 3.16 (right). It can be shown that all of them contain a triangle whose edges have the same color, so $R(3, 3) = 6$.

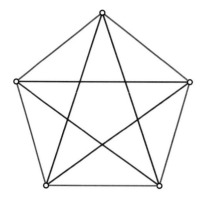
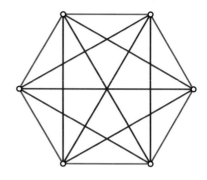

Fig. 3.16 Left: A 2-coloring of K_5 with no monochromatic K_3; Right: A 2-coloring of K_6 with several monochromatic K_3s

The numbers, $R(r,s)$, are called *Ramsey numbers*. It turns out that there are *very few* numbers r,s for which $R(r,s)$ is known because the number of colorings of a graph grows very fast! For example, there are

$$2^{43\times21} = 2^{903} > 1024^{90} > 10^{270}$$

2-colored complete graphs with 43 vertices, a huge number. In comparison, the universe is *only* approximately 14 billion years old, namely 14×10^9 years old.

For example, $R(4,4) = 18$, $R(4,5) = 25$, but $R(5,5)$ *is unknown*, although it can be shown that $43 \le R(5,5) \le 49$. Finding the $R(r,s)$, or at least some sharp bounds for them, is an open problem.

The notion of homomorphism and isomorphism also makes sense for undirected graphs. In order to adapt Definition 3.4, observe that any function $g\colon V_1 \to V_2$ can be extended in a natural way to a function from $V_1 \cup [V_1]^2$ to $V_2 \cup [V_2]^2$, also denoted g, so that

$$g(\{u,v\}) = \{g(u),g(v)\},$$

for all $\{u,v\} \in [V_1]^2$.

Definition 3.17. Given two graphs $G_1 = (V_1,E_1,st_1)$ and $G_2 = (V_2,E_2,st_2)$, a *homomorphism* (or *morphism*) $f\colon G_1 \to G_2$, *from* G_1 *to* G_2 is a pair $f = (f^v,f^e)$, with $f^v\colon V_1 \to V_2$ and $f^e\colon E_1 \to E_2$, preserving incidence, that is, for every edge $e \in E_1$, we have

$$st_2(f^e(e)) = f^v(st_1(e)).$$

These conditions can also be expressed by saying that the following diagram commutes.

$$E_1 \xrightarrow{\;f^e\;} E_2$$

$$st_1 \downarrow \qquad\qquad \downarrow st_2$$

$$V_1 \cup [V_1]^2 \xrightarrow[\;f^v\;]{} V_2 \cup [V_2]^2$$

As for directed graphs, we can compose homomorphisms of undirected graphs and the definition of an isomorphism of undirected graphs is the same as the definition of an isomorphism of digraphs. Definition 3.10 about various notions of subgraphs is immediately adapted to undirected graphs.

We now investigate the properties of a very important subclass of graphs, trees.

3.5 Trees and Arborescences

In this section, until further notice, we are dealing with undirected graphs. Given a graph G, edges having the property that their deletion increases the number of connected components of G play an important role and we would like to characterize such edges.

Definition 3.18. Given any graph $G = (V, E, st)$, any edge $e \in E$, whose deletion increases the number of connected components of G (i.e., $(V, E - \{e\}, st \restriction (E - \{e\}))$ has more connected components than G) is called a *bridge*.

For example, the edge $(v_4 v_5)$ in the graph shown in Figure 3.17 is a bridge.

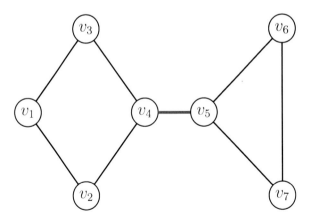

Fig. 3.17 A bridge in the graph G_7

Proposition 3.7. *Given any graph $G = (V, E, st)$, adjunction of a new edge e between u and v (this means that st is extended to st_e, with $st_e(e) = \{u, v\}$) to G has the following effect.*

1. *Either the number of components of G decreases by 1, in which case the edge e does not belong to any cycle of $G' = (V, E \cup \{e\}, st_e)$, or*
2. *The number of components of G is unchanged, in which case the edge e belongs to some cycle of $G' = (V, E \cup \{e\}, st_e)$.*

Proof. Two mutually exclusive cases are possible:

(a) The endpoints u and v (of e) belong to two disjoint connected components of G. In G', these components are merged. The edge e can't belong to a cycle of G' because the chain obtained by deleting e from this cycle would connect u and v in G, a contradiction.
(b) The endpoints u and v (of e) belong to the same connected component of G. Then, G' has the same connected components as G. Because u and v are connected, there is a simple chain from u to v (by Proposition 3.5) and by adding e to this simple chain, we get a cycle of G' containing e.   □

Corollary 3.5. *Given any graph $G = (V, E, st)$ an edge $e \in E$, is a bridge iff it does not belong to any cycle of G.*

Theorem 3.1. *Let G be a finite graph and let $m = |V| \geq 1$. The following properties hold.*

(i) If G is connected, then $|E| \geq m - 1$.
(ii) If G has no cycle, then $|E| \leq m - 1$.

Proof. We can build the graph G progressively by adjoining edges one at a time starting from the graph $(V, \emptyset)$, which has m connected components.

(i) Every time a new edge is added, the number of connected components decreases by at most 1. Therefore, it will take at least $m - 1$ steps to get a connected graph.

(ii) If G has no cycle, then every spanning graph has no cycle. Therefore, at every step, we are in case (1) of Proposition 3.7 and the number of connected components decreases by exactly 1. As G has at least one connected component, the number of steps (i.e., of edges) is at most $m - 1$.   □

In view of Theorem 3.1, it makes sense to define the following kind of graphs.

Definition 3.19. A *tree* is a graph that is connected and acyclic (i.e., has no cycles). A *forest* is a graph whose connected components are trees.

The picture of a tree is shown in Figure 3.18.
Our next theorem gives several equivalent characterizations of a tree.

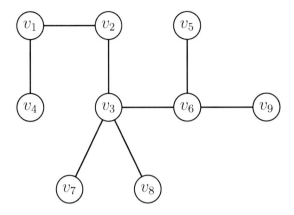

Fig. 3.18 A tree T_1

Theorem 3.2. *Let G be a finite graph with $m = |V| \geq 2$ nodes. The following properties characterize trees.*

(1) G is connected and acyclic.
(2) G is connected and minimal for this property (if we delete any edge of G, then the resulting graph is no longer connected).
(3) G is connected and has $m - 1$ edges.
(4) G is acyclic and maximal for this property (if we add any edge to G, then the resulting graph is no longer acyclic).
(5) G is acyclic and has $m - 1$ edges.
(6) Any two nodes of G are joined by a unique chain.

Proof. The implications

$$(1) \implies (3), (5)$$
$$(3) \implies (2)$$
$$(5) \implies (4)$$

all follow immediately from Theorem 3.1.

$(4) \implies (3)$. If G was not connected, we could add an edge between to disjoint connected components without creating any cycle in G, contradicting the maximality of G with respect to acyclicity. By Theorem 3.1, as G is connected and acyclic, it must have $m - 1$ edges.

$(2) \implies (6)$. As G is connected, there is a chain joining any two nodes of G. If, for two nodes u and v, we had two distinct chains from u to v, deleting any edge from one of these two chains would not destroy the connectivity of G contradicting the fact that G is minimal with respect to connectivity.

$(6) \implies (1)$. If G had a cycle, then there would be at least two distinct chains joining two nodes in this cycle, a contradiction.

The reader should then draw the directed graph of implications that we just established and check that this graph is strongly connected. Indeed, we have the cycle of implications

$$(1) \Longrightarrow (5) \Longrightarrow (4) \Longrightarrow (3) \Longrightarrow (2) \Longrightarrow (6) \Longrightarrow (1).$$

$\square$

Remark: The equivalence of (1) and (6) holds for infinite graphs too.

Corollary 3.6. *For any tree G adding a new edge e to G yields a graph G' with a unique cycle.*

Proof. Because G is a tree, all cycles of G' must contain e. If G' had two distinct cycles, there would be two distinct chains in G joining the endpoints of e, contradicting property (6) of Theorem 3.2. $\square$

Corollary 3.7. *Every finite connected graph possesses a spanning tree.*

Proof. This is a consequence of property (2) of Theorem 3.2. Indeed, if there is some edge $e \in E$, such that deleting e yields a connected graph G_1, we consider G_1 and repeat this deletion procedure. Eventually, we get a minimal connected graph that must be a tree. $\square$

An example of a spanning tree (shown in thicker lines) in a graph is shown in Figure 3.19.

An *endpoint* or *leaf* in a graph is a node of degree 1.

Proposition 3.8. *Every finite tree with $m \geq 2$ nodes has at least two endpoints.*

Proof. By Theorem 3.2, our tree has $m - 1$ edges and by the version of Proposition 3.1 for undirected graphs,

$$\sum_{u \in V} d_G(u) = 2(m - 1).$$

If we had $d_G(u) \geq 2$ except for a single node u_0, we would have

$$\sum_{u \in V} d_G(u) \geq 2m - 1,$$

contradicting the above. $\square$

Remark: A forest with m nodes and p connected components has $m - p$ edges. Indeed, if each connected component has m_i nodes, then the total number of edges is

$$(m_1 - 1) + (m_2 - 1) + \cdots + (m_p - 1) = m - p.$$

We now briefly consider directed versions of a tree.

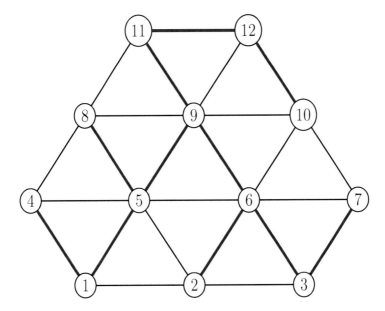

Fig. 3.19 A spanning tree

Definition 3.20. Given a digraph $G = (V, E, s, t)$, a node $a \in V$ is a *root* (respectively, *antiroot*) iff for every node $u \in V$, there is a path from a to u (respcetively, there is a path from u to a). A digraph with at least two nodes is an *arborescence with root a* iff

1. The node a is a root of G.
2. G is a tree (as an undirected graph).

A digraph with at least two nodes is an *antiarborescence with antiroot a* iff

1. The node a is an antiroot of G
2. G is a tree (as an undirected graph).

Note that orienting the edges in a tree does not necessarily yield an arborescence (or an antiarborescence). Also, if we reverse the orientation of the arcs of an arborescence we get an antiarborescence. An arborescence is shown in Figure 3.20.

There is a version of Theorem 3.2 giving several equivalent characterizations of an arborescence. The proof of this theorem is left as an exercise to the reader.

Theorem 3.3. *Let G be a finite digraph with $m = |V| \geq 2$ nodes. The following properties characterize arborescences with root a.*

(1) G is a tree (as undirected graph) with root a.
(2) For every $u \in V$, there is a unique path from a to u.
(3) G has a as a root and is minimal for this property (if we delete any edge of G, then a is not a root any longer).

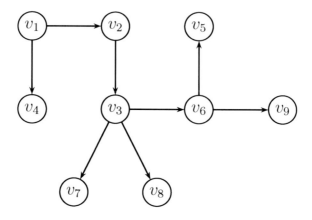

Fig. 3.20 An arborescence T_2

(4) G is connected (as undirected graph) and moreover

$$(*) \begin{cases} d_G^-(a) = 0 \\ d_G^-(u) = 1, \text{ for all } u \in V, u \neq a. \end{cases}$$

(5) G is acyclic (as undirected graph) and the properties $(*)$ are satisfied.
(6) G is acyclic (as undirected graph) and has a as a root.
(7) G has a as a root and has $m - 1$ arcs.

3.6 Minimum (or Maximum) Weight Spanning Trees

For a certain class of problems, it is necessary to consider undirected graphs (without loops) whose edges are assigned a "cost" or "weight."

Definition 3.21. A *weighted graph* is a finite graph without loops $G = (V, E, st)$, together with a function $c : E \to \mathbb{R}$, called a *weight function* (or *cost function*). We denote a weighted graph by (G, c). Given any set of edges $E' \subseteq E$, we define the *weight (or cost)* of E' by

$$c(E') = \sum_{e \in E'} c(e).$$

Given a weighted graph (G, c), an important problem is to find a spanning tree T such that $c(T)$ is maximum (or minimum). This problem is called the *maximal weight spanning tree* (respectively, *minimal weight spanning tree*). Actually, it is easy to see that any algorithm solving any one of the two problems can be converted to an algorithm solving the other problem. For example, if we can solve the maximal weight spanning tree, we can solve the minimal weight spanning tree by replacing every weight $c(e)$ by $-c(e)$, and by looking for a spanning tree T that is a maximal spanning tree, because

$$\min_{T \subseteq G} c(T) = -\max_{T \subseteq G} -c(T).$$

There are several algorithms for finding such spanning trees, including one due to Kruskal and another one due to Robert C. Prim. The fastest known algorithm at present is due to Bernard Chazelle (1999).

Because every spanning tree of a given graph $G = (V, E, st)$ has the same number of edges (namely, $|V| - 1$), adding the same constant to the weight of every edge does not affect the maximal nature a spanning tree, that is, the set of maximal weight spanning trees is preserved. Therefore, we may assume that all the weights are nonnegative.

In order to justify the correctness of Kruskal's algorithm, we need two definitions. Let (G, c) be any connected weighted graph with $G = (V, E, st)$ and let T be any spanning tree of G. For every edge $e \in E - T$, let C_e be the set of edges belonging to the unique chain in T joining the endpoints of e (the vertices in $st(e)$). For example, in the graph shown in Figure 3.21, the set $C_{\{8,11\}}$ associated with the edge $\{8, 11\}$ (shown as a dashed line) corresponds to the following set of edges (shown as dotted lines) in T,

$$C_{\{8,11\}} = \{\{8,5\}, \{5,9\}, \{9,11\}\}.$$

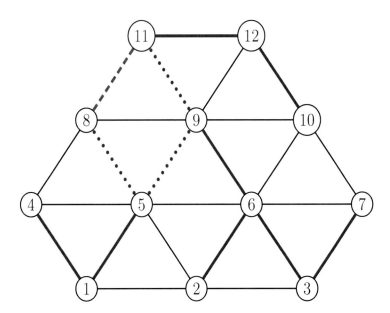

Fig. 3.21 The set C_e associated with an edge $e \in G - T$

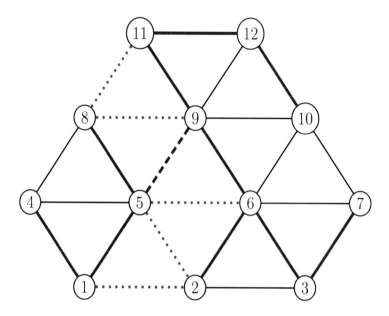

Fig. 3.22 The set $\Omega_{\{5,9\}}$ obtained by deleting the edge $\{5,9\}$ from the spanning tree.

Also, given any edge $e \in T$, observe that the result of deleting e yields a graph denoted $T - e$ consisting of two disjoint subtrees of T. We let Ω_e be the set of edges $e' \in G - T$, such that if $st(e') = \{u, v\}$, then u and v belong to the two distinct connected components of $T - \{e\}$. For example, in Figure 3.22, deleting the edge $\{5, 9\}$ yields the set of edges (shown as dotted lines)

$$\Omega_{\{5,9\}} = \{\{1,2\}, \{5,2\}, \{5,6\}, \{8,9\}, \{8,11\}\}.$$

Observe that in the first case, deleting any edge from C_e and adding the edge $e \in E - T$ yields a new spanning tree and in the second case, deleting any edge $e \in T$ and adding any edge in Ω_e also yields a new spanning tree. These observations are crucial ingredients in the proof of the following theorem.

Theorem 3.4. *Let (G, c) be any connected weighted graph and let T be any spanning tree of G. (1) The tree T is a maximal weight spanning tree iff any of the following (equivalent) conditions hold.*

(i) For every $e \in E - T$,

$$c(e) \leq \min_{e' \in C_e} c(e')$$

(ii) For every $e \in T$,

$$c(e) \geq \max_{e' \in \Omega_e} c(e').$$

(2) The tree T is a minimal weight spanning tree iff any of the following (equivalent) conditions hold.

(i) For every $e \in E - T$,

$$c(e) \geq \max_{e' \in C_e} c(e')$$

(ii) For every $e \in T$,

$$c(e) \leq \min_{e' \in \Omega_e} c(e').$$

Proof. (1) First, assume that T is a maximal weight spanning tree. Observe that

(a) For any $e \in E - T$ and any $e' \in C_e$, the graph $T' = (V, (T \cup \{e\}) - \{e'\})$ is acyclic and has $|V| - 1$ edges, so it is a spanning tree. Then, (i) must hold, as otherwise we would have $c(T') > c(T)$, contradicting the maximality of T.
(b) For any $e \in T$ and any $e' \in \Omega_e$, the graph $T' = (V, (T \cup \{e'\}) - \{e\})$ is connected and has $|V| - 1$ edges, so it is a spanning tree. Then, (ii) must hold, as otherwise we would have $c(T') > c(T)$, contradicting the maximality of T.

Let us now assume that (i) holds. We proceed by contradiction. Let T be a spanning tree satisfying condition (i) and assume there is another spanning tree T' with $c(T') > c(T)$. There are only finitely many spanning trees of G, therefore we may assume that T' is maximal. Consider any edge $e \in T' - T$ and let $st(e) = \{u, v\}$. In T, there is a unique chain C_e joining u and v and this chain must contain some edge $e' \in T$ joining the two connected components of $T' - e$; that is, $e' \in \Omega_e$. As (i) holds, we get $c(e) \leq c(e')$. However, as T' is maximal, (ii) holds (as we just proved), so $c(e) \geq c(e')$. Therefore, we get

$$c(e) = c(e').$$

Consequently, if we form the graph $T_2 = (T' \cup \{e'\}) - \{e\})$, we see that T_2 is a spanning tree having some edge from T and $c(T_2) = c(T')$. We can repeat this process of edge substitution with T_2 and T and so on. Ultimately, we obtain the tree T with the weight $c(T') > c(T)$, which is absurd. Therefore, T is indeed maximal.

Finally, assume that (ii) holds. The proof is analogous to the previous proof: We pick some edge $e' \in T - T'$ and e is some edge in $\Omega_{e'}$ belonging to the chain joining the endpoints of e' in T'.

(2) The proof of (2) is analogous to the proof of (1) but uses 2(i) and 2(ii) instead of 1(i) and 1(ii). $\quad\square$

We are now in the position to present a version of Kruskal's algorithm and to prove its correctness.

Here is a version of Kruskal's algorithm for finding a minimal weight spanning tree using criterion 2(i). Let n be the number of edges of the weighted graph (G, c), where $G = (V, E, st)$.

function *Kruskal*$((G, c)$: weighted graph): tree
  begin
    Sort the edges in nondecreasing order of weights:
    $c(e_1) \leq c(e_2) \leq \cdots \leq c(e_n)$;

Fig. 3.23 Joseph Kruskal, 1928–

$T := \emptyset;$
for $i := 1$ **to** n **do**
    if $(V, T \cup \{e_i\})$ is acyclic **then** $T := T \cup \{e_i\}$
    endif
  endfor;
  $Kruskal := T$
end

We admit that the above description of Kruskal's algorithm is a bit sketchy as we have not explicitly specified how we check that adding an edge to a tree preserves acyclicity. On the other hand, it is quite easy to prove the correctness of the above algorithm. It is not difficult to refine the above "naive" algorithm to make it totally explicit but this involves a good choice of data structures. We leave these considerations to an algorithms course.

Clearly, the graph T returned by the algorithm is acyclic, but why is it connected? Well, suppose T is not connected and consider two of its connected components, say T_1 and T_2. Being acyclic and connected, T_1 and T_2 are trees. Now, as G itself is connected, for any node of T_1 and any node of T_2, there is some chain connecting these nodes. Consider such a chain C, of minimal length. Then, as T_1 is a tree, the first edge e_j of C cannot belong to T_1 because otherwise we would get an even shorter chain connecting T_1 and T_2 by deleting e_j. Furthermore, e_j does not belong to any other connected component of T, as these connected components are pairwise disjoint. But then, $T + e_j$ is acyclic, which means that when we considered the addition of edge e_j to the current graph $T^{(j)}$, the test should have been positive and e_j should have been added to $T^{(j)}$. Therefore, T is connected and so it is a spanning tree. Now observe that as the edges are sorted in nondecreasing order of weight, condition 2(i) is enforced and by Theorem 3.4, T is a minimal weight spanning tree.

We can easily design a version of Kruskal's algorithm based on Condition 2(ii). This time, we sort the edges in nonincreasing order of weights and, starting with G, we attempt to delete each edge e_j as long as the remaining graph is still connected. We leave the design of this algorithm as an exercise to the reader.

Prim's algorithm is based on a rather different observation. For any node, $v \in V$, let U_v be the set of edges incident with v that are not loops,

$$U_v = \{e \in E \mid v \in st(e), \, st(e) \in [V]^2\}.$$

Choose in U_v some edge of minimum weight that we (ambiguously) denote by $e(v)$.

Proposition 3.9. *Let (G, c) be a connected weighted graph with $G = (V, E, st)$. For every vertex $v \in V$, there is a minimum weight spanning tree T so that $e(v) \in T$.*

Proof. Let T' be a minimum weight spanning tree of G and assume that $e(v) \notin T'$. Let C be the chain in T' that joins the endpoints of $e(v)$ and let e be the edge of C that is incident with v. Then, the graph $T'' = (V, (T' \cup \{e(v)\}) - \{e\})$ is a spanning tree of weight less than or equal to the weight of T' and as T' has minimum weight, so does T''. By construction, $e(v) \in T''$. $\square$

Prim's algorithm uses an edge-contraction operation described below:

Definition 3.22. Let $G = (V, E, st)$ be a graph, and let $e \in E$ be some edge that is not a loop; that is, $st(e) = \{u, v\}$, with $u \neq v$. The graph $C_e(G)$ obtained by *contracting the edge e* is the graph obtained by merging u and v into a single node and deleting e. More precisely, $C_e(G) = ((V - \{u, v\}) \cup \{w\}, E - \{e\}, st_e)$, where w is any new node not in V and where

1. $st_e(e') = st(e')$ iff $u \notin st(e')$ and $v \notin st(e')$.
2. $st_e(e') = \{w, z\}$ iff $st(e') = \{u, z\}$, with $z \notin st(e)$.
3. $st_e(e') = \{z, w\}$ iff $st(e') = \{z, v\}$, with $z \notin st(e)$.
4. $st_e(e') = w$ iff $st(e') = \{u, v\}$.

Proposition 3.10. *Let $G = (V, E, st)$ be a graph. For any edge, $e \in E$, the graph G is a tree iff $C_e(G)$ is a tree.*

Proof. Proposition 3.10 follows from Theorem 3.2. Observe that G is connected iff $C_e(G)$ is connected. Moreover, if G is a tree, the number of nodes of $C_e(G)$ is $n_e = |V| - 1$ and the number of edges of $C_e(G)$ is $m_e = |E| - 1$. Because $|E| = |V| - 1$, we get $m_e = n_e - 1$ and $C_e(G)$ is a tree. Conversely, if $C_e(G)$ is a tree, then $m_e = n_e - 1$, $|V| = n_e + 1$ and $|E| = m_e + 1$, so $m = n - 1$ and G is a tree. $\square$

Here is a "naive" version of Prim's algorithm.

function *Prim*$((G = (V, E, st), c)$: weighted graph): tree
  begin
    $T := \emptyset$;
    while $|V| \geq 2$ **do**
      pick any vertex $v \in V$;
      pick any edge (not a loop), e, in U_v of minimum weight;
      $T := T \cup \{e\}$; $G := C_e(G)$
    endwhile;
    Prim $:= T$
  end

The correctness of Prim's algorithm is an immediate consequence of Proposition 3.9 and Proposition 3.10; the details are left to the reader.

3.7 Summary

This chapter deals with the concepts of directed and undirected graphs and some of their basic properties, in particular, connectivity. Trees are characterized in various ways and methods for finding (minimal weight) spanning trees are briefly studied.

- We begin with a problem motivating the use of directed graphs.
- We define *directed graphs* using *source* and *taget* functions from *edges* to *vertices*.
- We define *simple* directed graphs.
- We define *adjacency* and *incidence*.
- We define the *outer half-degree*, *inner half-degree*, and the *degree* of a vertex.
- We define a *regular* graph.
- We define *homomorphisms* and *isomorphisms* of directed graphs.
- We define the notion of (*open or closed*) *path* (or *walk*) in a directed graph.
- We define *e-simple* paths and *simple* paths.
- We prove that every nonnull path contains a simple subpath.
- We define the *concatenation* of paths.
- We define when two nodes are *strongly connected* and the *strongly connected components* (SCCs) of a directed graph. We give a simple algorithm for computing the SCCs of a directed graph.
- We define *circuits* and *simple circuits*.
- We prove some basic properties of circuits and simple circuits.
- We define the *reduced graph* of a directed graph and prove that it contains no circuits.
- We define *subgraphs*, *induced subgraphs*, *spanning subgraphs*, *partial graphs* and *partial subgraphs*.
- Next we consider *undirected graphs*.
- We define a notion of undirected path called a *chain*.
- We define *e-simple* chains and *simple* chains.
- We define when two nodes are *connected* and the *connected components* of a graph.
- We define undirected circuits, called *cycles* and *simple cycles*
- We define *undirected graphs* in terms of a function from the set of edges to the union of the set of vertices and the set of two-element subsets of vertices.
- We revisit the notion of *chain* in the framework of undirected graphs.
- We define the *degree* of a node in an undirected graph.
- We define the *complete graph* K_n on n vertices.
- We state a version of *Ramsey's theorem* and define *Ramsey numbers*.
- We define *homomorphisms* and *isomorphisms* of undirected graphs.
- We define the notion of a *bridge* in an undirected graph and give a characterization of a bridge in terms of cycles.
- We prove a basic relationship between the number of vertices and the number of edges in a finite undirected graph G having to do with the fact that either G is connected or G has no cycle.

- We define *trees* and *forests*.
- We give several characterizations of a tree.
- We prove that every connected graph possesses a spanning tree.
- We define a *leaf* or *endpoint*.
- We prove that every tree with at least two nodes has at least two leaves.
- We define a *root* and an *antiroot* in a directed graph.
- We define an *arborescence* (with a root or an antiroot).
- We state a characterization of arborescences.
- We define (undirected) *weighted graphs*.
- We prove a theorem characterizing *maximal weight spanning trees* (and *minimal weight spanning trees*).
- We present *Kruskal's algorithm* for finding a minimal weight spanning tree.
- We define *edge contraction*.
- We present *Prim's algorithm* for finding a minimal weight spanning tree.

Problems

3.1. (a) Give the list of all directed simple graphs with two nodes.
(b) Give the list of all undirected simple graphs with two nodes.

3.2. Prove that in a party with an odd number of people, there is always a person who knows an even number of others. Here we assume that the relation "knowing" is symmetric (i.e., if A knows B, then B knows A). Also, there may be pairs of people at the party who don't know each other or even people who don't know anybody else so "even" includes zero.

3.3. What is the maximum number of edges that an undirected simple graph with 10 nodes can have?

3.4. Prove that every undirected simple graph with $n \geq 2$ nodes and more than $(n-1)(n-2)/2$ edges is connected.

3.5. Prove that if $f = (f^e, f^v)$ is a graph isomorphism, then both f^e and f^v are bijections. Assume that $f = (f^e, f^v)$ is a graph homomorphism and that both f^e and f^v are bijections. Must f be a graph isomorphism?

3.6. If G_1 and G_2 are isomorphic finite directed graphs, then prove that for every $k \geq 0$, the number of nodes u in G_1 such that $d_{G_1}^-(u) = k$, is equal to the number of nodes $v \in G_2$, such that $d_{G_2}^-(v) = k$ (respectively, the number of nodes u in G_1 such that $d_{G_1}^+(u) = k$, is equal to the number of nodes $v \in G_2$, such that $d_{G_2}^+(v) = k$). Give a counterexample showing that the converse property is false.

3.7. Prove that every undirected simple graph with at least two nodes has two nodes with the same degree.

3.8. An undirected graph G is *h-connected* ($h \geq 1$) iff the result of deleting any $h-1$ vertices and the edges adjacent to these vertices does not disconnect G. An *articulation point* u in G is a vertex whose deletion increases the number of connected components. Prove that if G hs $n \geq 3$ nodes, then the following properties are equivalent.

(1) G is 2-connected.
(2) G is connected and has no articulation point.
(3) For every pair of vertices (u,v) in G, there is a simple cycle passing through u and v.
(4) For every vertex u in G and every edge $e \in G$, there is a simple cycle passing through u containing e.
(5) For every pair of edges (e,f) in G, there is a simple cycle containing e and f.
(6) For every triple of vertices (a,b,c) in G, there is a chain from a to b passing through c.
(7) For every triple of vertices (a,b,c) in G, there is a chain from a to b not passing through c.

3.9. Give an algorithm for finding the connected components of an undirected finite graph.

3.10. If $G = (V,E)$ is an undirected simple graph, then its *complement* is the graph, $\overline{G} = (V,\overline{E})$; that is, an edge, $\{u,v\}$, is an edge of $\overline{G}$ iff it is not an edge of G.
  (a) Prove that either G or $\overline{G}$ is connected.
  (b) Give an example of an undirected simple graph with nine nodes that is isomorphic to its complement.

3.11. Let $G = (V,E)$ be an undirected graph. Let E' be the set of edges in any cycle in G. Then, every vertex of the partial graph (V,E') has even degree.

3.12. A directed graph G is *quasi-strongly connected* iff for every pair of nodes (a,b) there is some node c in G such that there is a path from c to a and a path from c to b. Prove that G is quasi-strongly connected iff G has a root.

3.13. A directed graph $G = (V,E,s,t)$ is

1. *Injective* iff $d_G^-(u) \leq 1$, for all $u \in V$.
2. *Functional* iff $d_G^+(u) \leq 1$, for all $u \in V$.

  (a) Prove that an injective graph is quasi-strongly connected iff it is connected (as an undirected graph).
  (b) Prove that an undirected simple graph G can be oriented to form either an injective graph or a functional graph iff every connected component of G has at most one cycle.

3.14. Design a version of Kruskal's algorithm based on condition 2(ii) of Theorem 3.4.

3.15. (a) List all (unoriented) trees with four nodes and then five nodes.

(b) Recall that the *complete graph* K_n with n vertices ($n \geq 2$) is the simple undirected graph whose edges are all two-element subsets $\{i, j\}$, with $i, j \in \{1, 2, \ldots, n\}$ and $i \neq j$. List all spanning trees of the complete graphs K_2 (one tree), K_3 (3 trees), and K_4 (16 trees).

Remark: It can be shown that the number of spanning trees of K_n is n^{n-2}, a formula due to Cayley (1889); see Problem 4.24.

3.16. Prove that the graph K_5 with the coloring shown on Figure 3.16 (left) does not contain any complete subgraph on three vertices whose edges are all of the same color. Prove that for every edge coloring of the graph K_6 using two colors (say red and blue), there is a complete subgraph on three vertices whose edges are all of the same color.

References

1. Claude Berge. *Graphs and Hypergraphs*. Amsterdam: Elsevier North-Holland, first edition, 1973.
2. Béla Bollobas. *Modern Graph Theory*. GTM No. 184. New York: Springer Verlag, first edition, 1998.
3. Reinhard Diestel. *Graph Theory*. GTM No. 173. New York: Springer Verlag, third edition, 2005.
4. Frank Harary. *Graph Theory*. Reading, MA: Addison Wesley, first edition, 1971.
5. Michel Sakarovitch. *Optimisation Combinatoire, Méthodes mathématiques et algorithmiques. Graphes et Programmation Linéaire*. Paris: Hermann, first edition, 1984.

Chapter 4
Some Counting Problems; Multinomial Coefficients, The Principle of Inclusion–Exclusion, Sylvester's Formula, The Sieve Formula

4.1 Counting Permutations and Functions

In this section, we consider some simple counting problems. Let us begin with permutations. Recall that a *permutation* of a set A is any bijection between A and itself. If A is a finite set with n elements, we mentioned earlier (without proof) that A has $n!$ permutations, where the *factorial function*, $n \mapsto n!$ $(n \in \mathbb{N})$, is given recursively by:

$$0! = 1$$
$$(n+1)! = (n+1)n!.$$

The reader should check that the existence of the function $n \mapsto n!$ can be justified using the recursion theorem (Theorem 2.1).

Proposition 4.1. *The number of permutations of a set of n elements is n!.*

Proof. We prove that if A and B are any two finite sets of the same cardinality n, then the number of bijections between A and B is $n!$. Now, in the special case where $B = A$, we get our theorem.

The proof is by induction on n. For $n = 0$, the empty set has one bijection (the empty function). So, there are $0! = 1$ permutations, as desired.

Assume inductively that if A and B are any two finite sets of the same cardinality, n, then the number of bijections between A and B is $n!$. If A and B are sets with $n+1$ elements, then pick any element $a \in A$, and write $A = A' \cup \{a\}$, where $A' = A - \{a\}$ has n elements. Now, any bijection $f: A \to B$ must assign some element of B to a and then $f \upharpoonright A'$ is a bijection between A' and $B' = B - \{f(a)\}$. By the induction hypothesis, there are $n!$ bijections between A' and B'. There are $n+1$ ways of picking $f(a)$ in B, thus the total number of bijections between A and B is $(n+1)n! = (n+1)!$, establishing the induction hypothesis. $\square$

Let us also count the number of functions between two finite sets.

J. Gallier, *Discrete Mathematics*, Universitext,
DOI 10.1007/978-1-4419-8047-2_4, © Springer Science+Business Media, LLC 2011

Proposition 4.2. *If A and B are finite sets with $|A| = m$ and $|B| = n$, then the set of function B^A from A to B has n^m elements.*

Proof. We proceed by induction on m. For $m = 0$, we have $A = \emptyset$, and the only function is the empty function. In this case, $n^0 = 1$ and the base case holds.

Assume the induction hypothesis holds for m and assume $|A| = m + 1$. Pick any element $a \in A$ and let $A' = A - \{a\}$, a set with m elements. Any function $f: A \to B$ assigns an element $f(a) \in B$ to a and $f \restriction A'$ is a function from A' to B. By the induction hypothesis, there are n^m functions from A' to B. There are n ways of assigning $f(a) \in B$ to a, thus there are $n \cdot n^m = n^{m+1}$ functions from A to B, establishing the induction hypothesis. $\square$

As a corollary, we determine the cardinality of a finite power set.

Corollary 4.1. *For any finite set A, if $|A| = n$, then $|2^A| = 2^n$.*

Proof. By proposition 2.5, there is a bijection between 2^A and the set of functions $\{0, 1\}^A$. Because $|\{0, 1\}| = 2$, we get $|2^A| = |\{0, 1\}^A| = 2^n$, by Proposition 4.2. $\square$

Computing the value of the factorial function for a few inputs, say $n = 1, 2 \ldots, 10$, shows that it grows very fast. For example,

$$10! = 3,628,800.$$

Is it possible to quantify how fast the factorial grows compared to other functions, say n^n or e^n? Remarkably, the answer is yes. A beautiful formula due to James Stirling (1692–1770) tells us that

$$n! \sim \sqrt{2\pi n} \left(\frac{n}{e}\right)^n,$$

which means that

$$\lim_{n \to \infty} \frac{n!}{\sqrt{2\pi n} \left(\frac{n}{e}\right)^n} = 1.$$

Here, of course,

$$e = 1 + \frac{1}{1!} + \frac{1}{2!} + \frac{1}{3!} + \cdots + \frac{1}{n!} + \cdots,$$

the base of the natural logarithm. It is even possible to estimate the error. It turns out that

$$n! = \sqrt{2\pi n} \left(\frac{n}{e}\right)^n e^{\lambda_n},$$

where

$$\frac{1}{12n + 1} < \lambda_n < \frac{1}{12n},$$

a formula due to Jacques Binet (1786–1856).

Let us introduce some notation used for comparing the rate of growth of functions. We begin with the "big oh" notation.

Fig. 4.1 Jacques Binet, 1786–1856

Given any two functions, $f \colon \mathbb{N} \to \mathbb{R}$ and $g \colon \mathbb{N} \to \mathbb{R}$, we say that f *is* $O(g)$ *(or* $f(n)$ *is* $O(g(n))$*)* iff there is some $N > 0$ and a constant $c > 0$ such that

$$|f(n)| \le c|g(n)|, \text{ for all } n \ge N.$$

In other words, for n large enough, $|f(n)|$ is bounded by $c|g(n)|$. We sometimes write $n >> 0$ to indicate that n is "large."

For example, λ_n is $O(1/12n)$. By abuse of notation, we often write $f(n) = O(g(n))$ even though this does not make sense.

The "big omega" notation means the following: f *is* $\Omega(g)$ *(or* $f(n)$ *is* $\Omega(g(n))$*)* iff there is some $N > 0$ and a constant $c > 0$ such that

$$|f(n)| \ge c|g(n)|, \text{ for all } n \ge N.$$

The reader should check that $f(n)$ is $O(g(n))$ iff $g(n)$ is $\Omega(f(n))$. We can combine O and Ω to get the "big theta" notation: f *is* $\Theta(g)$ *(or* $f(n)$ *is* $\Theta(g(n))$*)* iff there is some $N > 0$ and some constants $c_1 > 0$ and $c_2 > 0$ such that

$$c_1|g(n)| \le |f(n)| \le c_2|g(n)|, \text{ for all } n \ge N.$$

Finally, the "little oh" notation expresses the fact that a function f has much slower growth than a function g. We say that f *is* $o(g)$ *(or* $f(n)$ *is* $o(g(n))$*)* iff

$$\lim_{n \to \infty} \frac{f(n)}{g(n)} = 0.$$

For example, $\sqrt{n}$ is $o(n)$.

4.2 Counting Subsets of Size k; Binomial and Multinomial Coefficients

Let us now count the number of subsets of cardinality k of a set of cardinality n, with $0 \leq k \leq n$. Denote this number by $\binom{n}{k}$ (say "n choose k"). Actually, in the proposition below, it is more convenient to assume that $k \in \mathbb{Z}$.

Proposition 4.3. *For all $n \in \mathbb{N}$ and all $k \in \mathbb{Z}$, if $\binom{n}{k}$ denotes the number of subsets of cardinality k of a set of cardinality n, then*

$$\binom{0}{0} = 1$$

$$\binom{n}{k} = 0 \quad \text{if} \quad k \notin \{0, 1, \ldots, n\}$$

$$\binom{n}{k} = \binom{n-1}{k} + \binom{n-1}{k-1} \quad (n \geq 1, 0 \leq k \leq n).$$

Proof. Obviously, when k is "out of range," that is, when $k \notin \{0, 1, \ldots, n\}$, we have

$$\binom{n}{k} = 0.$$

Next, assume that $0 \leq k \leq n$. Clearly, we may assume that our set is $[n] = \{1, \ldots, n\}$ ($[0] = \emptyset$). If $n = 0$, we have

$$\binom{0}{0} = 1,$$

because the empty set is the only subset of size 0.

If $n \geq 1$, we need to consider the cases $k = 0$ and $k = n$ separately. If $k = 0$, then the only subset of $[n]$ with 0 elements is the empty set, so

$$\binom{n}{0} = 1 = \binom{n-1}{0} + \binom{n-1}{-1} = 1 + 0,$$

inasmuch as $\binom{n-1}{0} = 1$ and $\binom{n-1}{-1} = 0$. If $k = n$, then the only subset of $[n]$ with n elements is $[n]$ itself, so

$$\binom{n}{n} = 1 = \binom{n-1}{n} + \binom{n-1}{n-1} = 0 + 1,$$

because $\binom{n-1}{n} = 0$ and $\binom{n-1}{n-1} = 1$.

If $1 \leq k \leq n-1$, then there are two kinds of subsets of $\{1, \ldots, n\}$ having k elements: those containing 1, and those not containing 1. Now, there are as many subsets of k elements from $\{1, \ldots, n\}$ containing 1 as there are subsets of $k - 1$ elements from $\{2, \ldots, n\}$, namely $\binom{n-1}{k-1}$, and there are as many subsets of k elements from $\{1, \ldots, n\}$ not containing 1 as there are subsets of k elements from $\{2, \ldots, n\}$, namely $\binom{n-1}{k}$. Thus, the number of subsets of $\{1,\dots,n\}$ consisting of k elements is $\binom{n-1}{k} + \binom{n-1}{k-1}$, which is equal to $\binom{n}{k}$. □

The numbers $\binom{n}{k}$ are also called *binomial coefficients*, because they arise in the expansion of the binomial expression $(a+b)^n$, as we show shortly. The binomial coefficients can be computed inductively using the formula

$$\binom{n}{k} = \binom{n-1}{k} + \binom{n-1}{k-1}$$

(sometimes known as *Pascal's recurrence formula*) by forming what is usually called *Pascal's triangle*, which is based on the recurrence for $\binom{n}{k}$; see Table 4.1.

n	$\binom{n}{0}$	$\binom{n}{1}$	$\binom{n}{2}$	$\binom{n}{3}$	$\binom{n}{4}$	$\binom{n}{5}$	$\binom{n}{6}$	$\binom{n}{7}$	$\binom{n}{8}$	$\binom{n}{9}$	$\binom{n}{10}$	$\cdots$
0	1											
1	1	1										
2	1	2	1									
3	1	3	3	1								
4	1	4	6	4	1							
5	1	5	10	10	5	1						
6	1	6	15	20	15	6	1					
7	1	7	21	35	35	21	7	1				
8	1	8	28	56	70	56	28	8	1			
9	1	9	36	84	126	126	84	36	9	1		
10	1	10	45	120	210	252	210	120	45	10	1	
$\vdots$	$\vdots$	$\vdots$	$\vdots$	$\vdots$	$\vdots$	$\vdots$	$\vdots$	$\vdots$	$\vdots$	$\vdots$	$\vdots$	$\vdots$

Table 4.1 Pascal's Triangle

We can also give the following explicit formula for $\binom{n}{k}$ in terms of the factorial function.

Proposition 4.4. *For all $n, k \in \mathbb{N}$, with $0 \leq k \leq n$, we have*

$$\binom{n}{k} = \frac{n!}{k!(n-k)!}.$$

Proof. Left as an exercise to the reader (use induction on n and Pascal's recurrence formula). □

Then, it is very easy to see that we have the *symmetry identity*

$$\binom{n}{k} = \binom{n}{n-k} = \frac{n(n-1)\cdots(n-k+1)}{k(k-1)\cdots 2\cdot 1}.$$

Remarks:

(1) The binomial coefficients were already known in the twelfth century by the Indian scholar Bhaskra. Pascal's triangle was taught back in 1265 by the Persian philosopher, Nasir-Ad-Din.

Fig. 4.2 Blaise Pascal, 1623–1662

(2) The formula given in Proposition 4.4 suggests generalizing the definition of the binomial coefficients to upper indices taking *real* values. Indeed, for all $r \in \mathbb{R}$ and all integers $k \in \mathbb{Z}$ we can set

$$\binom{r}{k} = \begin{cases} \dfrac{r^{\underline{k}}}{k!} = \dfrac{r(r-1)\cdots(r-k+1)}{k(k-1)\cdots 2\cdot 1} & \text{if } k \geq 0 \\ 0 & \text{if } k < 0. \end{cases}$$

Note that the expression in the numerator, $r^{\underline{k}}$, stands for the product of the k terms

$$\overbrace{r(r-1)\cdots(r-k+1)}^{k \text{ terms}}.$$

By convention, the value of this expression is 1 when $k = 0$, so that $\binom{r}{0} = 1$. The expression $\binom{r}{k}$ can be viewed as a polynomial of degree k in r. The generalized binomial coefficients allow for a useful extension of the binomial formula (see next) to real exponents. However, beware that the symmetry identity fails when r is not a natural number and that the formula in Proposition 4.4 (in terms of the factorial function) only makes sense for natural numbers.

We now prove the "binomial formula" (also called "binomial theorem").

Proposition 4.5. *(Binomial Formula) For all $n \in \mathbb{N}$ and for all reals $a, b \in \mathbb{R}$, (or more generally, any two commuting variables a, b, i.e., satisfying $ab = ba$), we have the formula:*

$$(a+b)^n = a^n + \binom{n}{1}a^{n-1}b + \cdots + \binom{n}{k}a^{n-k}b^k + \cdots + \binom{n}{n-1}ab^{n-1} + b^n.$$

The above can be written concisely as

$$(a+b)^n = \sum_{k=0}^{n} \binom{n}{k} a^{n-k} b^k.$$

Proof. We proceed by induction on n. For $n = 0$, we have $(a+b)^0 = 1$ and the sum on the right hand side is also 1, inasmuch as $\binom{0}{0} = 1$.

Assume inductively that the formula holds for n. Because

$$(a+b)^{n+1} = (a+b)^n (a+b),$$

using the induction hypothesis, we get

$$
\begin{aligned}
(a+b)^{n+1} &= (a+b)^n (a+b) \\
&= \left(\sum_{k=0}^{n} \binom{n}{k} a^{n-k} b^k \right) (a+b) \\
&= \sum_{k=0}^{n} \binom{n}{k} a^{n+1-k} b^k + \sum_{k=0}^{n} \binom{n}{k} a^{n-k} b^{k+1} \\
&= a^{n+1} + \sum_{k=1}^{n} \binom{n}{k} a^{n+1-k} b^k + \sum_{k=0}^{n-1} \binom{n}{k} a^{n-k} b^{k+1} + b^{n+1} \\
&= a^{n+1} + \sum_{k=1}^{n} \binom{n}{k} a^{n+1-k} b^k + \sum_{k=1}^{n} \binom{n}{k-1} a^{n+1-k} b^k + b^{n+1} \\
&= a^{n+1} + \sum_{k=1}^{n} \left(\binom{n}{k} + \binom{n}{k-1} \right) a^{n+1-k} b^k + b^{n+1} \\
&= \sum_{k=0}^{n+1} \binom{n+1}{k} a^{n+1-k} b^k,
\end{aligned}
$$

where we used Proposition 4.3 to go from the next to the last line to the last line. This establishes the induction step and thus, proves the binomial formula. $\square$

Remark: The binomial formula can be generalized to the case where the exponent r is a real number (even negative). This result is usually known as the *binomial theorem* or *Newton's generalized binomial theorem*. Formally, the binomial theorem states that

$$(a+b)^r = \sum_{k=0}^{\infty} \binom{r}{k} a^{r-k} b^k, \quad r \in \mathbb{N} \text{ or } |b/a| < 1.$$

Observe that when r is not a natural number, the right-hand side is an infinite sum and the condition $|b/a| < 1$ ensures that the series converges. For example, when $a = 1$ and $r = 1/2$, if we rename b as x, we get

$$(1+x)^{\frac{1}{2}} = \sum_{k=0}^{\infty} \binom{\frac{1}{2}}{k} x^k$$

$$= 1 + \sum_{k=1}^{\infty} \frac{1}{k!} \frac{1}{2} \left(\frac{1}{2} - 1 \right) \left(\frac{1}{2} - 2 \right) \cdots \left(\frac{1}{2} - k + 1 \right) x^k$$

$$= 1 + \sum_{k=1}^{\infty} (-1)^{k-1} \frac{1 \cdot 3 \cdot 5 \cdots (2k-3)}{2 \cdot 4 \cdot 6 \cdots 2k} x^k$$

$$= 1 + \sum_{k=1}^{\infty} \frac{(-1)^{k-1}(2k)!}{(2k-1)(k!)^2 2^{2k}} x^k$$

$$= 1 + \sum_{k=1}^{\infty} \frac{(-1)^{k-1}}{2^{2k}(2k-1)} \binom{2k}{k} x^k$$

$$= 1 + \sum_{k=1}^{\infty} \frac{(-1)^{k-1}}{2^{2k-1}} \frac{1}{k} \binom{2k-2}{k-1} x^k$$

which converges if $|x| < 1$. The first few terms of this series are

$$(1+x)^{\frac{1}{2}} = 1 + \frac{1}{2}x - \frac{1}{8}x^2 + \frac{1}{16}x^3 - \frac{5}{128}x^4 + \cdots,$$

For $r = -1$, we get the familiar geometric series

$$\frac{1}{1+x} = 1 - x + x^2 - x^3 + \cdots + (-1)^k x^k + \cdots,$$

which converges if $|x| < 1$.

We also stated earlier that the number of injections between a set with m elements and a set with n elements, where $m \le n$, is given by $n!/(n-m)!$ and we now prove it.

Proposition 4.6. *The number of injections between a set A with m elements and a set B with n elements, where $m \le n$, is given by $n!/(n-m)! = n(n-1) \cdots (n-m+1)$.*

Proof. We proceed by induction on $m \le n$. If $m = 0$, then $A = \emptyset$ and there is only one injection, namely the empty function from $\emptyset$ to B. Because

$$\frac{n!}{(n-0)!} = \frac{n!}{n!} = 1,$$

the base case holds.

Assume the induction hypothesis holds for m and consider a set A with $m+1$ elements, where $m+1 \le n$. Pick any element $a \in A$ and let $A' = A - \{a\}$, a set with m elements. Any injection $f: A \to B$ assigns some element $f(a) \in B$ to a and then $f \upharpoonright A'$ is an injection from A' to $B' = B - \{f(a)\}$, a set with $n-1$ elements. By the induction hypothesis, there are

$$\frac{(n-1)!}{(n-1-m)!}$$

injections from A' to B'. There are n ways of picking $f(a)$ in B, therefore the number of injections from A to B is

$$n\frac{(n-1)!}{(n-1-m)!} = \frac{n!}{(n-(m+1))!},$$

establishing the induction hypothesis. □

Counting the number of surjections between a set with n elements and a set with p elements, where $n \geq p$, is harder. We state the following formula without giving a proof right now. Finding a proof of this formula is an interesting exercise. We give a quick proof using the principle of inclusion–exclusion in Section 4.4.

Proposition 4.7. *The number of surjections S_{np} between a set A with n elements and a set B with p elements, where $n \geq p$, is given by*

$$S_{np} = p^n - \binom{p}{1}(p-1)^n + \binom{p}{2}(p-2)^n + \cdots + (-1)^{p-1}\binom{p}{p-1}.$$

Remarks:

1. It can be shown that S_{np} satisfies the following peculiar version of Pascal's recurrence formula,

$$S_{np} = p(S_{n-1\,p} + S_{n-1\,p-1}), \qquad p \geq 2,$$

   and, of course, $S_{n1} = 1$ and $S_{np} = 0$ if $p > n$. Using this recurrence formula and the fact that $S_{nn} = n!$, simple expressions can be obtained for $S_{n+1\,n}$ and $S_{n+2\,n}$.

2. The numbers S_{np} are intimately related to the so-called *Stirling numbers of the second kind*, denoted $\left\{{n \atop p}\right\}$, $S(n,p)$, or $S_n^{(p)}$, which count the number of partitions of a set of n elements into p nonempty pairwise disjoint blocks (see Section 5.6). In fact,

$$S_{np} = p!\left\{{n \atop p}\right\}.$$

The Stirling numbers $\left\{{n \atop p}\right\}$ satisfy a recurrence equation that is another variant of Pascal's recurrence formula:

$$\left\{{n \atop 1}\right\} = 1$$

$$\left\{{n \atop n}\right\} = 1$$

$$\left\{{n \atop p}\right\} = \left\{{n-1 \atop p-1}\right\} + p\left\{{n-1 \atop p}\right\} \qquad (1 \leq p < n).$$

The total numbers of partitions of a set with $n \geq 1$ elements is given by the *Bell number*,

$$b_n = \sum_{p=1}^{n} \left\{ {n \atop p} \right\}.$$

There is a recurrence formula for the Bell numbers but it is complicated and not very useful because the formula for b_{n+1} involves all the previous Bell numbers.

Fig. 4.3 Eric Temple Bell, 1883–1960 (left) and Donald Knuth, 1938– (right)

A good reference for all these special numbers is Graham, Knuth, and Patashnik [4], Chapter 6.

The binomial coefficients can be generalized as follows. For all $n, m, k_1, \ldots, k_m \in \mathbb{N}$, with $k_1 + \cdots + k_m = n$ and $m \geq 2$, we have the *multinomial coefficient*,

$$\binom{n}{k_1 \cdots k_m},$$

which counts the number of ways of splitting a set of n elements into an ordered sequence of m disjoint subsets, the ith subset having $k_i \geq 0$ elements. Such sequences of disjoint subsets whose union is $\{1, \ldots, n\}$ itself are sometimes called *ordered partitions*. Beware that some of the subsets in an ordered partition may be empty, so we feel that the terminology "partition" is confusing because as we show in Section 5.6, the subsets that form a partition are never empty. Note that when $m = 2$, the number of ways of splitting a set of n elements into two disjoint subsets where the first subset has k_1 elements and the second subset has $k_2 = n - k_1$ elements is precisely the number of subsets of size k_1 of a set of n elements; that is,

$$\binom{n}{k_1 \, k_2} = \binom{n}{k_1}.$$

Observe that the order of the m subsets matters. For example, for $n = 5$, $m = 4$, $k_1 = 2$, and $k_2 = k_3 = k_4 = 1$, the sequences of subsets $(\{1,2\}, \{3\}, \{4\}, \{5\})$, $(\{1,2\}, \{3\}, \{5\}, \{4\})$, $(\{1,2\}, \{5\}, \{3\}, \{4\})$, $(\{1,2\}, \{4\}, \{3\}, \{5\})$, $(\{1,2\}, \{4\}, \{5\}, \{3\})$, $(\{1,2\}, \{5\}, \{4\}, \{3\})$ are all different and they correspond to the same partition, $\{\{1,2\}, \{3\}, \{4\}, \{5\}\}$.

Proposition 4.8. *For all $n, m, k_1, \ldots, k_m \in \mathbb{N}$, with $k_1 + \cdots + k_m = n$ and $m \geq 2$, we have*

$$\binom{n}{k_1 \cdots k_m} = \frac{n!}{k_1! \cdots k_m!}.$$

Proof. There are $\binom{n}{k_1}$ ways of forming a subset of k_1 elements from the set of n elements; there are $\binom{n-k_1}{k_2}$ ways of forming a subset of k_2 elements from the remaining $n - k_1$ elements; there are $\binom{n-k_1-k_2}{k_3}$ ways of forming a subset of k_3 elements from the remaining $n - k_1 - k_2$ elements and so on; finally, there are $\binom{n-k_1-\cdots-k_{m-2}}{k_{m-1}}$ ways of forming a subset of k_{m-1} elements from the remaining $n - k_1 - \cdots - k_{m-2}$ elements and there remains a set of $n - k_1 - \cdots - k_{m-1} = k_m$ elements. This shows that

$$\binom{n}{k_1 \cdots k_m} = \binom{n}{k_1} \binom{n-k_1}{k_2} \cdots \binom{n-k_1-\cdots-k_{m-2}}{k_{m-1}}.$$

But then, using the fact that $k_m = n - k_1 - \cdots - k_{m-1}$, we get

$$\binom{n}{k_1 \cdots k_m} = \frac{n!}{k_1!(n-k_1)!} \frac{(n-k_1)!}{k_2!(n-k_1-k_2)!} \cdots \frac{(n-k_1-\cdots-k_{m-2})!}{k_{m-1}!(n-k_1-\cdots-k_{m-1})!}$$

$$= \frac{n!}{k_1! \cdots k_m!},$$

as claimed. $\square$

As in the binomial case, it is convenient to set

$$\binom{n}{k_1 \cdots k_m} = 0$$

if $k_i < 0$ or $k_i > n$, for any i, with $1 \leq i \leq m$. Then, Proposition 4.3 is generalized as follows.

Proposition 4.9. *For all $n, m, k_1, \ldots, k_m \in \mathbb{N}$, with $k_1 + \cdots + k_m = n$, $n \geq 1$ and $m \geq 2$, we have*

$$\binom{n}{k_1 \cdots k_m} = \sum_{i=1}^{m} \binom{n-1}{k_1 \cdots (k_i - 1) \cdots k_m}.$$

Proof. Note that we have $k_i - 1 = -1$ when $k_i = 0$. First, observe that

$$k_i \binom{n}{k_1 \cdots k_m} = n \binom{n-1}{k_1 \cdots (k_i - 1) \cdots k_m}$$

even if $k_i = 0$. This is because if $k_i \geq 1$, then

$$\binom{n}{k_1 \cdots k_m} = \frac{n!}{k_1! \cdots k_m!} = \frac{n}{k_i} \frac{(n-1)!}{k_1! \cdots (k_i-1)! \cdots k_m!} = \frac{n}{k_i} \binom{n-1}{k_1 \cdots (k_i - 1) \cdots k_m},$$

and so,

$$k_i \binom{n}{k_1 \cdots k_m} = n \binom{n-1}{k_1 \cdots (k_i - 1) \cdots k_m}.$$

With our convention that $\left(\begin{smallmatrix} n-1 \\ k_1 \cdots -1 \cdots k_m \end{smallmatrix} \right) = 0$, the above identity also holds when $k_i = 0$. Then, we have

$$\sum_{i=1}^{m} \binom{n-1}{k_1 \cdots (k_i - 1) \cdots k_m} = \left(\frac{k_1}{n} + \cdots + \frac{k_m}{n} \right) \binom{n}{k_1 \cdots k_m}$$
$$= \binom{n}{k_1 \cdots k_m},$$

because $k_1 + \cdots + k_m = n$. $\square$

Remark: Proposition 4.9 shows that Pascal's triangle generalizes to "higher dimensions," that is, to $m \geq 3$. Indeed, it is possible to give a geometric interpretation of Proposition 4.9 in which the multinomial coefficients corresponding to those $k_1, \ldots, k_m$ with $k_1 + \cdots + k_m = n$ lie on the hyperplane of equation $x_1 + \cdots + x_m = n$ in $\mathbb{R}^m$, and all the multinomial coefficients for which $n \leq N$, for any fixed N, lie in a generalized tetrahedron called a *simplex*. When $m = 3$, the multinomial coefficients for which $n \leq N$ lie in a tetrahedron whose faces are the planes of equations, $x = 0$; $y = 0$; $z = 0$; and $x + y + z = N$.

We also have the following generalization of Proposition 4.5.

Proposition 4.10. *(Multinomial Formula) For all* $n, m \in \mathbb{N}$ *with* $m \geq 2$, *for all pairwise commuting variables* $a_1, \ldots, a_m$, *we have*

$$(a_1 + \cdots + a_m)^n = \sum_{\substack{k_1, \ldots, k_m \geq 0 \\ k_1 + \cdots + k_m = n}} \binom{n}{k_1 \cdots k_m} a_1^{k_1} \cdots a_m^{k_m}.$$

Proof. We proceed by induction on n and use Proposition 4.9. The case $n = 0$ is trivially true.

Assume the induction hypothesis holds for $n \geq 0$, then we have

$$(a_1 + \cdots + a_m)^{n+1} = (a_1 + \cdots + a_m)^n (a_1 + \cdots + a_m)$$
$$= \left(\sum_{\substack{k_1, \ldots, k_m \\ k_1 + \cdots + k}} \binom{n}{k_1 \cdots k_m} a_1^{k_1} \cdots a_m^{k_m} \right) (a_1 + \cdots + a_m)$$
$$= \sum_{i=1}^{m} \sum_{\substack{k_1, \ldots, k \\ k_1 + \cdots + k}} \binom{n}{k_1 \cdots k_i \cdots k_m} a_1^{k_1} \cdots a_i^{k_i + 1} \cdots a_m^{k_m}$$
$$= \sum_{i=1}^{m} \sum_{\substack{k_1, \ldots, k_m \\ k_1 + \cdots + k}} \binom{n}{k_1 \cdots (k_i - 1) \cdots k_m} a_1^{k_1} \cdots a_i^{k_i} \cdots a_m^{k_m}.$$

We seem to hit a snag, namely, that k_i but recall that

$$\binom{n}{k_1 \cdots - 1 \cdots k_m} = 0,$$

so we have

$$(a_1 + \cdots + a_m)^{n+1} = \sum_{i=1}^{m} \sum_{\substack{k_1,\ldots,k_m \geq 0, k_i \geq 1 \\ k_1 + \cdots + k_m = n+1}} \binom{n}{k_1 \cdots (k_i - 1) \cdots k_m} a_1^{k_1} \cdots a_i^{k_i} \cdots a_m^{k_m}$$

$$= \sum_{i=1}^{m} \sum_{\substack{k_1,\ldots,k_m \geq 0, \\ k_1 + \cdots + k_m = n+1}} \binom{n}{k_1 \cdots (k_i - 1) \cdots k_m} a_1^{k_1} \cdots a_i^{k_i} \cdots a_m^{k_m}$$

$$= \sum_{\substack{k_1,\ldots,k_m \geq 0, \\ k_1 + \cdots + k_m = n+1}} \left(\sum_{i=1}^{m} \binom{n}{k_1 \cdots (k_i - 1) \cdots k_m} \right) a_1^{k_1} \cdots a_i^{k_i} \cdots a_m^{k_m}$$

$$= \sum_{\substack{k_1,\ldots,k_m \geq 0, \\ k_1 + \cdots + k_m = n+1}} \binom{n+1}{k_1 \cdots k_i \cdots k_m} a_1^{k_1} \cdots a_i^{k_i} \cdots a_m^{k_m},$$

where we used Proposition 4.9 to justify the last equation. Therefore, the induction step is proved and so is our proposition. $\Box$

How many terms occur on the right-hand side of the multinomial formula? After a moment of reflection, we see that this is the number of finite multisets of size n whose elements are drawn from a set of m elements, which is also equal to the number of m-tuples, $k_1, \ldots, k_m$, with $k_i \in \mathbb{N}$ and

$$k_1 + \cdots + k_m = n.$$

The following proposition is left an exercise.

Proposition 4.11. *The number of finite multisets of size $n \geq 0$ whose elements come from a set of size $m \geq 1$ is*

$$\binom{m+n-1}{n}.$$

4.3 Some Properties of the Binomial Coefficients

The binomial coefficients satisfy many remarkable identities.

If one looks at the Pascal triangle, it is easy to figure out what are the sums of the elements in any given row. It is also easy to figure out what are the sums of $n - m + 1$ consecutive elements in any given column (starting from the top and with $0 \leq m \leq n$).

What about the sums of elements on the diagonals? Again, it is easy to determine what these sums are. Here are the answers, beginning with the sums of the elements in a column.

(a) Sum of the first $n - m + 1$ elements in column m ($0 \le m \le n$).

For example, if we consider the sum of the first five (nonzero) elements in column $m = 3$ (so, $n = 7$), we find that

$$1 + 4 + 10 + 20 + 35 = 70,$$

where 70 is the entry on the next row and the next column. Thus, we conjecture that

$$\binom{m}{m} + \binom{m+1}{m} + \cdots + \binom{n-1}{m} + \binom{n}{m} = \binom{n+1}{m+1},$$

which is easily proved by induction.

$$n \quad \binom{n}{0} \ \binom{n}{1} \ \binom{n}{2} \ \binom{n}{3} \ \binom{n}{4} \ \binom{n}{5} \ \binom{n}{6} \ \binom{n}{7} \ \binom{n}{8} \ \cdots$$

```
0  1
1  1   1
2  1   2   1
3  1   3   3    1
4  1   4   6    4    1
5  1   5   10   10   5    1
6  1   6   15   20   15   6    1
7  1   7   21   35   35   21   7    1
8  1   8   28   56   70   56   28   8    1
   :   :   :    :    :    :    :    :    :    :
```

The above formula can be written concisely as

$$\sum_{k=m}^{n} \binom{k}{m} = \binom{n+1}{m+1},$$

or even as

$$\sum_{k=0}^{n} \binom{k}{m} = \binom{n+1}{m+1},$$

because $\binom{k}{m} = 0$ when $k < m$. It is often called the *upper summation formula* inasmuch as it involves a sum over an index k, appearing in the upper position of the binomial coefficient $\binom{k}{m}$.

(b) Sum of the elements in row n.

For example, if we consider the sum of the elements in row $n = 6$, we find that

$$1 + 6 + 15 + 20 + 15 + 6 + 1 = 64 = 2^6.$$

$$n \quad \binom{n}{0} \quad \binom{n}{1} \quad \binom{n}{2} \quad \binom{n}{3} \quad \binom{n}{4} \quad \binom{n}{5} \quad \binom{n}{6} \quad \binom{n}{7} \quad \binom{n}{8} \quad \cdots$$

```
0  1
1  1   1
2  1   2    1
3  1   3    3    1
4  1   4    6    4    1
5  1   5   10   10    5    1
6  1   6   15   20   15    6    1
7  1   7   21   35   35   21    7    1
8  1   8   28   56   70   56   28    8    1
   ⋮   ⋮    ⋮    ⋮    ⋮    ⋮    ⋮    ⋮    ⋮   ⋮   ⋮
```

Thus, we conjecture that

$$\binom{n}{0} + \binom{n}{1} + \cdots + \binom{n}{n-1} + \binom{n}{n} = 2^n.$$

This is easily proved by induction by setting $a = b = 1$ in the binomial formula for $(a+b)^n$.

Unlike the columns for which there is a formula for the partial sums, there is no closed-form formula for the partial sums of the rows. However, there is a closed-form formula for partial alternating sums of rows. Indeed, it is easily shown by induction that

$$\sum_{k=0}^{m} (-1)^k \binom{n}{k} = (-1)^m \binom{n-1}{m},$$

if $0 \le m \le n$. For example,

$$1 - 7 + 21 - 35 = -20.$$

Also, for $m = n$, we get

$$\sum_{k=0}^{n} (-1)^k \binom{n}{k} = 0.$$

(c) Sum of the first $n+1$ elements on the descending diagonal starting from row m.

For example, if we consider the sum of the first five elements starting from row $m = 3$ (so, $n = 4$), we find that

$$1 + 4 + 10 + 20 + 35 = 70,$$

the elements on the next row below the last element, 35.

n $\binom{n}{0}$ $\binom{n}{1}$ $\binom{n}{2}$ $\binom{n}{3}$ $\binom{n}{4}$ $\binom{n}{5}$ $\binom{n}{6}$ $\binom{n}{7}$ $\binom{n}{8}$ $\cdots$

n									
0	1								
1	1	1							
2	1	2	1						
3	1	3	3	1					
4	1	4	6	4	1				
5	1	5	10	10	5	1			
6	1	6	15	20	15	6	1		
7	1	7	21	35	35	21	7	1	
8	1	8	28	56	70	56	28	8	1
$\vdots$	$\vdots$	$\vdots$	$\vdots$	$\vdots$	$\vdots$	$\vdots$	$\vdots$	$\vdots$	$\vdots$

Thus, we conjecture that

$$\binom{m}{0} + \binom{m+1}{1} + \cdots + \binom{m+n}{n} = \binom{m+n+1}{n},$$

which is easily shown by induction. The above formula can be written concisely as

$$\sum_{k=0}^{n} \binom{m+k}{k} = \binom{m+n+1}{n}.$$

It is often called the *parallel summation formula* because it involves a sum over an index k appearing both in the upper and in the lower position of the binomial coefficient $\binom{m+k}{k}$.

(d) Sum of the elements on the ascending diagonal starting from row n.

n F_{n+1} $\binom{n}{0}$ $\binom{n}{1}$ $\binom{n}{2}$ $\binom{n}{3}$ $\binom{n}{4}$ $\binom{n}{5}$ $\binom{n}{6}$ $\binom{n}{7}$ $\binom{n}{8}$ $\cdots$

n	F_{n+1}									
0	1	1								
1	1	1	1							
2	2	1	2	1						
3	3	1	3	3	1					
4	5	1	4	6	4	1				
5	8	1	5	10	10	5	1			
6	13	1	6	15	20	15	6	1		
7	21	1	7	21	35	35	21	7	1	
8	34	1	8	28	56	70	56	28	8	1
$\vdots$	$\vdots$	$\vdots$	$\vdots$	$\vdots$	$\vdots$	$\vdots$	$\vdots$	$\vdots$	$\vdots$	$\vdots$

For example, the sum of the numbers on the diagonal starting on row 6 (in green), row 7 (in blue) and row 8 (in red) are:

$$1 + 6 + 5 + 1 = 13$$
$$4 + 10 + 6 + 1 = 21$$
$$1 + 10 + 15 + 7 + 1 = 34.$$

We recognize the Fibonacci numbers F_7, F_8, and F_9; what a nice surprise. Recall that $F_0 = 0$, $F_1 = 1$, and

$$F_{n+2} = F_{n+1} + F_n.$$

Thus, we conjecture that

$$F_{n+1} = \binom{n}{0} + \binom{n-1}{1} + \binom{n-2}{2} + \cdots + \binom{0}{n}.$$

The above formula can indeed be proved by induction, but we have to distinguish the two cases where n is even or odd.

We now list a few more formulae that are often used in the manipulations of binomial coefficients. They are among the "top ten binomial coefficient identities" listed in Graham, Knuth, and Patashnik [4]; see Chapter 5.

(e) The equation

$$\binom{n}{i}\binom{n-i}{k-i} = \binom{k}{i}\binom{n}{k},$$

holds for all n, i, k, with $0 \leq i \leq k \leq n$.

This is because we find that after a few calculations,

$$\binom{n}{i}\binom{n-i}{k-i} = \frac{n!}{i!(k-i)!(n-k)!} = \binom{k}{i}\binom{n}{k}.$$

Observe that the expression in the middle is really the trinomial coefficient

$$\binom{n}{i \; k-i \; n-k}.$$

For this reason, the equation (e) is often called *trinomial revision*.

For $i = 1$, we get

$$n\binom{n-1}{k-1} = k\binom{n}{k}.$$

So, if $k \neq 0$, we get the equation

$$\binom{n}{k} = \frac{n}{k}\binom{n-1}{k-1}, \qquad k \neq 0.$$

This equation is often called the *absorption identity*.

(f) The equation

$$\binom{m+p}{n} = \sum_{k=0}^{m}\binom{m}{k}\binom{p}{n-k}$$

holds for $m, n, p \geq 0$ such that $m + p \geq n$. This equation is usually known as *Vandermonde convolution*.

One way to prove this equation is to observe that $\binom{m+p}{n}$ is the coefficient of $a^{m+p-n} b^n$ in $(a + b)^{m+p} = (a + b)^m (a + b)^p$; a detailed proof is left as an exercise (see Problem 4.5). By making the change of variables $n = r + s$ and $k = r + i$, we get another version of Vandermonde convolution, namely:

$$\binom{m + p}{r + s} = \sum_{i = -r}^{s} \binom{m}{r + i} \binom{p}{s - i}$$

for $m, r, s, p \geq 0$ such that $m + p \geq r + s$.

An interesting special case of Vandermonde convolution arises when $m = p = n$. In this case, we get the equation

$$\binom{2n}{n} = \sum_{k=0}^{n} \binom{n}{k} \binom{n}{n - k}.$$

However, $\binom{n}{k} = \binom{n}{n-k}$, so we get

$$\sum_{k=0}^{n} \binom{n}{k}^2 = \binom{2n}{n},$$

that is, the sum of the squares of the entries on row n of the Pascal triangle is the middle element on row $2n$.

A summary of the top nine binomial coefficient identities is given in Table 4.2.

Remark: Going back to the generalized binomial coefficients $\binom{r}{k}$, where r is a real number, possibly negative, the following formula is easily shown.

$$\binom{r}{k} = (-1)^k \binom{k - r - 1}{k},$$

where $r \in \mathbb{R}$ and $k \in \mathbb{Z}$. When $k < 0$, both sides are equal to 0 and if $k = 0$ then both sides are equal to zero. If $r < 0$ and $k \geq 1$ then $k - r - 1 > 0$, so the formula shows how a binomial coefficient with negative upper index can be expessed as a binomial coefficient with positive index. For this reason, this formula is known as *negating the upper index*.

Next, we would like to better understand the growth pattern of the binomial coefficients. Looking at the Pascal triangle, it is clear that when $n = 2m$ is even, the central element $\binom{2m}{m}$ is the largest element on row $2m$ and when $n = 2m + 1$ is odd, the two central elements $\binom{2m+1}{m} = \binom{2m+1}{m+1}$ are the largest elements on row $2m + 1$. Furthermore, $\binom{n}{k}$ is strictly increasing until it reaches its maximal value and then it is strictly decreasing (with two equal maximum values when n is odd).

The above facts are easy to prove by considering the ratio

$$\binom{n}{k} = \frac{n!}{k!(n-k)!}, \qquad 0 \le k \le n \qquad \textit{factorial expansion}$$

$$\binom{n}{k} = \binom{n}{n-k}, \qquad 0 \le k \le n \qquad \textit{symmetry}$$

$$\binom{n}{k} = \frac{n}{k}\binom{n-1}{k-1}, \qquad k \ne 0 \qquad \textit{absorption}$$

$$\binom{n}{k} = \binom{n-1}{k} + \binom{n-1}{k-1}, \qquad 0 \le k \le n \qquad \textit{addition/induction}$$

$$\binom{n}{i}\binom{n-i}{k-i} = \binom{k}{i}\binom{n}{k}, \qquad 0 \le i \le k \le n \qquad \textit{trinomial revision}$$

$$(a+b)^n = \sum_{k=0}^{n}\binom{n}{k}a^{n-k}b^{k}, \qquad n \ge 0 \qquad \textit{binomial formula}$$

$$\sum_{k=0}^{n}\binom{m+k}{k} = \binom{m+n+1}{n}, \qquad m,n \ge 0 \qquad \textit{parallel summation}$$

$$\sum_{k=0}^{n}\binom{k}{m} = \binom{n+1}{m+1}, \qquad 0 \le m \le n \qquad \textit{upper summation}$$

$$\binom{m+p}{n} = \sum_{k=0}^{m}\binom{m}{k}\binom{p}{n-k} \qquad \begin{array}{c} m+p \ge n \\ m,n,p \ge 0 \end{array} \qquad \textit{Vandermonde convolution}$$

Table 4.2 Summary of Binomial Coefficient Identities

$$\binom{n}{k} \bigg/ \binom{n}{k+1} = \frac{n!}{k!(n-k)!}\frac{(k+1)!(n-k-1)!}{n!} = \frac{k+1}{n-k},$$

where $0 \le k \le n-1$. Because

$$\frac{k+1}{n-k} = \frac{2k-(n-1)}{n-k} + 1,$$

we see that if $n = 2m$, then

$$\binom{2m}{k} < \binom{2m}{k+1} \quad \text{if } k < m$$

and if $n = 2m+1$, then

$$\binom{2m+1}{k} < \binom{2m+1}{k+1} \quad \text{if } k < m.$$

By symmetry,

$$\binom{2m}{k} > \binom{2m}{k+1} \text{ if } k > m$$

and

$$\binom{2m+1}{k} > \binom{2m+1}{k+1} \text{ if } k > m+1.$$

It would be nice to have an estimate of how large is the maximum value of the largest binomial coefficient $\binom{n}{\lfloor n/2 \rfloor}$. The sum of the elements on row n is 2^n and there are $n+1$ elements on row n, therefore some rough bounds are

$$\frac{2^n}{n+1} \leq \binom{n}{\lfloor n/2 \rfloor} < 2^n$$

for all $n \geq 1$. Thus, we see that the middle element on row n grows very fast (exponentially). We can get a sharper estimate using Stirling's formula (see Section 4.1). We give such an estimate when $n = 2m$ is even, the case where n is odd being similar (see Problem 4.13). We have

$$\binom{2m}{m} = \frac{(2m)!}{(m!)^2},$$

and because by Stirling's formula,

$$n! \sim \sqrt{2\pi n} \left(\frac{n}{e}\right)^n,$$

we get

$$\binom{2m}{m} \sim \frac{2^{2m}}{\sqrt{\pi m}}.$$

The next question is to figure out how quickly $\binom{n}{k}$ drops from its maximum value, $\binom{n}{\lfloor n/2 \rfloor}$. Let us consider the case where $n = 2m$ is even, the case when n is odd being similar and left as an exercise (see Problem 4.14). We would like to estimate the ratio

$$\binom{2m}{m-t} \Big/ \binom{2m}{m},$$

where $0 \leq t \leq m$. Actually, it is more convenient to deal with the inverse ratio,

$$r(t) = \binom{2m}{m} \Big/ \binom{2m}{m-t} = \frac{(2m)!}{(m!)^2} \Big/ \frac{(2m)!}{(m-t)!(m+t)!} = \frac{(m-t)!(m+t)!}{(m!)^2}.$$

Observe that

$$r(t) = \frac{(m+t)(m+t-1)\cdots(m+1)}{m(m-1)\cdots(m-t+1)}.$$

The above expression is not easy to handle but if we take its (natural) logarithm, we can use basic inequalities about logarithms to get some bounds. We make use of the following proposition.

Proposition 4.12. *We have the inequalities*

$$1 - \frac{1}{x} \le \ln x \le x - 1,$$

for all $x \in \mathbb{R}$ with $x > 0$.

Proof. These inequalities are quite obvious if we plot the curves but a rigorous proof can be given using the power series expansion of the exponential function and the fact that $x \mapsto \log x$ is strictly increasing and that it is the inverse of the exponential. Recall that

$$e^x = \sum_{n=0}^{\infty} \frac{x^n}{n!},$$

for all $x \in \mathbb{R}$. First, we can prove that

$$x \le e^{x-1},$$

for all $x \in \mathbb{R}$. This is clear when $x < 0$ because $e^{x-1} > 0$ and if $x \ge 1$, then

$$e^{x-1} = 1 + x - 1 + \sum_{n=2}^{\infty} \frac{(x-1)^n}{n!} = x + C$$

with $C \ge 0$. When $0 \le x < 1$, we have $-1 \le x - 1 < 0$ and we still have

$$e^{x-1} = x + \sum_{n=2}^{\infty} \frac{(x-1)^n}{n!}.$$

In order to prove that the second term on the right-hand side is nonnegative, it suffices to prove that

$$\frac{(x-1)^{2n}}{(2n)!} + \frac{(x-1)^{2n+1}}{(2n+1)!} \ge 0,$$

for all $n \ge 1$, which amounts to proving that

$$\frac{(x-1)^{2n}}{(2n)!} \ge -\frac{(x-1)^{2n+1}}{(2n+1)!},$$

which (because $2n$ is even) is equivalent to

$$2n + 1 \ge 1 - x,$$

which holds, inasmuch as $0 \le x < 1$.

Now, because $x \le e^{x-1}$ for all $x \in \mathbb{R}$, taking logarithms, we get

$$\ln x \le x - 1,$$

for all $x > 0$ (recall that $\ln x$ is undefined if $x \leq 0$).

Next, if $x > 0$, applying the above formula to $1/x$, we get

$$\ln\left(\frac{1}{x}\right) \leq \frac{1}{x} - 1;$$

that is,

$$-\ln x \leq \frac{1}{x} - 1,$$

which yields

$$1 - \frac{1}{x} \leq \ln x,$$

as claimed. □

We are now ready to prove the following inequalities:

Proposition 4.13. *For every* $m \geq 0$ *and every* t, *with* $0 \leq t \leq m$, *we have the inequalities*

$$e^{-t^2/(m-t+1)} \leq \binom{2m}{m-t} \Big/ \binom{2m}{m} \leq e^{-t^2/(m+t)}.$$

This implies that

$$\binom{2m}{m-t} \Big/ \binom{2m}{m} \sim e^{-t^2/m},$$

for m *large and* $0 \leq t \leq m$.

Proof. Recall that

$$r(t) = \binom{2m}{m} \Big/ \binom{2m}{m-t} = \frac{(m+t)(m+t-1)\cdots(m+1)}{m(m-1)\cdots(m-t+1)}$$

and take logarithms. We get

$$\ln r(t) = \ln\left(\frac{m+t}{m}\right) + \ln\left(\frac{m+t-1}{m-1}\right) + \cdots + \ln\left(\frac{m+1}{m-t+1}\right)$$
$$= \ln\left(1 + \frac{t}{m}\right) + \ln\left(1 + \frac{t}{m-1}\right) + \cdots + \ln\left(1 + \frac{t}{m-t+1}\right).$$

By Proposition 4.12, we have $\ln(1+x) \leq x$ for $x > -1$, therefore we get

$$\ln r(t) \leq \frac{t}{m} + \frac{t}{m-1} + \cdots + \frac{t}{m-t+1}.$$

If we replace the denominators on the right-hand side by the smallest one, $m-t+1$, we get an upper bound on this sum, namely,

$$\ln r(t) \leq \frac{t^2}{m-t+1}.$$

Now, remember that $r(t)$ is the inverse of the ratio in which we are really interested. So, by exponentiating and then taking inverses, we get

$$e^{-t^2/(m-t+1)} \leq \binom{2m}{m-t} \Big/ \binom{2m}{m}.$$

Proposition 4.12 also says that $(x-1)/x \leq \ln(x)$ for $x > 0$, thus from

$$\ln r(t) = \ln\left(1 + \frac{t}{m}\right) + \ln\left(1 + \frac{t}{m-1}\right) + \cdots + \ln\left(1 + \frac{t}{m-t+1}\right),$$

we get

$$\frac{t}{m} \Big/ \frac{m+t}{m} + \frac{t}{m-1} \Big/ \frac{m+t-1}{m-1} + \cdots + \frac{t}{m-t+1} \Big/ \frac{m+1}{m-t+1} \leq \ln r(t);$$

that is,

$$\frac{t}{m+t} + \frac{t}{m+t-1} + \cdots + \frac{t}{m+1} \leq \ln r(t).$$

This time, if we replace the denominators on the left-hand side by the largest one, $m+t$, we get a lower bound, namely,

$$\frac{t^2}{m+t} \leq \ln r(t).$$

Again, if we exponentiate and take inverses, we get

$$\binom{2m}{m-t} \Big/ \binom{2m}{m} \leq e^{-t^2/(m+t)},$$

as claimed. Finally, because $m - t + 1 \leq m \leq m + t$, it is easy to see that

$$e^{-t^2/(m-t+1)} \leq e^{-t^2/m} \leq e^{-t^2/(m+t)},$$

so we deduce that

$$\binom{2m}{m-t} \Big/ \binom{2m}{m} \sim e^{-t^2/m},$$

for m large and $0 \leq t \leq m$, as claimed. $\square$

What is remarkable about Proposition 4.13 is that it shows that $\binom{2m}{m-t}$ varies according to the *Gaussian curve* (also known as *the bell curve*), $t \mapsto e^{-t^2/m}$, which is the probability density function of the *normal distribution* (or *Gaussian distribution*). If we make the change of variable $k = m - t$, we see that if $0 \leq k \leq 2m$, then

$$\binom{2m}{k} \sim e^{-(m-k)^2/m} \binom{2m}{m}.$$

If we plot this curve, we observe that it reaches its maximum for $k = m$ and that it decays very quickly as k varies away from m. It is an interesting exercise to plot a bar chart of the binomial coefficients and the above curve together, say for $m = 50$. One will find that the bell curve is an excellent fit.

Given some number $c > 1$, it sometimes desirable to find for which values of t does the inequality

$$\binom{2m}{m} \Big/ \binom{2m}{m-t} > c$$

hold. This question can be answered using Proposition 4.13.

Proposition 4.14. *For every constant $c > 1$ and every natural number $m \geq 0$, if $\sqrt{m \ln c} + \ln c \leq t \leq m$, then*

$$\binom{2m}{m} \Big/ \binom{2m}{m-t} > c$$

and if $0 \leq t \leq \sqrt{m \ln c} - \ln c \leq m$, then

$$\binom{2m}{m} \Big/ \binom{2m}{m-t} \leq c.$$

The proof uses the inequalities of Proposition 4.13 and is left as an exercise (see Problem 4.15). As an example, if $m = 1000$ and $c = 100$, we have

$$\binom{1000}{500} \Big/ \binom{1000}{500-(500-k)} > 100$$

or equivalently

$$\binom{1000}{k} \Big/ \binom{1000}{500} < \frac{1}{100}$$

when $500 - k \geq \sqrt{500 \ln 100} + \ln 100$, that is, when

$$k \leq 447.4.$$

It is also possible to give an upper on the partial sum

$$\binom{2m}{0} + \binom{2m}{1} + \cdots + \binom{2m}{k-1},$$

with $0 \leq k \leq m$ in terms of the ratio $c = \binom{2m}{k} \Big/ \binom{2m}{m}$. The following proposition is taken from Lovász, Pelikán, and Vesztergombi [5].

Proposition 4.15. *For any natural numbers m and k with $0 \leq k \leq m$, if we let $c = \binom{2m}{k} \Big/ \binom{2m}{m}$, then we have*

$$\binom{2m}{0} + \binom{2m}{1} + \cdots + \binom{2m}{k-1} < c\, 2^{2m-1}.$$

The proof of Proposition 4.15 is not hard; this is the proof of Lemma 3.8.2 in Lovász, Pelikán, and Vesztergombi [5]. This proposition implies an important result in (discrete) probability theory as explained in [5] (see Chapter 5).

Observe that 2^{2m} is the sum of all the entries on row $2m$. As an application, if $k \leq 447$, the sum of the first 447 numbers on row 1000 of the Pascal triangle makes up less than 0.5% of the total sum and similarly for the last 447 entries. Thus, the middle 107 entries account for 99% of the total sum.

4.4 The Principle of Inclusion–Exclusion, Sylvester's Formula, The Sieve Formula

We close this chapter with the proof of a powerful formula for determining the cardinality of the union of a finite number of (finite) sets in terms of the cardinalities of the various intersections of these sets. This identity variously attributed to Nicholas Bernoulli, de Moivre, Sylvester, and Poincaré, has many applications to counting problems and to probability theory. We begin with the "baby case" of two finite sets.

Fig. 4.4 Abraham de Moivre, 1667–1754 (left) and Henri Poincaré, 1854–1912 (right)

Proposition 4.16. *Given any two finite sets A and B, we have*

$$|A \cup B| = |A| + |B| - |A \cap B|.$$

Proof. This formula is intuitively obvious because if some element $a \in A \cup B$ belongs to both A and B then it is counted twice in $|A| + |B|$ and so we need to subtract its contribution to $A \cap B$. Now,

$$A \cup B = (A - (A \cap B)) \cup (A \cap B) \cup (B - (A \cap B)),$$

where the three sets on the right-hand side are pairwise disjoint. If we let $a = |A|$, $b = |B|$, and $c = |A \cap B|$, then it is clear that

$$|A - (A \cap B)| = a - c$$
$$|B - (A \cap B)| = b - c,$$

so we get

$$\begin{aligned}|A \cup B| &= |A - (A \cap B)| + |A \cap B| + |B - (A \cap B)| \\ &= a - c + c + b - c = a + b - c \\ &= |A| + |B| - |A \cap B|,\end{aligned}$$

as desired. One can also give a proof by induction on $n = |A \cup B|$. □

We generalize the formula of Proposition 4.16 to any finite collection of finite sets, $A_1, \ldots, A_n$. A moment of reflection shows that when $n = 3$, we have

$$|A \cup B \cup C| = |A| + |B| + |C| - |A \cap B| - |A \cap C| - |B \cap C| + |A \cap B \cap C|.$$

One of the obstacles in generalizing the above formula to n sets is purely notational. We need a way of denoting arbitrary intersections of sets belonging to a family of sets indexed by $\{1, \ldots, n\}$. We can do this by using indices ranging over subsets of $\{1, \ldots, n\}$, as opposed to indices ranging over integers. So, for example, for any nonempty subset $I \subseteq \{1, \ldots, n\}$, the expression $\bigcap_{i \in I} A_i$ denotes the intersection of all the subsets whose index i belongs to I.

Theorem 4.1. *(Principle of Inclusion–Exclusion) For any finite sequence $A_1, \ldots,$ A_n, of $n \geq 2$ subsets of a finite set X, we have*

$$\left| \bigcup_{k=1}^{n} A_k \right| = \sum_{\substack{I \subseteq \{1,\ldots,n\} \\ I \neq \emptyset}} (-1)^{(|I|-1)} \left| \bigcap_{i \in I} A_i \right|.$$

Proof. We proceed by induction on $n \geq 2$. The base case, $n = 2$, is exactly Proposition 4.16. Let us now consider the induction step. We can write

$$\bigcup_{k=1}^{n+1} A_k = \left(\bigcup_{k=1}^{n} A_k \right) \cup \{A_{n+1}\}$$

and so, by Proposition 4.16, we have

$$\left| \bigcup_{k=1}^{n+1} A_k \right| = \left| \left(\bigcup_{k=1}^{n} A_k \right) \cup \{A_{n+1}\} \right|$$

$$= \left| \bigcup_{k=1}^{n} A_k \right| + |A_{n+1}| - \left| \left(\bigcup_{k=1}^{n} A_k \right) \cap \{A_{n+1}\} \right|.$$

We can apply the induction hypothesis to the first term and we get

$$\left| \bigcup_{k=1}^{n} A_k \right| = \sum_{\substack{J \subseteq \{1,\dots,n\} \\ J \neq \emptyset}} (-1)^{(|J|-1)} \left| \bigcap_{j \in J} A_j \right|.$$

Using distributivity of intersection over union, we have

$$\left(\bigcup_{k=1}^{n} A_k \right) \cap \{A_{n+1}\} = \bigcup_{k=1}^{n} (A_k \cap A_{n+1}).$$

Again, we can apply the induction hypothesis and obtain

$$-\left| \bigcup_{k=1}^{n} (A_k \cap A_{n+1}) \right| = -\sum_{\substack{J \subseteq \{1,\dots,n\} \\ J \neq \emptyset}} (-1)^{(|J|-1)} \left| \bigcap_{j \in J} (A_j \cap A_{n+1}) \right|$$

$$= \sum_{\substack{J \subseteq \{1,\dots,n\} \\ J \neq \emptyset}} (-1)^{|J|} \left| \bigcap_{j \in J \cup \{n+1\}} A_j \right|$$

$$= \sum_{\substack{J \subseteq \{1,\dots,n\} \\ J \neq \emptyset}} (-1)^{(|J \cup \{n+1\}| - 1)} \left| \bigcap_{j \in J \cup \{n+1\}} A_j \right|.$$

Putting all this together, we get

$$\left| \bigcup_{k=1}^{n+1} A_k \right| = \sum_{\substack{J \subseteq \{1,\dots,n\} \\ J \neq \emptyset}} (-1)^{(|J|-1)} \left| \bigcap_{j \in J} A_j \right| + |A_{n+1}|$$

$$+ \sum_{\substack{J \subseteq \{1,\dots,n\} \\ J \neq \emptyset}} (-1)^{(|J \cup \{n+1\}| - 1)} \left| \bigcap_{j \in J \cup \{n+1\}} A_j \right|$$

$$= \sum_{\substack{J \subseteq \{1,\dots,n+1\} \\ J \neq \emptyset, n+1 \notin J}} (-1)^{(|J|-1)} \left| \bigcap_{j \in J} A_j \right| + \sum_{\substack{J \subseteq \{1,\dots,n+1\} \\ n+1 \in J}} (-1)^{(|J|-1)} \left| \bigcap_{j \in J} A_j \right|$$

$$= \sum_{\substack{I \subseteq \{1,\dots,n+1\} \\ I \neq \emptyset}} (-1)^{(|I|-1)} \left| \bigcap_{i \in I} A_i \right|,$$

establishing the induction hypothesis and finishing the proof. $\quad\square$

As an application of the inclusion–exclusion principle, let us prove the formula for counting the number of surjections from $\{1,\ldots,n\}$ to $\{1,\ldots,p\}$, with $p \leq n$, given in Proposition 4.7.

Recall that the total number of functions from $\{1,\ldots,n\}$ to $\{1,\ldots,p\}$ is p^n. The trick is to count the number of functions that are *not* surjective. Any such function has the property that its image misses one element from $\{1,\ldots,p\}$. So, if we let

$$A_i = \{f \colon \{1,\ldots,n\} \to \{1,\ldots,p\} \mid i \notin \mathrm{Im}(f)\},$$

we need to count $|A_1 \cup \cdots \cup A_p|$. But, we can easily do this using the inclusion–exclusion principle. Indeed, for any nonempty subset I of $\{1,\ldots,p\}$, with $|I| = k$, the functions in $\bigcap_{i \in I} A_i$ are exactly the functions whose range misses I. But, these are exactly the functions from $\{1,\ldots,n\}$ to $\{1,\ldots,p\} - I$ and there are $(p-k)^n$ such functions. Thus,

$$\left| \bigcap_{i \in I} A_i \right| = (p-k)^n.$$

As there are $\binom{p}{k}$ subsets $I \subseteq \{1,\ldots,p\}$ with $|I| = k$, the contribution of all k-fold intersections to the inclusion–exclusion principle is

$$\binom{p}{k}(p-k)^n.$$

Note that $A_1 \cap \cdots \cap A_p = \emptyset$, because functions have a nonempty image. Therefore, the inclusion–exclusion principle yields

$$|A_1 \cup \cdots \cup A_p| = \sum_{k=1}^{p-1} (-1)^{k-1} \binom{p}{k}(p-k)^n,$$

and so, the number of surjections S_{np} is

$$S_{np} = p^n - |A_1 \cup \cdots \cup A_p| = p^n - \sum_{k=1}^{p-1} (-1)^{k-1} \binom{p}{k}(p-k)^n$$

$$= \sum_{k=0}^{p-1} (-1)^k \binom{p}{k}(p-k)^n$$

$$= p^n - \binom{p}{1}(p-1)^n + \binom{p}{2}(p-2)^n + \cdots + (-1)^{p-1}\binom{p}{p-1},$$

which is indeed the formula of Proposition 4.7.

Another amusing application of the inclusion–exclusion principle is the formula giving the number p_n of permutations of $\{1,\ldots,n\}$ that leave no element fixed (i.e., $f(i) \neq i$, for all $i \in \{1,\ldots,n\}$). Such permutations are often called *derangements*. We get

$$p_n = n! \left(1 - \frac{1}{1!} + \frac{1}{2!} + \cdots + \frac{(-1)^k}{k!} + \cdots + \frac{(-1)^n}{n!} \right)$$

$$= n! - \binom{n}{1}(n-1)! + \binom{n}{2}(n-2)! + \cdots + (-1)^n.$$

Remark: We know (using the series expansion for e^x in which we set $x = -1$) that

$$\frac{1}{e} = 1 - \frac{1}{1!} + \frac{1}{2!} + \cdots + \frac{(-1)^k}{k!} + \cdots .$$

Consequently, the factor of $n!$ in the above formula for p_n is the sum of the first $n+1$ terms of $1/e$ and so,

$$\lim_{n \to \infty} \frac{p_n}{n!} = \frac{1}{e}.$$

It turns out that the series for $1/e$ converges very rapidly, so $p_n \approx n!/e$. The ratio $p_n/n!$ has an interesting interpretation in terms of probabilities. Assume n persons go to a restaurant (or to the theatre, etc.) and that they all check their coats. Unfortunately, the clerk loses all the coat tags. Then $p_n/n!$ is the probability that nobody will get her or his own coat back. As we just explained, this probability is roughly $1/e \approx 1/3$, a surprisingly large number.

The inclusion–exclusion principle can be easily generalized in a useful way as follows. Given a finite set X, let m be any given function $m \colon X \to \mathbb{R}_+$ and for any nonempty subset $A \subseteq X$, set

$$m(A) = \sum_{a \in A} m(a),$$

with the convention that $m(\emptyset) = 0$ (recall that $\mathbb{R}_+ = \{x \in \mathbb{R} \mid x \geq 0\}$). For any $x \in X$, the number $m(x)$ is called the *weight* (or *measure*) of x and the quantity $m(A)$ is often called the *measure of the set* A. For example, if $m(x) = 1$ for all $x \in A$, then $m(A) = |A|$, the cardinality of A, which is the special case that we have been considering. For any two subsets $A, B \subseteq X$, it is obvious that

$$m(A \cup B) = m(A) + m(B)$$
$$m(X - A) = m(X) - m(A)$$
$$m(\overline{A \cup B}) = m(\overline{A} \cap \overline{B})$$
$$m(\overline{A \cap B}) = m(\overline{A} \cup \overline{B}),$$

where $\overline{A} = X - A$. Then, we have the following version of Theorem 4.1.

Theorem 4.2. *(Principle of Inclusion–Exclusion, Version 2) Given any measure function $m \colon X \to \mathbb{R}_+$, for any finite sequence $A_1, \ldots, A_n$, of $n \geq 2$ subsets of a finite set X, we have*

$$m\left(\bigcup_{k=1}^{n} A_k\right) = \sum_{\substack{I \subseteq \{1,\dots,n\} \\ I \neq \emptyset}} (-1)^{(|I|-1)} \, m\left(\bigcap_{i \in I} A_i\right).$$

Proof. The proof is obtained from the proof of Theorem 4.1 by changing everywhere any expression of the form $|B|$ to $m(B)$. □

A useful corollary of Theorem 4.2 often known as Sylvester's formula is the following.

Fig. 4.5 James Joseph Sylvester, 1814–1897

Theorem 4.3. *(Sylvester's Formula) Given any measure* $m\colon X \to \mathbb{R}_+$, *for any finite sequence* $A_1,\dots,A_n$ *of* $n \geq 2$ *subsets of a finite set* X, *the measure of the set of elements of* X *that do not belong to any of the sets* A_i *is given by*

$$m\left(\bigcap_{k=1}^{n} \overline{A}_k\right) = m(X) + \sum_{\substack{I \subseteq \{1,\dots,n\} \\ I \neq \emptyset}} (-1)^{|I|} \, m\left(\bigcap_{i \in I} A_i\right).$$

Proof. Observe that

$$\bigcap_{k=1}^{n} \overline{A}_k = X - \bigcup_{k=1}^{n} A_k.$$

Consequently, using Theorem 4.2, we get

$$m\left(\bigcap_{k=1}^{n}\overline{A}_k\right) = m\left(X - \bigcup_{k=1}^{n}A_k\right)$$

$$= m(X) - m\left(\bigcup_{k=1}^{n}A_k\right)$$

$$= m(X) - \sum_{\substack{I\subseteq\{1,\dots,n\}\\ I\neq\emptyset}}(-1)^{(|I|-1)}\,m\left(\bigcap_{i\in I}A_i\right)$$

$$= m(X) + \sum_{\substack{I\subseteq\{1,\dots,n\}\\ I\neq\emptyset}}(-1)^{|I|}\,m\left(\bigcap_{i\in I}A_i\right),$$

establishing Sylvester's formula. □

Note that if we use the convention that when the index set I is empty then

$$\bigcap_{i\in\emptyset}A_i = X,$$

hence the term $m(X)$ can be included in the above sum by removing the condition that $I\neq\emptyset$ and this version of Sylvester's formula is written:

$$m\left(\bigcap_{k=1}^{n}\overline{A}_k\right) = \sum_{I\subseteq\{1,\dots,n\}}(-1)^{|I|}\,m\left(\bigcap_{i\in I}A_i\right).$$

Sometimes, it is also convenient to regroup terms involving subsets I having the same cardinality, and another way to state Sylvester's formula is as follows.

$$m\left(\bigcap_{k=1}^{n}\overline{A}_k\right) = \sum_{k=0}^{n}(-1)^k\sum_{\substack{I\subseteq\{1,\dots,n\}\\ |I|=k}}m\left(\bigcap_{i\in I}A_i\right). \qquad \text{(Sylvester's Formula)}$$

Finally, Sylvester's formula can be generalized to a formula usually known as the "sieve formula."

Theorem 4.4. *(Sieve Formula) Given any measure* $m\colon X \to \mathbb{R}_+$ *for any finite sequence* $A_1,\dots,A_n$ *of* $n\geq 2$ *subsets of a finite set* X, *the measure of the set of elements of* X *that belong to exactly* p *of the sets* A_i $(0\leq p\leq n)$ *is given by*

$$T_n^p = \sum_{k=p}^{n}(-1)^{k-p}\binom{k}{p}\sum_{\substack{I\subseteq\{1,\dots,n\}\\ |I|=k}}m\left(\bigcap_{i\in I}A_i\right).$$

Proof. Observe that the set of elements of X that belong to exactly p of the sets A_i (with $0\leq p\leq n$) is given by the expression

$$\bigcup_{\substack{I \subseteq \{1,\ldots,n\} \\ |I|=p}} \left(\bigcap_{i \in I} A_i \cap \bigcap_{j \notin I} \overline{A}_j \right).$$

For any subset $I \subseteq \{1,\ldots,n\}$, if we apply Sylvester's formula to $X = \bigcap_{i \in I} A_i$ and to the subsets $A_j \cap \bigcap_{i \in I} A_i$ for which $j \notin I$ (i.e., $j \in \{1,\ldots,n\} - I$), we get

$$m \left(\bigcap_{i \in I} A_i \cap \bigcap_{j \notin I} \overline{A}_j \right) = \sum_{\substack{J \subseteq \{1,\ldots,n\} \\ I \subseteq J}} (-1)^{|J|-|I|} m \left(\bigcap_{j \in J} A_j \right).$$

Hence,

$$T_n^p = \sum_{\substack{I \subseteq \{1,\ldots,n\} \\ |I|=p}} m \left(\bigcap_{i \in I} A_i \cap \bigcap_{j \notin I} \overline{A}_j \right)$$

$$= \sum_{\substack{I \subseteq \{1,\ldots,n\} \\ |I|=p}} \sum_{\substack{J \subseteq \{1,\ldots,n\} \\ I \subseteq J}} (-1)^{|J|-|I|} m \left(\bigcap_{j \in J} A_j \right)$$

$$= \sum_{\substack{J \subseteq \{1,\ldots,n\} \\ |J| \geq p}} \sum_{\substack{I \subseteq J \\ |I|=p}} (-1)^{|J|-|I|} m \left(\bigcap_{j \in J} A_j \right)$$

$$= \sum_{k=p}^{n} (-1)^{k-p} \binom{k}{p} \sum_{\substack{J \subseteq \{1,\ldots,n\} \\ |J|=k}} m \left(\bigcap_{j \in J} A_j \right),$$

establishing the sieve formula. □

Observe that Sylvester's formula is the special case of the sieve formula for which $p = 0$. The inclusion–exclusion principle (and its relatives) plays an important role in combinatorics and probability theory as the reader may verify by consulting any text on combinatorics. A classical reference on combinatorics is Berge [1]; a more recent one is Cameron [2]; more advanced references are van Lint and Wilson [8] and Stanley [7]. Another great (but deceptively tough) reference covering discrete mathematics and including a lot of combinatorics is Graham, Knuth, and Patashnik [4]. Conway and Guy [3] is another beautiful book that presents many fascinating and intriguing geometric and combinatorial properties of numbers in a very un-tertaining manner. For readers interested in geometry with a combinatriol flavor, Matousek [6] is a delightful (but more advanced) reference.

We are now ready to study special kinds of relations: partial orders and equiva-lence relations.

4.5 Summary

This chapter provided a very brief and elementary introduction to combinatorics. To be more precise, we considered various counting problems, such as counting the number of permutations of a finite set, the number of functions from one set to another, the number of injections from one set to another, the number of surjections from one set to another, the number of subsets of size k in a finite set of size n and the number of partitions of a set of size n into p blocks. This led us to the binomial (and the multinomial) coefficients and various properties of these very special numbers. We also presented various formulae for determining the size of the union of a finite collection of sets in terms of various intersections of these sets. We discussed the principle of inclusion–exclusion (PIE), Sylvester's formula, and the sieve formula.

- We review the notion of a *permutation* and the *factorial function* $(n \mapsto n!)$.
- We show that a set of size n has $n!$ permutations.
- We show that if A has m elements and B has n elements, then B^A (the set of functions from A to B) has n^m elements.
- We state *Stirling's formula*, as an estimation of the factorial function.
- We defined the "big oh" notation, the "big Ω" notation, the "big Θ" notation, and the "little oh" notation.
- We give recurrence relations for computing the number of subsets of size k of a set of size n (the "Pascal recurrence relations"); these are the *binomial coefficients* $\binom{n}{k}$.
- We give an explicit formula for $\binom{n}{k}$ and we prove the *binomial formula* (expressing $(a+b)^n$ in terms of the monomials $a^{n-k}b^k$).
- We give a formula for the number of injections from a finite set into another finite set.
- We state a formula for the number of surjections S_{np} from a finite set of n elements onto another finite set of p elements.
- We relate the S_{np} to the *Stirling numbers of the second kind* $\left\{ {n \atop p} \right\}$ that count the number of partitions of a set of n elements into p disjoint blocks.
- We define the *bell numbers*, which count the number of partitions of a finite set.
- We define the *multinomial coefficients* $\binom{n}{k_1 \cdots k_m}$ and give an explicit formula for these numbers.
- We prove the *multinomial formula* (expressing $(a_1 + \cdots + a_m)^n$).
- We prove some useful identities about the binomial coefficients summarized in Table 4.2.
- We estimate the value of the central (and largest) binomial coefficient $\binom{2m}{m}$ on row $2m$.
- We give bounds for the ratio $\binom{2m}{m-t} \big/ \binom{2m}{m}$ and show that it is approximately $e^{-t^2/m}$.
- We prove the formula for the *principle of inclusion–exclusion*.
- We apply this formula to derive a formula for S_{np}.
- We define *derangements* as permutations that leave no element fixed and give a formula for counting them.

- We generalize slightly the inclusion–exclusion principle by allowing finite sets with *weights* (defining a *measure* on the set).
- We prove *Sylvester's formula*.
- We prove the *sieve formula*.

Problems

4.1. Let S_{np} be the number of surjections from the set $\{1,\dots,n\}$ onto the set $\{1,\dots,p\}$, where $1 \le p \le n$. Observe that $S_{n1} = 1$.

(a) Recall that $n!$ (factorial) is defined for all $n \in \mathbb{N}$ by $0! = 1$ and $(n+1)! = (n+1)n!$. Also recall that $\binom{n}{k}$ (n choose k) is defined for all $n \in \mathbb{N}$ and all $k \in \mathbb{Z}$ as follows.

$$\binom{n}{k} = 0, \text{ if } k \notin \{0,\dots,n\}$$

$$\binom{0}{0} = 1$$

$$\binom{n}{k} = \binom{n-1}{k} + \binom{n-1}{k-1}, \text{ if } n \ge 1.$$

Prove by induction on n that

$$\binom{n}{k} = \frac{n!}{k!(n-k)!}.$$

(b) Prove that

$$\sum_{k=0}^{n} \binom{n}{k} = 2^n \quad (n \ge 0) \quad \text{and} \quad \sum_{k=0}^{n} (-1)^k \binom{n}{k} = 0 \quad (n \ge 1).$$

Hint. Use the *binomial formula*. For all $a, b \in \mathbb{R}$ and all $n \ge 0$,

$$(a+b)^n = \sum_{k=0}^{n} \binom{n}{k} a^{n-k} b^k.$$

(c) Prove that

$$p^n = S_{np} + \binom{p}{1} S_{np-1} + \binom{p}{2} S_{np-2} + \dots + \binom{p}{p-1}.$$

(d) For all $p \ge 1$ and all i, k, with $0 \le i \le k \le p$, prove that

$$\binom{p}{i}\binom{p-i}{k-i} = \binom{k}{i}\binom{p}{k}.$$

Use the above to prove that

$$\binom{p}{0}\binom{p}{k} - \binom{p}{1}\binom{p-1}{k-1} + \cdots + (-1)^k \binom{p}{k}\binom{p-k}{0} = 0.$$

(e) Prove that

$$S_{np} = p^n - \binom{p}{1}(p-1)^n + \binom{p}{2}(p-2)^n + \cdots + (-1)^{p-1}\binom{p}{p-1}.$$

Hint. Write all p equations given by (c) for $1, 2, \ldots, p-1, p$, multiply both sides of the equation involving $(p-k)^n$ by $(-1)^k \binom{p}{k}$, add up both sides of these equations, and use (b) to simplify the sum on the right-hand side.

4.2. (a) Let S_{np} be the number of surjections from a set of n elements onto a set of p elements, with $1 \le p \le n$. Prove that

$$S_{np} = p(S_{n-1\,p-1} + S_{n-1\,p}).$$

Hint. Adapt the proof of Pascal's recurrence formula.
  (b) Prove that

$$S_{n+1\,n} = \frac{n(n+1)!}{2}$$

and

$$S_{n+2\,n} = \frac{n(3n+1)(n+2)!}{24}.$$

Hint. First, show that $S_{nn} = n!$.
  (c) Let P_{np} be the number of partitions of a set of n elements into p blocks (equivalence classes), with $1 \le p \le n$. Note that P_{np} is usually denoted by

$$\begin{Bmatrix} n \\ p \end{Bmatrix}, \quad S(n, p) \quad \text{or} \quad S_n^{(p)}.$$

Prove that

$$\begin{Bmatrix} n \\ 1 \end{Bmatrix} = 1$$

$$\begin{Bmatrix} n \\ n \end{Bmatrix} = 1$$

$$\begin{Bmatrix} n \\ p \end{Bmatrix} = \begin{Bmatrix} n-1 \\ p-1 \end{Bmatrix} + p\begin{Bmatrix} n-1 \\ p \end{Bmatrix} \quad (1 \le p < n).$$

Hint. Fix the the first of the n elements, say a_1. There are two kinds of partitions: those in which $\{a_1\}$ is a block and those in which the block containing a_1 has at least two elements.

Construct the array of $\left\{{n \atop p}\right\}$s for $n, p \in \{1, \ldots, 6\}$.

(d) Prove that

$$S_{np} = p! P_{np}.$$

Deduce from the above that

$$P_{np} = \frac{1}{p!} \left(p^n - \binom{p}{1}(p-1)^n + \binom{p}{2}(p-2)^n + \cdots + (-1)^{p-1}\binom{p}{p-1} \right).$$

4.3. The *Fibonacci numbers* F_n are defined recursively as follows.

$$F_0 = 0$$
$$F_1 = 1$$
$$F_{n+2} = F_{n+1} + F_n, \quad n \geq 0.$$

For example, $0, 1, 1, 2, 3, 5, 8, 13, 21, 34, 55, \ldots$ are the first 11 Fibonacci numbers. Prove that

$$F_{n+1} = \binom{n}{0} + \binom{n-1}{1} + \binom{n-2}{2} + \cdots + \binom{0}{n}.$$

Hint. Use complete induction. Also, consider the two cases, n even and n odd.

4.4. Given any natural number, $n \geq 1$, let p_n denote the number of permutations $f: \{1, \ldots, n\} \to \{1, \ldots, n\}$ that leave no element fixed, that is, such that $f(i) \neq i$, for all $i \in \{1, \ldots, n\}$. Such permutations are sometimes called *derangements*. Note that $p_1 = 0$ and set $p_0 = 1$.

(a) Prove that

$$n! = p_n + \binom{n}{1} p_{n-1} + \binom{n}{2} p_{n-2} + \cdots + \binom{n}{n}.$$

Hint. For every permutation $f: \{1, \ldots, n\} \to \{1, \ldots, n\}$, let

$$Fix(f) = \{i \in \{1, \ldots, n\} \mid f(i) = i\}$$

be the set of elements left fixed by f. Prove that there are p_{n-k} permutations associated with any fixed set $Fix(f)$ of cardinality k.

(b) Prove that

$$p_n = n! \left(1 - \frac{1}{1!} + \frac{1}{2!} + \cdots + \frac{(-1)^k}{k!} + \cdots + \frac{(-1)^n}{n!} \right)$$
$$= n! - \binom{n}{1}(n-1)! + \binom{n}{2}(n-2)! + \cdots + (-1)^n.$$

Hint. Use the same method as in Problem 4.1.

Conclude from (b) that

$$\lim_{n \to \infty} \frac{p_n}{n!} = \frac{1}{e}.$$

Hint. Recall that

$$e^x = 1 + \frac{x}{1!} + \frac{x^2}{2!} + \cdots + \frac{x^n}{n!} + \cdots .$$

Remark: The ratio $p_n/n!$ has an interesting interpretation in terms of probabilities. Assume n persons go to a restaurant (or to the theatre, etc.) and that they all cherk their coats. Unfortunately, the cleck loses all the coat tags. Then, $p_n/n!$ is the probability that nobody will get her or his own coat back.

(c) Prove that

$$p_n = np_{n-1} + (-1)^n,$$

for all $n \geq 1$, with $p_0 = 1$.

Note that $n!$ is defined by $n! = n(n-1)!$. So, p_n is a sort of "weird factorial" with a strange corrective term $(-1)^n$.

4.5. Prove that if $m + p \geq n$ and $m, n, p \geq 0$, then

$$\binom{m+p}{n} = \sum_{k=0}^{m} \binom{m}{k}\binom{p}{n-k}.$$

Hint. Observe that $\binom{m+p}{n}$ is the coefficient of $a^{m+p-n}b^n$ in $(a+b)^{m+p} = (a+b)^m(a+b)^p$.

Show that the above implies that if $n \geq p$, then

$$\binom{m+p}{n} = \binom{m}{n-p}\binom{p}{p} + \binom{m}{n-p+1}\binom{p}{p-1}$$
$$+ \binom{m}{n-p+2}\binom{p}{p-2} + \cdots + \binom{m}{n}\binom{p}{0}$$

and if $n \leq p$ then

$$\binom{m+p}{n} = \binom{m}{0}\binom{p}{n} + \binom{m}{1}\binom{p}{n-1} + \binom{m}{2}\binom{p}{n-2} + \cdots + \binom{m}{n}\binom{p}{0}.$$

4.6. Prove that

$$\binom{0}{m} + \binom{1}{m} + \cdots + \binom{n}{m} = \binom{n+1}{m+1},$$

for all $m, n \in \mathbb{N}$ with $0 \leq m \leq n$.

4.7. Prove that

$$\binom{m}{0} + \binom{m+1}{1} + \cdots + \binom{m+n}{n} = \binom{m+n+1}{n}.$$

4.8. Prove that

$$\sum_{k=0}^{m}(-1)^k\binom{n}{k} = (-1)^m\binom{n-1}{m},$$

if $0 \le m \le n$.

4.9. (1) Prove that

$$\binom{r}{k} = (-1)^k\binom{k-r-1}{k},$$

where $r \in \mathbb{R}$ and $k \in \mathbb{Z}$ (*negating the upper index*).

(2) Use (1) and the identity of Problem 4.7 to prove that

$$\sum_{k=0}^{m}(-1)^k\binom{n}{k} = (-1)^m\binom{n-1}{m},$$

if $0 \le m \le n$.

4.10. Prove that

$$\sum_{k=0}^{n}\binom{n}{k}\binom{k}{m} = 2^{n-m}\binom{m}{k},$$

where $0 \le m \le n$.

4.11. Prove that

$$(1+x)^{-\frac{1}{2}} = 1 + \sum_{k=1}^{\infty}(-1)^k\frac{1\cdot3\cdot5\cdots(2k-1)}{2\cdot4\cdot6\cdots2k}x^k$$

$$= 1 + \sum_{k=1}^{\infty}\frac{(-1)^k(2k)!}{(k!)^22^{2k}}x^k$$

$$= 1 + \sum_{k=1}^{\infty}\frac{(-1)^k}{2^{2k}}\binom{2k}{k}x^k$$

if $|x| < 1$.

4.12. Prove that

$$\ln(1+x) \le x - \frac{x^2}{2} + \frac{x^3}{3},$$

for all $x \ge -1$.

4.13. If $n = 2m+1$, prove that

$$\binom{2m+1}{m} \sim \sqrt{\frac{2m+1}{2\pi m(m+1)}}\left(1+\frac{1}{2m}\right)^m\left(1-\frac{1}{2(m+1)}\right)^{m+1}2^{2m+1},$$

for m large and so,

$$\binom{2m+1}{m} \sim \sqrt{\frac{2m+1}{2\pi m(m+1)}}2^{2m+1},$$

for m large.

4.14. If $n = 2m + 1$, prove that

$$e^{-t(t+1)/(m+1-t)} \le \binom{2m+1}{m-t} \Big/ \binom{2m+1}{m} \le e^{-t(t+1)/(m+1+t)}$$

with $0 \le t \le m$. Deduce from this that

$$\binom{2m+1}{k} \Big/ \binom{2m+1}{m} \sim e^{1/(4(m+1))} e^{-(2m+1-2k)^2/(4(m+1))},$$

for m large and $0 \le k \le 2m + 1$.

4.15. Prove Proposition 4.14.
Hint. First, show that the function

$$t \mapsto \frac{t^2}{m+t}$$

is strictly increasing for $t \ge 0$.

4.16. (1) Prove that

$$\frac{1 - \sqrt{1 - 4x}}{2x} = 1 + \sum_{k=1}^{\infty} \frac{1}{k+1} \binom{2k}{k} x^k.$$

(2) The numbers

$$C_n = \frac{1}{n+1} \binom{2n}{n},$$

are known as the *Catalan numbers* ($n \ge 0$). The Catalan numbers are the solution of many counting problems in combinatorics. The Catalan sequence begins with

$$1, 1, 2, 5, 14, 42, 132, 429, 1430, 4862, 16796, \ldots.$$

Prove that
$$C_n = \binom{2n}{n} - \binom{2n}{n-1} = \frac{1}{2n+1} \binom{2n+1}{n}.$$

(3) Prove that $C_0 = 1$ and that

$$C_{n+1} = \frac{2(2n+1)}{n+2} C_n.$$

(4) Prove that C_n is the number of ways a convex polygon with $n + 2$ sides can be subdivided into triangles (triangulated) by connecting vertices of the polygon with (nonintersecting) line segments.
Hint. Observe that any triangulation of a convex polygon with $n + 2$ sides has $n - 1$ edges in addition to the sides of the polygon and thus, a total of $2n + 1$ edges. Prove that

$$(4n+2)C_n = (n+2)C_{n+1}.$$

(5) Prove that C_n is the number of full binary trees with $n+1$ leaves (a full binary tree is a tree in which every node has degree 0 or 2).

4.17. Which of the following expressions is the number of partitions of a set with $n \geq 1$ elements into two disjoint blocks:

$$(1)\ 2^n - 2 \quad (2)\ 2^{n-1} - 1.$$

Justify your answer.

4.18. Let H_n, called the nth *harmonic number*, be given by

$$H_n = 1 + \frac{1}{2} + \frac{1}{3} + \cdots + \frac{1}{n},$$

with $n \geq 1$.

(a) Prove that $H_n \notin \mathbb{N}$ for all $n \geq 2$; that is, H_n is not a whole number for all $n \geq 2$. *Hint.* First, prove that every sequence $1, 2, 3, \ldots, n$, with $n \geq 2$, contains a unique number of the form $2^k q$, with $k \geq 1$ as big as possible and q odd ($q = 1$ is possible), which means that for every other number of the form $2^{k'} q'$, with $2^{k'} q' \neq 2^k q$, $1 \leq 2^{k'} q' \leq n$, $k' \geq 1$ and q' odd, we must have $k' < k$. Then, prove that the numerator of H_n is odd and that the denominator of H_n is even, for all $n \geq 2$.

(b) Prove that

$$H_1 + H_2 + \cdots + H_n = (n+1)(H_{n+1} - 1) = (n+1)H_n - n.$$

(c) Prove that

$$\ln(n+1) \leq H_n,$$

for all $n \geq 1$.

Hint. Use the fact that

$$\ln(1+x) \leq x,$$

for all $x > -1$. Compute the sum

$$\sum_{k=1}^{n} \left(\frac{1}{k} - \ln\left(1 + \frac{1}{k}\right) \right).$$

Prove that

$$\ln(n) + \frac{1}{2n} \leq H_n.$$

Hint. We have

$$\ln(n+1) = \ln(n) + \ln\left(1 + \frac{1}{n}\right)$$

and use the fact that

$$\ln(1+x) \geq x - \frac{x^2}{2}.$$

for all x, where $0 \leq x \leq 1$ (in fact, for all $x \geq 0$).

(d) Prove that

$$H_n \leq 1 + \ln(n) + \frac{1}{2n}.$$

Hint. Compute the sum

$$\sum_{k=1}^{n} \left(\ln\left(1 + \frac{1}{k}\right) - \frac{1}{k} + \frac{1}{2k^2} \right).$$

Use previous hints and show that

$$\sum_{k=1}^{n} \frac{1}{k^2} \leq 2 - \frac{1}{n}.$$

(e) It is known that $\ln(1 + x)$ is given by the following convergent series for $|x| < 1$,

$$\ln(1 + x) = x - \frac{x^2}{2} + \frac{x^3}{3} + \cdots + (-1)^{n+1} \frac{x^n}{n} + \cdots .$$

Deduce from this that

$$\ln\left(\frac{x}{x-1}\right) = \frac{1}{x} + \frac{1}{2x^2} + \frac{1}{3x^3} + \cdots + \frac{1}{nx^n} + \cdots .$$

for all x with $|x| > 1$.

Let

$$H_n^{(r)} = \sum_{k=1}^{n} \frac{1}{k^r}.$$

If $r > 1$, it is known that each $H_n^{(r)}$ converges to a limit denoted $H_\infty^{(r)}$ or $\zeta(r)$, where ζ is *Riemann's zeta function* given by

$$\zeta(r) = \sum_{k=1}^{\infty} \frac{1}{k^r},$$

for all $r > 1$.

Fig. 4.6 G. F. Bernhard Riemann, 1826–1866

Prove that

$$\ln(n) = \sum_{k=2}^{n} \left(\frac{1}{k} + \frac{1}{2k^2} + \frac{1}{3k^3} + \cdots + \frac{1}{mk^m} + \cdots \right)$$

$$= (H_n - 1) + \frac{1}{2}(H_n^{(2)} - 1) + \frac{1}{3}(H_n^{(3)} - 1) + \cdots + \frac{1}{m}(H_n^{(m)} - 1) + \cdots$$

and therefore,

$$H_n - \ln(n) = 1 - \frac{1}{2}(H_n^{(2)} - 1) - \frac{1}{3}(H_n^{(3)} - 1) - \cdots - \frac{1}{m}(H_n^{(m)} - 1) - \cdots .$$

Remark: The right-hand side has the limit

$$\gamma = 1 - \frac{1}{2}(\zeta(2) - 1) - \frac{1}{3}(\zeta(3) - 1) - \cdots - \frac{1}{m}(\zeta(m) - 1) - \cdots$$

known as *Euler's constant* (or the *Euler–Mascheroni number*).

Fig. 4.7 Leonhard Euler, 1707–1783 (left) and Jacob Bernoulli, 1654–1705 (right)

It is known that
$$\gamma = 0.577215664901 \cdots$$

but we don't even know whether γ is irrational! It can be shown that

$$H_n = \ln(n) + \gamma + \frac{1}{2n} - \frac{1}{12n^2} + \frac{\varepsilon_n}{120n^4},$$

with $0 < \varepsilon_n < 1$.

4.19. The purpose of this problem is to derive a formula for the sum

$$S_k(n) = 1^k + 2^k + 3^k + \cdots + n^k$$

in terms of a polynomial in n (where $k, n \geq 1$ and $n \geq 0$, with the understanding that this sum is 0 when $n = 0$). Such a formula was derived by Jacob Bernoulli (1654–1705) and is expressed in terms of certain numbers now called *Bernoulli numbers*.

The Bernoulli numbers B^k are defined inductively by solving some equations listed below,

$$B^0 = 1$$
$$B^2 - 2B^1 + 1 = B^2$$
$$B^3 - 3B^2 + 3B^1 - 1 = B^3$$
$$B^4 - 4B^3 + 6B^2 - 4B^1 + 1 = B^4$$
$$B^5 - 5B^4 + 10B^3 - 10B^2 + 5B^1 - 1 = B^5$$

and, in general,

$$\sum_{i=0}^{k} \binom{k}{i} (-1)^i B^{k-i} = B^k, \; k \geq 2.$$

Because $B^1, \dots, B^{k-2}$ are known inductively, this equation can be used to compute B^{k-1}.

Remark: It should be noted that there is more than one definition of the Bernoulli numbers. There are two main versions that differ in the choice of B^1:

1. $B^1 = \frac{1}{2}$
2. $B^1 = -\frac{1}{2}$.

The first version is closer to Bernoulli's original definition and we find it more convenient for stating the identity for $S_k(n)$ but the second version is probably used more often and has its own advantages.

(a) Prove that the first 14 Bernoulli numbers are the numbers listed below:

n	0	1	2	3	4	5	6	7	8	9	10	11	12	13	14
B^n	1	$\frac{1}{2}$	$\frac{1}{6}$	0	$\frac{-1}{30}$	0	$\frac{1}{42}$	0	$\frac{-1}{30}$	0	$\frac{5}{66}$	0	$\frac{-691}{2730}$	0	$\frac{7}{6}$

Observe two patterns:

1. All Bernoulli numbers B^{2k+1}, with $k \geq 1$, appear to be zero.
2. The signs of the Bernoulli numbers B^n, alternate for $n \geq 2$.

The above facts are indeed true but not so easy to prove from the defining equations. However, they follow fairly easily from the fact that the generating function of the numbers

$$\frac{B^k}{k!}$$

can be computed explicitly in terms of the exponential function.

(b) Prove that

$$\frac{z}{1 - e^{-z}} = \sum_{k=0}^{\infty} B^k \frac{z^k}{k!}.$$

Hint. Expand $z/(1 - e^{-z})$ into a power series

$$\frac{z}{1-e^{-z}} = \sum_{k=0}^{\infty} b_k \frac{z^k}{k!}$$

near 0, multiply both sides by $1 - e^{-z}$, and equate the coefficients of z^{k+1}; from this, prove that $b_k = B^k$ for all $k \geq 0$.

Remark: If we define $B^1 = -\frac{1}{2}$, then we get

$$\frac{z}{e^z - 1} = \sum_{k=0}^{\infty} B^k \frac{z^k}{k!}.$$

(c) Prove that $B^{2k+1} = 0$, for all $k \geq 1$.

Hint. Observe that

$$\frac{z}{1-e^{-z}} - \frac{z}{2} = \frac{z(e^z + 1)}{2(e^z - 1)} = 1 + \sum_{k=2}^{\infty} B^k \frac{z^k}{k!}$$

is an even function (which means that it has the same value when we change z to $-z$).

(d) Define the *Bernoulli polynomial* $B_k(x)$ by

$$B_k(x) = \sum_{i=0}^{k} \binom{k}{i} x^{k-i} B^i,$$

for every $k \geq 0$. Prove that

$$B_{k+1}(n) - B_{k+1}(n-1) = (k+1)n^k,$$

for all $k \geq 0$ and all $n \geq 1$. Deduce from the above identities that

$$S_k(n) = \frac{1}{k+1}(B_{k+1}(n) - B_{k+1}(0)) = \frac{1}{k+1} \sum_{i=0}^{k} \binom{k+1}{i} n^{k+1-i} B^i,$$

an identity often known as *Bernoulli's formula*.

Hint. Expand $(n-1)^{k+1-i}$ using the binomial formula and use the fact that

$$\binom{m}{i}\binom{m-i}{j} = \binom{m}{i+j}\binom{i+j}{i}.$$

Remark: If we assume that $B^1 = -\frac{1}{2}$, then

$$B_{k+1}(n+1) - B_{k+1}(n) = (k+1)n^k.$$

Find explicit formulae for $S_4(n)$ and $S_5(n)$.

Extra Credit. It is reported that Euler computed the first 30 Bernoulli numbers. Prove that

$$B^{20} = \frac{-174611}{330}, \quad B^{32} = \frac{-7\,709\,321\,041\,217}{510}.$$

What does the prime 37 have to do with the numerator of B^{32}?

Remark: Because

$$\frac{z}{1-e^{-z}} - \frac{z}{2} = \frac{z(e^z+1)}{2(e^z-1)} = \frac{z}{2}\frac{e^{z/2}+e^{-z/2}}{e^{z/2}-e^{-z/2}} = \frac{z}{2}\coth\left(\frac{z}{2}\right),$$

where coth is the *hyperbolic tangent* function given by

$$\coth(z) = \frac{\cosh z}{\sinh z},$$

with

$$\cosh z = \frac{e^z+e^{-z}}{2}, \quad \sinh z = \frac{e^z-e^{-z}}{2}.$$

It follows that

$$z\coth z = \frac{2z}{1-e^{-2z}} - z = \sum_{k=0}^{\infty} B^{2k}\frac{(2z)^{2k}}{(2k)!} = \sum_{k=0}^{\infty} 4^k B^{2k}\frac{z^{2k}}{(2k)!}.$$

If we use the fact that

$$\sin z = -i\sinh iz, \quad \cos z = \cosh iz,$$

we deduce that $\cot z = \cos z/\sin z = i\coth iz$, which yields

$$z\cot z = \sum_{k=0}^{\infty} (-4)^k B^{2k}\frac{z^{2k}}{(2k)!}.$$

Now, Euler found the remarkable formula

$$z\cot z = 1 - 2\sum_{k=1}^{\infty} \frac{z^2}{k^2\pi^2 - z^2}.$$

By expanding the right-hand side of the above formula in powers of z^2 and equating the coefficients of z^{2k} in both series for $z\cot z$, we get the amazing formula:

$$\zeta(2k) = (-1)^{k-1}\frac{2^{2k-1}\pi^{2k}}{(2k)!}B^{2k},$$

for all $k \geq 1$, where $\zeta(r)$ is *Riemann's zeta function* given by

$$\zeta(r) = \sum_{n=1}^{\infty} \frac{1}{n^r},$$

for all $r > 1$. Therefore, we get

$$B^{2k} = \zeta(2k)(-1)^{k-1}\frac{(2k)!}{2^{2k-1}\pi^{2k}} = (-1)^{k-1}2(2k)!\sum_{n=1}^{\infty}\frac{1}{(2\pi n)^{2k}},$$

a formula due to due to Euler. This formula shows that the signs of the B^{2k} alternate for all $k \geq 1$. Using Stirling's formula, it also shows that

$$|B^{2k}| \sim 4\sqrt{\pi k}\left(\frac{k}{\pi e}\right)^{2k}$$

so B^{2k} tends to infinity rather quickly when k goes to infinity.

There is another remarkable relationship between the Euler constant γ, defined in Problem 4.18 and the Bernoulli numbers, namely the identity

$$\gamma = \frac{1}{2} + \sum_{k=1}^{\infty}\frac{B^{2k}}{2k}.$$

4.20. The purpose of this problem is to derive a recurrence formula for the sum

$$S_k(n) = 1^k + 2^2 + 3^k + \cdots + n^k.$$

Using the trick of writing $(n+1)^k$ as the "telescoping sum"

$$(n+1)^k = 1^k + (2^k - 1^k) + (3^k - 2^k) + \cdots + ((n+1)^k - n^k),$$

use the binomial formula to prove that

$$(n+1)^k = 1 + \sum_{j=0}^{k-1}\binom{k}{j}\sum_{i=1}^{n}i^j = 1 + \sum_{j=0}^{k-1}\binom{k}{j}S_j(n).$$

Deduce from the above formula the recurrence formula

$$(k+1)S_k(n) = (n+1)^{k+1} - 1 - \sum_{j=0}^{k-1}\binom{k+1}{j}S_j(n).$$

4.21. Given n cards and a table, we would like to create the largest possible overhang by stacking cards up over the table's edge, subject to the laws of gravity. To be more precise, we require the edges of the cards to be parallel to the edge of the table; see Figure 4.8. We assume that each card is 2 units long.

With a single card, obviously we get the maximum overhang when its center of gravity is just above the edge of the table. Because the center of gravity is in the middle of the card, we can create half of a cardlength, namely 1 unit, of overhang.

With two cards, a moment of thought reveals that we get maximum overhang when the center of gravity of the top card is just above the edge of the second card and the center of gravity of both cards combined is just above the edge of the table.

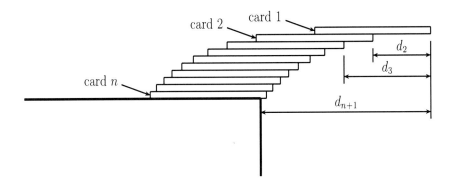

Fig. 4.8 Stack of overhanging cards

The joint center of gravity of two cards is in the middle of their common part, so we can achieve an additional half unit of overhang.

Given n cards, we find that we place the cards so that the center of gravity of the top k cards lies just above the edge of the $(k+1)$st card (which supports these top k cards). The table plays the role of the $(n+1)$st card. We can express this condition by defining the distance d_k from the extreme edge of the topmost card to the corresponding edge of the kth card from the top (see Figure 4.8). Note that $d_1 = 0$. In order for d_{k+1} to be the center of gravity of the first k cards, we must have

$$d_{k+1} = \frac{(d_1 + 1) + (d_2 + 2) + \cdots + (d_k + 1)}{k},$$

for $1 \le k \le n$. This is because the center of gravity of k objects having respective weights $w_1, \ldots, w_k$ and having respective centers of gravity at positions $x_1, \ldots, x_k$ is at position

$$\frac{w_1 x_1 + w_2 x_2 + \cdots + w_k x_k}{w_1 + w_2 + \cdots + w_k}.$$

Prove that the equations defining the d_{k+1} imply that

$$d_{k+1} = d_k + \frac{1}{k},$$

and thus, deduce that

$$d_{k+1} = H_k = 1 + \frac{1}{2} + \frac{1}{3} + \cdots + \frac{1}{k},$$

the kth Harmonic number (see Problem 4.18). Conclude that the total overhang with n cards is H_n.

Prove that it only takes four cards to achieve an overhang of one cardlength. What kind of overhang (in terms of cardlengths) is achieved with 52 cards? (See the end of Problem 4.18.)

4.22. Consider $n \geq 2$ lines in the plane. We say that these lines are in *general position* iff no two of them are parallel and no three pass through the same point. Prove that n lines in general position divide the plane into

$$\frac{n(n+1)}{2} + 1$$

regions.

4.23. (A deceptive induction, after Conway and Guy [3]) Place n distinct points on a circle and draw the line segments joining all pairs of these points. These line segments determine some regions inside the circle as shown in Figure 4.9 for five points. Assuming that the points are in general position, which means that no more than two line segments pass through any point inside the circle, we would like to compute the number of regions inside the circle. These regions are convex and their boundaries are line segments or possibly one circular arc.

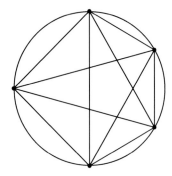

Fig. 4.9 Regions inside a circle

If we look at the first five circles in Figure 4.10, we see that the number of regions is

$$1, 2, 4, 8, 16.$$

Thus, it is reasonable to assume that with $n \geq 1$ points, there are $R = 2^{n-1}$ regions.

(a) Check that the circle with six points (the sixth circle in Figure 4.10) has 32 regions, confirming our conjecture.

(b) Take a closer look at the circle with six points on it. In fact, there are only 31 regions. Prove that the number of regions R corresponding to n points in general position is

$$R = \frac{1}{24}(n^4 - 6n^3 + 23n^2 - 18n + 24).$$

Thus, we get the following number of regions for $n = 1, \ldots, 14$:

$$n = 1\ 2\ 3\ 4\ 5\ 6\ 7\ 8\ 9\ \ 10\ \ 11\ \ 12\ \ 13\ \ \ 14$$
$$R = 1\ 2\ 4\ 8\ 16\ 31\ 57\ 99\ 163\ 256\ 386\ 562\ 794\ 1093$$

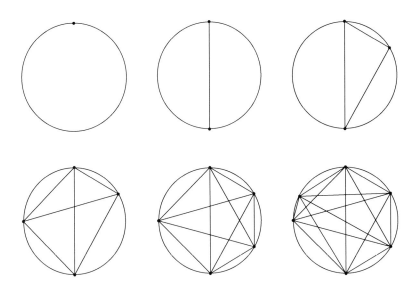

Fig. 4.10 Counting regions inside a circle

Hint. Label the points on the circle, $0, 1, \ldots, n-1$, in counterclockwise order. Next, design a procedure for assigning a unique label to every region. The region determined by the chord from 0 to $n-1$ and the circular arc from 0 to $n-1$ is labeled "empty". Every other region is labeled by a nonempty subset, S, of $\{0, 1, \ldots n-1\}$, where S has at most four elements as illustrated in Figure 4.11. The procedure for assigning labels to regions goes as follows.

For any quadruple of integers, a, b, c, d, with $0 < a < b < c < d \leq n-1$, the chords ac and bd intersect in a point that uniquely determines a region having this point as a vertex and lying to the right of the oriented line bd; we label this region $abcd$. In the special case where $a = 0$, this region, still lying to the right of the oriented line bd is labeled bcd. All regions that do not have a vertex on the circle are labeled that way. For any two integers c, d, with $0 < c < d \leq n-1$, there is a unique region having c as a vertex and lying to the right of the oriented line cd and we label it cd. In the special case where $c = 0$, this region, still lying to the right of the oriented line $0d$ is labeled d.

To understand the above procedure, label the regions in the six circles of Figure 4.10.

Use this labeling scheme to prove that the number of regions is

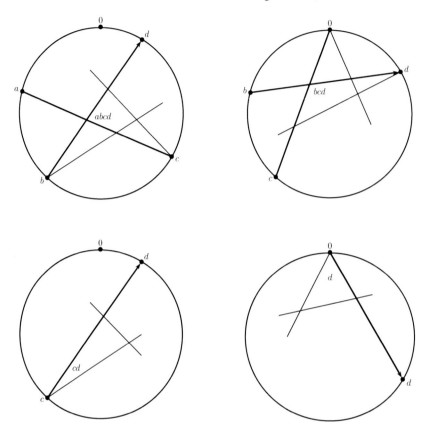

Fig. 4.11 Labeling the regions inside a circle

$$R = \binom{n-1}{0} + \binom{n-1}{1} + \binom{n-1}{2} + \binom{n-1}{3} + \binom{n-1}{4} = 1 + \binom{n}{2} + \binom{n}{4}.$$

(c) Prove again, using induction on n, that

$$R = 1 + \binom{n}{2} + \binom{n}{4}.$$

4.24. The *complete* graph K_n with n vertices ($n \geq 2$) is the simple undirected graph whose edges are all two-element subsets $\{i, j\}$, with $i, j \in \{1, 2, \ldots, n\}$ and $i \neq j$. The purpose of this problem is to prove that the number of spanning trees of K_n is n^{n-2}, a formula due to Cayley (1889).

(a) Let $T(n; d_1, \ldots, d_n)$ be the number of trees with $n \geq 2$ vertices $v_1, \ldots, v_n$, and degrees $d(v_1) = d_1, d(v_2) = d_2, \ldots, d(v_n) = d_n$, with $d_i \geq 1$. Prove that

$$T(n;d_1,\ldots,d_n) = \binom{n-2}{d_1-1\ d_2-1\ \cdots\ d_n-1}.$$

Hint. First, show that we must have

$$\sum_{i=1}^{n} d_i = 2(n-1).$$

We may assume that $d_1 \geq d_2 \geq \cdots \geq d_n$, with $d_n = 1$. Prove that

$$T(n;d_1,\ldots,d_n) = \sum_{\substack{1 \leq i \leq n \\ d_i \geq 2}}^{i} T(n-1;d_1,\ldots,d_i-1,\ldots,d_{n-1}).$$

Then, prove the formula by induction on n.

(b) Prove that $d_1,\ldots,d_n$, with $d_i \geq 1$, are degrees of a tree with n nodes iff

$$\sum_{i=1}^{n} d_i = 2(n-1).$$

(c) Use (a) and (b) to prove that the number of spanning trees of K_n is n^{n-2}.
Hint. Show that the number of spanning trees of K_n is

$$\sum_{\substack{d_1,\ldots,d_n \geq 1 \\ d_1+\cdots+d_n=2(n-1)}} \binom{n-2}{d_1-1\ d_2-1\ \cdots\ d_n-1}$$

and use the multinomial formula.

References

1. Claude Berge. *Principles of Combinatorics.* New York: Academic Press, first edition, 1971.
2. J. Cameron, Peter. *Combinatorics: Topics, Techniques, Algorithms.* Cambridge, UK: Cambridge University Press, first edition, 1994.
3. John H. Conway and Richard K. Guy, *The Book of Numbers.* Copernicus, New York: Springer-Verlag, first edition, 1996.
4. Ronald L. Graham, Donald E. Knuth, and Oren Patashnik. *Concrete Mathematics: A Foundation For Computer Science.* Reading, MA: Addison Wesley, second edition, 1994.
5. L. Lovász, J. Pelikán, and K. Vesztergombi. *Discrete Mathematics. Elementary and Beyond.* Undergraduate Texts in Mathematics. New York: Springer, first edition, 2003.
6. Jiri Matousek. *Lectures on Discrete Geometry.* GTM No. 212. New York: Springer Verlag, first edition, 2002.
7. Richard P. Stanley. *Enumerative Combinatorics, Vol. I.* Cambridge Studies in Advanced Mathematics, No. 49. Cambridge UK: Cambridge University Press, first edition, 1997.
8. J.H. van Lint and R.M. Wilson. *A Course in Combinatorics.* Cambridge UK: Cambridge University Press, second edition, 2001.

Chapter 5
Partial Orders, Lattices, Well-Founded Orderings, Unique Prime Factorization in $\mathbb{Z}$ and GCDs, Equivalence Relations, Fibonacci and Lucas Numbers, Public Key Cryptography and RSA, Distributive Lattices, Boolean Algebras, Heyting Algebras

5.1 Partial Orders

There are two main kinds of relations that play a very important role in mathematics and computer science:

1. Partial orders
2. Equivalence relations

In this section and the next few ones, we define partial orders and investigate some of their properties. As we show, the ability to use induction is intimately related to a very special property of partial orders known as well-foundedness.

Intuitively, the notion of order among elements of a set X captures the fact that some elements are bigger than others, perhaps more important, or perhaps that they carry more information. For example, we are all familiar with the natural ordering $\leq$ of the integers

$$\cdots, -3 \leq -2 \leq -1 \leq 0 \leq 1 \leq 2 \leq 3 \leq \cdots,$$

the ordering of the rationals (where

$$\frac{p_1}{q_1} \leq \frac{p_2}{q_2} \quad \text{iff} \quad \frac{p_2 q_1 - p_1 q_2}{q_1 q_2} \geq 0,$$

i.e., $p_2 q_1 - p_1 q_2 \geq 0$ if $q_1 q_2 > 0$ else $p_2 q_1 - p_1 q_2 \leq 0$ if $q_1 q_2 < 0$), and the ordering of the real numbers. In all of the above orderings, note that for any two numbers a and b, either $a \leq b$ or $b \leq a$. We say that such orderings are *total* orderings. A natural example of an ordering that is not total is provided by the subset ordering. Given a set X, we can order the subsets of X by the subset relation: $A \subseteq B$, where A, B are any subsets of X. For example, if $X = \{a, b, c\}$, we have $\{a\} \subseteq \{a, b\}$. However, note that neither $\{a\}$ is a subset of $\{b, c\}$ nor $\{b, c\}$ is a subset of $\{a\}$. We say that $\{a\}$ and $\{b, c\}$ are *incomparable*. Now, not all relations are partial orders, so which properties characterize partial orders? Our next definition gives us the answer.

J. Gallier, *Discrete Mathematics*, Universitext,
DOI 10.1007/978-1-4419-8047-2_5, © Springer Science+Business Media, LLC 2011

Definition 5.1. A binary relation $\leq$ on a set X is a *partial order* (or *partial ordering*) iff it is *reflexive, transitive*, and *antisymmetric*; that is:

(1) *(Reflexivity)*: $a \leq a$, for all $a \in X$.
(2) *(Transitivity)*: If $a \leq b$ and $b \leq c$, then $a \leq c$, for all $a, b, c \in X$.
(3) *(Antisymmetry)*: If $a \leq b$ and $b \leq a$, then $a = b$, for all $a, b \in X$.

A partial order is a *total order (ordering)* (or *linear order (ordering)*) iff for all $a, b \in X$, either $a \leq b$ or $b \leq a$. When neither $a \leq b$ nor $b \leq a$, we say that *a and b are incomparable*. A subset, $C \subseteq X$, is a *chain* iff $\leq$ induces a total order on C (so, for all $a, b \in C$, either $a \leq b$ or $b \leq a$). The *strict order (ordering)* $<$ *associated with* $\leq$ is the relation defined by: $a < b$ iff $a \leq b$ and $a \neq b$. If $\leq$ is a partial order on X, we say that the pair $\langle X, \leq \rangle$ is a *partially ordered set* or for short, a *poset*.

Remark: Observe that if $<$ is the strict order associated with a partial order $\leq$, then $<$ is transitive and *antireflexive*, which means that

(4) $a \not< a$, for all $a \in X$.

Conversely, let $<$ be a relation on X and assume that $<$ is transitive and antireflexive. Then, we can define the relation $\leq$ so that $a \leq b$ iff $a = b$ or $a < b$. It is easy to check that $\leq$ is a partial order and that the strict order associated with $\leq$ is our original relation, $<$.

Given a poset $\langle X, \leq \rangle$, by abuse of notation we often refer to $\langle X, \leq \rangle$ as the *poset* X, the partial order $\leq$ being implicit. If confusion may arise, for example, when we are dealing with several posets, we denote the partial order on X by $\leq_X$.

Here are a few examples of partial orders.

1. **The subset ordering**. We leave it to the reader to check that the subset relation $\subseteq$ on a set X is indeed a partial order. For example, if $A \subseteq B$ and $B \subseteq A$, where $A, B \subseteq X$, then $A = B$, because these assumptions are exactly those needed by the extensionality axiom.

2. **The natural order on** $\mathbb{N}$. Although we all know what the ordering of the natural numbers is, we should realize that if we stick to our axiomatic presentation where we defined the natural numbers as sets that belong to every inductive set (see Definition 1.11), then we haven't yet defined this ordering. However, this is easy to do because the natural numbers are sets. For any $m, n \in \mathbb{N}$, define $m \leq n$ as $m = n$ or $m \in n$. Then, it is not hard to check that this relation is a total order. (Actually, some of the details are a bit tedious and require induction; see Enderton [6], Chapter 4.)

3. **Orderings on strings**. Let $\Sigma = \{a_1, \ldots, a_n\}$ be an alphabet. The prefix, suffix, and substring relations defined in Section 2.11 are easily seen to be partial orders. However, these orderings are not total. It is sometimes desirable to have a total order on strings and, fortunately, the lexicographic order (also called dictionnary order) achieves this goal. In order to define the *lexicographic order* we assume that the symbols in Σ are totally ordered, $a_1 < a_2 < \cdots < a_n$. Then, given any two strings $u, v \in \Sigma^*$ we set

$$u \preceq v \quad \begin{cases} \text{if } v = uy, \text{ for some } y \in \Sigma^*, \text{ or} \\ \text{if } u = xa_iy, \ v = xa_jz, \\ \text{and } a_i < a_j, \text{ for some } x, y, z \in \Sigma^*. \end{cases}$$

In other words, either u is a prefix of v or else u and v share a common prefix x, and then there is a differing symbol, a_i in u and a_j in v, with $a_i < a_j$. It is fairly tedious to prove that the lexicographic order is a partial order. Moreover, the lexicographic order is a total order.

4. **The divisibility order on** $\mathbb{N}$. Let us begin by defining divisibility in $\mathbb{Z}$. Given any two integers, $a, b \in \mathbb{Z}$, with $b \neq 0$, we say that b *divides* a (*a is a multiple of b*) iff $a = bq$ for some $q \in \mathbb{Z}$. Such a q is called the *quotient of a and b*. Most number theory books use the notation $b \mid a$ to express that b divides a. For example, $4 \mid 12$ because $12 = 4 \cdot 3$ and $7 \mid -21$ because $-21 = 7 \cdot (-3)$ but 3 does not divide 16 because 16 is not an integer multiple of 3.

We leave the verification that the divisibility relation is reflexive and transitive as an easy exercise. What about antisymmetry? So, assume that $b \mid a$ and $a \mid b$ (thus, $a, b \neq 0$). This means that there exist $q_1, q_2 \in \mathbb{Z}$ so that

$$a = bq_1 \quad \text{and} \quad b = aq_2.$$

From the above, we deduce that $b = bq_1q_2$; that is,

$$b(1 - q_1q_2) = 0.$$

As $b \neq 0$, we conclude that

$$q_1q_2 = 1.$$

Now, let us restrict ourselves to $\mathbb{N}_+ = \mathbb{N} - \{0\}$, so that $a, b \geq 1$. It follows that $q_1, q_2 \in \mathbb{N}$ and in this case, $q_1q_2 = 1$ is only possible iff $q_1 = q_2 = 1$. Therefore, $a = b$ and the divisibility relation is indeed a partial order on $\mathbb{N}_+$. Why is divisibility not a partial order on $\mathbb{Z} - \{0\}$?

Given a poset $\langle X \leq \rangle$, if X is finite then there is a convenient way to describe the partial order $\leq$ on X using a graph. In preparation for that, we need a few preliminary notions.

Consider an arbitrary poset $\langle X \leq \rangle$ (not necessarily finite). Given any element $a \in X$, the following situations are of interest.

1. For **no** $b \in X$ do we have $b < a$. We say that a is a *minimal element* (of X).
2. There is some $b \in X$ so that $b < a$ and there is **no** $c \in X$ so that $b < c < a$. We say that b is an *immediate predecessor of a*.
3. For **no** $b \in X$ do we have $a < b$. We say that a is a *maximal element* (of X).
4. There is some $b \in X$ so that $a < b$ and there is **no** $c \in X$ so that $a < c < b$. We say that b is an *immediate successor of a*.

Note that an element may have more than one immediate predecessor (or more than one immediate successor).

If X is a finite set, then it is easy to see that every element that is not minimal has an immediate predecessor and any element that is not maximal has an immediate successor (why?). But if X is infinite, for example, $X = \mathbb{Q}$, this may not be the case. Indeed, given any two distinct rational numbers $a, b \in \mathbb{Q}$, we have

$$a < \frac{a+b}{2} < b.$$

Let us now use our notion of immediate predecessor to draw a diagram representing a finite poset $\langle X, \leq \rangle$. The trick is to draw a picture consisting of nodes and oriented edges, where the nodes are all the elements of X and where we draw an oriented edge from a to b iff a is an immediate predecessor of b. Such a diagram is called a *Hasse diagram* for $\langle X, \leq \rangle$. Observe that if $a < c < b$, then the diagram does **not** have an edge corresponding to the relation $a < b$. However, such information can be recovered from the diagram by following paths consisting of one or several consecutive edges. Similarly, the self-loops corresponding to the reflexive relations $a \leq a$ are omitted. A Hasse diagram is an economical representation of a finite poset and it contains the same amount of information as the partial order $\leq$.

The diagram associated with the partial order on the power set of the two-element set $\{a, b\}$ is shown in Figure 5.1.

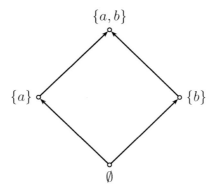

Fig. 5.1 The partial order of the power set $2^{\{a,b\}}$

The diagram associated with the partial order on the power set of the three-element set $\{a, b, c\}$ is shown in Figure 5.2.

Note that $\emptyset$ is a minimal element of the poset in Figure 5.2. (in fact, the smallest element) and $\{a, b, c\}$ is a maximal element (in fact, the greatest element). In this example, there is a unique minimal (respectively, maximal) element. A less trivial example with multiple minimal and maximal elements is obtained by deleting $\emptyset$ and $\{a, b, c\}$ and is shown in Figure 5.3.

Given a poset $\langle X, \leq \rangle$, observe that if there is some element $m \in X$ so that $m \leq x$ for all $x \in X$, then m is unique. Indeed, if m' is another element so that $m' \leq x$ for all

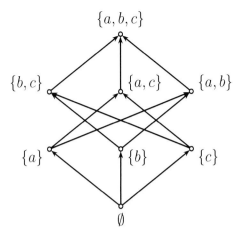

Fig. 5.2 The partial order of the power set $2^{\{a,b,c\}}$

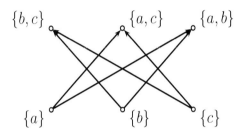

Fig. 5.3 Minimal and maximal elements in a poset

$x \in X$, then if we set $x = m'$ in the first case, we get $m \leq m'$ and if we set $x = m$ in the second case, we get $m' \leq m$, from which we deduce that $m = m'$, as claimed. Such an element m, is called the *smallest* or the *least element* of X. Similarly, an element $b \in X$, so that $x \leq b$ for all $x \in X$ is unique and is called the *greatest* or the *largest element* of X.

We summarize some of our previous definitions and introduce a few more useful concepts in the following.

Definition 5.2. Let $\langle X, \leq \rangle$ be a poset and let $A \subseteq X$ be any subset of X. An element $b \in X$ is a *lower bound of A* iff $b \leq a$ for all $a \in A$. An element $m \in X$ is an *upper bound of A* iff $a \leq m$ for all $a \in A$. An element $b \in X$ is the *least element of A* iff $b \in A$ and $b \leq a$ for all $a \in A$. An element $m \in X$ is the *greatest element of A* iff $m \in A$ and $a \leq m$ for all $a \in A$. An element $b \in A$ is *minimal in A* iff $a < b$ for no $a \in A$, or equivalently, if for all $a \in A$, $a \leq b$ implies that $a = b$. An element $m \in A$ is *maximal in A* iff $m < a$ for no $a \in A$, or equivalently, if for all $a \in A$, $m \leq a$ implies that $a = m$. An element $b \in X$ is the *greatest lower bound of A* iff the set of lower bounds of A is nonempty and if b is the greatest element of this set. An element $m \in X$ is the *least upper bound of A* iff the set of upper bounds of A is nonempty and if m is the least element of this set.

Remarks:

1. If b is a lower bound of A (or m is an upper bound of A), then b (or m) may not belong to A.
2. The least element of A is a lower bound of A that also belongs to A and the greatest element of A is an upper bound of A that also belongs to A. When $A = X$, the least element is often denoted $\bot$, sometimes 0, and the greatest element is often denoted $\top$, sometimes 1.
3. Minimal or maximal elements of A belong to A but they are not necessarily unique.
4. The greatest lower bound (or the least upper bound) of A may not belong to A. We use the notation $\bigwedge A$ for the greatest lower bound of A and the notation $\bigvee A$ for the least upper bound of A. In computer science, some people also use $\bigsqcup A$ instead of $\bigvee A$ and the symbol $\sqcup$ upside down instead of $\bigwedge$. When $A = \{a,b\}$, we write $a \wedge b$ for $\bigwedge \{a,b\}$ and $a \vee b$ for $\bigvee \{a,b\}$. The element $a \wedge b$ is called the *meet of a and b* and $a \vee b$ is the *join of a and b*. (Some computer scientists use $a \sqcap b$ for $a \wedge b$ and $a \sqcup b$ for $a \vee b$.)
5. Observe that if it exists, $\bigwedge \emptyset = \top$, the greatest element of X and if its exists, $\bigvee \emptyset = \bot$, the least element of X. Also, if it exists, $\bigwedge X = \bot$ and if it exists, $\bigvee X = \top$.

The reader should look at the posets in Figures 5.2 and 5.3 for examples of the above notions.

For the sake of completeness, we state the following fundamental result known as Zorn's lemma even though it is unlikely that we use it in this course. Zorn's lemma turns out to be equivalent to the axiom of choice. For details and a proof, the reader is referred to Suppes [16] or Enderton [6].

Fig. 5.4 Max Zorn, 1906–1993

Theorem 5.1. *(Zorn's Lemma) Given a poset* $\langle X, \leq \rangle$, *if every nonempty chain in X has an upper bound, then X has some maximal element.*

When we deal with posets, it is useful to use functions that are order preserving as defined next.

Definition 5.3. Given two posets $\langle X, \leq_X \rangle$ and $\langle Y, \leq_Y \rangle$, a function $f \colon X \to Y$ is *monotonic* (or *order preserving*) iff for all $a, b \in X$,

$$\text{if} \quad a \leq_X b \quad \text{then} \quad f(a) \leq_Y f(b).$$

5.2 Lattices and Tarski's Fixed-Point Theorem

We now take a closer look at posets having the property that every two elements have a meet and a join (a greatest lower bound and a least upper bound). Such posets occur a lot more often than we think. A typical example is the power set under inclusion, where meet is intersection and join is union.

Definition 5.4. A *lattice* is a poset in which any two elements have a meet and a join. A *complete lattice* is a poset in which any subset has a greatest lower bound and a least upper bound.

According to Part (5) of the remark just before Zorn's lemma, observe that a complete lattice must have a least element $\perp$ and a greatest element $\top$.

Remark: The notion of complete lattice is due to G. Birkhoff (1933). The notion of a lattice is due to Dedekind (1897) but his definition used properties (L1)–(L4) listed in Proposition 5.1. The use of meet and join in posets was first studied by C. S. Peirce (1880).

Fig. 5.5 J. W. Richard Dedekind, 1831–1916 (left), Garrett Birkhoff, 1911–1996 (middle) and Charles S. Peirce, 1839–1914 (right)

Figure 5.6 shows the lattice structure of the power set of $\{a, b, c\}$. It is actually a complete lattice.

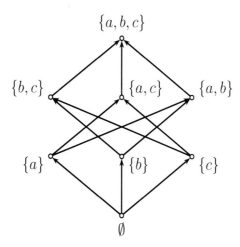

Fig. 5.6 The lattice $2^{\{a,b,c\}}$

It is easy to show that any finite lattice is a complete lattice and that a finite poset is a lattice iff it has a least element and a greatest element.

The poset $\mathbb{N}_+$ under the divisibility ordering is a lattice. Indeed, it turns out that the meet operation corresponds to *greatest common divisor* and the join operation corresponds to *least common multiple*. However, it is not a complete lattice. The power set of any set X is a complete lattice under the subset ordering. Indeed, one may verify immediately that for any collection $\mathscr{C}$ of subsets of X, the least upper bound of $\mathscr{C}$ is its union $\bigcup \mathscr{C}$ and the greatest lower bound of $\mathscr{C}$ is its intersection $\bigcap \mathscr{C}$. The least element of 2^X is $\emptyset$ and its greatest element is X itself.

The following proposition gathers some useful properties of meet and join.

Proposition 5.1. *If X is a lattice, then the following identities hold for all $a, b, c \in X$.*

L1	$a \vee b = b \vee a,$	$a \wedge b = b \wedge a$
L2	$(a \vee b) \vee c = a \vee (b \vee c),$	$(a \wedge b) \wedge c = a \wedge (b \wedge c)$
L3	$a \vee a = a,$	$a \wedge a = a$
L4	$(a \vee b) \wedge a = a,$	$(a \wedge b) \vee a = a.$

Properties (L1) correspond to commutativity, properties (L2) to associativity, properties (L3) to idempotence, and properties (L4) to absorption. Furthermore, for all $a, b \in X$, we have

$$a \leq b \quad \text{iff} \quad a \vee b = b \quad \text{iff} \quad a \wedge b = a,$$

called consistency.

Proof. The proof is left as an exercise to the reader. $\square$

Properties (L1)–(L4) are algebraic identities that were found by Dedekind (1897). A pretty symmetry reveals itself in these identities: they all come in pairs, one involving $\wedge$, the other involving $\vee$. A useful consequence of this symmetry is *duality*, namely, that each equation derivable from (L1)–(L4) has a dual statement obtained by exchanging the symbols $\wedge$ and $\vee$. What is even more interesting is that it is possible to use these properties to define lattices. Indeed, if X is a set together with two operations $\wedge$ and $\vee$ satisfying (L1)–(L4), we can define the relation $a \leq b$ by $a \vee b = b$ and then show that $\leq$ is a partial order such that $\wedge$ and $\vee$ are the corresponding meet and join. The first step is to show that

$$a \vee b = b \quad \text{iff} \quad a \wedge b = a.$$

If $a \vee b = b$, then substituting b for $a \vee b$ in (L4), namely

$$(a \vee b) \wedge a = a,$$

we get

$$b \wedge a = a,$$

which, by (L1), yields

$$a \wedge b = a,$$

as desired. Conversely, if $a \wedge b = a$, then by (L1) we have $b \wedge a = a$, and substituting a for $b \wedge a$ in the instance of (L4) where a and b are switched, namely

$$(b \wedge a) \vee b = b,$$

we get

$$a \vee b = b,$$

as claimed. Therefore, we can define $a \leq b$ as $a \vee b = b$ or equivalently as $a \wedge b = a$. After a little work, we obtain the following proposition.

Proposition 5.2. *Let X be a set together with two operations $\wedge$ and $\vee$ satisfying the axioms (L1)–(L4) of Proposition 5.1. If we define the relation $\leq$ by $a \leq b$ iff $a \vee b = b$ (equivalently, $a \wedge b = a$), then $\leq$ is a partial order and $(X, \leq)$ is a lattice whose meet and join agree with the original operations $\wedge$ and $\vee$.*

The following proposition shows that the existence of arbitrary least upper bounds (or arbitrary greatest lower bounds) is already enough ensure that a poset is a complete lattice.

Proposition 5.3. *Let $\langle X, \leq \rangle$ be a poset. If X has a greatest element $\top$, and if every nonempty subset A of X has a greatest lower bound $\bigwedge A$, then X is a complete lattice. Dually, if X has a least element $\bot$ and if every nonempty subset A of X has a least upper bound $\bigvee A$, then X is a complete lattice*

Proof. Assume X has a greatest element $\top$ and that every nonempty subset A of X has a greatest lower bound, $\bigwedge A$. We need to show that any subset S of X has a least upper bound. As X has a greatest element $\top$, the set U of upper bounds of S is nonempty and so, $m = \bigwedge U$ exists. We claim that $\bigwedge U = \bigvee S$ (i.e., m is the least upper bound of S). First, note that every element of S is a lower bound of U because U is the set of upper bounds of S. As $m = \bigwedge U$ is the greatest lower bound of U, we deduce that $s \leq m$ for all $s \in S$ (i.e., m is an upper bound of S). Next, if b is any upper bound for S, then $b \in U$ and as m is a lower bound of U (the greatest one), we have $m \leq b$ (i.e., m is the least upper bound of S). The other statement is proved by duality. $\square$

We are now going to prove a remarkable result due to A. Tarski (discovered in 1942, published in 1955). A special case (for power sets) was proved by B. Knaster (1928). First, we define fixed points.

Fig. 5.7 Alferd Tarski, 1902–1983

Definition 5.5. Let $\langle X, \leq \rangle$ be a poset and let $f \colon X \to X$ be a function. An element $x \in X$ is a *fixed point of f* (sometimes spelled *fixpoint*) iff

$$f(x) = x.$$

An element, $x \in X$, is a *least (respectively, greatest) fixed point of f* if it is a fixed point of f and if $x \leq y$ (resp. $y \leq x$) for every fixed point y of f.

Fixed points play an important role in certain areas of mathematics (e.g., topology, differential equations) and also in economics because they tend to capture the notion of stability or equilibrium.

We now prove the following pretty theorem due to Tarski and then immediately proceed to use it to give a very short proof of the Schröder–Bernstein theorem (Theorem 2.7).

Theorem 5.2. *(Tarski's Fixed-Point Theorem) Let $\langle X, \leq \rangle$ be a complete lattice and let $f \colon X \to X$ be any monotonic function. Then, the set F of fixed points of f is a complete lattice. In particular, f has a least fixed point,*

$$x_{\min} = \bigwedge \{ x \in X \mid f(x) \leq x \}$$

and a greatest fixed point

$$x_{\max} = \bigvee \{x \in X \mid x \le f(x)\}.$$

Proof.   We proceed in three steps.

Step 1. We prove that $x_{\max}$ is the largest fixed point of f.

Because $x_{\max}$ is an upper bound of $A = \{x \in X \mid x \le f(x)\}$ (the smallest one), we have $x \le x_{\max}$ for all $x \in A$. By monotonicity of f, we get $f(x) \le f(x_{\max})$ and because $x \in A$, we deduce

$$x \le f(x) \le f(x_{\max}) \quad \text{for all} \quad x \in A,$$

which shows that $f(x_{\max})$ is an upper bound of A. As $x_{\max}$ is the least upper bound of A, we get

$$x_{\max} \le f(x_{\max}). \tag{$*$}$$

Again, by monotonicity, from the above inequality, we get

$$f(x_{\max}) \le f(f(x_{\max})),$$

which shows that $f(x_{\max}) \in A$. As $x_{\max}$ is an upper bound of A, we deduce that

$$f(x_{\max}) \le x_{\max}. \tag{$**$}$$

But then, $(*)$ and $(**)$ yield

$$f(x_{\max}) = x_{\max},$$

which shows that $x_{\max}$ is a fixed point of f. If x is any fixed point of f, that is, if $f(x) = x$, we also have $x \le f(x)$; that is, $x \in A$. As $x_{\max}$ is the least upper bound of A, we have $x \le x_{\max}$, which proves that $x_{\max}$ is the greatest fixed point of f.

Step 2. We prove that $x_{\min}$ is the least fixed point of f.

This proof is dual to the proof given in Step 1.

Step 3. We know that the set of fixed points F of f has a least element and a greatest element, so by Proposition 5.3, it is enough to prove that any nonempty subset $S \subseteq F$ has a greatest lower bound. If we let

$$I = \{x \in X \mid x \le s \quad \text{for all} \quad s \in S \quad \text{and} \quad x \le f(x)\},$$

then we claim that $a = \bigvee I$ is a fixed point of f and that it is the greatest lower bound of S.

The proof that $a = \bigvee I$ is a fixed point of f is analogous to the proof used in Step 1. Because a is an upper bound of I, we have $x \le a$ for all $x \in I$. By monotonicity of f and the fact that $x \in I$, we get

$$x \le f(x) \le f(a).$$

Thus, $f(a)$ is an upper bound of I and so, as a is the least upper bound of I, we have

$$a \le f(a). \tag{$\dagger$}$$

By monotonicity of f, we get $f(a) \leq f(f(a))$. Now, to claim that $f(a) \in I$, we need to check that $f(a)$ is a lower bound of S. However, by definition of I, every element of S is an upper bound of I and because a is the least upper bound of I, we must have $a \leq s$ for all $s \in S$; that is, a is a lower bound of S. By monotonicity of f and the fact that S is a set of fixed points, we get

$$f(a) \leq f(s) = s, \text{ for all } s \in S,$$

which shows that $f(a)$ is a lower bound of S and thus, $f(a) \in I$, as contended. As a is an upper bound of I and $f(a) \in I$, we must have

$$f(a) \leq a, \tag{$\dagger\dagger$}$$

and together with $(\dagger)$, we conclude that $f(a) = a$; that is, a is a fixed point of f.

We already proved that a is a lower bound of S thus it only remains to show that if x is any fixed point of f and x is a lower bound of S, then $x \leq a$. But, if x is any fixed point of f then $x \leq f(x)$ and because x is also a lower bound of S, then $x \in I$. As a is an upper bound of I, we do get $x \leq a$. $\square$

It should be noted that the least upper bounds and the greatest lower bounds in F do not necessarily agree with those in X. In technical terms, F is generally not a sublattice of X.

Now, as promised, we use Tarski's fixed-point theorem to prove the Schröder–Bernstein theorem.

Theorem 2.7 *Given any two sets A and B, if there is an injection from A to B and an injection from B to A, then there is a bijection between A and B.*
Proof. Let $f: A \to B$ and $g: B \to A$ be two injections. We define the function $\varphi: 2^A \to 2^A$ by

$$\varphi(S) = A - g(B - f(S)),$$

for any $S \subseteq A$. Because of the two complementations, it is easy to check that φ is monotonic (check it). As 2^A is a complete lattice, by Tarski's fixed point theorem, the function φ has a fixed point; that is, there is some subset $C \subseteq A$ so that

$$C = A - g(B - f(C)).$$

By taking the complement of C in A, we get

$$A - C = g(B - f(C)).$$

Now, as f and g are injections, the restricted functions $f \upharpoonright C: C \to f(C)$ and $g \upharpoonright (B - f(C)): (B - f(C)) \to (A - C)$ are bijections. Using these functions, we define the function $h: A \to B$ as follows.

$$h(a) = \begin{cases} f(a) & \text{if } a \in C \\ (g \upharpoonright (B - f(C))^{-1}(a) & \text{if } a \notin C. \end{cases}$$

The reader may check that h is indeed a bijection. $\square$

The above proof is probably the shortest known proof of the Schröder–Bernstein theorem because it uses Tarski's fixed-point theorem, a powerful result. If one looks carefully at the proof, one realizes that there are two crucial ingredients:

1. The set C is closed under $g \circ f$; that is, $g \circ f(C) \subseteq C$.
2. $A - C \subseteq g(B)$.

Using these observations, it is possible to give a proof that circumvents the use of Tarski's theorem. Such a proof is given in Enderton [6], Chapter 6, and we give a sketch of this proof below.

Define a sequence of subsets C_n of A by recursion as follows.

$$C_0 = A - g(B)$$
$$C_{n+1} = (g \circ f)(C_n),$$

and set

$$C = \bigcup_{n \geq 0} C_n.$$

Clearly, $A - C \subseteq g(B)$ and because direct images preserve unions, $(g \circ f)(C) \subseteq C$. The definition of h is similar to the one used in our proof:

$$h(a) = \begin{cases} f(a) & \text{if } a \in C \\ (g \restriction (A - C))^{-1}(a) & \text{if } a \notin C. \end{cases}$$

When $a \notin C$, that is, $a \in A - C$, as $A - C \subseteq g(B)$ and g is injective, $g^{-1}(a)$ is indeed well-defined. As f and g are injective, so is g^{-1} on $A - C$. So, to check that h is injective, it is enough to prove that $f(a) = g^{-1}(b)$ with $a \in C$ and $b \notin C$ is impossible. However, if $f(a) = g^{-1}(b)$, then $(g \circ f)(a) = b$. Because $(g \circ f)(C) \subseteq C$ and $a \in C$, we get $b = (g \circ f)(a) \in C$, yet $b \notin C$, a contradiction. It is not hard to verify that h is surjective and therefore, h is a bijection between A and B. $\square$

The classical reference on lattices is Birkhoff [1]. We highly recommend this beautiful book (but it is not easy reading).

We now turn to special properties of partial orders having to do with induction.

5.3 Well-Founded Orderings and Complete Induction

Have you ever wondered why induction on $\mathbb{N}$ actually "works"? The answer, of course, is that $\mathbb{N}$ was defined in such a way that, by Theorem 1.4, it is the "smallest" inductive set. But this is not a very illuminating answer. The key point is that every nonempty subset of $\mathbb{N}$ has a least element. This fact is intuitively clear inasmuch as if we had some nonempty subset of $\mathbb{N}$ with no smallest element, then we could construct an infinite strictly decreasing sequence, $k_0 > k_1 > \cdots > k_n > \cdots$. But this is absurd, as such a sequence would eventually run into 0 and stop. It turns out that the deep reason why induction "works" on a poset is indeed that the poset ordering has a very special property and this leads us to the following definition.

Definition 5.6. Given a poset $\langle X, \leq \rangle$ we say that $\leq$ is a *well-order (well-ordering)* and that X is *well-ordered by* $\leq$ iff every nonempty subset of X has a least element.

When X is nonempty, if we pick any two-element subset $\{a, b\}$ of X, because the subset $\{a, b\}$ must have a least element, we see that either $a \leq b$ or $b \leq a$; that is, *every well-order is a total order*. First, let us confirm that $\mathbb{N}$ is indeed well-ordered.

Theorem 5.3. *(Well-Ordering of $\mathbb{N}$) The set of natural numbers $\mathbb{N}$ is well-ordered.*

Proof. Not surprisingly we use induction, but we have to be a little shrewd. Let A be any nonempty subset of $\mathbb{N}$. We prove by contradiction that A has a least element. So, suppose A does not have a least element and let $P(m)$ be the predicate

$$P(m) \equiv (\forall k \in \mathbb{N})(k < m \Rightarrow k \notin A),$$

which says that no natural number strictly smaller than m is in A. We prove by induction on m that $P(m)$ holds. But then, the fact that $P(m)$ holds for all m shows that $A = \emptyset$, a contradiction.

Let us now prove $P(m)$ by induction. The base case $P(0)$ holds trivially. Next, assume $P(m)$ holds; we want to prove that $P(m + 1)$ holds. Pick any $k < m + 1$. Then, either

(1) $k < m$, in which case, by the induction hypothesis, $k \notin A$; or
(2) $k = m$. By the induction hypothesis, $P(m)$ holds. Now, if m were in A, as $P(m)$ holds no $k < m$ would belong to A and m would be the least element of A, contradicting the assumption that A has no least element. Therefore, $m \notin A$.

Thus in both cases we proved that if $k < m + 1$, then $k \notin A$, establishing the induction hypothesis. This concludes the induction and the proof of Theorem 5.3. $\square$

Theorem 5.3 yields another induction principle which is often more flexible than our original induction principle. This principle, called *complete induction* (or sometimes *strong induction*), was already encountered in Section 2.3. It turns out that it is a special case of induction on a well-ordered set but it does not hurt to review it in the special case of the natural ordering on $\mathbb{N}$. Recall that $\mathbb{N}_+ = \mathbb{N} - \{0\}$.

Complete Induction Principle on $\mathbb{N}$.
In order to prove that a predicate $P(n)$ holds for all $n \in \mathbb{N}$ it is enough to prove that

(1) $P(0)$ holds (the base case).
(2) For every $m \in \mathbb{N}_+$, if $(\forall k \in \mathbb{N})(k < m \Rightarrow P(k))$ then $P(m)$.

As a formula, complete induction is stated as

$$P(0) \wedge (\forall m \in \mathbb{N}_+)[(\forall k \in \mathbb{N})(k < m \Rightarrow P(k)) \Rightarrow P(m)] \Rightarrow (\forall n \in \mathbb{N})P(n).$$

The difference between ordinary induction and complete induction is that in complete induction, the induction hypothesis $(\forall k \in \mathbb{N})(k < m \Rightarrow P(k))$ assumes that $P(k)$ holds for all $k < m$ and not just for $m - 1$ (as in ordinary induction), in order to deduce $P(m)$. This gives us more proving power as we have more knowledge in order to prove $P(m)$.

We have many occasions to use complete induction but let us first check that it is a valid principle. Even though we already sketched how the validity of complete induction is a consequence of the (ordinary) induction principle (Version 3) on $\mathbb{N}$ in Section 2.3 and we soon give a more general proof of the validity of complete induction for a well-ordering, we feel that it is helpful to give the proof in the case of $\mathbb{N}$ as a warm-up.

Theorem 5.4. *The complete induction principle for $\mathbb{N}$ is valid.*

Proof. Let $P(n)$ be a predicate on $\mathbb{N}$ and assume that $P(n)$ satisfies Conditions (1) and (2) of complete induction as stated above. We proceed by contradiction. So, assume that $P(n)$ fails for some $n \in \mathbb{N}$. If so, the set

$$F = \{n \in \mathbb{N} \mid P(n) = \textbf{false}\}$$

is nonempty. By Theorem 5.3, the set A has a least element m and thus

$$P(m) = \textbf{false}.$$

Now, we can't have $m = 0$, as we assumed that $P(0)$ holds (by (1)) and because m is the least element for which $P(m) = \textbf{false}$, we must have

$$P(k) = \textbf{true} \text{ for all } k < m.$$

But, this is exactly the premise in (2) and as we assumed that (2) holds, we deduce that
$$P(m) = \textbf{true},$$

contradicting the fact that we already know that $P(m) = \textbf{false}$. Therefore, $P(n)$ must hold for all $n \in \mathbb{N}$. $\square$

Remark: In our statement of the principle of complete induction, we singled out the base case (1), and consequently we stated the induction step (2) for every $m \in \mathbb{N}_+$, excluding the case $m = 0$, which is already covered by the base case. It is also possible to state the principle of complete induction in a more concise fashion as follows.

$$(\forall m \in \mathbb{N})[(\forall k \in \mathbb{N})(k < m \Rightarrow P(k)) \Rightarrow P(m)] \Rightarrow (\forall n \in \mathbb{N})P(n).$$

In the above formula, observe that when $m = 0$, which is now allowed, the premise $(\forall k \in \mathbb{N})(k < m \Rightarrow P(k))$ of the implication within the brackets is trivially true and so, $P(0)$ must still be established. In the end, exactly the same amount of work is required but some people prefer the second more concise version of the principle of complete induction. We feel that it would be easier for the reader to make the transition from ordinary induction to complete induction if we make explicit the fact that the base case must be established.

Let us illustrate the use of the complete induction principle by proving that every natural number factors as a product of primes. Recall that for any two natural numbers, $a, b \in \mathbb{N}$ with $b \neq 0$, we say that b divides a iff $a = bq$, for some $q \in \mathbb{N}$. In this case, we say that *a is divisible by b* and that *b is a factor of a*. Then, we say that a natural number $p \in \mathbb{N}$ is a *prime number* (for short, a *prime*) if $p \geq 2$ and if p is only divisible by itself and by 1. Any prime number but 2 must be odd but the converse is false. For example, $2, 3, 5, 7, 11, 13, 17$ are prime numbers, but 9 is not. There are infinitely many prime numbers but to prove this, we need the following theorem.

Theorem 5.5. *Every natural number $n \geq 2$ can be factored as a product of primes; that is, n can be written as a product $n = p_1^{m_1} \cdots p_k^{m_k}$, where the p_is are pairwise distinct prime numbers and $m_i \geq 1$ $(1 \leq i \leq k)$.*

Proof. We proceed by complete induction on $n \geq 2$. The base case, $n = 2$ is trivial, inasmuch as 2 is prime.

Consider any $n > 2$ and assume that the induction hypothesis holds; that is, every m with $2 \leq m < n$ can be factored as a product of primes. There are two cases.

(a) The number n is prime. Then, we are done.
(b) The number n is not a prime. In this case, n factors as $n = n_1 n_2$, where $2 \leq n_1, n_2 < n$. By the induction hypothesis, n_1 has some prime factorization and so does n_2. If $\{p_1, \ldots, p_k\}$ is the union of all the primes occurring in these factorizations of n_1 and n_2, we can write

$$n_1 = p_1^{i_1} \cdots p_k^{i_k} \quad \text{and} \quad n_2 = p_1^{j_1} \cdots p_k^{j_k},$$

where $i_h, j_h \geq 0$ and, in fact, $i_h + j_h \geq 1$, for $1 \leq h \leq k$. Consequently, n factors as the product of primes,

$$n = p_1^{i_1 + j_1} \cdots p_k^{i_k + j_k},$$

with $i_h + j_h \geq 1$, establishing the induction hypothesis. $\square$

For example, $21 = 3^1 \cdot 7^1$, $98 = 2^1 \cdot 7^2$, and $396 = 2^2 \cdot 3^3 \cdot 11$.

Remark: The prime factorization of a natural number is unique up to permutation of the primes $p_1, \ldots, p_k$ but this requires the Euclidean division lemma. However, we can prove right away that there are infinitely primes.

Theorem 5.6. *Given any natural number $n \geq 1$ there is a prime number p such that $p > n$. Consequently, there are infinitely many primes.*

Proof. Let $m = n! + 1$. If m is prime, we are done. Otherwise, by Theorem 5.5, the number m has a prime decomposition. We claim that $p > n$ for every prime p in this decomposition. If not, $2 \leq p \leq n$ and then p would divide both $n! + 1$ and $n!$, so p would divide 1, a contradiction. □

As an application of Theorem 5.3, we prove the Euclidean division lemma for the integers.

Theorem 5.7. *(Euclidean Division Lemma for $\mathbb{Z}$) Given any two integers $a, b \in \mathbb{Z}$, with $b \neq 0$, there is some unique integer $q \in \mathbb{Z}$ (the quotient) and some unique natural number $r \in \mathbb{N}$ (the remainder or residue), so that*

$$a = bq + r \quad with \quad 0 \leq r < |b|.$$

Proof. First, let us prove the existence of q and r with the required condition on r. We claim that if we show existence in the special case where $a, b \in \mathbb{N}$ (with $b \neq 0$), then we can prove existence in the general case. There are four cases:

1. If $a, b \in \mathbb{N}$, with $b \neq 0$, then we are done.
2. If $a \geq 0$ and $b < 0$, then $-b > 0$, so we know that there exist q, r with

$$a = (-b)q + r \quad with \quad 0 \leq r \leq -b - 1.$$

Then,
$$a = b(-q) + r \quad with \quad 0 \leq r \leq |b| - 1.$$

3. If $a < 0$ and $b > 0$, then $-a > 0$, so we know that there exist q, r with

$$-a = bq + r \quad with \quad 0 \leq r \leq b - 1.$$

Then,
$$a = b(-q) - r \quad with \quad 0 \leq r \leq b - 1.$$

If $r = 0$, we are done. Otherwise, $1 \leq r \leq b - 1$, which implies $1 \leq b - r \leq b - 1$, so we get

$$a = b(-q) - b + b - r = b(-(q+1)) + b - r \quad with \quad 0 \leq b - r \leq b - 1.$$

4. If $a < 0$ and $b < 0$, then $-a > 0$ and $-b > 0$, so we know that there exist q, r with

$$-a = (-b)q + r \quad with \quad 0 \leq r \leq -b - 1.$$

Then,
$$a = bq - r \quad with \quad 0 \leq r \leq -b - 1.$$

If $r = 0$, we are done. Otherwise, $1 \leq r \leq -b - 1$, which implies $1 \leq -b - r \leq -b - 1$, so we get

$$a = bq + b - b - r = b(q+1) + (-b - r) \quad with \quad 0 \leq -b - r \leq |b| - 1.$$

We are now reduced to proving the existence of q and r when $a, b \in \mathbb{N}$ with $b \neq 0$. Consider the set

$$R = \{a - bq \in \mathbb{N} \mid q \in \mathbb{N}\}.$$

Note that $a \in R$ by setting $q = 0$, because $a \in \mathbb{N}$. Therefore, R is nonempty. By Theorem 5.3, the nonempty set R has a least element r. We claim that $r \leq b - 1$ (of course, $r \geq 0$ as $R \subseteq \mathbb{N}$). If not, then $r \geq b$, and so $r - b \geq 0$. As $r \in R$, there is some $q \in \mathbb{N}$ with $r = a - bq$. But now, we have

$$r - b = a - bq - b = a - b(q + 1)$$

and as $r - b \geq 0$, we see that $r - b \in R$ with $r - b < r$ (because $b \neq 0$), contradicting the minimality of r. Therefore, $0 \leq r \leq b - 1$, proving the existence of q and r with the required condition on r.

We now go back to the general case where $a, b \in \mathbb{Z}$ with $b \neq 0$ and we prove uniqueness of q and r (with the required condition on r). So, assume that

$$a = bq_1 + r_1 = bq_2 + r_2 \quad \text{with} \quad 0 \leq r_1 \leq |b| - 1 \quad \text{and} \quad 0 \leq r_2 \leq |b| - 1.$$

Now, as $0 \leq r_1 \leq |b| - 1$ and $0 \leq r_2 \leq |b| - 1$, we have $|r_1 - r_2| < |b|$, and from $bq_1 + r_1 = bq_2 + r_2$, we get

$$b(q_2 - q_1) = r_1 - r_2,$$

which yields

$$|b||q_2 - q_1| = |r_1 - r_2|.$$

Because $|r_1 - r_2| < |b|$, we must have $r_1 = r_2$. Then, from $b(q_2 - q_1) = r_1 - r_2 = 0$, as $b \neq 0$, we get $q_1 = q_2$, which concludes the proof. $\square$

For example, $12 = 5 \cdot 2 + 2$, $200 = 5 \cdot 40 + 0$, and $42823 = 6409 \times 6 + 4369$. The remainder r in the Euclidean division, $a = bq + r$, of a by b, is usually denoted $a \bmod b$.

We now show that complete induction holds for a very broad class of partial orders called *well-founded orderings* that subsume well-orderings.

Definition 5.7. Given a poset $\langle X, \leq \rangle$, we say that $\leq$ is a *well-founded ordering (order)* and that X is *well founded* iff X has **no** infinite strictly decreasing sequence $x_0 > x_1 > x_2 > \cdots > x_n > x_{n+1} > \cdots$.

The following property of well-founded sets is fundamental.

Proposition 5.4. *A poset $\langle X, \leq \rangle$ is well founded iff every nonempty subset of X has a minimal element.*

Proof. First, assume that every nonempty subset of X has a minimal element. If we had an infinite strictly decreasing sequence, $x_0 > x_1 > x_2 > \cdots > x_n > \cdots$, then the set $A = \{x_n\}$ would have no minimal element, a contradiction. Therefore, X is well founded.

Now, assume that X is well founded. We prove that A has a minimal element by contradiction. So, let A be some nonempty subset of X and suppose A has no minimal element. This means that for every $a \in A$, there is some $b \in A$ with $a > b$. Using the axiom of choice (graph version), there is some function $g \colon A \to A$ with the property that

$$a > g(a), \text{ for all } a \in A.$$

Inasmuch as A is nonempty, we can pick some element, say $a \in A$. By the recursion Theorem (Theorem 2.1), there is a unique function $f \colon \mathbb{N} \to A$ so that

$$f(0) = a,$$
$$f(n+1) = g(f(n)) \text{ for all } n \in \mathbb{N}.$$

But then, f defines an infinite sequence $\{x_n\}$ with $x_n = f(n)$, so that $x_n > x_{n+1}$ for all $n \in \mathbb{N}$, contradicting the fact that X is well founded. $\square$

So, the seemingly weaker condition that there is **no** infinite strictly decreasing sequence in X is equivalent to the fact that every nonempty subset of X has a minimal element. If X is a total order, any minimal element is actually a least element and so we get the following.

Corollary 5.1. *A poset, $\langle X, \leq \rangle$, is well-ordered iff $\leq$ is total and X is well founded.*

Note that the notion of a well-founded set is more general than that of a well-ordered set, because a well-founded set is not necessarily totally ordered.

Remark: Suppose we can prove some property P by ordinary induction on $\mathbb{N}$. Then, I claim that P can also be proved by complete induction on $\mathbb{N}$. To see this, observe first that the base step is identical. Also, for all $m \in \mathbb{N}_+$, the implication

$$(\forall k \in \mathbb{N})(k < m \Rightarrow P(k)) \Rightarrow P(m-1)$$

holds and because the induction step (in ordinary induction) consists in proving for all $m \in \mathbb{N}_+$ that

$$P(m-1) \Rightarrow P(m)$$

holds, from this implication and the previous implication we deduce that for all $m \in \mathbb{N}_+$, the implication

$$(\forall k \in \mathbb{N})(k < m \Rightarrow P(k)) \Rightarrow P(m)$$

holds, which is exactly the induction step of the complete induction method. So, we see that complete induction on $\mathbb{N}$ subsumes ordinary induction on $\mathbb{N}$. The converse is also true but we leave it as a fun exercise. But now, by Theorem 5.3 (ordinary) induction on $\mathbb{N}$ implies that $\mathbb{N}$ is well-ordered and by Theorem 5.4, the fact that $\mathbb{N}$ is well-ordered implies complete induction on $\mathbb{N}$. We just showed that complete induction on $\mathbb{N}$ implies (ordinary) induction on $\mathbb{N}$, therefore we conclude that all three are equivalent; that is,

(ordinary) induction on $\mathbb{N}$ is valid
iff

complete induction on $\mathbb{N}$ is valid

iff

$\mathbb{N}$ is well-ordered.

These equivalences justify our earlier claim that the ability to do induction hinges on some key property of the ordering, in this case, that it is a well-ordering.

We finally come to the principle of *complete induction* (also called *transfinite induction* or *structural induction*), which, as we prove, is valid for all well-founded sets. Every well-ordered set is also well-founded, thus complete induction is a very general induction method.

Let $(X, \leq)$ be a well-founded poset and let P be a predicate on X (i.e., a function $P \colon X \to \{\mathbf{true}, \mathbf{false}\}$).

Principle of Complete Induction on a Well-Founded Set.
To prove that a property P holds for all $z \in X$, it suffices to show that, for every $x \in X$,

$(*)$ If x is minimal or $P(y)$ holds for all $y < x$,
$(**)$ Then $P(x)$ holds.

The statement $(*)$ is called the *induction hypothesis*, and the implication for all x, $(*)$ implies $(**)$ is called the *induction step*. Formally, the induction principle can be stated as:

$$(\forall x \in X)[(\forall y \in X)(y < x \Rightarrow P(y)) \Rightarrow P(x)] \Rightarrow (\forall z \in X)P(z) \qquad \text{(CI)}$$

Note that if x is minimal, then there is no $y \in X$ such that $y < x$, and $(\forall y \in X)(y < x \Rightarrow P(y))$ is true. Hence, we must show that $P(x)$ holds for every minimal element x. These cases are called the *base cases*.

Complete induction is not valid for arbitrary posets (see the problems) but holds for well-founded sets as shown in the following theorem.

Theorem 5.8. *The principle of complete induction holds for every well-founded set.*

Proof. We proceed by contradiction. Assume that (CI) is false. Then,

$$(\forall x \in X)[(\forall y \in X)(y < x \Rightarrow P(y)) \Rightarrow P(x)] \qquad (1)$$

holds and

$$(\forall z \in X)P(z) \qquad (2)$$

is false, that is, there is some $z \in X$ so that

$$P(z) = \mathbf{false}.$$

Hence, the subset F of X defined by

$$F = \{x \in X \mid P(x) = \mathbf{false}\}$$

is nonempty. Because X is well founded, by Proposition 5.4, F has some minimal element b. Because (1) holds for all $x \in X$, letting $x = b$, we see that

$$[(\forall y \in X)(y < b \Rightarrow P(y)) \Rightarrow P(b)] \tag{3}$$

holds. If b is also minimal in X, then there is no $y \in X$ such that $y < b$ and so,

$$(\forall y \in X)(y < b \Rightarrow P(y))$$

holds trivially and (3) implies that $P(b) = $ **true**, which contradicts the fact that $b \in F$. Otherwise, for every $y \in X$ such that $y < b$, $P(y) = $ **true**, because otherwise y would belong to F and b would not be minimal. But then,

$$(\forall y \in X)(y < b \Rightarrow P(y))$$

also holds and (3) implies that $P(b) = $ **true**, contradicting the fact that $b \in F$. Hence, complete induction is valid for well-founded sets. $\square$

As an illustration of well-founded sets, we define the *lexicographic ordering* on pairs. Given a partially ordered set $\langle X, \leq \rangle$, the *lexicographic ordering* $<<$ on $X \times X$ induced by $\leq$ is defined as follows. For all $x, y, x', y' \in X$,

$$(x, y) << (x', y') \quad \text{iff either}$$

$$\begin{aligned} x = x' \quad &\text{and} \quad y = y' \quad \text{or} \\ x < x' \quad &\text{or} \\ x = x' \quad &\text{and} \quad y < y'. \end{aligned}$$

We leave it as an exercise to check that $<<$ is indeed a partial order on $X \times X$. The following proposition is useful.

Proposition 5.5. *If $\langle X, \leq \rangle$ is a well-founded set, then the lexicographic ordering $<<$ on $X \times X$ is also well-founded.*

Proof. We proceed by contradiction. Assume that there is an infinite decreasing sequence $(\langle x_i, y_i \rangle)_i$ in $X \times X$. Then, either,

(1) There is an infinite number of distinct x_i, or
(2) There is only a finite number of distinct x_i.

In case (1), the subsequence consisting of these distinct elements forms a decreasing sequence in X, contradicting the fact that $\leq$ is well-founded. In case (2), there is some k such that $x_i = x_{i+1}$, for all $i \geq k$. By definition of $<<$, the sequence $(y_i)_{i \geq k}$ is a decreasing sequence in X, contradicting the fact that $\leq$ is well-founded. Hence, $<<$ is well-founded on $X \times X$. $\square$

As an illustration of the principle of complete induction, consider the following example in which it is shown that a function defined recursively is a total function.

Example (Ackermann's Function) The following function, $A: \mathbb{N} \times \mathbb{N} \to \mathbb{N}$, known as *Ackermann's function* is well known in recursive function theory for its extraordinary rate of growth. It is defined recursively as follows.

$$A(x,y) = \textbf{if } x = 0 \textbf{ then } y + 1$$
$$\textbf{else if } y = 0 \textbf{ then } A(x-1,1)$$
$$\textbf{else } A(x-1, A(x, y-1)).$$

We wish to prove that A is a total function. We proceed by complete induction over the lexicographic ordering on $\mathbb{N} \times \mathbb{N}$.

1. The base case is $x = 0$, $y = 0$. In this case, because $A(0,y) = y + 1$, $A(0,0)$ is defined and equal to 1.
2. The induction hypothesis is that for any (m,n), $A(m',n')$ is defined for all $(m',n') << (m,n)$, with $(m,n) \neq (m',n')$.
3. For the induction step, we have three cases:

   a. If $m = 0$, because $A(0,y) = y + 1$, $A(0,n)$ is defined and equal to $n+1$.
   b. If $m \neq 0$ and $n = 0$, because $(m-1,1) << (m,0)$ and $(m-1,1) \neq (m,0)$, by the induction hypothesis, $A(m-1,1)$ is defined, and so $A(m,0)$ is defined because it is equal to $A(m-1,1)$.
   c. If $m \neq 0$ and $n \neq 0$, because $(m,n-1) << (m,n)$ and $(m,n-1) \neq (m,n)$, by the induction hypothesis, $A(m,n-1)$ is defined. Because $(m-1,y) << (m,z)$ and $(m-1,y) \neq (m,z)$ no matter what y and z are, $(m-1, A(m,n-1)) << (m,n)$ and $(m-1, A(m,n-1)) \neq (m,n)$, and by the induction hypothesis, $A(m-1, A(m,n-1))$ is defined. But this is precisely $A(m,n)$, and so $A(m,n)$ is defined. This concludes the induction step.

Hence, $A(x,y)$ is defined for all $x,y \geq 0$. □

5.4 Unique Prime Factorization in $\mathbb{Z}$ and GCDs

In the previous section, we proved that every natural number $n \geq 2$ can be factored as a product of primes numbers. In this section, we use the Euclidean division lemma to prove that such a factorization is unique. For this, we need to introduce greatest common divisors (gcds) and prove some of their properties.

In this section, it is convenient to allow 0 to be a divisor. So, given any two integers, $a, b \in \mathbb{Z}$, we say that b *divides* a and that a *is a multiple of* b iff $a = bq$, for some $q \in \mathbb{Z}$. Contrary to our previous definition, $b = 0$ is allowed as a divisor. However, this changes very little because if 0 divides a, then $a = 0q = 0$; that is, *the only integer divisible by 0 is 0*. The notation $b \mid a$ is usually used to denote that b divides a. For example, $3 \mid 21$ because $21 = 2 \cdot 7$, $5 \mid -20$ because $-20 = 5 \cdot (-4)$ but 3 does not divide 20.

We begin by introducing a very important notion in algebra, that of an ideal due to Richard Dedekind, and prove a fundamental property of the ideals of $\mathbb{Z}$.

Fig. 5.8 Richard Dedekind, 1831–1916

Definition 5.8. An *ideal of* $\mathbb{Z}$ is any nonempty subset $\mathfrak{I}$ of $\mathbb{Z}$ satisfying the following two properties.

(ID1) If $a, b \in \mathfrak{I}$, then $b - a \in \mathfrak{I}$.
(ID2) If $a \in \mathfrak{I}$, then $ak \in \mathfrak{I}$ for every $k \in \mathbb{Z}$.

An ideal $\mathfrak{I}$ is a *principal ideal* if there is some $a \in \mathfrak{I}$, *called a generator*, such that $\mathfrak{I} = \{ak \mid k \in \mathbb{Z}\}$. The equality $\mathfrak{I} = \{ak \mid k \in \mathbb{Z}\}$ is also written as $\mathfrak{I} = a\mathbb{Z}$ or as $\mathfrak{I} = (a)$. The ideal $\mathfrak{I} = (0) = \{0\}$ is called the *null ideal*.

Note that if $\mathfrak{I}$ is an ideal, then $\mathfrak{I} = \mathbb{Z}$ iff $1 \in \mathfrak{I}$. Because by definition, an ideal $\mathfrak{I}$ is nonempty, there is some $a \in \mathfrak{I}$, and by (ID1) we get $0 = a - a \in \mathfrak{I}$. Then, for every $a \in \mathfrak{I}$, since $0 \in \mathfrak{I}$, by (ID1) we get $-a \in \mathfrak{I}$.

Theorem 5.9. *Every ideal* $\mathfrak{I}$ *of* $\mathbb{Z}$ *is a principal ideal; that is,* $\mathfrak{I} = m\mathbb{Z}$ *for some unique* $m \in \mathbb{N}$, *with* $m > 0$ *iff* $\mathfrak{I} \neq (0)$.

Proof. Note that $\mathfrak{I} = (0)$ iff $\mathfrak{I} = 0\mathbb{Z}$ and the theorem holds in this case. So, assume that $\mathfrak{I} \neq (0)$. Then, our previous observation that $-a \in \mathfrak{I}$ for every $a \in \mathfrak{I}$ implies that some positive integer belongs to $\mathfrak{I}$ and so, the set $\mathfrak{I} \cap \mathbb{N}_+$ is nonempty. As $\mathbb{N}$ is well ordered, this set has a smallest element, say $m > 0$. We claim that $\mathfrak{I} = m\mathbb{Z}$.
As $m \in \mathfrak{I}$, by (ID2), $m\mathbb{Z} \subseteq \mathfrak{I}$. Conversely, pick any $n \in \mathfrak{I}$. By the Euclidean division lemma, there are unique $q \in \mathbb{Z}$ and $r \in \mathbb{N}$ so that $n = mq + r$, with $0 \le r < m$. If $r > 0$, because $m \in \mathfrak{I}$, by (ID2), $mq \in \mathfrak{I}$, and by (ID1), we get $r = n - mq \in \mathfrak{I}$. Yet $r < m$, contradicting the minimality of m. Therefore, $r = 0$, so $n = mq \in m\mathbb{Z}$, establishing that $\mathfrak{I} \subseteq m\mathbb{Z}$ and thus, $\mathfrak{I} = m\mathbb{Z}$, as claimed. As to uniqueness, clearly $(0) \neq m\mathbb{Z}$ if $m \neq 0$, so assume $m\mathbb{Z} = m'\mathbb{Z}$, with $m > 0$ and $m' > 0$. Then, m divides m' and m' divides m, but we already proved earlier that this implies $m = m'$. $\square$

Theorem 5.9 is often phrased: $\mathbb{Z}$ is a *principal ideal domain*, for short, a *PID*. Note that the natural number m such that $\mathfrak{I} = m\mathbb{Z}$ is a divisor of every element in $\mathfrak{I}$.

Corollary 5.2. *For any two integers, $a,b \in \mathbb{Z}$, there is a unique natural number $d \in \mathbb{N}$, and some integers $u,v \in \mathbb{Z}$, so that d divides both a and b and*

$$ua + vb = d.$$

(The above is called the Bézout identity.) Furthermore, $d = 0$ iff $a = 0$ and $b = 0$.

Proof. It is immediately verified that

$$\mathfrak{I} = \{ha + kb \mid h,k \in \mathbb{Z}\}$$

is an ideal of $\mathbb{Z}$ with $a,b \in \mathfrak{I}$. Therefore, by Theorem 5.9, there is a unique $d \in \mathbb{N}$, so that $\mathfrak{I} = d\mathbb{Z}$. We already observed that d divides every number in $\mathfrak{I}$ so, as $a,b \in \mathfrak{I}$, we see that d divides a and b. If $d = 0$, as d divides a and b, we must have $a = b = 0$. Conversely, if $a = b = 0$, then $d = ua + bv = 0$. $\square$

Given any nonempty finite set of integers $S = \{a_1,\ldots,a_n\}$, it is easy to verify that the set
$$\mathfrak{I} = \{k_1 a_1 + \cdots + k_n a_n \mid k_1,\ldots,k_n \in \mathbb{Z}\}$$
is an ideal of $\mathbb{Z}$ and, in fact, the smallest (under inclusion) ideal containing S. This ideal is called the *ideal generated by S* and it is often denoted $(a_1,\ldots,a_n)$. Corollary 5.2 can be restated by saying that for any two distinct integers, $a,b \in \mathbb{Z}$, there is a unique natural number $d \in \mathbb{N}$, such that the ideal (a,b), generated by a and b is equal to the ideal $d\mathbb{Z}$ (also denoted (d)), that is,

$$(a,b) = d\mathbb{Z}.$$

This result still holds when $a = b$; in this case, we consider the ideal $(a) = (b)$. With a slight (but harmless) abuse of notation, when $a = b$, we also denote this ideal by (a,b).

Fig. 5.9 Étienne Bézout, 1730–1783

The natural number d of Corollary 5.2 divides both a and b. Moreover, every divisor of a and b divides $d = ua + vb$. This motivates the next definition.

Definition 5.9. Given any two integers $a, b \in \mathbb{Z}$, an integer $d \in \mathbb{Z}$ is a *greatest common divisor of a and b* (for short, a *gcd of a and b*) if d divides a and b and, for any integer, $h \in \mathbb{Z}$, if h divides a and b, then h divides d. We say that a and b are *relatively prime* if 1 is a gcd of a and b.

Remarks:

1. If $a = b = 0$ then any integer $d \in \mathbb{Z}$ is a divisor of 0. In particular, 0 divides 0. According to Definition 5.9, this implies $\gcd(0,0) = 0$. The ideal generated by 0 is the trivial ideal (0), so $\gcd(0,0) = 0$ is equal to the generator of the zero ideal, (0).

   If $a \neq 0$ or $b \neq 0$, then the ideal (a,b), generated by a and b is not the zero ideal and there is a unique integer, $d > 0$, such that

   $$(a,b) = d\mathbb{Z}.$$

   For any gcd d', of a and b, because d divides a and b we see that d must divide d'. As d' also divides a and b, the number d' must also divide d. Thus, $d = d'q'$ and $d' = dq$ for some $q, q' \in \mathbb{Z}$ and so, $d = dqq'$ which implies $qq' = 1$ (inasmuch as $d \neq 0$). Therefore, $d' = \pm d$. So, according to the above definition, when $(a,b) \neq (0)$, gcds are not unique. However, exactly one of d' or $-d'$ is positive and equal to the positive generator d, of the ideal (a,b). We refer to this positive gcd as "the" gcd of a and b and write $d = \gcd(a,b)$. Observe that $\gcd(a,b) = \gcd(b,a)$. For example, $\gcd(20,8) = 4$, $\gcd(1000,50) = 50$, $\gcd(42823,6409) = 17$, and $\gcd(5,16) = 1$.

2. Another notation commonly found for $\gcd(a,b)$ is (a,b), but this is confusing because (a,b) also denotes the ideal generated by a and b.

3. Observe that if $d = \gcd(a,b) \neq 0$, then d is indeed the largest positive common divisor of a and b because every divisor of a and b must divide d. However, we did not use this property as one of the conditions for being a gcd because such a condition does not generalize to other rings where a total order is not available. Another minor reason is that if we had used in the definition of a gcd the condition that $\gcd(a,b)$ should be the largest common divisor of a and b, as every integer divides 0, $\gcd(0,0)$ would be undefined.

4. If $a = 0$ and $b > 0$, then the ideal $(0,b)$, generated by 0 and b, is equal to the ideal $(b) = b\mathbb{Z}$, which implies $\gcd(0,b) = b$ and similarly, if $a > 0$ and $b = 0$, then $\gcd(a,0) = a$.

Let $p \in \mathbb{N}$ be a prime number. Then, note that for any other integer n, if p does not divide n, then $\gcd(p,n) = 1$, as the only divisors of p are 1 and p.

Proposition 5.6. *Given any two integers $a, b \in \mathbb{Z}$ a natural number $d \in \mathbb{N}$ is the greatest common divisor of a and b iff d divides a and b and if there are some integers, $u, v \in \mathbb{Z}$, so that*

$$ua + vb = d. \qquad \text{(Bézout identity)}$$

In particular, a and b are relatively prime iff there are some integers $u, v \in \mathbb{Z}$, so that

$$ua + vb = 1. \qquad \text{(Bézout identity)}$$

Proof. We already observed that half of Proposition 5.6 holds, namely if $d \in \mathbb{N}$ divides a and b and if there are some integers $u, v \in \mathbb{Z}$ so that $ua + vb = d$, then d is the gcd of a and b. Conversely, assume that $d = \gcd(a, b)$. If $d = 0$, then $a = b = 0$ and the proposition holds trivially. So, assume $d > 0$, in which case $(a, b) \neq (0)$. By Corollary 5.2, there is a unique $m \in \mathbb{N}$ with $m > 0$ that divides a and b and there are some integers $u, v \in \mathbb{Z}$ so that

$$ua + vb = m.$$

But now m is also the (positive) gcd of a and b, so $d = m$ and our proposition holds. Now, a and b are relatively prime iff $\gcd(a, b) = 1$ in which case the condition that $d = 1$ divides a and b is trivial. $\square$

The gcd of two natural numbers can be found using a method involving Euclidean division and so can the numbers u and v (see Problems 5.18 and 5.19). This method is based on the following simple observation.

Proposition 5.7. *If a, b are any two positive integers with $a \geq b$, then for every $k \in \mathbb{Z}$,*

$$\gcd(a, b) = \gcd(b, a - kb).$$

In particular,

$$\gcd(a, b) = \gcd(b, a - b) = \gcd(b, a + b),$$

and if $a = bq + r$ is the result of performing the Euclidean division of a by b, with $0 \leq r < b$, then

$$\gcd(a, b) = \gcd(b, r).$$

Proof. We claim that

$$(a, b) = (b, a - kb),$$

where (a, b) is the ideal generated by a and b and $(b, a - kb)$ is the ideal generated by b and $a - kb$. Recall that

$$(a, b) = \{k_1 a + k_2 b \mid k_1, k_2 \in \mathbb{Z}\},$$

and similarly for $(b, a - kb)$. Because $a = a - kb + kb$, we have $a \in (b, a - kb)$, so $(a, b) \subseteq (b, a - kb)$. Conversely, we have $a - kb \in (a, b)$ and so, $(b, a - kb) \subseteq (a, b)$. Therefore, $(a, b) = (b, a - kb)$, as claimed. But then, $(a, b) = (b, a - kb) = d\mathbb{Z}$ for a unique positive integer $d > 0$, and we know that

$$\gcd(a, b) = \gcd(b, a - kb) = d,$$

as claimed. The next two equations correspond to $k = 1$ and $k = -1$. When $a = bq + r$, we have $r = a - bq$, so the previous result applies with $k = q$. $\square$

Using the fact that $\gcd(a, 0) = a$, we have the following algorithm for finding the gcd of two natural numbers a, b, with $(a, b) \neq (0, 0)$.

Euclidean Algorithm for Finding the gcd.
The input consists of two natural numbers m, n, with $(m, n) \neq (0, 0)$.

begin
    $a := m; b := n;$
    if $a < b$ **then**
        $t := b; b := a; a := t;$ (swap a and b)
    while $b \neq 0$ **do**
        $r := a \bmod b;$ (divide a by b to obtain the remainder r)
        $a := b; b := r$
    endwhile;
    $\gcd(m, n) := a$
end

In order to prove the correctness of the above algorithm, we need to prove two facts:

1. The algorithm always terminates.
2. When the algorithm exits the while loop, the current value of a is indeed $\gcd(m, n)$.

The termination of the algorithm follows by induction on $\min\{m, n\}$. Without loss of generality, we may assume that $m \geq n$. If $n = 0$, then $b = 0$, the body of the while loop is not even entered and the algorithm stops. If $n > 0$, then $b > 0$, we divide m by n, obtaining $m = qn + r$, with $0 \leq r < n$ and we set a to n and b to r. Because $r < n$, we have $\min\{n, r\} = r < n = \min\{m, n\}$, and by the induction hypothesis, the algorithm terminates.

The correctness of the algorithm is an immediate consequence of Proposition 5.7. During any round through the while loop, the invariant $\gcd(a, b) = \gcd(m, n)$ is preserved, and when we exit the while loop, we have

$$a = \gcd(a, 0) = \gcd(m, n),$$

which proves that the current value of a when the algorithm stops is indeed $\gcd(m, n)$.

Let us run the above algorithm for $m = 42823$ and $n = 6409$. There are five division steps:

$$42823 = 6409 \times 6 + 4369$$
$$6409 = 4369 \times 1 + 2040$$
$$4369 = 2040 \times 2 + 289$$
$$2040 = 289 \times 7 + 17$$
$$289 = 17 \times 17 + 0,$$

so we find that

$$\gcd(42823, 6409) = 17.$$

You should also use your computation to find numbers x, y so that

$$42823x + 6409y = 17.$$

Check that $x = -22$ and $y = 147$ work.

The complexity of the Euclidean algorithm to compute the gcd of two natural numbers is quite interesting and has a long history. It turns out that Gabriel Lamé published a paper in 1844 in which he proved that if $m > n > 0$, then the number of divisions needed by the algorithm is bounded by $5\delta + 1$, where δ is the number of digits in n. For this, Lamé realized that the maximum number of steps is achieved by taking m and n to be two consecutive Fibonacci numbers (see Section 5.8). Dupré, in a paper published in 1845, improved the upper bound to $4.785\delta + 1$, also making use of the Fibonacci numbers. Using a variant of Euclidean division allowing negative remainders, in a paper published in 1841, Binet gave an algorithm with an even better bound: $(10/3)\delta + 1$. For more on these bounds, see Problems 5.18, 5.20, and 5.51. (It should observed that Binet, Lamé, and Dupré do not count the last division step, so the term $+1$ is not present in their upper bounds.)

The Euclidean algorithm can be easily adapted to also compute two integers, x and y, such that

$$mx + ny = \gcd(m, n);$$

see Problem 5.18. Such an algorithm is called the *extended Euclidean algorithm*. Another version of an algorithm for computing x and y is given in Problem 5.19.

What can be easily shown is the following proposition.

Proposition 5.8. *The number of divisions made by the Euclidean algorithm for gcd applied to two positive integers m, n, with $m > n$, is at most $\log_2 m + \log_2 n$.*

Proof. We claim that during every round through the while loop, we have

$$br < \frac{1}{2}ab.$$

Indeed, as $a \geq b$, we have $a = bq + r$, with $q \geq 1$ and $0 \leq r < b$, so $a \geq b + r > 2r$, and thus

$$br < \frac{1}{2}ab,$$

as claimed. But then, if the algorithm requires k divisions, we get

$$0 < \frac{1}{2^k}mn,$$

which yields $mn \geq 2^k$ and by taking logarithms, $k \leq \log_2 m + \log_2 n$. $\qquad\square$

The exact role played by the Fibonacci numbers in figuring out the complexity of the Euclidean algorithm for gcd is explored in Problem 5.51.

We now return to Proposition 5.6 as it implies a very crucial property of divisibility in any PID.

Proposition 5.9. *(Euclid's proposition) Let $a, b, c \in \mathbb{Z}$ be any integers. If a divides bc and a is relatively prime to b, then a divides c.*

Proof. From Proposition 5.6, a and b are relatively prime iff there exist some integers $u, v \in \mathbb{Z}$ such that

$$ua + vb = 1.$$

Then, we have

$$uac + vbc = c,$$

and because a divides bc, it divides both uac and vbc and so, a divides c. $\square$

Fig. 5.10 Euclid of Alexandria, about 325 BC–about 265 BC

In particular, if p is a prime number and if p divides ab, where $a, b \in \mathbb{Z}$ are nonzero, then either p divides a or p divides b because if p does not divide a, by a previous remark, then p and a are relatively prime, so Proposition 5.9 implies that p divides c.

Proposition 5.10. *Let $a, b_1, \ldots, b_m \in \mathbb{Z}$ be any integers. If a and b_i are relatively prime for all i, with $1 \le i \le m$, then a and $b_1 \cdots b_m$ are relatively prime.*

Proof. We proceed by induction on m. The case $m = 1$ is trivial. Let $c = b_2 \cdots b_m$. By the induction hypothesis, a and c are relatively prime. Let d be the gcd of a and $b_1 c$. We claim that d is relatively prime to b_1. Otherwise, d and b_1 would have some gcd $d_1 \ne 1$ which would divide both a and b_1, contradicting the fact that a and b_1 are relatively prime. Now, by Proposition 5.9, d divides $b_1 c$ and d and b_1 are relatively prime, thus d divides $c = b_2 \cdots b_m$. But then, d is a divisor of a and c, and because a and c are relatively prime, $d = 1$, which means that a and $b_1 \cdots b_m$ are relatively prime. $\square$

One of the main applications of the Euclidean algorithm is to find the inverse of a number in modular arithmetic, an essential step in the *RSA algorithm*, the first and still widely used algorithm for public-key cryptography.

Given any natural number $p \geq 1$, we can define a relation on $\mathbb{Z}$, called *congruence*, as follows.

$$n \equiv m \,(\mathrm{mod}\, p)$$

iff $p \mid n - m$; that is, iff $n = m + pk$, for some $k \in \mathbb{Z}$. We say that *m is a residue of n modulo p*.

The notation for congruence was introduced by Carl Friedrich Gauss (1777–1855), one of the greatest mathematicians of all time. Gauss contributed significantly to the theory of congruences and used his results to prove deep and fundamental results in number theory.

Fig. 5.11 Carl Friedrich Gauss, 1777–1855

If $n \geq 1$ and n and p are relatively prime, an *inverse of n modulo p* is a number $s \geq 1$ such that

$$ns \equiv 1 \,(\mathrm{mod}\, p).$$

Using Proposition 5.9 (Euclid's proposition), it is easy to see that that if s_1 and s_2 are both an inverse of n modulo p, then $s_1 \equiv s_2 \,(\mathrm{mod}\, p)$. Finding an inverse of n modulo p means finding some integers x, y, so that $nx = 1 + py$, that is, $nx - py = 1$, therefore we can find x and y using the extended Euclidean algorithm; see Problems 5.18 and 5.19. If $p = 1$, we can pick $x = 1$ and $y = n - 1$ and 1 is the smallest positive inverse of n modulo 1. Let us now assume that $p \geq 2$. Using Euclidean division (even if x is negative), we can write

$$x = pq + r,$$

where $1 \leq r < p$ ($r \neq 0$ because otherwise $p \geq 2$ would divide 1), so that

$$nx - py = n(pq + r) - py = nr - p(y - nq) = 1,$$

and r is the unique inverse of n modulo p such that $1 \leq r < p$.

We can now prove the uniqueness of prime factorizations in $\mathbb{N}$. The first rigorous proof of this theorem was given by Gauss.

Theorem 5.10. *(Unique Prime Factorization in $\mathbb{N}$) For every natural number $a \geq 2$, there exists a unique set $\{\langle p_1, k_1 \rangle, \ldots, \langle p_m, k_m \rangle\}$, where the p_is are distinct prime numbers and the k_is are (not necessarily distinct) integers, with $m \geq 1$, $k_i \geq 1$, so* *that*

$$a = p_1^{k_1} \cdots p_m^{k_m}.$$

Proof. The existence of such a factorization has already been proved in Theorem 5.5.

Let us now prove uniqueness. Assume that

$$a = p_1^{k_1} \cdots p_m^{k_m} \quad \text{and} \quad a = q_1^{h_1} \cdots q_n^{h_n}.$$

Thus, we have

$$p_1^{k_1} \cdots p_m^{k_m} = q_1^{h_1} \cdots q_n^{h_n}.$$

We prove that $m = n$, $p_i = q_i$, and $h_i = k_i$, for all i, with $1 \leq i \leq n$. The proof proceeds by induction on $h_1 + \cdots + h_n$.

If $h_1 + \cdots + h_n = 1$, then $n = 1$ and $h_1 = 1$. Then,

$$p_1^{k_1} \cdots p_m^{k_m} = q_1,$$

and because q_1 and the p_i are prime numbers, we must have $m = 1$ and $p_1 = q_1$ (a prime is only divisible by 1 or itself).

If $h_1 + \cdots + h_n \geq 2$, because $h_1 \geq 1$, we have

$$p_1^{k_1} \cdots p_m^{k_m} = q_1 q,$$

with

$$q = q_1^{h_1 - 1} \cdots q_n^{h_n},$$

where $(h_1 - 1) + \cdots + h_n \geq 1$ (and $q_1^{h_1 - 1} = 1$ if $h_1 = 1$). Now, if q_1 is not equal to any of the p_i, by a previous remark, q_1 and p_i are relatively prime, and by Proposition 5.10, q_1 and $p_1^{k_1} \cdots p_m^{k_m}$ are relatively prime. But this contradicts the fact that q_1 divides $p_1^{k_1} \cdots p_m^{k_m}$. Thus, q_1 is equal to one of the p_i. Without loss of generality, we can assume that $q_1 = p_1$. Then, as $q_1 \neq 0$, we get

$$p_1^{k_1 - 1} \cdots p_m^{k_m} = q_1^{h_1 - 1} \cdots q_n^{h_n},$$

where $p_1^{k_1 - 1} = 1$ if $k_1 = 1$, and $q_1^{h_1 - 1} = 1$ if $h_1 = 1$. Now, $(h_1 - 1) + \cdots + h_n < h_1 + \cdots + h_n$, and we can apply the induction hypothesis to conclude that $m = n$, $p_i = q_i$ and $h_i = k_i$, with $1 \leq i \leq n$. $\quad \square$

Theorem 5.10 is a basic but very important result of number theory and it has many applications. It also reveals the importance of the primes as the building blocks of all numbers.

Remark: Theorem 5.10 also applies to any nonzero integer $a \in \mathbb{Z} - \{-1, +1\}$, by adding a suitable sign in front of the prime factorization. That is, we have a unique prime factorization of the form

$$a = \pm p_1^{k_1} \cdots p_m^{k_m}.$$

Theorem 5.10 shows that $\mathbb{Z}$ is a *unique factorization domain*, for short, a *UFD*. Such rings play an important role because every nonzero element that is not a unit (i.e., which is not invertible) has a unique factorization (up to some unit factor) into so-called *irreducible elements* which generalize the primes.

Readers who would like to learn more about number theory are strongly advised to read Silverman's delightful and very "friendly" introductory text [15]. Another excellent but more advanced text is Davenport [3] and an even more comprehensive book (and a classic) is Niven, Zuckerman, and Montgomery [12]. For those interested in the history of number theory (up to Gauss), we highly recommend Weil [17], a fascinating book (but no easy reading).

In the next section, we give a beautiful application of the pigeonhole principle to number theory due to Dirichlet (1805–1949).

5.5 Dirichlet's Diophantine Approximation Theorem

The pigeonhole principle (see Section 2.9) was apparently first stated explicitly by Dirichlet in 1834. Dirichlet used the pigeonhole principle (under the name *Schubfachschluß*) to prove a fundamental theorem about the approximation of irrational numbers by fractions (rational numbers). The proof is such a beautiful illustration

Fig. 5.12 Johan Peter Gustav Lejeune Dirichlet, 1805–1859

of the use of the pigeonhole principle that we can't resist presenting it. Recall that a real number $\alpha \in \mathbb{R}$ is *irrational* iff it cannot be written as a fraction $p/q \in \mathbb{Q}$.

Theorem 5.11. *(Dirichlet) For every positive irrational number $\alpha > 0$, there are infinitely many pairs of positive integers, (x, y), such that $\gcd(x, y) = 1$ and*

$$|x - y\alpha| < \frac{1}{y}.$$

Proof. Pick any positive integer m such that $m \geq 1/\alpha$, and consider the numbers

$$0, \alpha, 2\alpha, 3\alpha, \cdots, m\alpha.$$

We can write each number in the above list as the sum of a whole number (a natural number) and a decimal real part, between 0 and 1, say

$$0 = N_0 + F_0$$
$$\alpha = N_1 + F_1$$
$$2\alpha = N_2 + F_2$$
$$3\alpha = N_3 + F_3$$
$$\vdots$$
$$m\alpha = N_m + F_m,$$

with $N_0 = F_0 = 0$, $N_i \in \mathbb{N}$, and $0 \le F_i < 1$, for $i = 1,\ldots,m$. Observe that there are $m+1$ numbers $F_0,\ldots,F_m$. Consider the m "boxes" consisting of the intervals

$$\left\{ t \in \mathbb{R} \;\middle|\; \frac{i}{m} \le t < \frac{i+1}{m} \right\}, \qquad 0 \le i \le m-1.$$

There are $m+1$ numbers F_i, and only m intervals, thus by the pigeonhole principle, two of these numbers must be in the same interval, say F_i and F_j, for $i < j$. As

$$\frac{i}{m} \le F_i, F_j < \frac{i+1}{m},$$

we must have

$$|F_i - F_j| < \frac{1}{m}$$

and because $i\alpha = N_i + F_i$ and $j\alpha = N_j + F_j$, we conclude that

$$|i\alpha - N_i - (j\alpha - N_j)| < \frac{1}{m};$$

that is,

$$|N_j - N_i - (j-i)\alpha| < \frac{1}{m}.$$

Note that $1 \le j - i \le m$ and so, if $N_j - N_i = 0$, then

$$\alpha < \frac{1}{(j-i)m} \le \frac{1}{m},$$

which contradicts the hypothesis $m \ge 1/\alpha$. Therefore, $x = N_j - N_i > 0$ and $y = j - i > 0$ are positive integers such that $y \le m$ and

$$|x - y\alpha| < \frac{1}{m}.$$

If $\gcd(x,y) = d > 1$, then write $x = dx'$, $y = dy'$, and divide both sides of the above inequality by d to obtain

$$|x' - y'\alpha| < \frac{1}{md} < \frac{1}{m}$$

with $\gcd(x', y') = 1$ and $y' < m$. In either case, we proved that there exists a pair of positive integers (x, y), with $y \leq m$ and $\gcd(x, y) = 1$ such that

$$|x - y\alpha| < \frac{1}{m}.$$

However, $y \leq m$, so we also have

$$|x - y\alpha| < \frac{1}{m} \leq \frac{1}{y},$$

as desired.

Suppose that there are only finitely many pairs (x, y) satisfying $\gcd(x, y) = 1$ and

$$|x - y\alpha| < \frac{1}{y}.$$

In this case, there are finitely many values for $|x - y\alpha|$ and thus, the minimal value of $|x - y\alpha|$ is achieved for some (x_0, y_0). Furthermore, as α is irrational, we have $0 < |x_0 - y_0\alpha|$. However, if we pick m large enough, we can find (x, y) such that $\gcd(x, y) = 1$ and

$$|x - y\alpha| < \frac{1}{m} < |x_0 - y_0\alpha|,$$

contradicting the minimality of $|x_0 - y_0\alpha|$. Therefore, there are infinitely many pairs (x, y), satisfying the theorem. $\square$

Note that Theorem 5.11 yields rational approximations for α, because after division by y, we get

$$\left|\frac{x}{y} - \alpha\right| < \frac{1}{y^2}.$$

For example,

$$\frac{355}{113} = 3.1415929204,$$

a good approximation of

$$\pi = 3.1415926535\ldots$$

The fraction

$$\frac{103993}{33102} = 3.1415926530$$

is even better.

Remark: Actually, Dirichlet proved his approximation theorem for irrational numbers of the form $\sqrt{D}$, where D is a positive integer that is not a perfect square, but a trivial modification of his proof applies to any (positive) irrational number. One should consult Dirichlet's original proof in Dirichlet [5], Supplement VIII. This book was actually written by R. Dedekind in 1863 based on Dirichlet's lectures, after Dirichlet's death. It is considered as one of the most important mathematics book of the nineteenth century and it is a model of exposition for its clarity.

Theorem 5.11 only gives a brute-force method for finding x and y, namely, given y, we pick x to be the integer closest to $y\alpha$. There are better ways for finding rational approximations based on *continued fractions*; see Silverman [15], Davenport [3], or Niven, Zuckerman, and Montgomery [12].

It should also be noted that Dirichlet made another clever use of the pigeonhole principle to prove that the equation (known as *Pell's equation*)

$$x^2 - Dy^2 = 1,$$

where D is a positive integer that is not a perfect square, has some solution (x,y), where x and y are positive integers. Such equations had been considered by Fermat around the 1640s and long before that by the Indian mathematicians, Brahmagupta (598–670) and Bhaskaracharya (1114–1185). Surprisingly, the solution with the smallest x can be very large. For example, the smallest (positive) solution of

$$x^2 - 61y^2 = 1$$

is $(x_1, y_1) = (1766319049, 226153980)$.

It can also be shown that Pell's equation has infinitely many solutions (in positive integers) and that these solutions can be expressed in terms of the smallest solution. For more on Pell's equation, see Silverman [15] and Niven, Zuckerman, and Montgomery [12].

We now take a well-deserved break from partial orders and induction and study equivalence relations, an equally important class of relations.

5.6 Equivalence Relations and Partitions

Equivalence relations basically generalize the identity relation. Technically, the definition of an equivalence relation is obtained from the definition of a partial order (Definition 5.1) by changing the third condition, antisymmetry, to *symmetry*.

Definition 5.10. A binary relation R on a set X is an *equivalence relation* iff it is *reflexive, transitive*, and *symmetric*, that is:

(1) (*Reflexivity*): aRa, for all $a \in X$
(2) (*Transitivity*): If aRb and bRc, then aRc, for all $a, b, c \in X$
(3) (*symmetry*): If aRb, then bRa, for all $a, b \in X$

Here are some examples of equivalence relations.

1. The identity relation id_X on a set X is an equivalence relation.
2. The relation $X \times X$ is an equivalence relation.

3. Let S be the set of students in CIS160. Define two students to be equivalent iff they were born the same year. It is trivial to check that this relation is indeed an equivalence relation.
4. Given any natural number $p \geq 1$, recall that we can define a relation on $\mathbb{Z}$ as follows,

$$n \equiv m \,(\text{mod } p)$$

iff $p \mid n - m$; that is, $n = m + pk$, for some $k \in \mathbb{Z}$. It is an easy exercise to check that this is indeed an equivalence relation called *congruence modulo p*.
5. Equivalence of propositions is the relation defined so that $P \equiv Q$ iff $P \Rightarrow Q$ and $Q \Rightarrow P$ are both provable (say, classically). It is easy to check that logical equivalence is an equivalence relation.
6. Suppose $f\colon X \to Y$ is a function. Then, we define the relation $\equiv_f$ on X by

$$x \equiv_f y \quad \text{iff} \quad f(x) = f(y).$$

It is immediately verified that $\equiv_f$ is an equivalence relation. Actually, we show that every equivalence relation arises in this way, in terms of (surjective) functions.

The crucial property of equivalence relations is that they *partition* their domain X into pairwise disjoint nonempty blocks. Intuitively, they carve out X into a bunch of puzzle pieces.

Definition 5.11. Given an equivalence relation R on a set X for any $x \in X$, the set

$$[x]_R = \{y \in X \mid xRy\}$$

is the *equivalence class of x*. Each equivalence class $[x]_R$ is also denoted $\bar{x}_R$ and the subscript R is often omitted when no confusion arises. The set of equivalence classes of R is denoted by X/R. The set X/R is called the *quotient of X by R* or *quotient of X modulo R*. The function, $\pi\colon X \to X/R$, given by

$$\pi(x) = [x]_R, \ x \in X,$$

is called the *canonical projection* (or *projection*) of X onto X/R.

Every equivalence relation is reflexive, that is, xRx for every $x \in X$, therefore observe that $x \in [x]_R$ for any $x \in R$; that is, every equivalence class is *nonempty*. It is also clear that the projection $\pi\colon X \to X/R$ is surjective. The main properties of equivalence classes are given by the following.

Proposition 5.11. *Let R be an equivalence relation on a set X. For any two elements $x, y \in X$ we have*

$$xRy \quad \text{iff} \quad [x] = [y].$$

Moreover, the equivalence classes of R satisfy the following properties.

(1) $[x] \neq \emptyset$, for all $x \in X$

(2) If $[x] \neq [y]$ *then* $[x] \cap [y] = \emptyset$
(3) $X = \bigcup_{x \in X} [x]$.

Proof. First, assume that $[x] = [y]$. We observed that by reflexivity, $y \in [y]$. As $[x] = [y]$, we get $y \in [x]$ and by definition of $[x]$, this means that xRy.

Next, assume that xRy. Let us prove that $[y] \subseteq [x]$. Pick any $z \in [y]$; this means that yRz. By transitivity, we get xRz; that is, $z \in [x]$, proving that $[y] \subseteq [x]$. Now, as R is symmetric, xRy implies that yRx and the previous argument yields $[x] \subseteq [y]$. Therefore, $[x] = [y]$, as needed.

Property (1) follows from the fact that $x \in [x]$ (by reflexivity).

Let us prove the contrapositive of (2). So, assume $[x] \cap [y] \neq \emptyset$. Thus, there is some z so that $z \in [x]$ and $z \in [y]$; that is,

$$xRz \text{ and } yRz.$$

By symmetry, we get zRy and by transitivity, xRy. But then, by the first part of the proposition, we deduce $[x] = [y]$, as claimed.

The third property follows again from the fact that $x \in [x]$. $\square$

A useful way of interpreting Proposition 5.11 is to say that the equivalence classes of an equivalence relation form a partition, as defined next.

Definition 5.12. Given a set X, a *partition of* X is any family $\Pi = \{X_i\}_{i \in I}$, of subsets of X such that

(1) $X_i \neq \emptyset$, for all $i \in I$ (each X_i is nonempty)
(2) If $i \neq j$ then $X_i \cap X_j = \emptyset$ (the X_i are pairwise disjoint)
(3) $X = \bigcup_{i \in I} X_i$ (the family is exhaustive).

Each set X_i is called a *block* of the partition.

In the example where equivalence is determined by the same year of birth, each equivalence class consists of those students having the same year of birth. Let us now go back to the example of congruence modulo p (with $p > 0$) and figure out what are the blocks of the corresponding partition. Recall that

$$m \equiv n \pmod{p}$$

iff $m - n = pk$ for some $k \in \mathbb{Z}$. By the division theorem (Theorem 5.7), we know that there exist some unique q, r, with $m = pq + r$ and $0 \leq r \leq p - 1$. Therefore, for every $m \in \mathbb{Z}$,

$$m \equiv r \pmod{p} \text{ with } 0 \leq r \leq p - 1,$$

which shows that there are p equivalence classes, $[0], [1], \ldots, [p-1]$, where the equivalence class $[r]$ (with $0 \leq r \leq p - 1$) consists of all integers of the form $pq + r$, where $q \in \mathbb{Z}$, that is, those integers whose residue modulo p is r.

Proposition 5.11 defines a map from the set of equivalence relations on X to the set of partitions on X. Given any set X, let $\text{Equiv}(X)$ denote the set of equivalence relations on X and let $\text{Part}(X)$ denote the set of partitions on X. Then, Proposition 5.11 defines the function $\Pi\colon \text{Equiv}(X) \to \text{Part}(X)$ given by,

$$\Pi(R) = X/R = \{[x]_R \mid x \in X\},$$

where R is any equivalence relation on X. We also write Π_R instead of $\Pi(R)$.

There is also a function $\mathscr{R}\colon \text{Part}(X) \to \text{Equiv}(X)$ that assigns an equivalence relation to a partition as shown by the next proposition.

Proposition 5.12. *For any partition* $\Pi = \{X_i\}_{i \in I}$ *on a set* X, *the relation* $\mathscr{R}(\Pi)$ *defined by*

$$x\mathscr{R}(\Pi)y \text{ iff } (\exists i \in I)(x,y \in X_i),$$

is an equivalence relation whose equivalence classes are exactly the blocks X_i.

Proof. We leave this easy proof as an exercise to the reader.   □

Putting Propositions 5.11 and 5.12 together we obtain the useful fact that there is a bijection between $\text{Equiv}(X)$ and $\text{Part}(X)$. Therefore, in principle, it is a matter of taste whether we prefer to work with equivalence relations or partitions. In computer science, it is often preferable to work with partitions, but not always.

Proposition 5.13. *Given any set* X *the functions* $\Pi\colon \text{Equiv}(X) \to \text{Part}(X)$ *and* $\mathscr{R}\colon \text{Part}(X) \to \text{Equiv}(X)$ *are mutual inverses; that is,*

$$\mathscr{R} \circ \Pi = \text{id} \quad and \quad \Pi \circ \mathscr{R} = \text{id}.$$

Consequently, there is a bijection between the set $\text{Equiv}(X)$ *of equivalence relations on* X *and the set* $\text{Part}(X)$ *of partitions on* X.

Proof. This is a routine verification left to the reader.   □

Now, if $f\colon X \to Y$ is a surjective function, we have the equivalence relation $\equiv_f$ defined by

$$x \equiv_f y \text{ iff } f(x) = f(y).$$

It is clear that the equivalence class of any $x \in X$ is the inverse image $f^{-1}(f(x))$, of $f(x) \in Y$. Therefore, there is a bijection between $X/\equiv_f$ and Y. Thus, we can identify f and the projection π, from X onto $X/\equiv_f$. If f is not surjective, note that f is surjective onto $f(X)$ and so, we see that f can be written as the composition

$$f = i \circ \pi,$$

where $\pi\colon X \to f(X)$ is the canonical projection and $i\colon f(X) \to Y$ is the *inclusion function* mapping $f(X)$ into Y (i.e., $i(y) = y$, for every $y \in f(X)$).

Given a set X, the inclusion ordering on $X \times X$ defines an ordering on binary relations on X, namely,

$$R \leq S \quad \text{iff} \quad (\forall x,y \in X)(xRy \Rightarrow xSy).$$

When $R \leq S$, we say that R *refines* S. If R and S are equivalence relations and $R \leq S$, we observe that every equivalence class of R is contained in some equivalence class of S. Actually, in view of Proposition 5.11, we see that *every equivalence class of S is the union of equivalence classes of R*. We also note that id_X is the least equivalence relation on X and $X \times X$ is the largest equivalence relation on X. This suggests the following question: Is $\text{Equiv}(X)$ a lattice under refinement?

The answer is yes. It is easy to see that the meet of two equivalence relations is $R \cap S$, their intersection. But beware, their join is not $R \cup S$, because in general, $R \cup S$ is not transitive. However, there is a least equivalence relation containing R and S, and this is the join of R and S. This leads us to look at various closure properties of relations.

5.7 Transitive Closure, Reflexive and Transitive Closure, Smallest Equivalence Relation

Let R be any relation on a set X. Note that R is reflexive iff $id_X \subseteq R$. Consequently, the smallest reflexive relation containing R is $id_X \cup R$. This relation is called the *reflexive closure of R*.

Note that R is transitive iff $R \circ R \subseteq R$. This suggests a way of making the smallest transitive relation containing R (if R is not already transitive). Define R^n by induction as follows.

$$R^0 = id_X$$
$$R^{n+1} = R^n \circ R.$$

Definition 5.13. Given any relation R on a set X, the *transitive closure of R* is the relation R^+ given by

$$R^+ = \bigcup_{n \geq 1} R^n.$$

The *reflexive and transitive closure of R* is the relation R^*, given by

$$R^* = \bigcup_{n \geq 0} R^n = id_X \cup R^+.$$

The proof of the following proposition is left an an easy exercise.

Proposition 5.14. *Given any relation R on a set X, the relation R^+ is the smallest transitive relation containing R and R^* is the smallest reflexive and transitive relation containing R.*

If R is reflexive, then it is easy to see that $R \subseteq R^2$ and so, $R^k \subseteq R^{k+1}$ for all $k \geq 0$. From this, we can show that if X is a finite set, then there is a smallest k so that

$R^k = R^{k+1}$. In this case, R^k is the reflexive and transitive closure of R. If X has n elements it can be shown that $k \leq n - 1$.

Note that a relation R is symmetric iff $R^{-1} = R$. As a consequence, $R \cup R^{-1}$ is the smallest symmetric relation containing R. This relation is called the *symmetric closure of R*. Finally, given a relation R, what is the smallest equivalence relation containing R? The answer is given by

Proposition 5.15. *For any relation R on a set X, the relation*

$$(R \cup R^{-1})^*$$

is the smallest equivalence relation containing R.

5.8 Fibonacci and Lucas Numbers; Mersenne Primes

We have encountered the Fibonacci numbers (after Leonardo Fibonacci, also known as *Leonardo of Pisa*, 1170–1250) in Section 2.3. These numbers show up unexpectedly in many places, including algorithm design and analysis, for example, Fibonacci heaps. The Lucas numbers (after Edouard Lucas, 1842–1891) are closely related to the Fibonacci numbers. Both arise as special instances of the recurrence relation

$$u_{n+2} = u_{n+1} + u_n, \ n \geq 0,$$

where u_0 and u_1 are some given initial values.

Fig. 5.13 Leonardo Pisano Fibonacci, 1170–1250 (left) and F. Edouard Lucas, 1842–1891 (right)

The *Fibonacci sequence* (F_n) arises for $u_0 = 0$ and $u_1 = 1$ and the *Lucas sequence* (L_n) for $u_0 = 2$ and $u_1 = 1$. These two sequences turn out to be intimately related and they satisfy many remarkable identities. The Lucas numbers play a role in testing for primality of certain kinds of numbers of the form $2^p - 1$, where p is a prime, known as *Mersenne numbers*. In turns out that the largest known primes so far are Mersenne numbers and large primes play an important role in cryptography.

It is possible to derive a closed-form formula for both F_n and L_n using some simple linear algebra.

Observe that the recurrence relation

$$u_{n+2} = u_{n+1} + u_n$$

yields the recurrence

$$\begin{pmatrix} u_{n+1} \\ u_n \end{pmatrix} = \begin{pmatrix} 1 & 1 \\ 1 & 0 \end{pmatrix} \begin{pmatrix} u_n \\ u_{n-1} \end{pmatrix}$$

for all $n \geq 1$, and so,

$$\begin{pmatrix} u_{n+1} \\ u_n \end{pmatrix} = \begin{pmatrix} 1 & 1 \\ 1 & 0 \end{pmatrix}^n \begin{pmatrix} u_1 \\ u_0 \end{pmatrix}$$

for all $n \geq 0$. Now, the matrix

$$A = \begin{pmatrix} 1 & 1 \\ 1 & 0 \end{pmatrix}$$

has characteristic polynomial, $\lambda^2 - \lambda - 1$, which has two real roots

$$\lambda = \frac{1 \pm \sqrt{5}}{2}.$$

Observe that the larger root is the famous *golden ratio*, often denoted

$$\varphi = \frac{1 + \sqrt{5}}{2} = 1.618033988749 \cdots$$

and that

$$\frac{1 - \sqrt{5}}{2} = -\varphi^{-1}.$$

Inasmuch as A has two distinct eigenvalues, it can be diagonalized and it is easy to show that

$$A = \begin{pmatrix} 1 & 1 \\ 1 & 0 \end{pmatrix} = \frac{1}{\sqrt{5}} \begin{pmatrix} \varphi & -\varphi^{-1} \\ 1 & 1 \end{pmatrix} \begin{pmatrix} \varphi & 0 \\ 0 & -\varphi^{-1} \end{pmatrix} \begin{pmatrix} 1 & \varphi^{-1} \\ -1 & \varphi \end{pmatrix}.$$

It follows that

$$\begin{pmatrix} u_{n+1} \\ u_n \end{pmatrix} = \frac{1}{\sqrt{5}} \begin{pmatrix} \varphi & -\varphi^{-1} \\ 1 & 1 \end{pmatrix} \begin{pmatrix} (\varphi^{-1} u_0 + u_1)\varphi^n \\ (\varphi u_0 - u_1)(-\varphi^{-1})^n \end{pmatrix},$$

and so,

$$u_n = \frac{1}{\sqrt{5}} \left((\varphi^{-1} u_0 + u_1)\varphi^n + (\varphi u_0 - u_1)(-\varphi^{-1})^n \right),$$

for all $n \geq 0$.

For the Fibonacci sequence, $u_0 = 0$ and $u_1 = 1$, so

$$F_n = \frac{1}{\sqrt{5}} \left(\varphi^n - (-\varphi^{-1})^n \right) = \frac{1}{\sqrt{5}} \left[\left(\frac{1 + \sqrt{5}}{2} \right)^n - \left(\frac{1 - \sqrt{5}}{2} \right)^n \right],$$

a formula established by Jacques Binet (1786–1856) in 1843 and already known to Euler, Daniel Bernoulli, and de Moivre. Because

$$\frac{\varphi^{-1}}{\sqrt{5}} = \frac{\sqrt{5}-1}{2\sqrt{5}} < \frac{1}{2},$$

we see that F_n is the closest integer to $\varphi^n/\sqrt{5}$ and that

$$F_n = \left\lfloor \frac{\varphi^n}{\sqrt{5}} + \frac{1}{2} \right\rfloor.$$

It is also easy to see that

$$F_{n+1} = \varphi F_n + (-\varphi^{-1})^n,$$

which shows that the ratio F_{n+1}/F_n approaches φ as n goes to infinity.

For the Lucas sequence, $u_0 = 2$ and $u_1 = 1$, so

$$\varphi^{-1} u_0 + u_1 = 2\frac{(\sqrt{5}-1)}{2} + 1 = \sqrt{5},$$

$$\varphi u_0 - u_1 = 2\frac{(1+\sqrt{5})}{2} - 1 = \sqrt{5}$$

and we get

$$L_n = \varphi^n + (-\varphi^{-1})^n = \left(\frac{1+\sqrt{5}}{2}\right)^n + \left(\frac{1-\sqrt{5}}{2}\right)^n.$$

Because

$$\varphi^{-1} = \frac{\sqrt{5}-1}{2} < 0.62$$

it follows that L_n is the closest integer to φ^n.

When $u_0 = u_1$, because $\varphi - \varphi^{-1} = 1$, we get

$$u_n = \frac{u_0}{\sqrt{5}}\left(\varphi^{n+1} - (-\varphi^{-1})^{n+1}\right);$$

that is,

$$u_n = u_0 F_{n+1}.$$

Therefore, from now on, we assume that $u_0 \neq u_1$. It is easy to prove the following by induction.

Proposition 5.16. *The following identities hold.*

$$F_0^2 + F_1^2 + \cdots + F_n^2 = F_n F_{n+1}$$
$$F_0 + F_1 + \cdots + F_n = F_{n+2} - 1$$
$$F_2 + F_4 + \cdots + F_{2n} = F_{2n+1} - 1$$

$$F_1 + F_3 + \cdots + F_{2n+1} = F_{2n+2}$$

$$\sum_{k=0}^{n} k F_k = n F_{n+2} - F_{n+3} + 2$$

for all $n \geq 0$ (with the third sum interpreted as F_0 for $n = 0$).

Following Knuth (see [7]), the third and fourth identities yield the identity

$$F_{(n \bmod 2)+2} + \cdots + F_{n-2} + F_n = F_{n+1} - 1,$$

for all $n \geq 2$.

The above can be used to prove the *Zeckendorf representation* of the natural numbers (see Knuth [7], Chapter 6).

Proposition 5.17. *(Zeckendorf's Representation) Every natural number $n \in \mathbb{N}$ with $n > 0$, has a unique representation of the form*

$$n = F_{k_1} + F_{k_2} + \cdots + F_{k_r},$$

with $k_i \geq k_{i+1} + 2$ for $i = 1, \ldots, r - 1$ and $k_r \geq 2$.

For example,

$$30 = 21 + 8 + 1$$
$$= F_8 + F_6 + F_2$$

and

$$1000000 = 832040 + 121393 + 46368 + 144 + 55$$
$$= F_{30} + F_{26} + F_{24} + F_{12} + F_{10}.$$

The fact that

$$F_{n+1} = \varphi F_n + (-\varphi^{-1})^n$$

and the Zeckendorf representation lead to an amusing method for converting between kilometers and miles (see [7], Section 6.6). Indeed, φ is nearly the number of kilometers in a mile (the exact number is 1.609344 and $\varphi = 1.618033$). It follows that a distance of F_{n+1} kilometers is very nearly a distance of F_n miles,

Thus, to convert a distance d expressed in kilometers into a distance expressed in miles, first find the Zeckendorf representation of d and then shift each F_{k_i} in this representation to $F_{k_i - 1}$. For example,

$$30 = 21 + 8 + 1 = F_8 + F_6 + F_2$$

so the corresponding distance in miles is

$$F_7 + F_6 + F_1 = 13 + 5 + 1 = 19.$$

The "exact" distance in miles is 18.64 miles.

We can prove two simple formulas for obtaining the Lucas numbers from the Fibonacci numbers and vice-versa:

Proposition 5.18. *The following identities hold:*

$$L_n = F_{n-1} + F_{n+1}$$
$$5F_n = L_{n-1} + L_{n+1},$$

for all $n \geq 1$.

The Fibonacci sequence begins with
$$0, 1, 1, 2, 3, 5, 8, 13, 21, 34, 55, 89, 144, 233, 377, 610$$
and the Lucas sequence begins with
$$2, 1, 3, 4, 7, 11, 18, 29, 47, 76, 123, 199, 322, 521, 843, 1364.$$
Notice that $L_n = F_{n-1} + F_{n+1}$ is equivalent to

$$2F_{n+1} = F_n + L_n.$$

It can also be shown that
$$F_{2n} = F_n L_n,$$

for all $n \geq 1$.

The proof proceeds by induction but one finds that it is necessary to prove an auxiliary fact.

Proposition 5.19. *For any fixed $k \geq 1$ and all $n \geq 0$, we have*

$$F_{n+k} = F_k F_{n+1} + F_{k-1} F_n.$$

The reader can also prove that

$$L_n L_{n+2} = L_{n+1}^2 + 5(-1)^n$$
$$L_{2n} = L_n^2 - 2(-1)^n$$
$$L_{2n+1} = L_n L_{n+1} - (-1)^n$$
$$L_n^2 = 5F_n^2 + 4(-1)^n.$$

Using the matrix representation derived earlier, the following can be shown.

Proposition 5.20. *The sequence given by the recurrence*

$$u_{n+2} = u_{n+1} + u_n$$

satisfies the equation:

$$u_{n+1} u_{n-1} - u_n^2 = (-1)^{n-1}(u_0^2 + u_0 u_1 - u_1^2).$$

For the Fibonacci sequence, where $u_0 = 0$ and $u_1 = 1$, we get the *Cassini identity* (after Jean-Dominique Cassini, also known as Giovanni Domenico Cassini, 1625–1712),

$$F_{n+1}F_{n-1} - F_n^2 = (-1)^n, \qquad n \geq 1.$$

The above identity is a special case of *Catalan's identity*,

$$F_{n+r}F_{n-r} - F_n^2 = (-1)^{n-r+1}F_r^2, \qquad n \geq r,$$

due to Eugène Catalan (1814–1894).

Fig. 5.14 Jean-Dominique Cassini, 1748–1845 (left) and Eugène Charles Catalan, 1814–1984 (right)

For the Lucas numbers, where $u_0 = 2$ and $u_1 = 1$ we get

$$L_{n+1}L_{n-1} - L_n^2 = 5(-1)^{n-1}, \qquad n \geq 1.$$

In general, we have

$$u_k u_{n+1} + u_{k-1} u_n = u_1 u_{n+k} + u_0 u_{n+k-1},$$

for all $k \geq 1$ and all $n \geq 0$.

For the Fibonacci sequence, where $u_0 = 0$ and $u_1 = 1$, we just re-proved the identity

$$F_{n+k} = F_k F_{n+1} + F_{k-1} F_n.$$

For the Lucas sequence, where $u_0 = 2$ and $u_1 = 1$, we get

$$\begin{aligned}
L_k L_{n+1} + L_{k-1} L_n &= L_{n+k} + 2L_{n+k-1} \\
&= L_{n+k} + L_{n+k-1} + L_{n+k-1} \\
&= L_{n+k+1} + L_{n+k-1} \\
&= 5F_{n+k};
\end{aligned}$$

that is,

$$L_k L_{n+1} + L_{k-1} L_n = L_{n+k+1} + L_{n+k-1} = 5F_{n+k},$$

for all $k \geq 1$ and all $n \geq 0$.

The identity

$$F_{n+k} = F_k F_{n+1} + F_{k-1} F_n$$

plays a key role in the proof of various divisibility properties of the Fibonacci numbers. Here are two such properties.

Proposition 5.21. *The following properties hold.*

1. F_n divides F_{mn}, for all $m, n \geq 1$.
2. $\gcd(F_m, F_n) = F_{\gcd(m,n)}$, for all $m, n \geq 1$.

An interesting consequence of this divisibility property is that if F_n is a prime and $n > 4$, then n must be a prime. Indeed, if $n \geq 5$ and n is not prime, then $n = pq$ for some integers p, q (possibly equal) with $p \geq 2$ and $q \geq 3$, so F_q divides $F_{pq} = F_n$ and becaue $q \geq 3$, $F_q \geq 2$ and F_n is not prime. For $n = 4$, $F_4 = 3$ is prime. However, there are prime numbers $n \geq 5$ such that F_n is not prime, for example, $n = 19$, as $F_{19} = 4181 = 37 \times 113$ is not prime.

The gcd identity can also be used to prove that for all m, n with $2 < n < m$, if F_n divides F_m, then n divides m, which provides a converse of our earlier divisibility property.

The formulae

$$2F_{m+n} = F_m L_n + F_n L_m$$
$$2L_{m+n} = L_m L_n + 5F_m F_n$$

are also easily established using the explicit formulae for F_n and L_n in terms of φ and φ^{-1}.

The Fibonacci sequence and the Lucas sequence contain primes but it is unknown whether they contain infinitely many primes. Here are some facts about Fibonacci and Lucas primes taken from *The Little Book of Bigger Primes*, by Paulo Ribenboim [13].

As we proved earlier, if F_n is a prime and $n \neq 4$, then n must be a prime but the converse is false. For example,

$$F_3, F_4, F_5, F_7, F_{11}, F_{13}, F_{17}, F_{23}$$

are prime but $F_{19} = 4181 = 37 \times 113$ is not a prime. One of the largest prime Fibonacci numbers is F_{81839}. This number has $17, 103$ digits. Concerning the Lucas numbers, we prove shortly that if L_n is an odd prime and n is not a power of 2, then n is a prime. Again, the converse is false. For example,

$$L_0, L_2, L_4, L_5, L_7, L_8, L_{11}, L_{13}, L_{16}, L_{17}, L_{19}, L_{31}$$

are prime but $L_{23} = 64079 = 139 \times 461$ is not a prime. Similarly, $L_{32} = 4870847 = 1087 \times 4481$ is not prime. One of the largest Lucas primes is L_{51169}.

Generally, divisibility properties of the Lucas numbers are not easy to prove because there is no simple formula for L_{m+n} in terms of other L_ks. Nevertheless, we can prove that if $n, k \geq 1$ and k is odd, then L_n divides L_{kn}. This is not necessarily true if k is even. For example, $L_4 = 7$ and $L_8 = 47$ are prime. The trick is that when k is odd, the binomial expansion of $L_n^k = (\varphi^n + (-\varphi^{-1})^n)^k$ has an even number of terms and these terms can be paired up. Indeed, if k is odd, say $k = 2h + 1$, we have the formula

$$L_n^{2h+1} = L_{(2h+1)n} + \binom{2h+1}{1}(-1)^n L_{(2h-1)n} + \binom{2h+1}{2}(-1)^{2n} L_{(2h-3)n} + \cdots$$
$$+ \binom{2h+1}{h}(-1)^{hn} L_n.$$

By induction on h, we see that L_n divides $L_{(2h+1)n}$ for all $h \geq 0$. Consequently, if $n \geq 2$ is not prime and not a power of 2, then either $n = 2^i q$ for some odd integer, $q \geq 3$, and some $i \geq 1$ and thus, $L_{2^i} \geq 3$ divides L_n, or $n = pq$ for some odd integers (possibly equal), $p \geq 3$ and $q \geq 3$, and so, $L_p \geq 4$ (and $L_q \geq 4$) divides L_n. Therefore, if L_n is an odd prime (so $n \neq 1$, because $L_1 = 1$) then either n is a power of 2 or n is prime.

Remark: When k is even, say $k = 2h$, the "middle term," $\binom{2h}{h}(-1)^{hn}$, in the binomial expansion of $L_n^{2h} = (\varphi^n + (-\varphi^{-1})^n)^{2h}$ stands alone, so we get

$$L_n^{2h} = L_{2hn} + \binom{2h}{1}(-1)^n L_{(2h-2)n} + \binom{2h}{2}(-1)^{2n} L_{(2h-4)n} + \cdots$$
$$+ \binom{2h}{h-1}(-1)^{(h-1)n} L_{2n} + \binom{2h}{h}(-1)^{hn}.$$

Unfortunately, the above formula seems of little use to prove that L_{2hn} is divisible by L_n. Note that the last term is always even inasmuch as

$$\binom{2h}{h} = \frac{(2h)!}{h!h!} = \frac{2h}{h}\frac{(2h-1)!}{(h-1)!h!} = 2\binom{2h-1}{h}.$$

It should also be noted that not every sequence (u_n) given by the recurrence

$$u_{n+2} = u_{n+1} + u_n$$

and with $\gcd(u_0, u_1) = 1$ contains a prime number. According to Ribenboim [13], Graham found an example in 1964 but it turned out to be incorrect. Later, Knuth gave correct sequences (see *Concrete Mathematics* [7], Chapter 6), one of which began with

$$u_0 = 62638280004239857$$
$$u_1 = 49463435743205655.$$

We just studied some properties of the sequences arising from the recurrence relation

$$u_{n+2} = u_{n+1} + u_n.$$

Lucas investigated the properties of the more general recurrence relation

$$u_{n+2} = Pu_{n+1} - Qu_n,$$

where $P, Q \in \mathbb{Z}$ are any integers with $P^2 - 4Q \neq 0$, in two seminal papers published in 1878.

We can prove some of the basic results about these Lucas sequences quite easily using the matrix method that we used before. The recurrence relation

$$u_{n+2} = Pu_{n+1} - Qu_n$$

yields the recurrence

$$\begin{pmatrix} u_{n+1} \\ u_n \end{pmatrix} = \begin{pmatrix} P & -Q \\ 1 & 0 \end{pmatrix} \begin{pmatrix} u_n \\ u_{n-1} \end{pmatrix}$$

for all $n \geq 1$, and so,

$$\begin{pmatrix} u_{n+1} \\ u_n \end{pmatrix} = \begin{pmatrix} P & -Q \\ 1 & 0 \end{pmatrix}^n \begin{pmatrix} u_1 \\ u_0 \end{pmatrix}$$

for all $n \geq 0$. The matrix

$$A = \begin{pmatrix} P & -Q \\ 1 & 0 \end{pmatrix}$$

has the characteristic polynomial $-(P - \lambda)\lambda + Q = \lambda^2 - P\lambda + Q$, which has the discriminant $D = P^2 - 4Q$. If we assume that $P^2 - 4Q \neq 0$, the polynomial $\lambda^2 - P\lambda + Q$ has two distinct roots:

$$\alpha = \frac{P + \sqrt{D}}{2}, \qquad \beta = \frac{P - \sqrt{D}}{2}.$$

Obviously,

$$\alpha + \beta = P$$
$$\alpha\beta = Q$$
$$\alpha - \beta = \sqrt{D}.$$

The matrix A can be diagonalized as

$$A = \begin{pmatrix} P & -Q \\ 1 & 0 \end{pmatrix} = \frac{1}{\alpha - \beta} \begin{pmatrix} \alpha & \beta \\ 1 & 1 \end{pmatrix} \begin{pmatrix} \alpha & 0 \\ 0 & \beta \end{pmatrix} \begin{pmatrix} 1 & -\beta \\ -1 & \alpha \end{pmatrix}.$$

Thus, we get

$$\begin{pmatrix} u_{n+1} \\ u_n \end{pmatrix} = \frac{1}{\alpha - \beta} \begin{pmatrix} \alpha & \beta \\ 1 & 1 \end{pmatrix} \begin{pmatrix} (-\beta u_0 + u_1)\alpha^n \\ (\alpha u_0 - u_1)\beta^n \end{pmatrix}$$

and so,

$$u_n = \frac{1}{\alpha - \beta} \left((-\beta u_0 + u_1)\alpha^n + (\alpha u_0 - u_1)\beta^n \right).$$

Actually, the above formula holds for $n = 0$ only if $\alpha \neq 0$ and $\beta \neq 0$, that is, iff $Q \neq 0$. If $Q = 0$, then either $\alpha = 0$ or $\beta = 0$, in which case the formula still holds if we assume that $0^0 = 1$.

For $u_0 = 0$ and $u_1 = 1$, we get a generalization of the Fibonacci numbers,

$$U_n = \frac{\alpha^n - \beta^n}{\alpha - \beta}$$

and for $u_0 = 2$ and $u_1 = P$, because

$$-\beta u_0 + u_1 = -2\beta + P = -2\beta + \alpha + \beta = \alpha - \beta$$

and

$$\alpha u_0 - u_1 = 2\alpha - P = 2\alpha - (\alpha + \beta) = \alpha - \beta,$$

we get a generalization of the Lucas numbers,

$$V_n = \alpha^n + \beta^n.$$

The orginal Fibonacci and Lucas numbers correspond to $P = 1$ and $Q = -1$. The vectors $\binom{0}{1}$ and $\binom{2}{P}$ are linearly independent, therefore every sequence arising from the recurrence relation

$$u_{n+2} = P u_{n+1} - Q u_n$$

is a unique linear combination of the sequences (U_n) and (V_n).

It is possible to prove the following generalization of the Cassini identity.

Proposition 5.22. *The sequence defined by the recurrence*

$$u_{n+2} = P u_{n+1} - Q u_n$$

(with $P^2 - 4Q \neq 0$) satisfies the identity:

$$u_{n+1} u_{n-1} - u_n^2 = Q^{n-1}(-Q u_0^2 + P u_0 u_1 - u_1^2).$$

For the U-sequence, $u_0 = 0$ and $u_1 = 1$, so we get

$$U_{n+1} U_{n-1} - U_n^2 = -Q^{n-1}.$$

For the V-sequence, $u_0 = 2$ and $u_1 = P$, so we get

$$V_{n+1} V_{n-1} - V_n^2 = Q^{n-1} D,$$

where $D = P^2 - 4Q$.

Because $\alpha^2 - Q = \alpha(\alpha - \beta)$ and $\beta^2 - Q = -\beta(\alpha - \beta)$, we easily get formulae expressing U_n in terms of the V_ks and vice versa.

Proposition 5.23. *We have the following identities relating the U_n and the V_n,*

$$V_n = U_{n+1} - QU_{n-1}$$
$$DU_n = V_{n+1} - QV_{n-1},$$

for all $n \geq 1$.

The following identities are also easy to derive.

$$U_{2n} = U_n V_n$$
$$V_{2n} = V_n^2 - 2Q^n$$
$$U_{m+n} = U_m U_{n+1} - QU_n U_{m-1}$$
$$V_{m+n} = V_m V_n - Q^n V_{m-n}.$$

Lucas numbers play a crucial role in testing the primality of certain numbers of the form $N = 2^p - 1$, called *Mersenne numbers*. A Mersenne number which is prime is called a *Mersenne prime*.

Fig. 5.15 Marin Mersenne, 1588–1648

First, let us show that if $N = 2^p - 1$ is prime, then p itself must be a prime. This is because if $p = ab$ is a composite, with $a, b \geq 2$, as

$$2^p - 1 = 2^{ab} - 1 = (2^a - 1)(1 + 2^a + 2^{2a} + \cdots + 2^{(b-1)a}),$$

then $2^a - 1 > 1$ divides $2^p - 1$, a contradiction.

For $p = 2, 3, 5, 7$ we see that $3 = 2^2 - 1$, $7 = 2^3 - 1$, $31 = 2^5 - 1$, $127 = 2^7 - 1$ are indeed prime.

However, the condition that the exponent p be prime is not sufficient for $N = 2^p - 1$ to be prime, because for $p = 11$, we have $2^{11} - 1 = 2047 = 23 \times 89$. Mersenne (1588–1648) stated in 1644 that $N = 2^p - 1$ is prime when

$$p = 2, 3, 5, 7, 13, 17, 19, 31, 67, 127, 257.$$

Mersenne was wrong about $p = 67$ and $p = 257$, and he missed $p = 61, 89$, and 107. Euler showed that $2^{31} - 1$ was indeed prime in 1772 and at that time, it was known that $2^p - 1$ is indeed prime for $p = 2, 3, 5, 7, 13, 17, 19, 31$.

Then came Lucas. In 1876, Lucas, proved that $2^{127} - 1$ was prime. Lucas came up with a method for testing whether a Mersenne number is prime, later rigorously proved correct by Lehmer, and known as the *Lucas–Lehmer test*. This test does not require the actual computation of $N = 2^p - 1$ but it requires an efficient method for squaring large numbers (less that N) and a way of computing the residue modulo $2^p - 1$ just using p.

A version of the Lucas–Lehmer test uses the Lucas sequence given by the recurrence

$$V_{n+2} = 2V_{n+1} + 2V_n,$$

starting from $V_0 = V_1 = 2$. This corresponds to $P = 2$ and $Q = -2$. In this case, $D = 12$ and it is easy to see that $\alpha = 1 + \sqrt{3}$, $\beta = 1 - \sqrt{3}$, so

$$V_n = (1 + \sqrt{3})^n + (1 - \sqrt{3})^n.$$

This sequence starts with

$$2, 2, 8, 20, 56, \ldots.$$

Here is the first version of the Lucas–Lehmer test for primality of a Mersenne number.

Fig. 5.16 Derrick Henry Lehmer, 1905–1991

Theorem 5.12. *Lucas–Lehmer test (Version 1) The number $N = 2^p - 1$ is prime for any odd prime p iff N divides $V_{2^{p-1}}$.*

A proof of the Lucas–Lehmer test can be found in *The Little Book of Bigger Primes* [13]. Shorter proofs exist and are available on the Web but they require some knowledge of algebraic number theory. The most accessible proof that we are aware of (it only uses the quadratic reciprocity law) is given in Volume 2 of Knuth [8]; see Section 4.5.4. Note that the test does not apply to $p = 2$ because $3 = 2^2 - 1$ does not divide $V_2 = 8$ but that's not a problem.

The numbers $V_{2^{p-1}}$ get large very quickly but if we observe that

$$V_{2n} = V_n^2 - 2(-2)^n,$$

we may want to consider the sequence S_n, given by

$$S_{n+1} = S_n^2 - 2,$$

starting with $S_0 = 4$. This sequence starts with

$$4, 14, 194, 37643, 1416317954, \ldots .$$

Then, it turns out that

$$V_{2^k} = S_{k-1} 2^{2^{k-1}},$$

for all $k \geq 1$. It is also easy to see that

$$S_k = (2 + \sqrt{3})^{2^k} + (2 - \sqrt{3})^{2^k}.$$

Now, $N = 2^p - 1$ is prime iff N divides $V_{2^{p-1}}$ iff $N = 2^p - 1$ divides $S_{p-2} 2^{2^{p-2}}$ iff N divides S_{p-2} (because if N divides $2^{2^{p-2}}$, then N is not prime).

Thus, we obtain an improved version of the Lucas–Lehmer test for primality of a Mersenne number.

Theorem 5.13. *Lucas–Lehmer test (Version 2) The number, $N = 2^p - 1$, is prime for any odd prime p iff*

$$S_{p-2} \equiv 0 \pmod{N}.$$

The test does not apply to $p = 2$ because $3 = 2^2 - 1$ does not divide $S_0 = 4$ but that's not a problem.

The above test can be performed by computing a sequence of residues mod N, using the recurrence $S_{n+1} = S_n^2 - 2$, starting from 4.

As of January 2009, only 46 Mersenne primes were known. The largest one was found in August 2008 by mathematicians at UCLA. This is

$$M_{46} = 2^{43112609} - 1,$$

and it has $12,978,189$ digits. It is an open problem whether there are infinitely many Mersenne primes.

Going back to the second version of the Lucas–Lehmer test, because we are computing the sequence of S_ks modulo N, the squares being computed never exceed $N^2 = 2^{2p}$. There is also a clever way of computing $n \bmod 2^p - 1$ without actually performing divisions if we express n in binary. This is because

$$n \equiv (n \bmod 2^p) + \lfloor n/2^p \rfloor \pmod{2^p - 1}.$$

But now, if n is expressed in binary, $(n \bmod 2^p)$ consists of the p rightmost (least significant) bits of n and $\lfloor n/2^p \rfloor$ consists of the bits remaining as the head of the string obtained by deleting the rightmost p bits of n. Thus, we can compute the remainder modulo $2^p - 1$ by repeating this process until at most p bits remain. Observe that if n is a multiple of $2^p - 1$, the algorithm will produce $2^p - 1$ in binary as opposed to 0 but this exception can be handled easily. For example,

$$916 \bmod 2^5 - 1 = 111001010 0_2 \ (\bmod \ 2^5 - 1)$$
$$= 10100_2 + 11100_2 \ (\bmod \ 2^5 - 1)$$
$$= 110000_2 \ (\bmod \ 2^5 - 1)$$
$$= 10000_2 + 1_2 \ (\bmod \ 2^5 - 1)$$
$$= 10001_2 \ (\bmod \ 2^5 - 1)$$
$$= 10001_2$$
$$= 17.$$

The Lucas–Lehmer test applied to $N = 127 = 2^7 - 1$ yields the following steps, if we denote $S_k \bmod 2^p - 1$ by r_k.

$r_0 = 4$,

$r_1 = 4^2 - 2 = 14 \ (\bmod \ 127)$; that is, $r_1 = 14$.

$r_2 = 14^2 - 2 = 194 \ (\bmod \ 127)$; that is, $r_2 = 67$.

$r_3 = 67^2 - 2 = 4487 \ (\bmod \ 127)$; that is, $r_3 = 42$.

$r_4 = 42^2 - 2 = 1762 \ (\bmod \ 127)$; that is, $r_4 = 111$.

$r_5 = 111^2 - 2 = 12319 \ (\bmod \ 127)$; that is, $r_5 = 0$.

As $r_5 = 0$, the Lucas–Lehmer test confirms that $N = 127 = 2^7 - 1$ is indeed prime.

5.9 Public Key Cryptography; The RSA System

Ever since written communication was used, people have been interested in trying to conceal the content of their messages from their adversaries. This has led to the development of techniques of secret communication, a science known as *cryptography*.

The basic situation is that one party, A, say Albert, wants to send a message to another party, J, say Julia. However, there is a danger that some ill-intentioned third party, Machiavelli, may intercept the message and learn things that he is not supposed to know about and as a result, do evil things. The original message, understandable to all parties, is known as the *plain text*. To protect the content of the message, Albert *encrypts* his message. When Julia receives the encrypted message, she must *decrypt* it in order to be able to read it. Both Albert and Julia share some information that Machiavelli does not have, a *key*. Without a key, Machiavelli, is incapable of decrypting the message and thus, to do harm.

There are many schemes for generating keys to encrypt and decrypt messages. We are going to describe a method involving *public and private keys* known as the *RSA Cryptosystem*, named after its inventors, Ronald Rivest, Adi Shamir, and Leonard Adleman (1978), based on ideas by Diffie and Hellman (1976). We highly recommend reading the orginal paper by Rivest, Shamir, and Adleman [14]. It is beautifully written and easy to follow. A very clear, but concise exposition can also be found in Koblitz [9]. An encyclopedic coverage of cryptography can be found in Menezes, van Oorschot, and Vanstone's *Handbook* [11].

The RSA system is widely used in practice, for example in SSL (Secure Socket Layer), which in turn is used in https (secure http). Any time you visit a "secure site" on the Internet (to read e-mail or to order merchandise), your computer generates a public key and a private key for you and uses them to make sure that your credit card number and other personal data remain secret. Interestingly, although one might think that the mathematics behind such a scheme is very advanced and complicated, this is not so. In fact, little more than the material of Section 5.4 is needed. Therefore, in this section, we are going to explain the basics of RSA.

The first step is to convert the plain text of characters into an integer. This can be done easily by assigning distinct integers to the distinct characters, for example, by converting each character to its ASCII code. From now on, we assume that this conversion has been performed.

The next and more subtle step is to use modular arithmetic. We pick a (large) positive integer m and perform arithmetic modulo m. Let us explain this step in more detail.

Recall that for all $a, b \in \mathbb{Z}$, we write $a \equiv b \pmod{m}$ iff $a - b = km$, for some $k \in \mathbb{Z}$, and we say that a and b are congruent modulo m. We already know that congruence is an equivalence relation but it also satisfies the following properties.

Proposition 5.24. *For any positive integer m, for all $a_1, a_2, b_1, b_2 \in \mathbb{Z}$, the following properties hold. If $a_1 \equiv b_1 \pmod{m}$ and $a_2 \equiv b_2 \pmod{m}$, then*

(1) $a_1 + a_2 \equiv b_1 + b_2 \pmod{m}$.
(2) $a_1 - a_2 \equiv b_1 - b_2 \pmod{m}$.
(3) $a_1 a_2 \equiv b_1 b_2 \pmod{m}$.

Proof. We only check (3), leaving (1) and (2) as easy exercises. Because $a_1 \equiv b_1 \pmod{m}$ and $a_2 \equiv b_2 \pmod{m}$, we have $a_1 = b_1 + k_1 m$ and $a_2 = b_2 + k_2 m$, for some $k_1, k_2 \in \mathbb{Z}$, and so

$$a_1 a_2 = (b_1 + k_1 m)(b_2 + k_2 m) = b_1 b_2 + (b_1 k_2 + k_1 b_2 + k_1 m k_2)m,$$

which means that $a_1 a_2 \equiv b_1 b_2 \pmod{m}$. A more elegant proof consists in observing that

$$a_1 a_2 - b_1 b_2 = a_1(a_2 - b_2) + (a_1 - b_1)b_2$$
$$= (a_1 k_2 + k_1 b_2)m,$$

as claimed.   □

Proposition 5.24 allows us to define addition, subtraction, and multiplication on equivalence classes modulo m. If we denote by $\mathbb{Z}/m\mathbb{Z}$ the set of equivalence classes modulo m and if we write $\bar{a}$ for the equivalence class of a, then we define

$$\bar{a} + \bar{b} = \overline{a+b}$$
$$\bar{a} - \bar{b} = \overline{a-b}$$
$$\bar{a}\bar{b} = \overline{ab}.$$

The above make sense because $\overline{a+b}$ does not depend on the representatives chosen in the equivalence classes $\bar{a}$ and $\bar{b}$, and similarly for $\overline{a-b}$ and $\overline{ab}$. Of course, each equivalence class $\bar{a}$ contains a unique representative from the set of remainders $\{0, 1, \ldots, m-1\}$, modulo m, so the above operations are completely determined by $m \times m$ tables. Using the arithmetic operations of $\mathbb{Z}/m\mathbb{Z}$ is called *modular arithmetic*.

For an arbitrary m, the set $\mathbb{Z}/m\mathbb{Z}$ is an algebraic structure known as a *ring*. Addition and subtraction behave as in $\mathbb{Z}$ but multiplication is stranger. For example, when $m = 6$,

$$2 \cdot 3 = 0$$
$$3 \cdot 4 = 0,$$

inasmuch as $2 \cdot 3 = 6 \equiv 0 \,(\mathrm{mod}\, 6)$, and $3 \cdot 4 = 12 \equiv 0 \,(\mathrm{mod}\, 6)$. Therefore, it is not true that every nonzero element has a multiplicative inverse. However, we know from Section 5.4 that a nonzero integer a has a multiplicative inverse iff $\gcd(a, m) = 1$ (use the Bézout identity). For example,

$$5 \cdot 5 = 1,$$

because $5 \cdot 5 = 25 \equiv 1 \,(\mathrm{mod}\, 6)$.

As a consequence, when m is a prime number, every nonzero element not divisible by m has a multiplicative inverse. In this case, $\mathbb{Z}/m\mathbb{Z}$ is more like $\mathbb{Q}$; it is a *finite field*. However, note that in $\mathbb{Z}/m\mathbb{Z}$ we have

$$\underbrace{1 + 1 + \cdots + 1}_{m \text{ times}} = 0$$

(because $m \equiv 0 \,(\mathrm{mod}\, m)$), a phenomenom that does not happen in $\mathbb{Q}$ (or $\mathbb{R}$).

The RSA method uses modular arithmetic. One of the main ingredients of public key cryptography is that one should use an encryption function, $f \colon \mathbb{Z}/m\mathbb{Z} \to \mathbb{Z}/m\mathbb{Z}$, which is easy to compute (i.e., can be computed efficiently) but such that its inverse f^{-1} is practically impossible to compute unless one has *special additional information*. Such functions are usually referred to as *trapdoor one-way functions*. Remarkably, *exponentiation modulo* m, that is, the function, $x \mapsto x^e \bmod m$, is a trapdoor one-way function for suitably chosen m and e.

Thus, we claim the following.

(1) Computing $x^e \bmod m$ can be done efficiently .
(2) Finding x such that

$$x^e \equiv y \,(\mathrm{mod}\, m)$$

with $0 \leq x, y \leq m - 1$, is hard, unless one has extra information about m. The function that finds an eth root modulo m is sometimes called a *discrete logarithm*.

We explain shortly how to compute $x^e \bmod m$ efficiently using the *square and multiply* method also known as *repeated squaring*.

As to the second claim, actually, no proof has been given yet that this function is a one-way function but, so far, this has not been refuted either.

Now, what's the trick to make it a trapdoor function?

What we do is to pick two distinct large prime numbers, p and q (say over 200 decimal digits), which are "sufficiently random" and we let

$$m = pq.$$

Next, we pick a random e, with $1 < e < (p-1)(q-1)$, relatively prime to $(p-1)(q-1)$.

Because $\gcd(e, (p-1)(q-1)) = 1$, we know from the discussion just before Theorem 5.10 that there is some d with $1 < d < (p-1)(q-1)$, such that $ed \equiv 1 \pmod{(p-1)(q-1)}$.

Then, we claim that to find x such that

$$x^e \equiv y \pmod{m},$$

we simply compute $y^d \bmod m$, and this can be done easily, as we claimed earlier. The reason why the above "works" is that

$$x^{ed} \equiv x \pmod{m}, \qquad\qquad (*)$$

for all $x \in \mathbb{Z}$, which we prove later.

Setting up RSA

In, summary to set up RSA for Albert (A) to receive encrypted messages, perform the following steps.

1. Albert generates two distinct large and sufficiently random primes, p_A and q_A. They are kept secret.
2. Albert computes $m_A = p_A q_A$. This number called the *modulus* will be made public.
3. Albert picks at random some e_A, with $1 < e_A < (p_A - 1)(q_A - 1)$, so that $\gcd(e_A, (p_A - 1)(q_A - 1)) = 1$. The number e_A is called the *encryption key* and it will also be public.
4. Albert computes the inverse, $d_A = e_A^{-1}$ modulo m_A, of e_A. This number is kept secret. The pair (d_A, m_A) is Albert's *private key* and d_A is called the *decryption key*.
5. Albert publishes the pair (e_A, m_A) as his *public key*.

Encrypting a Message

Now, if Julia wants to send a message, x, to Albert, she proceeds as follows. First, she splits x into chunks, $x_1, \ldots, x_k$, each of length at most $m_A - 1$, if necessary (again, I assume that x has been converted to an integer in a preliminary step). Then she looks up Albert's public key (e_A, m_A) and she computes

$$y_i = E_A(x_i) = x_i^{e_A} \bmod m_A,$$

for $i = 1, \ldots, k$. Finally, she sends the sequence $y_1, \ldots, y_k$ to Albert. This encrypted message is known as the *cyphertext*. The function E_A is Albert's *encryption function*.

Decrypting a Message

In order to decrypt the message $y_1, \ldots, y_k$ that Julia sent him, Albert uses his private key (d_A, m_A) to compute each

$$x_i = D_A(y_i) = y_i^{d_A} \bmod m_A,$$

and this yields the sequence $x_1, \ldots, x_k$. The function D_A is Albert's *decryption function*.

Similarly, in order for Julia to receive encrypted messages, she must set her own public key (e_J, m_J) and private key (d_J, m_J) by picking two distinct primes p_J and q_J and e_J, as explained earlier.

The beauty of the scheme is that the sender only needs to know the public key of the recipient to send a message but an eavesdropper is unable to decrypt the encoded message unless he somehow gets his hands on the secret key of the receiver.

Let us give a concrete illustration of the RSA scheme using an example borrowed from Silverman [15] (Chapter 18). We write messages using only the 26 upper-case letters A, B, ..., Z, encoded as the integers A = 11, B = 12, ..., Z = 36. It would be more convenient to have assigned a number to represent a blank space but to keep things as simple as possible we do not do that.

Say Albert picks the two primes $p_A = 12553$ and $q_A = 13007$, so that $m_A = p_A q_A = 163,276,871$ and $(p_A - 1)(q_A - 1) = 163,251,312$. Albert also picks $e_A = 79921$, relatively prime to $(p_A - 1)(q_A - 1)$ and then finds the inverse d_A, of e_A modulo $(p_A - 1)(q_A - 1)$ using the extended Euclidean algorithm (more details are given in Section 5.11) which turns out to be $d_A = 145,604,785$. One can check that

$$145,604,785 \cdot 79921 - 71282 \cdot 163,251,312 = 1,$$

which confirms that d_A is indeed the inverse of e_A modulo $163,251,312$.

Now, assume that Albert receives the following message, broken in chunks of at most nine digits, because $m_A = 163,276,871$ has nine digits.

145387828   47164891   152020614   27279275   35356191.

Albert decrypts the above messages using his private key (d_A, m_A), where $d_A = 145,604,785$, using the repeated squaring method (described in Section 5.11) and finds that

$$145387828^{145,604,785} \equiv 30182523 \,(\mathrm{mod}\, 163,276,871)$$

$$47164891^{145,604,785} \equiv 26292524 \,(\mathrm{mod}\, 163,276,871)$$

$$152020614^{145,604,785} \equiv 19291924 \,(\mathrm{mod}\, 163,276,871)$$

$$27279275^{145,604,785} \equiv 30282531 \,(\mathrm{mod}\, 163,276,871)$$

$$35356191^{145,604,785} \equiv 122215 \,(\mathrm{mod}\, 163,276,871)$$

which yields the message

$$30182523 \; 26292524 \; 19291924 \; 30282531 \; 122215,$$

and finally, translating each two-digit numeric code to its corresponding character, to the message

$$\mathrm{T\,H\,O\,M\,P\,S\,O\,N\,I\,S\,I\,N\,T\,R\,O\,U\,B\,L\,E}$$

or, in more readable format

Thompson is in trouble

It would be instructive to encrypt the decoded message

$$30182523 \; 26292524 \; 19291924 \; 30282531 \; 122215$$

using the public key $e_A = 79921$. If everything goes well, we should get our original message

$$145387828 \qquad 47164891 \qquad 152020614 \qquad 27279275 \qquad 35356191$$

back.

Let us now explain in more detail how the RSA system works and why it is correct.

5.10 Correctness of The RSA System

We begin by proving the correctness of the inversion formula $(*)$. For this, we need a classical result known as *Fermat's little theorem*.

This result was first stated by Fermat in 1640 but apparently no proof was published at the time and the first known proof was given by Leibnitz (1646–1716). This is basically the proof suggested in Problem 5.14. A different proof was given by Ivory in 1806 and this is the proof that w give here. It has the advantage that it can be easily generalized to Euler's version (1760) of Fermat's little theorem.

Theorem 5.14. *(Fermat's Little Theorem) If p is any prime number, then the following two equivalent properties hold.*

Fig. 5.17 Pierre de Fermat, 1601–1665

(1) For every integer, $a \in \mathbb{Z}$, if a is not divisible by p, then we have

$$a^{p-1} \equiv 1 \,(\mathrm{mod}\,p).$$

(2) For every integer, $a \in \mathbb{Z}$, we have

$$a^{p} \equiv a \,(\mathrm{mod}\,p).$$

Proof. (1) Consider the integers

$$a, 2a, 3a, \ldots, (p-1)a$$

and let

$$r_1, r_2, r_3, \ldots, r_{p-1}$$

be the sequence of remainders of the division of the numbers in the first sequence by p. Because $\gcd(a, p) = 1$, none of the numbers in the first sequence is divisible by p, so $1 \le r_i \le p - 1$, for $i = 1, \ldots, p - 1$. We claim that these remainders are all distinct. If not, then say $r_i = r_j$, with $1 \le i < j \le p - 1$. But then, because

$$ai \equiv r_i \,(\mathrm{mod}\,p)$$

and

$$aj \equiv r_j \,(\mathrm{mod}\,p),$$

we deduce that

$$aj - ai \equiv r_j - r_i \,(\mathrm{mod}\,p),$$

and because $r_i = r_j$, we get,

$$a(j - i) \equiv 0 \,(\mathrm{mod}\,p).$$

This means that p divides $a(j-i)$, but $\gcd(a,p)=1$ so, by Euclid's proposition (Proposition 5.9), p must divide $j-i$. However $1 \le j-i < p-1$, so we get a contradiction and the remainders are indeed all distinct.

There are $p-1$ distinct remainders and they are all nonzero, therefore we must have

$$\{r_1, r_2, \ldots, r_{p-1}\} = \{1, 2, \ldots, p-1\}.$$

Using Property (3) of congruences (see Proposition 5.24), we get

$$a \cdot 2a \cdot 3a \cdots (p-1)a \equiv 1 \cdot 2 \cdot 3 \cdots (p-1) \,(\mathrm{mod}\, p);$$

that is,

$$(a^{p-1} - 1) \cdot (p-1)! \equiv 0 \,(\mathrm{mod}\, p).$$

Again, p divides $(a^{p-1} - 1) \cdot (p-1)!$, but because p is relatively prime to $(p-1)!$, it must divide $a^{p-1} - 1$, as claimed.

(2) If $\gcd(a,p) = 1$, we proved in (1) that

$$a^{p-1} \equiv 1 \,(\mathrm{mod}\, p),$$

from which we get

$$a^p \equiv a \,(\mathrm{mod}\, p),$$

because $a \equiv a \,(\mathrm{mod}\, p)$. If a is divisible by p, then $a \equiv 0 \,(\mathrm{mod}\, p)$, which implies $a^p \equiv 0 \,(\mathrm{mod}\, p)$, and thus, that

$$a^p \equiv a \,(\mathrm{mod}\, p).$$

Therefore, (2) holds for all $a \in \mathbb{Z}$ and we just proved that (1) implies (2). Finally, if (2) holds and if $\gcd(a,p) = 1$, as p divides $a^p - a = a(a^{p-1} - 1)$, it must divide $a^{p-1} - 1$, which shows that (1) holds and so, (2) implies (1). $\square$

It is now easy to establish the correctness of RSA.

Proposition 5.25. *For any two distinct prime numbers p and q, if e and d are any two positive integers such that*

1. $1 < e, d < (p-1)(q-1)$,
2. $ed \equiv 1 \,(\mathrm{mod}\,(p-1)(q-1))$,

then for every $x \in \mathbb{Z}$ we have

$$x^{ed} \equiv x \,(\mathrm{mod}\, pq).$$

Proof. Because p and q are two distinct prime numbers, by Euclid's proposition it is enough to prove that both p and q divide $x^{ed} - x$. We show that $x^{ed} - x$ is divisible by p, the proof of divisibility by q being similar.

By condition (2), we have

$$ed = 1 + (p-1)(q-1)k,$$

with $k \geq 1$, inasmuch as $1 < e, d < (p-1)(q-1)$. Thus, if we write $h = (q-1)k$, we have $h \geq 1$ and

$$
\begin{aligned}
x^{ed} - x &\equiv x^{1+(p-1)h} - x \,(\mathrm{mod}\,p)\\
&\equiv x((x^{p-1})^h - 1)\,(\mathrm{mod}\,p)\\
&\equiv x(x^{p-1} - 1)((x^{p-1})^{h-1} + (x^{p-1})^{h-2} + \cdots + 1)\,(\mathrm{mod}\,p)\\
&\equiv (x^p - x)((x^{p-1})^{h-1} + (x^{p-1})^{h-2} + \cdots + 1)\,(\mathrm{mod}\,p)\\
&\equiv 0\,(\mathrm{mod}\,p),
\end{aligned}
$$

because $x^p - x \equiv 0\,(\mathrm{mod}\,p)$, by Fermat's little theorem. □

Remark: Of course, Proposition 5.25 holds if we allow $e = d = 1$, but this not interesting for encryption. The number $(p-1)(q-1)$ turns out to be the number of positive integers less than pq that are relatively prime to pq. For any arbitrary positive integer, m, the number of positive integers less than m that are relatively prime to m is given by the *Euler ϕ function* (or *Euler totient*), denoted ϕ (see Problems 5.23 and 5.27 or Niven, Zuckerman, and Montgomery [12], Section 2.1, for basic properties of ϕ).

Fermat's little theorem can be generalized to what is known as *Euler's formula* (see Problem 5.23): For every integer a, if $\gcd(a, m) = 1$, then

$$
a^{\phi(m)} \equiv 1\,(\mathrm{mod}\,m).
$$

Because $\phi(pq) = (p-1)(q-1)$, when $\gcd(x, \phi(pq)) = 1$, Proposition 5.25 follows from Euler's formula. However, that argument does not show that Proposition 5.25 holds when $\gcd(x, \phi(pq)) > 1$ and a special argument is required in this case.

It can be shown that if we replace pq by a positive integer m that is square-free (does not contain a square factor) and if we assume that e and d are chosen so that $1 < e, d < \phi(m)$ and $ed \equiv 1\,(\mathrm{mod}\,\phi(m))$, then

$$
x^{ed} \equiv x\,(\mathrm{mod}\,m)
$$

for all $x \in \mathbb{Z}$ (see Niven, Zuckerman, and Montgomery [12], Section 2.5, Problem 4).

We see no great advantage in using this fancier argument and this is why we used the more elementary proof based on Fermat's little theorem.

Proposition 5.25 immediately implies that the decrypting and encrypting RSA functions D_A and E_A are mutual inverses for any A. Furthermore, E_A is easy to compute but, without extra information, namely, the trapdoor d_A, it is practically impossible to compute $D_A = E_A^{-1}$. That D_A is hard to compute without a trapdoor is related to the fact that factoring a large number, such as m_A, into its factors p_A and q_A is hard. Today, it is practically impossible to factor numbers over 300 decimal digits long. Although no proof has been given so far, it is believed that factoring will remain a hard problem. So, even if in the next few years it becomes possible to factor 300-digit numbers, it will still be impossible to factor 400-digit numbers.

RSA has the peculiar property that it depends both on the fact that primality testing is easy but that factoring is hard. What a stroke of genius!

5.11 Algorithms for Computing Powers and Inverses Modulo m

First, we explain how to compute $x^n \bmod m$ efficiently, where $n \geq 1$. Let us first consider computing the nth power x^n of some positive integer. The idea is to look at the parity of n and to proceed recursively. If n is even, say $n = 2k$, then

$$x^n = x^{2k} = (x^k)^2,$$

so, compute x^k recursively and then square the result. If n is odd, say $n = 2k + 1$, then

$$x^n = x^{2k+1} = (x^k)^2 \cdot x,$$

so, compute x^k recursively, square it, and multiply the result by x.

What this suggests is to write $n \geq 1$ in binary, say

$$n = b_\ell \cdot 2^\ell + b_{\ell-1} \cdot 2^{\ell-1} + \cdots + b_1 \cdot 2^1 + b_0,$$

where $b_i \in \{0,1\}$ with $b_\ell = 1$ or, if we let $J = \{j \mid b_j = 1\}$, as

$$n = \sum_{j \in J} 2^j.$$

Then we have

$$x^n \equiv x^{\sum_{j \in J} 2^j} = \prod_{j \in J} x^{2^j} \bmod m.$$

This suggests computing the residues r_j such that

$$x^{2^j} \equiv r_j \pmod{m},$$

because then,

$$x^n \equiv r_\ell \cdots r_0 \pmod{m},$$

where we can compute this latter product modulo m two terms at a time.

For example, say we want to compute $999^{179} \bmod 1763$. First, we observe that

$$179 = 2^7 + 2^5 + 2^4 + 2^1 + 1,$$

and we compute the powers modulo 1763:

$$999^{2^1} \equiv 143 \,(\mathrm{mod}\,1763)$$
$$999^{2^2} \equiv 143^2 \equiv 1056 \,(\mathrm{mod}\,1763)$$
$$999^{2^3} \equiv 1056^2 \equiv 920 \,(\mathrm{mod}\,1763)$$
$$999^{2^4} \equiv 920^2 \equiv 160 \,(\mathrm{mod}\,1763)$$
$$999^{2^5} \equiv 160^2 \equiv 918 \,(\mathrm{mod}\,1763)$$
$$999^{2^6} \equiv 918^2 \equiv 10 \,(\mathrm{mod}\,1763)$$
$$999^{2^7} \equiv 10^2 \equiv 100 \,(\mathrm{mod}\,1763).$$

Consequently,

$$999^{179} \equiv 999 \cdot 143 \cdot 160 \cdot 918 \cdot 100 \,(\mathrm{mod}\,1763)$$
$$\equiv 54 \cdot 160 \cdot 918 \cdot 100 \,(\mathrm{mod}\,1763)$$

$$\equiv 1588 \cdot 918 \cdot 100 \,(\mathrm{mod}\,1763)$$
$$\equiv 1546 \cdot 100 \,(\mathrm{mod}\,1763)$$
$$\equiv 1219 \,(\mathrm{mod}\,1763),$$

and we find that
$$999^{179} \equiv 1219 \,(\mathrm{mod}\,1763).$$

Of course, it would be impossible to exponentiate 999^{179} first and then reduce modulo 1763. As we can see, the number of multiplications needed is $O(\log_2 n)$, which is quite good.

The above method can be implemented without actually converting n to base 2. If n is even, say $n = 2k$, then $n/2 = k$ and if n is odd, say $n = 2k+1$, then $(n-1)/2 = k$, so we have a way of dropping the unit digit in the binary expansion of n and shifting the remaining digits one place to the right without explicitly computing this binary expansion. Here is an algorithm for computing $x^n \bmod m$, with $n \geq 1$, using the *repeated squaring* method.

An Algorithm to Compute $x^n \bmod m$ Using Repeated Squaring

```
begin
    u := 1; a := x;
    while n > 1 do
        if even(n) then e := 0 else e := 1;
        if e = 1 then u := a · u mod m;
        a := a² mod m; n := (n − e)/2
    endwhile;
    u := a · u mod m
end
```

The final value of u is the result. The reason why the algorithm is correct is that after j rounds through the while loop, $a = x^{2^j} \bmod m$ and

$$u = \prod_{i \in J \mid i < j} x^{2^i} \bmod m,$$

with this product interpreted as 1 when $j = 0$.

Observe that the while loop is only executed $n - 1$ times to avoid squaring once more unnecessarily and the last multiplication $a \cdot u$ is performed outside of the while loop. Also, if we delete the reductions modulo m, the above algorithm is a fast method for computing the nth power of an integer x and the time speed-up of not performing the last squaring step is more significant. We leave the details of the proof that the above algorithm is correct as an exercise.

Let us now consider the problem of computing efficiently the inverse of an integer a, modulo m, provided that $\gcd(a, m) = 1$.

We mentioned in Section 5.4 how the extended Euclidean algorithm can be used to find some integers x, y, such that

$$ax + by = \gcd(a, b),$$

where a and b are any two positive integers. The details are worked out in Problem 5.18 and another version is explored in Problem 5.19. In our situation, $a = m$ and $b = a$ and we only need to find y (we would like a positive integer).

When using the Euclidean algorithm for computing $\gcd(m, a)$, with $2 \leq a < m$, we compute the following sequence of quotients and remainders.

$$m = aq_1 + r_1$$
$$a = r_1 q_2 + r_2$$
$$r_1 = r_2 q_3 + r_3$$
$$\vdots$$
$$r_{k-1} = r_k q_{k+1} + r_{k+1}$$
$$\vdots$$
$$r_{n-3} = r_{n-2} q_{n-1} + r_{n-1}$$
$$r_{n-2} = r_{n-1} q_n + 0,$$

with $n \geq 3$, $0 < r_1 < b$, $q_k \geq 1$, for $k = 1, \ldots, n$, and $0 < r_{k+1} < r_k$, for $k = 1, \ldots, n-2$. Observe that $r_n = 0$. If $n = 2$, we have just two divisions,

$$m = aq_1 + r_1$$
$$a = r_1 q_2 + 0,$$

with $0 < r_1 < b$, $q_1, q_2 \geq 1$, and $r_2 = 0$. Thus, it is convenient to set $r_{-1} = m$ and $r_0 = a$.

In Problem 5.18, it is shown that if we set

$$x_{-1} = 1$$
$$y_{-1} = 0$$
$$x_0 = 0$$
$$y_0 = 1$$
$$x_{i+1} = x_{i-1} - x_i q_{i+1}$$
$$y_{i+1} = y_{i-1} - y_i q_{i+1},$$

for $i = 0, \ldots, n - 2$, then

$$mx_{n-1} + ay_{n-1} = \gcd(m, a) = r_{n-1},$$

and so, if $\gcd(m, a) = 1$, then $r_{n-1} = 1$ and we have

$$ay_{n-1} \equiv 1 \pmod{m}.$$

Now, y_{n-1} may be greater than m or negative but we already know how to deal with that from the discussion just before Theorem 5.10. This suggests reducing modulo m during the recurrence and we are led to the following recurrence.

$$y_{-1} = 0$$
$$y_0 = 1$$
$$z_{i+1} = y_{i-1} - y_i q_{i+1}$$
$$y_{i+1} = z_{i+1} \bmod m \quad \text{if} \quad z_{i+1} \geq 0$$
$$y_{i+1} = m - ((-z_{i+1}) \bmod m) \quad \text{if} \quad z_{i+1} < 0,$$

for $i = 0, \ldots, n - 2$.

It is easy to prove by induction that

$$ay_i \equiv r_i \pmod{m}$$

for $i = 0, \ldots, n - 1$ and thus, if $\gcd(a, m) > 1$, then a does not have an inverse modulo m, else

$$ay_{n-1} \equiv 1 \pmod{m}$$

and y_{n-1} is the inverse of a modulo m such that $1 \leq y_{n-1} < m$, as desired. Note that we also get $y_0 = 1$ when $a = 1$.

We leave this proof as an exercise (see Problem 5.58). Here is an algorithm obtained by adapting the algorithm given in Problem 5.18.

An Algorithm for Computing the Inverse of a Modulo m

Given any natural number a with $1 \leq a < m$ and $\gcd(a, m) = 1$, the following algorithm returns the inverse of a modulo m as y.

```
begin
    y := 0; v := 1; g := m; r := a;
    pr := r; q := ⌊g/pr⌋; r := g − prq; (divide g by pr, to get g = prq + r)
    if r = 0 then
        y := 1; g := pr
    else
        r = pr;
        while r ≠ 0 do
            pr := r; pv := v;
            q := ⌊g/pr⌋; r := g − prq; (divide g by pr, to get g = prq + r)
            v := y − pvq;
            if v < 0 then
                v := m − ((−v) mod m)
            else
                v = v mod m
            endif
            g := pr; y := pv
        endwhile;
    endif;
    inverse(a) := y
end
```

For example, we used the above algorithm to find that $d_A = 145,604,785$ is the inverse of $e_A = 79921$ modulo $(p_A - 1)(q_A - 1) = 163,251,312$.

The remaining issues are how to choose large random prime numbers p, q, and how to find a random number e, which is relatively prime to $(p-1)(q-1)$. For this, we rely on a deep result of number theory known as the *prime number theorem*.

5.12 Finding Large Primes; Signatures; Safety of RSA

Roughly speaking, the prime number theorem ensures that the density of primes is high enough to guarantee that there are many primes with a large specified number of digits. The relevant function is the *prime counting function* $\pi(n)$.

Definition 5.14. The *prime counting function* π is the function defined so that

$$\pi(n) = \text{number of prime numbers } p, \text{ such that } p \leq n,$$

for every natural number $n \in \mathbb{N}$.

Obviously, $\pi(0) = \pi(1) = 0$. We have $\pi(10) = 4$ because the primes no greater than 10 are $2, 3, 5, 7$ and $\pi(20) = 8$ because the primes no greater than 20 are $2, 3, 5, 7, 11, 13, 17, 19$. The growth of the function π was studied by Legendre,

Gauss, Chebyshev, and Riemann between 1808 and 1859. By then, it was conjectured that

$$\pi(n) \sim \frac{n}{\ln(n)},$$

for n large, which means that

$$\lim_{n \to \infty} \pi(n) \bigg/ \frac{n}{\ln(n)} = 1.$$

However, a rigorous proof was not found until 1896. Indeed, in 1896, Jacques

Fig. 5.18 Pafnuty Lvovich Chebyshev, 1821–1894 (left), Jacques Salomon Hadamard, 1865–1963 (middle), and Charles Jean de la Vallée Poussin, 1866–1962 (right)

Hadamard and Charles de la Vallée-Poussin independendly gave a proof of this "most wanted theorem," using methods from complex analysis. These proofs are difficult and although more elementary proofs were given later, in particular by Erdös and Selberg (1949), those proofs are still quite hard. Thus, we content ourselves with a statement of the theorem.

Fig. 5.19 Paul Erdös, 1913–1996 (left), Atle Selberg, 1917–2007 (right)

Theorem 5.15. *(Prime Number Theorem) For n large, the number of primes $\pi(n)$ no larger than n is approximately equal to $n/\ln(n)$, which means that*

$$\lim_{n \to \infty} \pi(n) \bigg/ \frac{n}{\ln(n)} = 1.$$

For a rather detailed account of the history of the prime number theorem (for short, *PNT*), we refer the reader to Ribenboim [13] (Chapter 4).

As an illustration of the use of the PNT, we can estimate the number of primes with 200 decimal digits. Indeed this is the difference of the number of primes up to 10^{200} minus the number of primes up to 10^{199}, which is approximately

$$\frac{10^{200}}{200 \ln 10} - \frac{10^{199}}{199 \ln 10} \approx 1.95 \cdot 10^{197}.$$

Thus, we see that there is a huge number of primes with 200 decimal digits. The number of natural numbers with 200 digits is $10^{200} - 10^{199} = 9 \cdot 10^{199}$, thus the proportion of 200-digit numbers that are prime is

$$\frac{1.95 \cdot 10^{197}}{9 \cdot 10^{199}} \approx \frac{1}{460}.$$

Consequently, among the natural numbers with 200 digits, roughly one in every 460 is a prime.

Beware that the above argument is not entirely rigorous because the prime number theorem only yields an approximation of $\pi(n)$ but sharper estimates can be used to say how large n should be to guarantee a prescribed error on the probability, say 1%.

The implication of the above fact is that if we wish to find a random prime with 200 digits, we pick at random some natural number with 200 digits and test whether it is prime. If this number is not prime, then we discard it and try again, and so on. On the average, after 460 trials, a prime should pop up.

This leads us the question: How do we test for primality?

Primality testing has also been studied for a long time. Remarkably, Fermat's little theorem yields a test for nonprimality. Indeed, if $p > 1$ fails to divide $a^{p-1} - 1$ for some natural number a, where $2 \leq a \leq p - 1$, then p cannot be a prime. The simplest a to try is $a = 2$. From a practical point of view, we can compute $a^{p-1} \bmod p$ using the method of repeated squaring and check whether the remainder is 1.

But what if p fails the Fermat test? Unfortunately, there are natural numbers p, such that p divides $2^{p-1} - 1$ and yet, p is composite. For example $p = 341 = 11 \cdot 31$ is such a number.

Actually, 2^{340} being quite big, how do we check that $2^{340} - 1$ is divisible by 341?

We just have to show that $2^{340} - 1$ is divisible by 11 and by 31. We can use Fermat's little theorem. Because 11 is prime, we know that 11 divides $2^{10} - 1$. But,

$$2^{340} - 1 = (2^{10})^{34} - 1 = (2^{10} - 1)((2^{10})^{33} + (2^{10})^{32} + \cdots + 1),$$

so $2^{340} - 1$ is also divisible by 11.

As to divisibility by 31, observe that $31 = 2^5 - 1$, and

$$2^{340} - 1 = (2^5)^{68} - 1 = (2^5 - 1)((2^5)^{67} + (2^5)^{66} + \cdots + 1),$$

so $2^{340} - 1$ is also divisible by 31.

A number p that is not a prime but behaves like a prime in the sense that p divides $2^{p-1} - 1$, is called a *pseudo-prime*. Unfortunately, the Fermat test gives a "false positive" for pseudo-primes.

Rather than simply testing whether $2^{p-1} - 1$ is divisible by p, we can also try whether $3^{p-1} - 1$ is divisible by p and whether $5^{p-1} - 1$ is divisible by p, and so on.

Unfortunately, there are composite natural numbers p, such that p divides $a^{p-1} - 1$, for all positive natural numbers a with $\gcd(a, p) = 1$. Such numbers are known as *Carmichael numbers*. The smallest Carmichael number is $p = 561 = 3 \cdot 11 \cdot 17$. The reader should try proving that, in fact, $a^{560} - 1$ is divisible by 561 for every positive natural number a, such that $\gcd(a, 561) = 1$, using the technique that we used to prove that 341 divides $2^{340} - 1$.

Fig. 5.20 Robert Daniel Carmichael, 1879–1967

It turns out that there are infinitely many Carmichael numbers. Again, for a thorough introduction to primality testing, pseudo-primes, Carmichael numbers, and more, we highly recommend Ribenboim [13] (Chapter 2). An excellent (but more terse) account is also given in Koblitz [9] (Chapter V).

Still, what do we do about the problem of false positives? The key is to switch to *probabilistic methods*. Indeed, if we can design a method that is guaranteed to give a false positive with probablity less than 0.5, then we can repeat this test for randomly chosen as and reduce the probability of false positive considerably. For example, if we repeat the experiment 100 times, the probability of false positive is less than $2^{-100} < 10^{-30}$. This is probably less than the probability of hardware failure.

Various probabilistic methods for primality testing have been designed. One of them is the Miller–Rabin test, another the APR test, and yet another the Solovay–Strassen test. Since 2002, it has been known that primality testing can be done in polynomial time. This result is due to Agrawal, Kayal, and Saxena and known as the AKS test solved a long-standing problem; see Dietzfelbinger [4] and Crandall and Pomerance [2] (Chapter 4). Remarkably, Agrawal and Kayal worked on this problem for their senior project in order to complete their bachelor's degree. It remains to be seen whether this test is really practical for very large numbers.

A very important point to make is that these primality testing methods *do not* provide a factorization of m when m is composite. This is actually a crucial ingredient for the security of the RSA scheme. So far, it appears (and it is hoped) that *factoring* an integer is a much harder problem than testing for primality and all known methods are incapable of factoring natural numbers with over 300 decimal digits (it would take centuries).

For a comprehensive exposition of the subject of primality-testing, we refer the reader to Crandall and Pomerance [2] (Chapter 4) and again, to Ribenboim [13] (Chapter 2) and Koblitz [9] (Chapter V).

Going back to the RSA method, we now have ways of finding the large random primes p and q by picking at random some 200-digit numbers and testing for primality. Rivest, Shamir, and Adleman also recommend to pick p and q so that they differ by a few decimal digits, that both $p - 1$ and $q - 1$ should contain large prime factors and that $\gcd(p - 1, q - 1)$ should be small. The public key, e, relatively prime to $(p - 1)(q - 1)$ can also be found by a similar method: Pick at random a number, $e < (p - 1)(q - 1)$, which is large enough (say, greater than $\max\{p, q\}$) and test whether $\gcd(e, (p - 1)(q - 1)) = 1$, which can be done quickly using the extended Euclidean algorithm. If not, discard e and try another number, and so on. It is easy to see that such an e will be found in no more trials than it takes to find a prime; see Lovász, Pelikán, and Vesztergombi [10] (Chapter 15), which contains one of the simplest and clearest presentations of RSA that we know of. Koblitz [9] (Chapter IV) also provides some details on this topic as well as Menezes, van Oorschot, and Vanstone's *Handbook* [11].

If Albert receives a message coming from Julia, how can he be sure that this message does not come from an imposter? Just because the message is signed "Julia" does not mean that it comes from Julia; it could have been sent by someone else pretending to be Julia, inasmuch as all that is needed to send a message to Albert is Albert's public key, which is known to everybody. This leads us to the issue of *signatures*.

There are various schemes for adding a signature to an encrypted message to ensure that the sender of a message is really who he or she claims to be (with a high degree of confidence). The trick is to make use of the the sender's keys. We propose two scenarios.

1. The sender, Julia, encrypts the message x to be sent with *her own private key*, (d_J, m_J), creating the message $D_J(x) = y_1$. Then, Julia adds her signature, "Julia", at the end of the message y_1, encrypts the message "y_1 Julia" using *Albert's public key*, (e_A, m_A), creating the message $y_2 = E_A(y_1 \text{ Julia})$, and finally sends the message y_2 to Albert.

   When Albert receives the encrypted message y_2 claiming to come from *Julia*, first he decrypts the message using *his private key* (d_A, m_A). He will see an encrypted message, $D_A(y_2) = y_1$ Julia, with the legible signature, *Julia*. He will then delete the signature from this message and decrypt the message y_1 using *Julia's public key* (e_J, m_J), getting $x = E_J(y_1)$. Albert will know whether someone else faked this message if the result is garbage. Indeed, only Julia could have encrypted the original message x with her private key, which is only known to her. An eavesdropper who is pretending to be Julia would not know Julia's private key and so, would not have encrypted the original message to be sent using Julia's secret key.

2. The sender, Julia, first adds her signature, "Julia", to the message x to be sent and then, she encrypts the message "x Julia" with *Albert's public key* (e_A, m_A), creating the message $y_1 = E_A(x \text{ Julia})$. Julia also encrypts the original message x using *her private key* (d_J, m_J) creating the message $y_2 = D_J(x)$, and finally she sends the pair of messages (y_1, y_2).

   When Albert receives a pair of messages (y_1, y_2), claiming to have been sent by Julia, first Albert decrypts y_1 using *his private key* (d_A, m_A), getting the message $D_A(y_1) = x \text{ Julia}$. Albert finds the signature, Julia, and then decrypts y_2 using *Julia's public key* (e_J, m_J), getting the message $x' = E_J(y_2)$. If $x = x'$, then Albert has serious assurance that the sender is indeed Julia and not an imposter.

The last topic that we would like to discuss is the *security* of the RSA scheme. This is a difficult issue and many researchers have worked on it. As we remarked earlier, the security of RSA hinges on the fact that factoring is hard. It has been shown that if one has a method for breaking the RSA scheme (namely, to find the secret key d), then there is a probabilistic method for finding the factors p and q, of $m = pq$ (see Koblitz [9], Chapter IV, Section 2, or Menezes, van Oorschot, and Vanstone [11], Section 8.2.2). If p and q are chosen to be large enough, factoring $m = pq$ will be practically impossible and so it is unlikely that RSA can be cracked. However, there may be other attacks and, at present, there is no proof that RSA is fully secure.

Observe that because $m = pq$ is known to everybody, if somehow one can learn $N = (p-1)(q-1)$, then p and q can be recovered. Indeed $N = (p-1)(q-1) = pq - (p+q) + 1 = m - (p+q) + 1$ and so,

$$pq = m$$
$$p + q = m - N + 1,$$

and p and q are the roots of the quadratic equation

$$X^2 - (m - N + 1)X + m = 0.$$

Thus, a line of attack is to try to find the value of $(p-1)(q-1)$. For more on the security of RSA, see Menezes, van Oorschot, and Vanstone's *Handbook* [11].

5.13 Distributive Lattices, Boolean Algebras, Heyting Algebras

If we go back to one of our favorite examples of a lattice, namely, the power set 2^X of some set X, we observe that it is more than a lattice. For example, if we look at Figure 5.6, we can check that the two identities D1 and D2 stated in the next definition hold.

Definition 5.15. We say that a lattice X is a *distributive lattice* if (D1) and (D2) hold:

$$D1 \qquad a \wedge (b \vee c) = (a \wedge b) \vee (a \wedge c)$$
$$D2 \qquad a \vee (b \wedge c) = (a \vee b) \wedge (a \vee c).$$

Remark: Not every lattice is distributive but many lattices of interest are distributive.

It is a bit surprising that in a lattice (D1) and (D2) are actually equivalent, as we now show. Suppose (D1) holds, then

$$
\begin{aligned}
(a \vee b) \wedge (a \vee c) &= ((a \vee b) \wedge a) \vee ((a \vee b) \wedge c) & \text{(D1)} \\
&= a \vee ((a \vee b) \wedge c) & \text{(L4)} \\
&= a \vee ((c \wedge (a \vee b)) & \text{(L1)} \\
&= a \vee ((c \wedge a) \vee (c \wedge b)) & \text{(D1)} \\
&= a \vee ((a \wedge c) \vee (b \wedge c)) & \text{(L1)} \\
&= (a \vee (a \wedge c)) \vee (b \wedge c) & \text{(L2)} \\
&= ((a \wedge c) \vee a) \vee (b \wedge c) & \text{(L1)} \\
&= a \vee (b \wedge c) & \text{(L4)}
\end{aligned}
$$

which is (D2). Dually, (D2) implies (D1).

The reader should prove that every totally ordered poset is a distributive lattice. The lattice $\mathbb{N}_+ = \mathbb{N} - \{0\}$ under the divisibility ordering also turns out to be a distributive lattice.

Another useful fact about distributivity is that in any lattice

$$a \wedge (b \vee c) \geq (a \wedge b) \vee (a \wedge c).$$

This is because in any lattice, $a \wedge (b \vee c) \geq a \wedge b$ and $a \wedge (b \vee c) \geq a \wedge c$. Therefore, in order to establish distributivity in a lattice it suffices to show that

$$a \wedge (b \vee c) \leq (a \wedge b) \vee (a \wedge c).$$

Another important property of distributive lattices is the following.

Proposition 5.26. *In a distributive lattice X, if $z \wedge x = z \wedge y$ and $z \vee x = z \vee y$, then $x = y$ (for all $x, y, z \in X$).*

Proof. We have

$$
\begin{aligned}
x &= (x \vee z) \wedge x & \text{(L4)} \\
&= x \wedge (z \vee x) & \text{(L1)} \\
&= x \wedge (z \vee y) & \\
&= (x \wedge z) \vee (x \wedge y) & \text{(D1)}
\end{aligned}
$$

$$= (z \wedge x) \vee (x \wedge y) \qquad \text{(L1)}$$
$$= (z \wedge y) \vee (x \wedge y)$$
$$= (y \wedge z) \vee (y \wedge x) \qquad \text{(L1)}$$
$$= y \wedge (z \vee x) \qquad \text{(D1)}$$
$$= y \wedge (z \vee y)$$
$$= (y \vee z) \wedge y \qquad \text{(L1)}$$
$$= y; \qquad \text{(L4)}$$

that is, $x = y$, as claimed. $\square$

The power set lattice has yet some additional properties having to do with complementation. First, the power lattice 2^X has a least element $0 = \emptyset$ and a greatest element, $1 = X$. If a lattice X has a least element 0 and a greatest element 1, the following properties are clear: For all $a \in X$, we have

$$a \wedge 0 = 0 \qquad a \vee 0 = a$$
$$a \wedge 1 = a \qquad a \vee 1 = 1.$$

More important, for any subset $A \subseteq X$ we have the complement $\overline{A}$ of A in X, which satisfies the identities:

$$A \cup \overline{A} = X, \qquad A \cap \overline{A} = \emptyset.$$

Moreover, we know that the de Morgan identities hold. The generalization of these properties leads to what is called a complemented lattice.

Fig. 5.21 Augustus de Morgan, 1806–1871

Definition 5.16. Let X be a lattice and assume that X has a least element 0 and a greatest element 1 (we say that X is a *bounded lattice*). For any $a \in X$, a *complement of a* is any element $b \in X$, so that

$$a \vee b = 1 \quad \text{and} \quad a \wedge b = 0.$$

If every element of X has a complement, we say that X is a *complemented lattice*.

Remarks:

1. When $0 = 1$, the lattice X collapses to the degenerate lattice consisting of a single element. As this lattice is of little interest, from now on, we always assume that $0 \neq 1$.
2. In a complemented lattice, complements are generally not unique. However, as the next proposition shows, this is the case for distributive lattices.

Proposition 5.27. *Let X be a lattice with least element 0 and greatest element 1. If X is distributive, then complements are unique if they exist. Moreover, if b is the complement of a, then a is the complement of b.*

Proof. If a has two complements, b_1 and b_2, then $a \wedge b_1 = 0$, $a \wedge b_2 = 0$, $a \vee b_1 = 1$, and $a \vee b_2 = 1$. By Proposition 5.26, we deduce that $b_1 = b_2$; that is, a has a unique complement.

By commutativity, the equations

$$a \vee b = 1 \quad \text{and} \quad a \wedge b = 0$$

are equivalent to the equations

$$b \vee a = 1 \quad \text{and} \quad b \wedge a = 0,$$

which shows that a is indeed a complement of b. By uniqueness, a is *the* complement of b. □

In view of Proposition 5.27, if X is a complemented distributive lattice, we denote the complement of any element, $a \in X$, by $\bar{a}$. We have the identities

$$a \vee \bar{a} = 1$$
$$a \wedge \bar{a} = 0$$
$$\bar{\bar{a}} = a.$$

We also have the following proposition about the de Morgan laws.

Proposition 5.28. *Let X be a lattice with least element 0 and greatest element 1. If X is distributive and complemented, then the de Morgan laws hold:*

$$\overline{a \vee b} = \bar{a} \wedge \bar{b}$$
$$\overline{a \wedge b} = \bar{a} \vee \bar{b}.$$

Proof. We prove that

$$\overline{a \vee b} = \bar{a} \wedge \bar{b},$$

leaving the dual identity as an easy exercise. Using the uniqueness of complements, it is enough to check that $\bar{a} \wedge \bar{b}$ works, that is, satisfies the conditions of Definition 5.16. For the first condition, we have

$$
\begin{aligned}
(a \vee b) \vee (\bar{a} \wedge \bar{b}) &= ((a \vee b) \vee \bar{a}) \wedge ((a \vee b) \vee \bar{b}) \\
&= (a \vee (b \vee \bar{a})) \wedge (a \vee (b \vee \bar{b})) \\
&= (a \vee (\bar{a} \vee b)) \wedge (a \vee 1) \\
&= ((a \vee \bar{a}) \vee b) \wedge 1 \\
&= (1 \vee b) \wedge 1 \\
&= 1 \wedge 1 = 1.
\end{aligned}
$$

For the second condition, we have

$$
\begin{aligned}
(a \vee b) \wedge (\bar{a} \wedge \bar{b}) &= (a \wedge (\bar{a} \wedge \bar{b})) \vee (b \wedge (\bar{a} \wedge \bar{b})) \\
&= ((a \wedge \bar{a}) \wedge \bar{b}) \vee (b \wedge (\bar{b} \wedge \bar{a})) \\
&= (0 \wedge \bar{b}) \vee ((b \wedge \bar{b}) \wedge \bar{a}) \\
&= 0 \vee (0 \wedge \bar{a}) \\
&= 0 \vee 0 = 0.
\end{aligned}
$$

$\square$

All this leads to the definition of a Boolean lattice.

Definition 5.17. A *Boolean lattice* is a lattice with a least element 0, a greatest element 1, and which is distributive and complemented.

Of course, every power set is a Boolean lattice, but there are Boolean lattices that are not power sets. Putting together what we have done, we see that a Boolean lattice is a set X with two special elements, 0, 1, and three operations $\wedge$, $\vee$, and $a \mapsto \bar{a}$ satisfying the axioms stated in the following.

Proposition 5.29. *If X is a Boolean lattice, then the following equations hold for all $a, b, c \in X$.*

L1	$a \vee b = b \vee a,$	$a \wedge b = b \wedge a$
L2	$(a \vee b) \vee c = a \vee (b \vee c),$	$(a \wedge b) \wedge c = a \wedge (b \wedge c)$
L3	$a \vee a = a,$	$a \wedge a = a$
L4	$(a \vee b) \wedge a = a,$	$(a \wedge b) \vee a = a$
D1-D2	$a \wedge (b \vee c) = (a \wedge b) \vee (a \wedge c),$	$a \vee (b \wedge c) = (a \vee b) \wedge (a \vee c)$
LE	$a \vee 0 = a,$	$a \wedge 0 = 0$
GE	$a \vee 1 = 1,$	$a \wedge 1 = a$
C	$a \vee \bar{a} = 1,$	$a \wedge \bar{a} = 0$
I	$\bar{\bar{a}} = a$	
dM	$\overline{a \vee b} = \bar{a} \wedge \bar{b},$	$\overline{a \wedge b} = \bar{a} \vee \bar{b}.$

Conversely, if X is a set together with two special elements 0, 1, *and three operations* $\wedge$, $\vee$, *and* $a \mapsto \bar{a}$ *satisfying the axioms above, then it is a Boolean lattice under the ordering given by* $a \leq b$ *iff* $a \vee b = b$.

In view of Proposition 5.29, we make the following definition.

Definition 5.18. A set X together with two special elements 0, 1 and three operations $\wedge$, $\vee$, and $a \mapsto \bar{a}$ satisfying the axioms of Proposition 5.29 is called a *Boolean algebra*.

Proposition 5.29 shows that the notions of a Boolean lattice and of a Boolean algebra are equivalent. The first one is order-theoretic and the second one is algebraic.

Remarks:

1. As the name indicates, Boolean algebras were invented by G. Boole (1854). One of the first comprehensive accounts is due to E. Schröder (1890–1895).
2. The axioms for Boolean algebras given in Proposition 5.29 are not independent. There is a set of independent axioms known as the *Huntington axioms* (1933).

Fig. 5.22 George Boole, 1815–1864 (left) and Ernst Schröder 1841–1902 (right)

Let p be any integer with $p \geq 2$. Under the division ordering, it turns out that the set $\mathrm{Div}(p)$ of divisors of p is a distributive lattice. In general not every integer $k \in \mathrm{Div}(p)$ has a complement but when it does $\bar{k} = p/k$. It can be shown that $\mathrm{Div}(p)$ is a Boolean algebra iff p is not divisible by any square integer (an integer of the form m^2, with $m > 1$).

Classical logic is also a rich source of Boolean algebras. Indeed, it is easy to show that logical equivalence is an equivalence relation and, as homework problems, you have shown (with great pain) that all the axioms of Proposition 5.29 are provable equivalences (where $\vee$ is disjunction and $\wedge$ is conjunction, $\bar{P} = \neg P$; i.e., negation, $0 = \bot$ and $1 = \top$) (see Problems 1.8, 1.18, 1.28). Furthermore, again, as homework problems (see Problems 1.18–1.20), you have shown that logical equivalence is compatible with $\vee, \wedge, \neg$ in the following sense. If $P_1 \equiv Q_1$ and $P_2 \equiv Q_2$, then

$$(P_1 \vee P_2) \equiv (Q_1 \vee Q_2)$$
$$(P_1 \wedge P_2) \equiv (Q_1 \wedge Q_2)$$
$$\neg P_1 \equiv \neg Q_1.$$

Consequently, for any set T of propositions we can define the relation $\equiv_T$ by

$$P \equiv_T Q \text{ iff } T \vdash P \equiv Q,$$

that is, iff $P \equiv Q$ is provable from T (as explained in Section 1.11). Clearly, $\equiv_T$ is an equivalence relation on propositions and so, we can define the operations $\vee, \wedge$, and $-$ on the set of equivalence classes $\mathbf{B}_T$ of propositions as follows.

$$[P] \vee [Q] = [P \vee Q]$$
$$[P] \wedge [Q] = [P \wedge Q]$$
$$\overline{[P]} = [\neg P].$$

We also let $0 = [\bot]$ and $1 = [\top]$. Then, we get the Boolean algebra $\mathbf{B}_T$ called the *Lindenbaum algebra* of T.

It also turns out that Boolean algebras are just what's needed to give truth-value semantics to classical logic. Let B be any Boolean algebra. A *truth assignment* is any function v from the set $\mathbf{PS} = \{\mathbf{P}_1, \mathbf{P}_2, \ldots\}$ of propositional symbols to B. Then, we can recursively evaluate the truth value $P_B[v]$ in B of any proposition P with respect to the truth assignment v as follows.

$$(\mathbf{P}_i)_B[v] = v(P)$$
$$\bot_B[v] = 0$$
$$\top_B[v] = 1$$
$$(P \vee Q)_B[v] = P_B[v] \vee P_B[v]$$
$$(P \wedge Q)_B[v] = P_B[v] \wedge P_B[v]$$
$$(\neg P)_B[v] = \overline{P[v]_B}.$$

In the equations above, on the right-hand side, $\vee$ and $\wedge$ are the lattice operations of the Boolean algebra B. We say that a proposition P is *valid in the Boolean algebra B (or B-valid)* if $P_B[v] = 1$ for all truth assignments v. We say that P is *(classically) valid* if P is B-valid in all Boolean algebras B. It can be shown that every provable proposition is valid. This property is called *soundness*. Conversely, if P is valid, then it is provable. This second property is called *completeness*. Actually completeness holds in a much stronger sense: If a proposition is valid in the two-element Boolean algebra $\{0, 1\}$, then it is provable.

One might wonder if there are certain kinds of algebras similar to Boolean algebras well suited for intuitionistic logic. The answer is yes: such algebras are called *Heyting algebras*.

In our study of intuitionistic logic, we learned that negation is not a primary connective but instead it is defined in terms of implication by $\neg P = P \Rightarrow \bot$. This

Fig. 5.23 Arend Heyting, 1898–1980

suggests adding to the two lattice operations $\vee$ and $\wedge$ a new operation $\rightarrow$, that will behave like $\Rightarrow$. The trick is, what kind of axioms should we require on $\rightarrow$ to "capture" the properties of intuitionistic logic? Now, if X is a lattice with 0 and 1, given any two elements $a, b \in X$, experience shows that $a \rightarrow b$ should be the largest element c, such that $c \wedge a \leq b$. This leads to

Definition 5.19. A lattice X with 0 and 1 is a *Heyting lattice* iff it has a third binary operation $\rightarrow$ such that

$$c \wedge a \leq b \text{ iff } c \leq (a \rightarrow b)$$

for all $a, b, c \in X$. We define the *negation (or pseudo-complement) of a* as $\bar{a} = (a \rightarrow 0)$.

At first glance, it is not clear that a Heyting lattice is distributive but in fact, it is. The following proposition (stated without proof) gives an algebraic characterization of Heyting lattices which is useful to prove various properties of Heyting lattices.

Proposition 5.30. *Let X be a lattice with 0 and 1 and with a binary operation $\rightarrow$. Then, X is a Heyting lattice iff the following equations hold for all $a, b, c \in X$.*

$$a \rightarrow a = 1$$
$$a \wedge (a \rightarrow b) = a \wedge b$$
$$b \wedge (a \rightarrow b) = b$$
$$a \rightarrow (b \wedge c) = (a \rightarrow b) \wedge (a \rightarrow c).$$

A lattice with 0 and 1 and with a binary operation, $\rightarrow$, satisfying the equations of Proposition 5.30 is called a *Heyting algebra*. So, we see that Proposition 5.30 shows that the notions of Heyting lattice and Heyting algebra are equivalent (this is analogous to Boolean lattices and Boolean algebras).

The reader will notice that these axioms are propositions that were shown to be provable intuitionistically in homework problems. The proof of Proposition 5.30 is not really difficult but it is a bit tedious so we omit it. Let us simply show that the fourth equation implies that for any fixed $a \in X$, the map $b \mapsto (a \rightarrow b)$ is monotonic. So, assume $b \leq c$; that is, $b \wedge c = b$. Then, we get

$$a \to b = a \to (b \wedge c) = (a \to b) \wedge (a \to c),$$

which means that $(a \to b) \leq (a \to c)$, as claimed.

The following theorem shows that every Heyting algebra is distributive, as we claimed earlier. This theorem also shows "how close" to a Boolean algebra a Heyting algebra is.

Theorem 5.16. *(a) Every Heyting algebra is distributive.*
(b) A Heyting algebra X is a Boolean algebra iff $\bar{\bar{a}} = a$ for all $a \in X$.

Proof. (a) From a previous remark, to show distributivity, it is enough to show the inequality

$$a \wedge (b \vee c) \leq (a \wedge b) \vee (a \wedge c).$$

Observe that from the property characterizing $\to$, we have

$$b \leq a \to (a \wedge b) \quad \text{iff} \quad b \wedge a \leq a \wedge b$$

which holds, by commutativity of $\wedge$. Thus, $b \leq a \to (a \wedge b)$ and similarly, $c \leq a \to (a \wedge c)$.

Recall that for any fixed a, the map $x \mapsto (a \to x)$ is monotonic. Because $a \wedge b \leq (a \wedge b) \vee (a \wedge c)$ and $a \wedge c \leq (a \wedge b) \vee (a \wedge c)$, we get

$$a \to (a \wedge b) \leq a \to ((a \wedge b) \vee (a \wedge c)) \quad \text{and} \quad a \to (a \wedge c) \leq a \to ((a \wedge b) \vee (a \wedge c)).$$

These two inequalities imply $(a \to (a \wedge b)) \vee (a \to (a \wedge c)) \leq a \to ((a \wedge b) \vee (a \wedge c))$, and because we also have $b \leq a \to (a \wedge b)$ and $c \leq a \to (a \wedge c)$, we deduce that

$$b \vee c \leq a \to ((a \wedge b) \vee (a \wedge c)),$$

which, using the fact that $(b \vee c) \wedge a = a \wedge (b \vee c)$, means that

$$a \wedge (b \vee c) \leq (a \wedge b) \vee (a \wedge c),$$

as desired.

(b) We leave this part as an exercise. The trick is to see that the de Morgan laws hold and to apply one of them to $a \wedge \bar{a} = 0$.   □

Remarks:

1. Heyting algebras were invented by A. Heyting in 1930. Heyting algebras are sometimes known as "Brouwerian lattices".
2. Every Boolean algebra is automatically a Heyting algebra: Set $a \to b = \bar{a} \vee b$.
3. It can be shown that every finite distributive lattice is a Heyting algebra.

We conclude this brief exposition of Heyting algebras by explaining how they provide a truth-value semantics for intuitionistic logic analogous to the truth-value semantics that Boolean algebras provide for classical logic.

As in the classical case, it is easy to show that intuitionistic logical equivalence is an equivalence relation and you have shown (with great pain) that all the axioms of Heyting algebras are intuitionistically provable equivalences (where $\vee$ is disjunction, $\wedge$ is conjunction, and $\rightarrow$ is $\Rightarrow$). Furthermore, you have also shown that intuitionistic logical equivalence is compatible with $\vee, \wedge, \Rightarrow$ in the following sense. If $P_1 \equiv Q_1$ and $P_2 \equiv Q_2$, then

$$(P_1 \vee P_2) \equiv (Q_1 \vee Q_2)$$
$$(P_1 \wedge P_2) \equiv (Q_1 \wedge Q_2)$$
$$(P_1 \Rightarrow P_2) \equiv (Q_1 \Rightarrow Q_2).$$

Consequently, for any set T of propositions we can define the relation $\equiv_T$ by

$$P \equiv_T Q \text{ iff } T \vdash P \equiv Q,$$

that is iff $P \equiv Q$ is provable intuitionistically from T (as explained in Section 1.11). Clearly, $\equiv_T$ is an equivalence relation on propositions and we can define the operations $\vee, \wedge$, and $\rightarrow$ on the set of equivalence classes $\mathbf{H}_T$ of propositions as follows.

$$[P] \vee [Q] = [P \vee Q]$$
$$[P] \wedge [Q] = [P \wedge Q]$$
$$[P] \rightarrow [Q] = [P \Rightarrow Q].$$

We also let $0 = [\bot]$ and $1 = [\top]$. Then, we get the Heyting algebra $\mathbf{H}_T$ called the *Lindenbaum algebra* of T, as in the classical case.

Now, let H be any Heyting algebra. By analogy with the case of Boolean algebras, a *truth assignment* is any function v from the set $\mathbf{PS} = \{\mathbf{P}_1, \mathbf{P}_2, \ldots\}$ of propositional symbols to H. Then, we can recursively evaluate the truth value $P_H[v]$ in H of any proposition P, with respect to the truth assignment v as follows.

$$(\mathbf{P}_i)_H[v] = v(P)$$
$$\bot_H[v] = 0$$
$$\top_H[v] = 1$$
$$(P \vee Q)_H[v] = P_H[v] \vee P_H[v]$$
$$(P \wedge Q)_H[v] = P_H[v] \wedge P_H[v]$$
$$(P \Rightarrow Q)_H[v] = (P_H[v] \rightarrow P_H[v])$$
$$(\neg P)_H[v] = (P_H[v] \rightarrow 0).$$

In the equations above, on the right-hand side, $\vee, \wedge$, and $\rightarrow$ are the operations of the Heyting algebra H. We say that a proposition P is *valid in the Heyting algebra H (or H-valid)* if $P_H[v] = 1$ for all truth assignments, v. We say that P is *HA-valid (or intuitionistically valid)* if P is H-valid in all Heyting algebras H. As in the classical case, it can be shown that every intuitionistically provable proposition is HA-valid. This property is called *soundness*. Conversely, if P is HA-valid, then it is intuitionistically provable. This second property is called *completeness*. A stronger completeness result actually holds: if a proposition is H-valid in all *finite* Heyting algebras H, then it is intuitionistically provable. As a consequence, if a proposition is *not* provable intuitionistically, then it can be falsified in some finite Heyting algebra.

Remark: If X is any set, a *topology on X* is a family $\mathcal{O}$ of subsets of X satisfying the following conditions.

(1) $\emptyset \in \mathcal{O}$ and $X \in \mathcal{O}$.
(2) For every family (even infinite), $(U_i)_{i \in I}$, of sets $U_i \in \mathcal{O}$, we have $\bigcup_{i \in I} U_i \in \mathcal{O}$.
(3) For every *finite* family, $(U_i)_{1 \leq i \leq n}$, of sets $U_i \in \mathcal{O}$, we have $\bigcap_{1 \leq i \leq n} U_i \in \mathcal{O}$.

Every subset in $\mathcal{O}$ is called an *open subset* of X (in the topology $\mathcal{O}$). The pair $\langle X, \mathcal{O} \rangle$ is called a *topological space*. Given any subset A of X, the union of all open subsets contained in A is the largest open subset of A and is denoted $\overset{\circ}{A}$.

Given a topological space $\langle X, \mathcal{O} \rangle$, we claim that $\mathcal{O}$ with the inclusion ordering is a Heyting algebra with $0 = \emptyset$; $1 = X$; $\vee = \cup$ (union); $\wedge = \cap$ (intersection); and with

$$(U \rightarrow V) = \overset{\overset{\circ}{\frown}}{(X - U) \cup V}.$$

(Here, $X - U$ is the complement of U in X.) In this Heyting algebra, we have

$$\overline{U} = \overset{\overset{\circ}{\frown}}{X - U}.$$

Because $X - U$ is usually not open, we generally have $\overline{\overline{U}} \neq U$. Therefore, we see that topology yields another supply of Heyting algebras.

5.14 Summary

In this chapter, we introduce two important kinds of relations, partial orders and equivalence relations, and we study some of their main properties. Equivalence relations induce partitions and are, in a precise sense, equivalent to them. The ability to use induction to prove properties of the elements of a partially ordered set is related to a property known as *well-foundedness*. We investigate quite thoroughly induction principles valid for well-ordered sets and, more generally, well-founded sets. As an application, we prove the unique prime factorization theorem for the integers. Section 5.8 on Fibonacci and Lucas numbers and the use of Lucas numbers to test a Mersenne number for primality should be viewed as a lovely illustration of complete induction and as an incentive for the reader to take a deeper look into the fascinating and mysterious world of prime numbers and more generally, number theory. Section 5.9 on public key cryptography and the RSA system is a wonderful application of the notions presented in Section 5.4, gcd and versions of Euclid's algorithm, and another excellent motivation for delving further into number theory. An excellent introduction to the theory of prime numbers with a computational emphasis is Crandall and Pomerance [2] and a delightful and remarkably clear introduction to number theory can be found in Silverman [15]. We also investigate the properties of partially ordered sets where the partial order has some extra properties. For example, we briefly study lattices, complete lattices, Boolean algebras, and Heyting algebras. Regarding complete lattices, we prove a beautiful theorem due to Tarski (Tarski's fixed-point theorem) and use it to give a very short proof of the Schröder–Bernstein theorem (Theorem 2.7).

- We begin with the definition of a *partial order*.
- Next, we define *total orders*, *chains*, *strict orders*, and *posets*.
- We define a *minimal element*, an *immediate predecessor*, a *maximal element*, and an *immediate successor*.
- We define the *Hasse diagram* of a poset.
- We define a *lower bound*, and *upper bound*, a *least element*, a *greatest element*, a *greatest lower bound*, and a *least upper bound*.
- We define a *meet* and a *join*.
- We state *Zorn's lemma*.
- We define *monotonic* functions.
- We define *lattices* and *complete lattices*.
- We prove some basic properties of lattices and introduce *duality*.
- We define *fixed points* as well as *least* and *greatest* fixed points.
- We state and prove *Tarski's fixed-point theorem*.
- As a consequence of Tarski's fixed-point theorem we give a short proof of the *Schröder–Bernstein theorem* (Theorem 2.7).
- We define a *well order* and show that $\mathbb{N}$ is well ordered.
- We revisit *complete induction* on $\mathbb{N}$ and prove its validity.
- We define *prime numbers* and we apply complete induction to prove that every natural number $n \geq 2$ can be factored as a product of primes.
- We prove that there are infinitely many primes.
- We use the fact that $\mathbb{N}$ is well ordered to prove the correctness of Euclidean division.
- We define *well-founded orderings*.
- We characterize well-founded orderings in terms of minimal elements.
- We define the principle of *complete induction on a well-founded set* and prove its validity.
- We define the *lexicographic ordering* on pairs.
- We give the example of *Ackermann's function* and prove that it is a total function.
- We define *divisibility* on $\mathbb{Z}$ (the integers).
- We define *ideals* and *prime ideals* of $\mathbb{Z}$.
- We prove that every ideal of $\mathbb{Z}$ is a principal ideal.
- We prove the *Bézout identity*.
- We define *greatest common divisors* (*gcds*) and *relatively prime* numbers.
- We characterize gcds in terms of the Bézout identity.

- We describe the Euclidean algorithm for computing the gcd and prove its correctness.
- We prove *Euclid's proposition*.
- We prove *unique prime factorization* in $\mathbb{N}$.
- We prove Dirichlet's diophantine approximation theorem, a great application of the pigeonhole principle
- We define *equivalence relations, equivalence classes, quotient sets,* and the *canonical projection.*
- We define *partitions* and *blocks* of a partition.
- We define a bijection between equivalence relations and partitions (on the same set).
- We define when an equivalence relation is a *refinement* of another equivalence relation.
- We define the *reflexive closure*, the *transitive closure*, and the *reflexive and transitive closure* of a relation.
- We characterize the smallest equivalence relation containing a relation.
- We define the *Fibonacci numbers* F_n and the *Lucas numbers* L_n, and investigate some of their properties, including explicit formulae for F_n and L_n.
- We state the *Zeckendorf representation* of natural numbers in terms of Fibonacci numbers.
- We give various versions of the *Cassini identity*
- We define a generalization of the Fibonacci and the Lucas numbers and state some of their properties.
- We define *Mersenne numbers* and *Mersenne primes*.
- We state two versions of the *Lucas–Lehmer test* to check whether a Mersenne number is a prime.
- We introduce some basic notions of *cryptography: encryption, decryption,* and *keys.*
- We define *modular arithmetic* in $\mathbb{Z}/m\mathbb{Z}$.
- We define the notion of a *trapdoor one-way function.*
- We claim that *exponentiation modulo m* is a trapdoor one-way function; its inverse is the *discrete logarithm.*
- We explain how to set up the *RSA scheme*; we describe *public keys* and *private keys.*
- We describe the procedure to *encrypt* a message using RSA and the procedure to *decrypt* a message using RSA.
- We prove *Fermat's little theorem.*
- We prove the correctness of the RSA scheme.
- We describe an algorithm for computing $x^n \bmod m$ using *repeated squaring* and give an example.
- We give an explicit example of an RSA scheme and an explicit example of the decryption of a message.
- We explain how to modify the extended Euclidean algorithm to find the inverse of an integer a modulo m (assuming $\gcd(a, m) = 1$).

- We define the *prime counting function*, $\pi(n)$, and state the *prime number theorem* (or *PNT*).
- We use the PNT to estimate the proportion of primes among positive integers with 200 decimal digits ($1/460$).
- We discuss briefly *primality testing* and the Fermat test.
- We define *pseudo-prime* numbers and *Carmichael* numbers.
- We mention *probabilistic methods* for primality testing.
- We stress that *factoring* integers is a hard problem, whereas primality testing is much easier and in theory, can be done in polynomial time.
- We discuss briefly scenarios for *signatures*.
- We briefly discuss the *security* of RSA, which hinges on the fact that factoring is hard.
- We define *distributive lattices* and prove some properties about them.
- We define *complemented lattices* and prove some properties about them.
- We define *Boolean lattices*, state some of their properties, and define *Boolean algebras*.
- We discuss the *Boolean-valued semantics* of classical logic.
- We define the *Lindenbaum algebra* of a set of propositions.
- We define *Heyting lattices* and prove some properties about them and define *Heyting algebras*.
- We show that every Heyting algebra is distributive and characterize when a Heyting algebra is a Boolean algebra.
- We discuss the *semantics* of intuitionistic logic in terms of Heyting algebras (*HA-validity*).
- We conclude with the definition of a *topological space* and show how the open sets form a Heyting algebra.

Problems

5.1. Give a proof for Proposition 5.1.

5.2. Give a proof for Proposition 5.2.

5.3. Draw the Hasse diagram of all the (positive) divisors of 60, where the partial ordering is the division ordering (i.e., $a \leq b$ iff a divides b). Does every pair of elements have a meet and a join?

5.4. Check that the lexicographic ordering on strings is indeed a total order.

5.5. Check that the function $\varphi \colon 2^A \to 2^A$ used in the proof of Theorem 2.7, is indeed monotonic. Check that the function $h \colon A \to B$ constructed during the proof of Theorem 2.7, is indeed a bijection.

5.6. Give an example of a poset in which complete induction fails.

5.7. Prove that the lexicographic ordering $<<$ on pairs is indeed a partial order.

5.8. Prove that the set
$$\mathfrak{I} = \{ha + kb \mid h, k \in \mathbb{Z}\}$$
used in the proof of Corollary 5.2 is indeed an ideal.

5.9. Prove by complete induction that
$$u_n = 3(3^n - 2^n)$$
is the solution of the recurrence relations:
$$u_0 = 0$$
$$u_1 = 3$$
$$u_{n+2} = 5u_{n+1} - 6u_n,$$
for all $n \geq 0$.

5.10. Consider the recurrence relation
$$u_{n+2} = 3u_{n+1} - 2u_n.$$
For $u_0 = 0$ and $u_1 = 1$, we obtain the sequence (U_n) and for $u_0 = 2$ and $u_1 = 3$, we obtain the sequence (V_n).
    (1) Prove that
$$U_n = 2^n - 1$$
$$V_n = 2^n + 1,$$
for all $n \geq 0$.
    (2) Prove that if U_n is a prime number, then n must be a prime number.
Hint. Use the fact that
$$2^{ab} - 1 = (2^a - 1)(1 + 2^a + 2^{2a} + \cdots + 2^{(b-1)a}).$$

Remark: The numbers of the form $2^p - 1$, where p is prime are known as *Mersenne numbers*. It is an open problem whether there are infinitely many Mersenne primes.

    (3) Prove that if V_n is a prime number, then n must be a power of 2; that is, $n = 2^m$, for some natural number m.
Hint. Use the fact that
$$a^{2k+1} + 1 = (a + 1)(a^{2k} - a^{2k-1} + a^{2k-2} + \cdots + a^2 - a + 1).$$

Remark: The numbers of the form $2^{2^m} + 1$ are known as *Fermat numbers*. It is an open problem whether there are infinitely many Fermat primes.

5.11. Find the smallest natural number n such that the remainder of the division of n by k is $k-1$, for $k = 2, 3, 4, \ldots, 10$.

5.12. Prove that if z is a real zero of a polynomial equation of the form

$$z^n + a_{n-1}z^{n-1} + \cdots + a_1 z + a_0 = 0,$$

where $a_0, a_1, \ldots, a_{n-1}$ are integers and z is not an integer, then z must be irrational.

5.13. Prove that for every integer $k \geq 2$ there is some natural number n so that the k consecutive numbers, $n+1, \ldots, n+k$, are all composite (not prime).
Hint. Consider sequences starting with $(k+1)! + 2$.

5.14. Let p be any prime number. (1) Prove that for every k, with $1 \leq k \leq p-1$, the prime p divides $\binom{p}{k}$.
Hint. Observe that

$$k\binom{p}{k} = p\binom{p-1}{k-1}.$$

(2) Prove that for every natural number a, if p is prime then p divides $a^p - a$.
Hint. Use induction on a.

Deduce *Fermat's little theorem*: For any prime p and any natural number a, if p does not divide a, then p divides $a^{p-1} - 1$.

5.15. If one wants to prove a property $P(n)$ of the natural numbers, rather than using induction, it is sometimes more convenient to use the method of *proof by smallest counterexample*. This is a method that proceeds by contradiction as follows.

1. If P is false, then we know from Theorem 5.3 that there is a smallest $k \in \mathbb{N}$ such that $P(k)$ is false; this k is the *smallest counterexample*.
2. Next, we prove that $k \neq 0$. This is usually easy and it is a kind of basis step.
3. Because $k \neq 0$, the number $k-1$ is a natural number and $P(k-1)$ must hold because k is the smallest counterexample. Then, use this fact and the fact that $P(k)$ is false to derive a contradiction.

Use the method of proof by smallest counterexample to prove that every natural number is either odd or even.

5.16. Prove that for any two positive natural numbers a and m, if $\gcd(a, m) > 1$, then

$$a^{m-1} \not\equiv 1 \pmod{m}.$$

5.17. Let a, b be any two positive integers. (1) Prove that if a is even and b odd, then

$$\gcd(a, b) = \gcd\left(\frac{a}{2}, b\right).$$

(2) Prove that if both a and b are even, then

$$\gcd(a, b) = 2\gcd\left(\frac{a}{2}, \frac{b}{2}\right).$$

5.18. Let a,b be any two positive integers and assume $a \geq b$. When using the Euclidean alorithm for computing the gcd, we compute the following sequence of quotients and remainders.

$$a = bq_1 + r_1$$
$$b = r_1 q_2 + r_2$$
$$r_1 = r_2 q_3 + r_3$$
$$\vdots$$
$$r_{k-1} = r_k q_{k+1} + r_{k+1}$$
$$\vdots$$
$$r_{n-3} = r_{n-2} q_{n-1} + r_{n-1}$$
$$r_{n-2} = r_{n-1} q_n + 0,$$

with $n \geq 3, 0 < r_1 < b, q_k \geq 1$, for $k = 1,\ldots,n$, and $0 < r_{k+1} < r_k$, for $k = 1,\ldots,n-2$. Observe that $r_n = 0$.

If $n = 1$, we have a single division,

$$a = bq_1 + 0,$$

with $r_1 = 0$ and $q_1 \geq 1$ and if $n = 2$, we have two divisions,

$$a = bq_1 + r_1$$
$$b = r_1 q_2 + 0$$

with $0 < r_1 < b, q_1, q_2 \geq 1$ and $r_2 = 0$. Thus, it is convenient to set $r_{-1} = a$ and $r_0 = b$, so that the first two divisions are also written as

$$r_{-1} = r_0 q_1 + r_1$$
$$r_0 = r_1 q_2 + r_2.$$

(1) Prove (using Proposition 5.7) that $r_{n-1} = \gcd(a,b)$.
(2) Next, we prove that some integers x, y such that

$$ax + by = \gcd(a,b) = r_{n-1}$$

can be found as follows:

If $n = 1$, then $a = bq_1$ and $r_0 = b$, so we set $x = 1$ and $y = -(q_1 - 1)$.
If $n \geq 2$, we define the sequence (x_i, y_i) for $i = 0,\ldots,n-1$, so that

$$x_0 = 0, \ y_0 = 1, \ x_1 = 1, \ y_1 = -q_1$$

and, if $n \geq 3$, then

$$x_{i+1} = x_{i-1} - x_i q_{i+1}, \ y_{i+1} = y_{i-1} - y_i q_{i+1},$$

for $i = 1, \ldots, n - 2$.

Prove that if $n \geq 2$, then

$$ax_i + by_i = r_i,$$

for $i = 0, \ldots, n - 1$ (recall that $r_0 = b$) and thus, that

$$ax_{n-1} + by_{n-1} = \gcd(a, b) = r_{n-1}.$$

(3) When $n \geq 2$, if we set $x_{-1} = 1$ and $y_{-1} = 0$ in addition to $x_0 = 0$ and $y_0 = 1$, then prove that the recurrence relations

$$x_{i+1} = x_{i-1} - x_i q_{i+1}, \; y_{i+1} = y_{i-1} - y_i q_{i+1},$$

are valid for $i = 0, \ldots, n - 2$.

Remark: Observe that r_{i+1} is given by the formula

$$r_{i+1} = r_{i-1} - r_i q_{i+1}.$$

Thus, the three sequences, (r_i), (x_i), and (y_i) all use the same recurrence relation,

$$w_{i+1} = w_{i-1} - w_i q_{i+1},$$

but they have different initial conditions: The sequence r_i starts with $r_{-1} = a, r_0 = b$, the sequence x_i starts with $x_{-1} = 1, x_0 = 0$, and the sequence y_i starts with $y_{-1} = 0, y_0 = 1$.

(4) Consider the following version of the gcd algorithm that also computes integers x, y, so that

$$mx + ny = \gcd(m, n),$$

where m and n are positive integers.

Extended Euclidean Algorithm

```
begin
    x := 1; y := 0; u := 0; v := 1; g := m; r := n;
    if m < n then
        t := g; g := r; r := t; (swap g and r)
    pr := r; q := ⌊g/pr⌋; r := g − prq; (divide g by r, to get g = prq + r)
    if r = 0 then
        x := 1; y := −(q − 1); g := pr
    else
        r = pr;
        while r ≠ 0 do
            pr := r; pu := u; pv := v;
            q := ⌊g/pr⌋; r := g − prq; (divide g by pr, to get g = prq + r)
            u := x − puq; v := y − pvq;
            g := pr; x := pu; y := pv
        endwhile;
```

endif;
gcd$(m,n) := g$;
if $m < n$ **then** $t := x$; $x = y$; $y = t$ (swap x and y)
end

Prove that the above algorithm is correct, that is, it always terminates and computes x, y so that

$$mx + ny = \gcd(m,n),$$

5.19. As in Problem 5.18, let a, b be any two positive integers and assume $a \geq b$. Consider the sequence of divisions,

$$r_{i-1} = r_i q_{i+1} + r_{i+1},$$

with $r_{-1} = a$, $r_0 = b$, with $0 \leq i \leq n-1$, $n \geq 1$, and $r_n = 0$. We know from Problem 5.18 that

$$\gcd(a,b) = r_{n-1}.$$

In this problem, we give another algorithm for computing two numbers x and y so that

$$ax + by = \gcd(a,b),$$

that proceeds from the bottom up (we proceed by "'back-substitution"). Let us illustate this in the case where $n = 4$. We have the four divisions:

$$a = bq_1 + \mathbf{r_1}$$
$$b = r_1 q_2 + \mathbf{r_2}$$
$$r_1 = r_2 q_3 + \mathbf{r_3}$$
$$r_2 = r_3 q_3 + 0,$$

with $r_3 = \gcd(a,b)$.
From the third equation, we can write

$$r_3 = r_1 - r_2 q_3. \tag{3}$$

From the second equation, we get

$$r_2 = b - r_1 q_2,$$

and by substituting the right-hand side for r_2 in (3), we get

$$r_3 = b - (b - r_1 q_2)q_3 = -bq_3 + r_1(1 + q_2 q_3);$$

that is,

$$r_3 = -bq_3 + r_1(1 + q_2 q_3). \tag{2}$$

From the first equation, we get

$$r_1 = a - bq_1,$$

and by substituting the right-hand side for r_2 in (2), we get

$$r_3 = -bq_3 + (a - bq_1)(1 + q_2q_3) = a(1 + q_2q_3) - b(q_3 + q_1(1 + q_2q_3));$$

that is,

$$r_3 = a(1 + q_2q_3) - b(q_3 + q_1(1 + q_2q_3)), \tag{1}$$

which yields $x = 1 + q_2q_3$ and $y = q_3 + q_1(1 + q_2q_3)$.

In the general case, we would like to find a sequence s_i for $i = 0, \ldots, n$ such that

$$r_{n-1} = r_{i-1}s_{i+1} + r_is_i, \tag{$*$}$$

for $i = n - 1, \ldots, 0$. For such a sequence, for $i = 0$, we have

$$\gcd(a, b) = r_{n-1} = r_{-1}s_1 + r_0s_0 = as_1 + bs_0,$$

so s_1 and s_0 are solutions of our problem.

The equation $(*)$ must hold for $i = n - 1$, namely,

$$r_{n-1} = r_{n-2}s_n + r_{n-1}s_{n-1},$$

therefore we should set $s_n = 0$ and $s_{n-1} = 1$.

(1) Prove that $(*)$ is satisfied if we set

$$s_{i-1} = -q_is_i + s_{i+1},$$

for $i = n - 1, \ldots, 0$.

(2) Write an algorithm computing the sequence (s_i) as in (1) and compare its performance with the extended Euclidean algorithm of Problem 5.18. Observe that the computation of the sequence (s_i) requires saving all the quotients $q_1, \ldots, q_{n-1}$, so the new algorithm will require more memory when the number of steps n is large.

5.20. In a paper published in 1841, Binet described a variant of the Euclidean algorithm for computing the gcd which runs faster than the standard algorithm. This algorithm makes use of a variant of the division algorithm that allows negative remainders. Let a, b be any two positive integers and assume $a > b$. In the usual division, we have

$$a = bq + r,$$

where $0 \leq r < b$; that is, the remainder r is nonnegative. If we replace q by $q + 1$, we get

$$a = b(q + 1) - (b - r),$$

where $1 \leq b - r \leq b$. Now, if $r > \lfloor b/2 \rfloor$, then $b - r < \lfloor b/2 \rfloor$, so by using a negative remainder, we can always write

$$a = bq \pm r,$$

with $0 \leq r \leq \lfloor b/2 \rfloor$. The proof of Proposition 5.7 also shows that

$$\gcd(a,b) = \gcd(b,r).$$

As in Problem 5.18 we can compute the following sequence of quotients and remainders:

$$a = bq_1' \pm r_1'$$
$$b = r_1'q_2' \pm r_2'$$
$$r_1' = r_2'q_3' \pm r_3'$$

$$\vdots$$

$$r_{k-1}' = r_k'q_{k+1}' \pm r_{k+1}'$$

$$\vdots$$

$$r_{n-3}' = r_{n-2}'q_{n-1}' \pm r_{n-1}'$$
$$r_{n-2}' = r_{n-1}'q_n' + 0,$$

with $n \geq 3$, $0 < r_1' \leq \lfloor b/2 \rfloor$, $q_k' \geq 1$, for $k = 1, \ldots, n$, and $0 < r_{k+1}' \leq \lfloor r_k'/2 \rfloor$, for $k = 1, \ldots, n-2$. Observe that $r_n' = 0$.

If $n = 1$, we have a single division,

$$a = bq_1' + 0,$$

with $r_1' = 0$ and $q_1' \geq 1$ and if $n = 2$, we have two divisions,

$$a = bq_1' \pm r_1'$$
$$b = r_1'q_2' + 0$$

with $0 < r_1' \leq \lfloor b/2 \rfloor$, $q_1', q_2' \geq 1$, and $r_2' = 0$. As in Problem 5.18, we set $r_{-1}' = a$ and $r_0' = b$.

(1) Prove that

$$r_{n-1}' = \gcd(a,b).$$

(2) Prove that

$$b \geq 2^{n-1}r_{n-1}'.$$

Deduce from this that

$$n \leq \frac{\log(b) - \log(r_{n-1})}{\log(2)} + 1 \leq \frac{10}{3}\log(b) + 1 \leq \frac{10}{3}\delta + 1,$$

where δ is the number of digits in b (the logarithms are in base 10).

Observe that this upper bound is better than Lamé's bound, $n \leq 5\delta + 1$ (see Problem 5.51).

(3) Consider the following version of the gcd algorithm using Binet's method. The input is a pair of positive integers, (m, n).

begin
  $a := m;\ b := n;$
  if $a < b$ **then**
    $t := b;\ b := a;\ a := t;$ (swap a and b)
  while $b \neq 0$ **do**
    $r := a \bmod b;$ (divide a by b to obtain the remainder r)
    if $2r > b$ **then** $r := b - r;$
    $a := b;\ b := r$
  endwhile;
  $\gcd(m,n) := a$
end

Prove that the above algorithm is correct; that is, it always terminates and it outputs $a = \gcd(m,n)$.

5.21. In this problem, we investigate a version of the extended Euclidean algorithm (see Problem 5.18) for Binet's method described in Problem 5.20.

Let a,b be any two positive integers and assume $a > b$. We define sequences, q_i, r_i, q'_i, and r'_i inductively, where the q_i and r_i denote the quotients and remainders in the usual Euclidean division and the q'_i and r'_i denote the quotient and remainders in the modified division allowing negative remainders. The sequences r_i and r'_i are defined starting from $i = -1$ and the sequence q_i and q'_i starting from $i = 1$. All sequences end for some $n \geq 1$.

We set $r_{-1} = r'_{-1} = a, r_0 = r'_0 = b$, and for $0 \leq i \leq n - 1$, we have

$$r'_{i-1} = r'_i q_{i+1} + r_{i+1},$$

the result of the usual Euclidean division, where if $n = 1$, then $r_1 = r'_1 = 0$ and $q_1 = q'_1 \geq 1$, else if $n \geq 2$, then $1 \leq r_{i+1} < r_i$, for $i = 0, \ldots, n-2$, $q_i \geq 1$, for $i = 1, \ldots, n$, $r_n = 0$, and with

$$q'_{i+1} = \begin{cases} q_{i+1} & \text{if } 2r_{i+1} \leq r'_i \\ q_{i+1} + 1 & \text{if } 2r_{i+1} > r'_i \end{cases}$$

and

$$r'_{i+1} = \begin{cases} r_{i+1} & \text{if } 2r_{i+1} \leq r'_i \\ r'_i - r_{i+1} & \text{if } 2r_{i+1} > r'_i, \end{cases}$$

for $i = 0, \ldots, n - 1$.

(1) Check that

$$r'_{i-1} = \begin{cases} r'_i q'_{i+1} + r'_{i+1} & \text{if } 2r_{i+1} \leq r'_i \\ r'_i q'_{i+1} - r'_{i+1} & \text{if } 2r_{i+1} > r'_i \end{cases}$$

and prove that

$$r'_{n-1} = \gcd(a,b).$$

(2) If $n \geq 2$, define the sequences, x_i and y_i inductively as follows:
$x_{-1} = 1, x_0 = 0, y_{-1} = 0, y_0 = 1,$

$$x_{i+1} = \begin{cases} x_{i-1} - x_i q'_{i+1} & \text{if } 2r_{i+1} \leq r'_i \\ x_i q'_{i+1} - x_{i-1} & \text{if } 2r_{i+1} > r'_i \end{cases}$$

and

$$y_{i+1} = \begin{cases} y_{i-1} - y_i q'_{i+1} & \text{if } 2r_{i+1} \leq r'_i \\ y_i q'_{i+1} - y_{i-1} & \text{if } 2r_{i+1} > r'_i, \end{cases}$$

for $i = 0, \ldots, n-2$.

Prove that if $n \geq 2$, then

$$ax_i + by_i = r'_i,$$

for $i = -1, \ldots, n-1$ and thus,

$$ax_{n-1} + by_{n-1} = \gcd(a,b) = r'_{n-1}.$$

(3) Design an algorithm combining the algorithms proposed in Problems 5.18 and 5.20.

5.22. (1) Let m_1, m_2 be any two positive natural numbers and assume that m_1 and m_2 are relatively prime.

Prove that for any pair of integers a_1, a_2 there is some integer x such that the following two congruences hold simultaneously.

$$x \equiv a_1 \pmod{m_1}$$
$$x \equiv a_2 \pmod{m_2}.$$

Furthermore, prove that if x and y are any two solutions of the above system, then $x \equiv y \pmod{m_1 m_2}$, so x is unique if we also require that $0 \leq x < m_1 m_2$.

Hint. By the Bézout identity (Proposition 5.6), there exist some integers, y_1, y_2, so that

$$m_1 y_1 + m_2 y_2 = 1.$$

Prove that $x = a_1 m_2 y_2 + a_2 m_1 y_1 = a_1 (1 - m_1 y_1) + a_2 m_1 y_1 = a_1 m_2 y_2 + a_2 (1 - m_2 y_2)$ works. For the second part, prove that if m_1 and m_2 both divide b and if $\gcd(m_1, m_2) = 1$, then $m_1 m_2$ divides b.

(2) Let $m_1, m_2, \ldots, m_n$ be any $n \geq 2$ positive natural numbers and assume that the m_i are pairwise relatively prime, which means that m_i and m_j are relatively prime for all $i \neq j$.

Prove that for any n integers $a_1, a_2, \ldots, a_n$, there is some integer x such that the following n congruences hold simultaneously.

$$x \equiv a_1 \pmod{m_1}$$
$$x \equiv a_2 \pmod{m_2}$$
$$\vdots$$
$$x \equiv a_n \pmod{m_n}.$$

Furthermore, prove that if x and y are any two solutions of the above system, then $x \equiv y \pmod{m}$, where $m = m_1 m_2 \cdots m_n$, so x is unique if we also require that $0 \leq x < m$. The above result is known as the *Chinese remainder theorem*.

Hint. Use induction on n. First, prove that m_1 and $m_2 \cdots m_n$ are relatively prime (because the m_i are pairwise relatively prime). By (1), there exists some z_1 so that

$$z_1 \equiv 1 \pmod{m_1}$$
$$z_1 \equiv 0 \pmod{m_2 \cdots m_n}.$$

By the induction hypothesis, there exists $z_2, \ldots, z_n$, so that

$$z_i \equiv 1 \pmod{m_i}$$
$$z_i \equiv 0 \pmod{m_j}$$

for all $i = 2, \ldots, n$ and all $j \neq i$, with $2 \leq j \leq n$; show that

$$x = a_1 z_1 + a_2 z_2 + \cdots + a_n z_n$$

works.

(3) Let $m = m_1 \cdots m_n$ and let $M_i = m/m_i = \prod_{j=1, j \neq i}^{n} m_j$, for $i = 1, \ldots, n$. As in (2), we know that m_i and M_i are relatively prime, thus by Bézout (or the extended Euclidean algorithm), we can find some integers u_i, v_i so that

$$m_i u_i + M_i v_i = 1,$$

for $i = 1, \ldots, n$. If we let $z_i = M_i v_i = m v_i / m_i$, then prove that

$$x = a_1 z_1 + \cdots + a_n z_n$$

is a solution of the system of congruences.

5.23. The *Euler ϕ-function* (or *totient*) is defined as follows. For every positive integer m, $\phi(m)$ is the number of integers, $n \in \{1, \ldots, m\}$, such that m is relatively prime to n. Observe that $\phi(1) = 1$.

(1) Prove the following fact. For every positive integer a, if a and m are relatively prime, then

$$a^{\phi(m)} \equiv 1 \pmod{m};$$

that is, m divides $a^{\phi(m)} - 1$.

Hint. Let $s_1, \ldots, s_k$ be the integers, $s_i \in \{1, \ldots, m\}$, such that s_i is relatively prime to m ($k = \phi(m)$). Let $r_1, \ldots, r_k$ be the remainders of the divisions of $s_1 a, s_2 a, \ldots, s_k a$ by m (so, $s_i a = m q_i + r_i$, with $0 \leq r_i < m$).

(i) Prove that $\gcd(r_i, m) = 1$, for $i = 1, \ldots, k$.
(ii) Prove that $r_i \neq r_j$ whenever $i \neq j$, so that

$$\{r_1, \ldots, r_k\} = \{s_1, \ldots, s_k\}.$$

Use (i) and (ii) to prove that

$$a^k s_1 \cdots s_k \equiv s_1 \cdots s_k \pmod{m}$$

and use this to conclude that

$$a^{\phi(m)} \equiv 1 \pmod{m}.$$

(2) Prove that if p is prime, then $\phi(p) = p - 1$ and thus, Fermat's little theorem is a special case of (1).

5.24. Prove that if p is a prime, then for every integer x we have $x^2 \equiv 1 \pmod{p}$ iff $x \equiv \pm 1 \pmod{p}$.

5.25. For any two positive integers a, m prove that $\gcd(a, m) = 1$ iff there is some integer x so that $ax \equiv 1 \pmod{m}$.

5.26. Prove that if p is a prime, then

$$(p - 1)! \equiv -1 \pmod{p}.$$

This result is known as *Wilson's theorem*.

Hint. The cases $p = 2$ and $p = 3$ are easily checked, so assume $p \geq 5$. Consider any integer a, with $1 \leq a \leq p - 1$. Show that $\gcd(a, p) = 1$. Then, by the result of Problem 5.25, there is a unique integer $\bar{a}$ such that $1 \leq \bar{a} \leq p - 1$ and $a\bar{a} \equiv 1 \pmod{p}$. Furthermore, a is the unique integer such that $1 \leq a \leq p - 1$ and $\bar{a}a \equiv 1 \pmod{p}$. Thus, the numbers in $\{1, \ldots, p - 1\}$ come in pairs $a, \bar{a}$ such that $\bar{a}a \equiv 1 \pmod{p}$. However, one must be careful because it may happen that $a = \bar{a}$, which is equivalent to $a^2 \equiv 1 \pmod{p}$. By Problem 5.24, this happens iff $a \equiv \pm 1 \pmod{p}$, iff $a = 1$ or $a = p - 1$. By pairing residues modulo p, prove that

$$\prod_{a=2}^{p-2} a \equiv 1 \pmod{p}$$

and use this to prove that

$$(p - 1)! \equiv -1 \pmod{p}.$$

5.27. Let ϕ be the Euler-ϕ function defined in Problem 5.23.
(1) Prove that for every prime p and any integer $k \geq 1$ we have $\phi(p^k) = p^{k-1}(p - 1)$.
(2) Prove that for any two positive integers m_1, m_2, if $\gcd(m_1, m_2) = 1$, then

$$\phi(m_1 m_2) = \phi(m_1)\phi(m_2).$$

Hint. For any integer $m \geq 1$, let

$$\mathscr{R}(m) = \{n \in \{1, \ldots, m\} \mid \gcd(m, n) = 1\}.$$

Let $m = m_1 m_2$. For every $n \in \mathscr{R}(m)$, if a_1 is the remainder of the division of n by m_1 and similarly if a_2 is the remainder of the division of n by m_2, then prove that

$\gcd(a_1, m_1) = 1$ and $\gcd(a_2, m_2) = 1$. Consequently, we get a function $\theta \colon \mathscr{R}(m) \to \mathscr{R}(m_1) \times \mathscr{R}(m_1)$, given by $\theta(n) = (a_1, a_2)$.

Prove that for every pair $(a_1, a_2) \in \mathscr{R}(m_1) \times \mathscr{R}(m_1)$, there is a unique $n \in \mathscr{R}(m)$, so that $\theta(n) = (a_1, a_2)$ (Use the Chinese remainder theorem; see Problem 5.22). Conclude that θ is a bijection. Use the bijection θ to prove that

$$\phi(m_1 m_2) = \phi(m_1)\phi(m_2).$$

(3) Use (1) and (2) to prove that for every integer $n \geq 2$, if $n = p_1^{k_1} \cdots p_r^{k_r}$ is the prime factorization of n, then

$$\phi(n) = p_1^{k_1-1} \cdots p_r^{k_r-1}(p_1 - 1) \cdots (p_r - 1) = n\left(1 - \frac{1}{p_1}\right) \cdots \left(1 - \frac{1}{p_r}\right).$$

5.28. Prove that the function, $f \colon \mathbb{N} \times \mathbb{N} \to \mathbb{N}$, given by

$$f(m, n) = 2^m(2n + 1) - 1$$

is a bijection.

5.29. Let $S = \{a_1, \ldots, a_n\}$ be any nonempty set of n positive natural numbers. Prove that there is a nonempty subset of S whose sum is divisible by n.
Hint. Consider the numbers, $b_1 = a_1$, $b_2 = a_1 + a_2$, $\ldots$, $b_n = a_1 + a_2 + \cdots + a_n$.

5.30. Establish the formula

$$A = \begin{pmatrix} 1 & 1 \\ 1 & 0 \end{pmatrix} = \frac{1}{\sqrt{5}} \begin{pmatrix} \varphi & -\varphi^{-1} \\ 1 & 1 \end{pmatrix} \begin{pmatrix} \varphi & 0 \\ 0 & -\varphi^{-1} \end{pmatrix} \begin{pmatrix} 1 & \varphi^{-1} \\ -1 & \varphi \end{pmatrix}$$

with $\varphi = (1 + \sqrt{5})/2$ given in Section 5.8 and use it to prove that

$$\begin{pmatrix} u_{n+1} \\ u_n \end{pmatrix} = \frac{1}{\sqrt{5}} \begin{pmatrix} \varphi & -\varphi^{-1} \\ 1 & 1 \end{pmatrix} \begin{pmatrix} (\varphi^{-1}u_0 + u_1)\varphi^n \\ (\varphi u_0 - u_1)(-\varphi^{-1})^n \end{pmatrix}.$$

5.31. If (F_n) denotes the Fibonacci sequence, prove that

$$F_{n+1} = \varphi F_n + (-\varphi^{-1})^n.$$

5.32. Prove the identities in Proposition 5.16, namely:

$$F_0^2 + F_1^2 + \cdots + F_n^2 = F_n F_{n+1}$$
$$F_0 + F_1 + \cdots + F_n = F_{n+2} - 1$$
$$F_2 + F_4 + \cdots + F_{2n} = F_{2n+1} - 1$$
$$F_1 + F_3 + \cdots + F_{2n+1} = F_{2n+2}$$
$$\sum_{k=0}^{n} kF_k = nF_{n+2} - F_{n+3} + 2$$

for all $n \geq 0$ (with the third sum interpreted as F_0 for $n = 0$).

5.33. Consider the undirected graph (*fan*) with $n + 1$ nodes and $2n - 1$ edges, with $n \geq 1$, shown in Figure 5.24

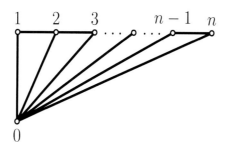

Fig. 5.24 A fan

The purpose of this problem is to prove that the number of spanning subtrees of this graph is F_{2n}, the $2n$th Fibonacci number.

(1) Prove that

$$1 + F_2 + F_4 + \cdots + F_{2n} = F_{2n+1}$$

for all $n \geq 0$, with the understanding that the sum on the left-hand side is 1 when $n = 0$ (as usual, F_k denotes the kth Fibonacci number, with $F_0 = 0$ and $F_1 = 1$).

(2) Let s_n be the number of spanning trees in the fan on $n + 1$ nodes ($n \geq 1$). Prove that $s_1 = 1$ and that $s_2 = 3$.

There are two kinds of spannings trees:

(a) Trees where there is no edge from node n to node 0.
(b) Trees where there is an edge from node n to node 0.

Prove that in case (a), the node n is connected to $n - 1$ and that in this case, there are s_{n-1} spanning subtrees of this kind; see Figure 5.25.

Observe that in case (b), there is some $k \leq n$ such that the edges between the nodes $n, n - 1, \ldots, k$ are in the tree but the edge from k to $k - 1$ is *not* in the tree and that none of the edges from 0 to any node in $\{n - 1, \ldots, k\}$ are in this tree; see Figure 5.26.

Furthermore, prove that if $k = 1$, then there is a single tree of this kind (see Figure 5.27) and if $k > 1$, then there are

$$s_{n-1} + s_{n-2} + \cdots + s_1$$

trees of this kind.

(3) Deduce from (2) that

$$s_n = s_{n-1} + s_{n-1} + s_{n-2} + \cdots + s_1 + 1,$$

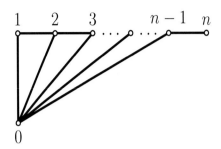

Fig. 5.25 Spanning trees of type (a)

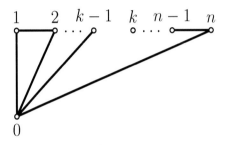

Fig. 5.26 Spanning trees of type (b) when $k > 1$

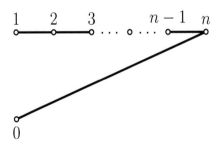

Fig. 5.27 Spanning tree of type (b) when $k = 1$

with $s_1 = 1$. Use (1) to prove that

$$s_n = F_{2n},$$

for all $n \geq 1$.

5.34. Prove the Zeckendorf representation of natural numbers, that is, Proposition 5.17.

Hint. For the existence part, prove by induction on $k \geq 2$ that a decomposition of the required type exists for all $n \leq F_k$ (with $n \geq 1$). For the uniqueness part, first prove that

$$F_{(n \bmod 2)+2} + \cdots + F_{n-2} + F_n = F_{n+1} - 1,$$

for all $n \geq 2$.

5.35. Prove Proposition 5.18 giving identities relating the Fibonacci numbers and the Lucas numbers:

$$L_n = F_{n-1} + F_{n+1}$$
$$5F_n = L_{n-1} + L_{n+1},$$

for all $n \geq 1$.

5.36. Prove Proposition 5.19; that is, for any fixed $k \geq 1$ and all $n \geq 0$, we have

$$F_{n+k} = F_k F_{n+1} + F_{k-1} F_n.$$

Use the above to prove that

$$F_{2n} = F_n L_n,$$

for all $n \geq 1$.

5.37. Prove the following identities.

$$L_n L_{n+2} = L_{n+1}^2 + 5(-1)^n$$
$$L_{2n} = L_n^2 - 2(-1)^n$$
$$L_{2n+1} = L_n L_{n+1} - (-1)^n$$
$$L_n^2 = 5F_n^2 + 4(-1)^n.$$

5.38. (a) Prove Proposition 5.20; that is,

$$u_{n+1} u_{n-1} - u_n^2 = (-1)^{n-1} (u_0^2 + u_0 u_1 - u_1^2).$$

(b) Prove the *Catalan identity*,

$$F_{n+r} F_{n-r} - F_n^2 = (-1)^{n-r+1} F_r^2, \qquad n \geq r.$$

5.39. Prove that any sequence defined by the recurrence

$$u_{n+2} = u_{n+1} + u_n$$

satisfies the following equation,

$$u_k u_{n+1} + u_{k-1} u_n = u_1 u_{n+k} + u_0 u_{n+k-1},$$

for all $k \geq 1$ and all $n \geq 0$.

5.40. Prove Proposition 5.21; that is,

1. F_n divides F_{mn}, for all $m, n \geq 1$.
2. $\gcd(F_m, F_n) = F_{\gcd(m,n)}$, for all $m, n \geq 1$.

Hint. For the first statement, use induction on $m \geq 1$. To prove the second statetement, first prove that

$$\gcd(F_n, F_{n+1}) = 1$$

for all $n \geq 1$. Then, prove that

$$\gcd(F_m, F_n) = \gcd(F_{m-n}, F_n).$$

5.41. Prove the formulae

$$2F_{m+n} = F_m L_n + F_n L_m$$
$$2L_{m+n} = L_m L_n + 5 F_m F_n.$$

5.42. Prove that

$$L_n^{2h+1} = L_{(2h+1)n} + \binom{2h+1}{1}(-1)^n L_{(2h-1)n} + \binom{2h+1}{2}(-1)^{2n} L_{(2h-3)n} + \cdots$$
$$+ \binom{2h+1}{h}(-1)^{hn} L_n.$$

5.43. Prove that

$$A = \begin{pmatrix} P & -Q \\ 1 & 0 \end{pmatrix} = \frac{1}{\alpha - \beta} \begin{pmatrix} \alpha & \beta \\ 1 & 1 \end{pmatrix} \begin{pmatrix} \alpha & 0 \\ 0 & \beta \end{pmatrix} \begin{pmatrix} 1 & -\beta \\ -1 & \alpha \end{pmatrix},$$

where

$$\alpha = \frac{P + \sqrt{D}}{2}, \qquad \beta = \frac{P - \sqrt{D}}{2}$$

and then prove that

$$\begin{pmatrix} u_{n+1} \\ u_n \end{pmatrix} = \frac{1}{\alpha - \beta} \begin{pmatrix} \alpha & \beta \\ 1 & 1 \end{pmatrix} \begin{pmatrix} (-\beta u_0 + u_1)\alpha^n \\ (\alpha u_0 - u_1)\beta^n \end{pmatrix}.$$

5.44. Prove Proposition 5.22; that is, the sequence defined by the recurrence

$$u_{n+2} = P u_{n+1} - Q u_n$$

(with $P^2 - 4Q \neq 0$) satisfies the identity:

$$u_{n+1} u_{n-1} - u_n^2 = Q^{n-1}(-Q u_0^2 + P u_0 u_1 - u_1^2).$$

5.45. Prove the following identities relating the U_n and the V_n;

$$V_n = U_{n+1} - QU_{n-1}$$
$$DU_n = V_{n+1} - QV_{n-1},$$

for all $n \geq 1$. Then, prove that

$$U_{2n} = U_n V_n$$
$$V_{2n} = V_n^2 - 2Q^n$$
$$U_{m+n} = U_m U_{n+1} - QU_n U_{m-1}$$
$$V_{m+n} = V_m V_n - Q^n V_{m-n}.$$

5.46. Consider the recurrence

$$V_{n+2} = 2V_{n+1} + 2V_n,$$

starting from $V_0 = V_1 = 2$. Prove that

$$V_n = (1 + \sqrt{3})^n + (1 - \sqrt{3})^n.$$

5.47. Consider the sequence S_n given by

$$S_{n+1} = S_n^2 - 2,$$

starting with $S_0 = 4$. Prove that

$$V_{2^k} = S_{k-1} 2^{2^{k-1}},$$

for all $k \geq 1$ and that

$$S_k = (2 + \sqrt{3})^{2^k} + (2 - \sqrt{3})^{2^k}.$$

5.48. Prove that

$$n \equiv (n \bmod 2^p) + \lfloor n/2^p \rfloor \pmod{2^p - 1}.$$

5.49. The Cassini identity,

$$F_{n+1} F_{n-1} - F_n^2 = (-1)^n, \qquad n \geq 1,$$

is the basis of a puzzle due to Lewis Carroll. Consider a square chess-board consisting of $8 \times 8 = 64$ squares and cut it up into four pieces using the Fibonacci numbers, $3, 5, 8$, as indicated by the bold lines in Figure 5.28 (a). Then, reassemble these four pieces into a rectangle consisting of $5 \times 13 = 65$ squares as shown in Figure 5.28 (b). Again, note the use of the Fibonacci numbers: $3, 5, 8, 13$. However, the original square has 64 small squares and the final rectangle has 65 small squares. Explain what's wrong with this apparent paradox.

5.50. The generating function of a sequence (u_n) is the power series

$$F(z) = \sum_{n=0}^{\infty} u_n z^n.$$

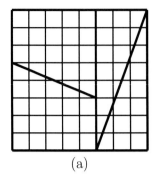
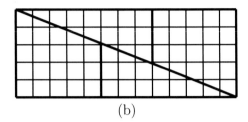

(a)   (b)

Fig. 5.28 (a) A square of 64 small squares. (b) A rectangle of 65 small squares

If the sequence (u_n) is defined by the recurrence relation

$$u_{n+2} = Pu_{n+1} - Qu_n$$

then prove that

$$F(z) = \frac{u_0 + (u_1 - Pu_0)z}{1 - Pz + Qz^2}.$$

For the Fibonacci-style sequence $u_0 = 0$, $u_1 = 1$, so we have

$$F_{\text{Fib}}(z) = \frac{z}{1 - Pz + Qz^2}$$

and for the Lucas-style sequence $u_0 = 2$, $u_1 = 1$, so we have

$$F_{\text{Luc}}(z) = \frac{2 + (1 - 2P)z}{1 - Pz + Qz^2}.$$

If $Q \neq 0$, prove that

$$F(z) = \frac{1}{\alpha - \beta}\left(\frac{-\beta u_0 + u_1}{1 - \alpha z} + \frac{\alpha u_0 - u_1}{1 - \beta z}\right).$$

Prove that the above formula for $F(z)$ yields, again,

$$u_n = \frac{1}{\alpha - \beta}\left((-\beta u_0 + u_1)\alpha^n + (\alpha u_0 - u_1)\beta^n\right).$$

Prove that the above formula is still valid for $Q = 0$, provided we assume that $0^0 = 1$.

5.51. (1) Prove that the Euclidean algorithm for gcd applied to two consecutive Fibonacci numbers F_n and F_{n+1} (with $n \geq 2$) requires $n - 1$ divisions.

(2) Prove that the Euclidean algorithm for gcd applied to two consecutive Lucas numbers L_n and L_{n+1} (with $n \geq 1$) requires n divisions.

(3) Prove that if $a > b \geq 1$ and if the Euclidean algorithm for gcd applied to a and b requires n divisions, then $a \geq F_{n+2}$ and $b \geq F_{n+1}$.

(4) Using the explicit formula for F_{n+1} and by taking logarithms in base 10, use (3) to prove that

$$n < 4.785\delta + 1,$$

where δ is the number of digits in b (Dupré's bound). This is slightly better than Lamé's bound, $n \leq 5\delta + 1$.

5.52. Let R and S be two relations on a set X. (1) Prove that if R and S are both reflexive, then $R \circ S$ is reflexive.

(2) Prove that if R and S are both symmetric and if $R \circ S = S \circ R$, then $R \circ S$ is symmetric.

(3) Prove that R is transitive iff $R \circ R \subseteq R$. Prove that if R and S are both transitive and if $R \circ S = S \circ R$, then $R \circ S$ is transitive.

Can the hypothesis $R \circ S = S \circ R$ be omitted?

(4) Prove that if R and S are both equivalence relations and if $R \circ S = S \circ R$, then $R \circ S$ is the smallest equivalence relation containing R and S.

5.53. Prove Proposition 5.12.

5.54. Prove Proposition 5.13.

5.55. Prove Proposition 5.14.

5.56. Prove Proposition 5.15

5.57. (1) Prove the correctness of the algorithm for computing $x^n \bmod m$ using repeated squaring.

(2) Use your algorithm to check that the message sent to Albert has been decrypted correctly and then encrypt the decrypted message and check that it is identical to the original message.

5.58. Recall the recurrence relations given in Section 5.9 to compute the inverse modulo m of an integer a such that $1 \leq a < m$ and $\gcd(m, a) = 1$:

$$y_{-1} = 0$$
$$y_0 = 1$$
$$z_{i+1} = y_{i-1} - y_i q_{i+1}$$
$$y_{i+1} = z_{i+1} \bmod m \quad \text{if} \quad z_{i+1} \geq 0$$
$$y_{i+1} = m - ((-z_{i+1}) \bmod m) \quad \text{if} \quad z_{i+1} < 0,$$

for $i = 0, \ldots, n - 2$.

(1) Prove by induction that

$$ay_i \equiv r_i \pmod{m}$$

for $i = 0, \ldots, n - 1$ and thus, that

$$ay_{n-1} \equiv 1 \,(\mathrm{mod}\, m),$$

with $1 \le y_{n-1} < m$, as desired.

(2) Prove the correctness of the algorithm for computing the inverse of an element modulo m proposed in Section 5.9.

(3) Design a faster version of this algorithm using "Binet's trick" (see Problem 5.20 and Problem 5.21).

5.59. Prove that $a^{560} - 1$ is divisible by 561 for every positive natural number, a, such that $\gcd(a, 561) = 1$.
Hint. Because $561 = 3 \cdot 11 \cdot 17$, it is enough to prove that $3 \mid (a^{560} - 1)$ for all positive integers a such that a is not a multiple of 3, that $11 \mid (a^{560} - 1)$ for all positive integers a such that a is not a multiple of 11, and that $17 \mid (a^{560} - 1)$ for all positive integers a such that a is not a multiple of 17.

5.60. Prove that 161038 divides $2^{161038} - 2$, yet $2^{161037} \equiv 80520 \,(\mathrm{mod}\, 161038)$.

This example shows that it would be undesirable to define a pseudo-prime as a positive natural number n that divides $2^n - 2$.

5.61. (a) Consider the sequence defined recursively as follows.

$$U_0 = 0$$
$$U_1 = 2$$
$$U_{n+2} = 6U_{n+1} - U_n, \ n \ge 0.$$

Prove the following identity,

$$U_{n+2}U_n = U_{n+1}^2 - 4,$$

for all $n \ge 0$.

(b) Consider the sequence defined recursively as follows:

$$V_0 = 1$$
$$V_1 = 3$$
$$V_{n+2} = 6V_{n+1} - V_n, \ n \ge 0.$$

Prove the following identity,

$$V_{n+2}V_n = V_{n+1}^2 + 8,$$

for all $n \ge 0$.

(c) Prove that

$$V_n^2 - 2U_n^2 = 1,$$

for all $n \ge 0$.

Hint. Use (a) and (b). You may also want to prove by simultaneous induction that

$$V_n^2 - 2U_n^2 = 1$$
$$V_n V_{n-1} - 2U_n U_{n-1} = 3,$$

for all $n \geq 1$.

5.62. Consider the sequences (U_n) and (V_n), given by the recurrence relations

$$U_0 = 0$$
$$V_0 = 1$$
$$U_1 = y_1$$
$$V_1 = x_1$$
$$U_{n+2} = 2x_1 U_{n+1} - U_n$$
$$V_{n+2} = 2x_1 V_{n+1} - V_n,$$

for any two positive integers x_1, y_1.

(1) If x_1 and y_1 are solutions of the (Pell) equation

$$x^2 - dy^2 = 1,$$

where d is a positive integer that is not a perfect square, then prove that

$$V_n^2 - dU_n^2 = 1$$
$$V_n V_{n-1} - dU_n U_{n-1} = x_1,$$

for all $n \geq 1$.

(2) Verify that

$$U_n = \frac{(x_1 + y_1\sqrt{d})^n - (x_1 - y_1\sqrt{d})^n}{2\sqrt{d}}$$
$$V_n = \frac{(x_1 + y_1\sqrt{d})^n + (x_1 - y_1\sqrt{d})^n}{2}.$$

Deduce from this that

$$V_n + U_n\sqrt{d} = (x_1 + y_1\sqrt{d})^n.$$

(3) Prove that the U_ns and V_ns also satisfy the following simultaneous recurrence relations:

$$U_{n+1} = x_1 U_n + y_1 V_n$$
$$V_{n+1} = dy_1 U_n + x_1 V_n,$$

for all $n \geq 0$. Use the above to prove that

$$V_{n+1} + U_{n+1}\sqrt{d} = (V_n + U_n\sqrt{d})(x_1 + y_1\sqrt{d})$$
$$V_{n+1} - U_{n+1}\sqrt{d} = (V_n - U_n\sqrt{d})(x_1 - y_1\sqrt{d})$$

for all $n \geq 0$ and then that

$$V_n + U_n\sqrt{d} = (x_1 + y_1\sqrt{d})^n$$
$$V_n - U_n\sqrt{d} = (x_1 - y_1\sqrt{d})^n$$

for all $n \geq 0$. Use the above to give another proof of the formulae for U_n and V_n in (2).

Remark: It can be shown that *Pell's equation*,

$$x^2 - dy^2 = 1,$$

where d is not a perfect square, always has solutions in positive integers. If (x_1, y_1) is the solution with smallest $x_1 > 0$, then every solution is of the form (V_n, U_n), where U_n and V_n are defined in (1). Curiously, the "smallest solution" (x_1, y_1) can involve some very large numbers. For example, it can be shown that the smallest positive solution of

$$x^2 - 61y^2 = 1$$

is $(x_1, y_1) = (1766319049, 226153980)$.

5.63. Prove that every totally ordered poset is a distributive lattice. Prove that the lattice $\mathbb{N}_+$ under the divisibility ordering is a distributive lattice.

5.64. Prove part (b) of Proposition 5.16.

5.65. Prove that every finite distributive lattice is a Heyting algebra.

References

1. Garrett Birkhoff. *Lattice Theory*. Colloquium Publications, Vol. XXV. Providence, RI: AMS, third edition, 1973.
2. Richard Crandall and Carl Pomerance. *Prime Numbers. A Computational Perspective*. New York: Springer, second edition, 2005.
3. H. Davenport. *The Higher Arithmetic. An Introduction to the Theory of Numbers*. Cambridge, UK: Cambridge University Press, eighth edition, 2008.
4. Martin Dietzfelbinger. *Primality Testing in Polynomial Time: From Randomized Algorithms to "Primes Is in P"*. LNCS No. 3000. New York: Springer Verlag, first edition, 2004.
5. Peter Gustav Lejeune Dirichlet. *Lectures on Number Theory*, volume 16 of *History of Mathematics*. Providence, RI: AMS, first edition, 1999.
6. Herbert B. Enderton. *Elements of Set Theory*. New York: Academic Press, first edition, 1977.
7. Ronald L. Graham, Donald E. Knuth, and Oren Patashnik. *Concrete Mathematics: A Foundation For Computer Science*. Reading, MA: Addison Wesley, second edition, 1994.

8. Donald E. Knuth. *The Art of Computer Programming, Volume 2: Seminumerical Algorithms.* Reading, MA: Addison Wesley, third edition, 1997.

9. Neal Koblitz. *A Course in Number Theory and Cryptography.* GTM No. 114. New York: Springer Verlag, second edition, 1994.

10. L. Lovász, J. Pelikán, and K. Vesztergombi. *Discrete Mathematics. Elementary and Beyond.* Undergraduate Texts in Mathematics. New York: Springer, first edition, 2003.

11. Alfred J. Menezes, Paul C. van Oorschot, and Scott A. Vanstone. *Handbook of Applied Cryptography.* Boca Raton, FL: CRC Press, fifth edition, 2001.

12. Ivan Niven, Herbert S. Zuckerman, and Hugh L. Montgomery. *An Introduction to the Theory of Numbers.* New York: Wiley, fifth edition, 1991.

13. Paulo Ribenboim. *The Little Book of Bigger Primes.* New York: Springer-Verlag, second edition, 2004.

14. R.L. Rivest, A. Shamir, and L. Adleman. A method for obtaining digital signatures and public-key cryptosystems. *Communications of the ACM,* 21(2):120–126, 1978.

15. Joseph H. Silverman. *A Friendly Introduction to Number Theory.* Upper Saddle River, NJ: Prentice Hall, third edition, 2006.

16. Patrick Suppes. *Axiomatic Set Theory.* New York: Dover, first edition, 1972.

17. André Weil. *Number Theory. An Approach Through History from Hammurapi to Legendre.* Boston: Birkhauser, first edition, 1987.

Chapter 6
Graphs, Part II: More Advanced Notions

6.1 Γ-Cycles, Cocycles, Cotrees, Flows, and Tensions

In this section, we take a closer look at the structure of cycles in a finite graph G. It turns out that there is a dual notion to that of a cycle, the notion of a *cocycle*. Assuming any orientation of our graph, it is possible to associate a vector space $\mathscr{F}$ with the set of cycles in G, another vector space $\mathscr{T}$ with the set of cocycles in G, and these vector spaces are mutually orthogonal (for the usual inner product). Furthermore, these vector spaces do not depend on the orientation chosen, up to isomorphism. In fact, if G has m nodes, n edges, and p connected components, we prove that $\dim \mathscr{F} = n - m + p$ and $\dim \mathscr{T} = m - p$. These vector spaces are the *flows* and the *tensions* of the graph G, and these notions are important in combinatorial optimization and the study of networks. This chapter assumes some basic knowledge of linear algebra.

Recall that if G is a directed graph, then a *cycle* C is a closed e-simple chain, which means that C is a sequence of the form $C = (u_0, e_1, u_1, e_2, u_2, \ldots, u_{n-1}, e_n, u_n)$, where $n \geq 1$; $u_i \in V$; $e_i \in E$ and

$$u_0 = u_n; \quad \{s(e_i), t(e_i)\} = \{u_{i-1}, u_i\}, \ 1 \leq i \leq n \text{ and } e_i \neq e_j \text{ for all } i \neq j.$$

The cycle C induces the sets C^+ and C^- where C^+ consists of the edges whose orientation agrees with the order of traversal induced by C and where C^- consists of the edges whose orientation is the inverse of the order of traversal induced by C. More precisely,

$$C^+ = \{e_i \in C \mid s(e_i) = u_{i-1}, t(e_i) = u_i\}$$

and

$$C^- = \{e_i \in C \mid s(e_i) = u_i, t(e_i) = u_{i-1}\}.$$

For the rest of this section, we assume that G is a finite graph and that its edges are

J. Gallier, *Discrete Mathematics*, Universitext,
DOI 10.1007/978-1-4419-8047-2_6, © Springer Science+Business Media, LLC 2011

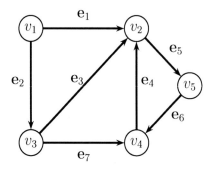

Fig. 6.1 Graph G_8

named, $\mathbf{e}_1, \ldots, \mathbf{e}_n$[1].

Definition 6.1. Given any finite directed graph G with n edges, with every cycle C is associated a *representative vector* $\gamma(C) \in \mathbb{R}^n$, defined so that for every i, with $1 \leq i \leq n$,

$$\gamma(C)_i = \begin{cases} +1 & \text{if } \mathbf{e}_i \in C^+ \\ -1 & \text{if } \mathbf{e}_i \in C^- \\ 0 & \text{if } \mathbf{e}_i \notin C. \end{cases}$$

For example, if $G = G_8$ is the graph of Figure 6.1, the cycle

$$C = (v_3, e_7, v_4, e_6, v_5, e_5, v_2, e_1, v_1, e_2, v_3)$$

corresponds to the vector

$$\gamma(C) = (-1, 1, 0, 0, -1, -1, 1).$$

Observe that distinct cycles may yield the same representative vector unless they are simple cycles. For example, the cycles

$$C_1 = (v_2, e_5, v_5, e_6, v_4, e_4, v_2, e_1, v_1, e_2, v_3, e_3, v_2)$$

and

$$C_2 = (v_2, e_1, v_1, e_2, v_3, e_3, v_2, e_5, v_5, e_6, v_4, e_4, v_2)$$

yield the same representative vector

$$\gamma = (-1, 1, 1, 1, 1, 1, 0).$$

In order to obtain a bijection between representative vectors and "cycles", we introduce the notion of a "Γ-cycle" (some authors redefine the notion of cycle and call "cycle" what we call a Γ-cycle, but we find this practice confusing).

---

[1] We use boldface notation for the edges in E in order to avoid confusion with the edges occurring in a cycle or in a chain; those are denoted in italic.

Definition 6.2. Given a finite directed graph $G = (V, E, s, t)$, a Γ-*cycle* is any set of edges $\Gamma = \Gamma^+ \cup \Gamma^-$ such that there is some cycle C in G with $\Gamma^+ = C^+$ and $\Gamma^- = C^-$; we say that the cycle C *induces the* Γ-*cycle*, Γ. The *representative vector* $\gamma(\Gamma)$ (for short, γ) associated with Γ is the vector $\gamma(C)$ from Definition 6.1, where C is any cycle inducing Γ. We say that a Γ-cycle Γ is a Γ-*circuit* iff either $\Gamma^+ = \emptyset$ or $\Gamma^- = \emptyset$ and that Γ is *simple* iff Γ arises from a simple cycle.

Remarks:

1. Given a Γ-cycle $\Gamma = \Gamma^+ \cup \Gamma^-$ we have the subgraphs $G^+ = (V, \Gamma^+, s, t)$ and $G^- = (V, \Gamma^-, s, t)$. Then, for every $u \in V$, we have

$$d_{G^+}^+(u) - d_{G^+}^-(u) - d_{G^-}^+(u) + d_{G^-}^-(u) = 0.$$

2. If Γ is a simple Γ-cycle, then every vertex of the graph (V, Γ, s, t) has degree 0 or 2.

3. When the context is clear and no confusion may arise, we often drop the "Γ" in Γ-cycle and simply use the term "cycle".

Proposition 6.1. *If G is any finite directed graph, then any Γ-cycle Γ is the disjoint union of simple Γ-cycles.*

Proof. This is an immediate consequence of Proposition 3.6. $\square$

Corollary 6.1. *If G is any finite directed graph, then any Γ-cycle Γ is simple iff it is minimal, that is, if there is no Γ-cycle Γ' such that $\Gamma' \subseteq \Gamma$ and $\Gamma' \neq \Gamma$.*

We now consider a concept that turns out to be dual to the notion of Γ-cycle.

Definition 6.3. Let G be a finite directed graph $G = (V, E, s, t)$ with n edges. For any subset of nodes $Y \subseteq V$, define the sets of edges $\Omega^+(Y)$ and $\Omega^-(Y)$ by

$$\Omega^+(Y) = \{e \in E \mid s(e) \in Y, t(e) \notin Y\}$$
$$\Omega^-(Y) = \{e \in E \mid s(e) \notin Y, t(e) \in Y\}$$
$$\Omega(Y) = \Omega^+(Y) \cup \Omega^-(Y).$$

Any set Ω of edges of the form $\Omega = \Omega(Y)$, for some set of nodes $Y \subseteq V$, is called a *cocycle* (or *cutset*). With every cocycle Ω we associate the *representative vector* $\omega(\Omega) \in \mathbb{R}^n$ defined so that

$$\omega(\Omega)_i = \begin{cases} +1 & \text{if } \mathbf{e}_i \in \Omega^+ \\ -1 & \text{if } \mathbf{e}_i \in \Omega^- \\ 0 & \text{if } \mathbf{e}_i \notin \Omega, \end{cases}$$

with $1 \leq i \leq n$. We also write $\omega(Y)$ for $\omega(\Omega)$ when $\Omega = \Omega(Y)$. If either $\Omega^+(Y) = \emptyset$ or $\Omega^-(Y) = \emptyset$, then Ω is called a *cocircuit* and a *simple cocycle* (or *bond*) is a minimal cocycle (i.e., there is no cocycle Ω' such that $\Omega' \subseteq \Omega$ and $\Omega' \neq \Omega$).

In the graph G_8 of Figure 6.1,

$$\Omega = \{e_5\} \cup \{e_1, e_2, e_6\}$$

is a cocycle induced by the set of nodes $Y = \{v_2, v_3, v_4\}$ and it corresponds to the vector

$$\omega(\Omega) = (-1, -1, 0, 0, 1, -1, 0).$$

This is not a simple cocycle because

$$\Omega' = \{e_5\} \cup \{e_6\}$$

is also a cocycle (induced by $Y' = \{v_1, v_2, v_3, v_4\}$). Observe that Ω' is a minimal cocycle, so it is a simple cocycle. Observe that the inner product

$$\gamma(C_1) \cdot \omega(\Omega) = (-1, 1, 1, 1, 1, 1, 0) \cdot (-1, -1, 0, 0, 1, -1, 0)$$
$$= 1 - 1 + 0 + 0 + 1 - 1 + 0 = 0$$

is zero. This is a general property that we prove shortly.

Observe that a cocycle Ω is the set of edges of G that join the vertices in a set Y to the vertices in its complement $V - Y$. Consequently, deletion of all the edges in Ω increases the number of connected components of G. We say that Ω is a *cutset* of G. Generally, a set of edges $K \subseteq E$ is a *cutset* of G if the graph $(V, E - K, s, t)$ has more connected components than G.

It should be noted that a cocycle $\Omega = \Omega(Y)$ may coincide with the set of edges of some cycle Γ. For example, in the graph displayed in Figure 6.2, the cocycle $\Omega = \Omega(\{1, 3, 5, 7\})$, shown in thicker lines, is equal to the set of edges of the cycle,

$$(1,2), (2,3), (3,4), (4,1), (5,6), (6,7), (7,8), (8,5).$$

If the edges of the graph are listed in the order

$$(1,2), (2,3), (3,4), (4,1), (5,6), (6,7), (7,8), (8,5), (1,5), (2,6), (3,7), (4,8)$$

the reader should check that the vectors

$$\gamma = (1, 1, 1, 1, 1, 1, 1, 1, 0, 0, 0, 0) \in \mathscr{F}$$

and

$$\omega = (1, -1, 1, -1, 1, -1, 1, -1, 0, 0, 0, 0) \in \mathscr{T}$$

correspond to Γ and Ω, respectively.

We now give several characterizations of simple cocycles.

Proposition 6.2. *Given a finite directed graph* $G = (V, E, s, t)$ *a set of edges* $S \subseteq E$ *is a simple cocycle iff it is a minimal cutset.*

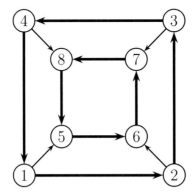

Fig. 6.2 A cocycle Ω equal to the edge set of a cycle Γ

Proof. We already observed that every cocycle is a cutset. Furthermore, we claim that every cutset contains a cocyle. To prove this, it is enough to consider a minimal cutset S and to prove the following satement.

Claim. Any minimal cutset S is the set of edges of G that join two nonempty sets of vertices Y_1 and Y_2 such that

(i) $Y_1 \cap Y_2 = \emptyset$.
(ii) $Y_1 \cup Y_2 = C$, some connected component of G.
(iii) The subgraphs G_{Y_1} and G_{Y_2}, induced by Y_1 and Y_2 are connected.

Indeed, if S is a minimal cutset, it disconnects a unique connected component of G, say C. Let $C_1, \ldots, C_k$ be the connected components of the graph $C - S$, obtained from C by deleting the edges in S. Adding any edge $e \in S$ to $C - S$ must connect two components of C because otherwise $S - \{e\}$ would disconnect C, contradicting the minimality of C. Furthermore, $k = 2$, because otherwise, again, $S - \{e\}$ would disconnect C. Then, if Y_1 is the set of nodes of C_1 and Y_2 is the set of nodes of C_2, it is clear that the claim holds.

Now, if S is a minimal cutset, the above argument shows that S contains a cocyle and this cocycle must be simple (i.e., minimal as a cocycle) as it is a cutset. Conversely, if S is a simple cocycle (i.e., minimal as a cocycle), it must be a minimal cutset because otherwise, S would contain a strictly smaller cutset which would then contain a cocycle strictly contained in S. $\quad\square$

Proposition 6.3. *Given a finite directed graph $G = (V, E, s, t)$ a set of edges $S \subseteq E$ is a simple cocycle iff S is the set of edges of G that join two nonempty sets of vertices Y_1 and Y_2 such that*

(i) $Y_1 \cap Y_2 = \emptyset$.
(ii) $Y_1 \cup Y_2 = C$, some connected component of G.
(iii) The subgraphs G_{Y_1} and G_{Y_2}, induced by Y_1 and Y_2 are connected.

Proof. It is clear that if S satisfies (i)–(iii), then S is a minimal cutset and by Proposition 6.3, it is a simple cocycle.

Let us first assume that G is connected and that $S = \Omega(Y)$ is a simple cocycle; that is, is minimal as a cocycle. If we let $Y_1 = Y$ and $Y_2 = X - Y_1$, it is clear that (i) and (ii) are satisfied. If G_{Y_1} or G_{Y_2} is not connected, then if Z is a connected component of one of these two graphs, we see that $\Omega(Z)$ is a cocycle strictly contained in $S = \Omega(Y_1)$, a contradiction. Therefore, (iii) also holds. If G is not connected, as S is a minimal cocycle it is a minimal cutset, and so it is contained in some connected component C of G and we apply the above argument to C. □

The following proposition is the analogue of Proposition 6.1 for cocycles.

Proposition 6.4. *Given a finite directed graph $G = (V, E, s, t)$, every cocycle $\Omega = \Omega(Y)$ is the disjoint union of simple cocycles.*

Proof. We give two proofs.

Proof 1: (Claude Berge) Let $Y_1, \ldots, Y_k$ be the connected components of the subgraph of G induced by Y. Then, it is obvious that

$$\Omega(Y) = \Omega(Y_1) \cup \cdots \cup \Omega(Y_k),$$

where the $\Omega(Y_i)$ are pairwise disjoint. So, it is enough to show that each $\Omega(Y_i)$ is the union of disjoint simple cycles.

Let C be the connected component of G that contains Y_i and let $C_1, \ldots, C_m$ be the connected components of the subgraph $C - Y$, obtained from C by deleting the nodes in Y_i and the edges incident to these nodes. Observe that the set of edges that are deleted when the nodes in Y_i are deleted is the union of $\Omega(Y_i)$ and the edges of the connected subgraph induced by Y_i. As a consequence, we see that

$$\Omega(Y_i) = \Omega(C_1) \cup \cdots \cup \Omega(C_m),$$

where $\Omega(C_k)$ is the set of edges joining C_k and nodes from Y_i in the connected subgraph induced by the nodes in $Y_i \cup \bigcup_{j \neq k} C_j$. By Proposition 6.3, the set $\Omega(C_k)$ is a simple cocycle and it is clear that the sets $\Omega(C_k)$ are pairwise disjoint inasmuch as the C_k are disjoint.

Proof 2: (Michel Sakarovitch)   Let $\Omega = \Omega(Y)$ be a cocycle in G. Now, Ω is a cutset and we can pick some minimal cocycle $\Omega_1 = \Omega(Z)$ contained in Ω. We proceed by induction on $|\Omega - \Omega_1|$. If $\Omega = \Omega_1$, we are done. Otherwise, we claim that $E_1 = \Omega - \Omega_1$ is a cutset in G. If not, let e be any edge in E_1; we may assume that $a = s(e) \in Y$ and $b = t(e) \in V - Y$. As E_1 is not a cutset, there is a chain C from a to b in $(V, E - E_1, s, t)$ and as Ω is a cutset, this chain must contain some edge e' in Ω, so $C = C_1(x, e', y)C_2$, where C_1 is a chain from a to x and C_2 is a chain from y to b. Then, because C has its edges in $E - E_1$ and $E_1 = \Omega - \Omega_1$, we must have $e' \in \Omega_1$. We may assume that $x = s(e') \in Z$ and $y = t(e') \in V - Z$. But, we have the chain $C_1^R(a, e, b)C_2^R$ joining x and y in $(V, E - \Omega_1)$, a contradiction. Therefore, E_1 is indeed a cutset of G. Now, there is some minimal cocycle Ω_2 contained in E_1. If $\Omega_2 = E_1$, we are done. Otherwise, if we let $E_2 = E_1 - \Omega_2$, we can show as we just did that E_2 is a cutset of G with $|E_2| < |E_1|$. Thus, we finish the proof by applying the induction hypothesis to E_2. $\square$

We now prove the key property of orthogonality between cycles and cocycles.

Proposition 6.5. *Given any finite directed graph* $G = (V, E, s, t)$, *if* $\gamma = \gamma(C)$ *is the representative vector of any* Γ-cycle $\Gamma = \Gamma(C)$ *and* $\omega = \omega(Y)$ *is the representative vector of any cocycle,* $\Omega = \Omega(Y)$, *then*

$$\gamma \cdot \omega = \sum_{i=1}^{n} \gamma_i \omega_i = 0;$$

that is, γ *and* ω *are orthogonal. (Here,* $|E| = n$.)

Proof. Recall that $\Gamma = C^+ \cup C^-$, where C is a cycle in G, say

$$C = (u_0, e_1, u_1, \ldots, u_{k-1}, e_k, u_k), \quad \text{with} \quad u_k = u_0.$$

Then, by definition, we see that

$$\gamma \cdot \omega = |C^+ \cap \Omega^+(Y)| - |C^+ \cap \Omega^-(Y)| - |C^- \cap \Omega^+(Y)| + |C^- \cap \Omega^-(Y)|. \quad (*)$$

As we traverse the cycle C, when we traverse the edge e_i between u_{i-1} and u_i ($1 \leq i \leq k$), we note that

$$e_i \in (C^+ \cap \Omega^+(Y)) \cup (C^- \cap \Omega^-(Y)) \quad \text{iff} \quad u_{i-1} \in Y, u_i \in V - Y$$
$$e_i \in (C^+ \cap \Omega^-(Y)) \cup (C^- \cap \Omega^+(Y)) \quad \text{iff} \quad u_{i-1} \in V - Y, u_i \in Y.$$

In other words, every time we traverse an edge coming out from Y, its contribution to $(*)$ is $+1$ and every time we traverse an edge coming into Y its contribution to $(*)$ is -1. After traversing the cycle C entirely, we must have come out from Y as many times as we came into Y, so these contributions must cancel out. $\square$

Note that Proposition 6.5 implies that $|\Gamma \cap \Omega|$ is even.

Definition 6.4. Given any finite digraph $G = (V, E, s, t)$, where $E = \{e_1, \ldots, e_n\}$, the subspace $\mathscr{F}(G)$ of $\mathbb{R}^n$ spanned by all vectors $\gamma(\Gamma)$, where Γ is any Γ-cycle, is called the *cycle space of* G or *flow space of* G and the subspace $\mathscr{T}(G)$ of $\mathbb{R}^n$ spanned by all vectors $\omega(\Omega)$, where Ω is any cocycle, is called the *cocycle space of* G or *tension space of* G (or *cut space of* G).

When no confusion is possible, we write $\mathscr{F}$ for $\mathscr{F}(G)$ and $\mathscr{T}$ for $\mathscr{T}(G)$. Thus, $\mathscr{F}$ is the space consisting of all linear combinations $\sum_{i=1}^{k} \alpha_i \gamma_i$ of representative vectors of Γ-cycles γ_i, and $\mathscr{T}$ is the the space consisting of all linear combinations $\sum_{i=1}^{k} \alpha_i \omega_i$ of representative vectors of cocycles ω_i with $\alpha_i \in \mathbb{R}$. Proposition 6.5 says that the spaces $\mathscr{F}$ and $\mathscr{T}$ are mutually orthogonal. Observe that $\mathbb{R}^n$ is isomorphic to the vector space of functions $f : E \to \mathbb{R}$. Consequently, a vector $f = (f_1, \ldots, f_n) \in \mathbb{R}^n$ may be viewed as a function from $E = \{\mathbf{e}_1, \ldots, \mathbf{e}_n\}$ to $\mathbb{R}$ and it is sometimes convenient to write $f(\mathbf{e}_i)$ instead of f_i.

Remark: The seemingly odd terminology "flow space" and "tension space" is explained later.

Our next goal is be to determine the dimensions of $\mathscr{F}$ and $\mathscr{T}$ in terms of the number of edges, the number of nodes, and the number of connected components of G, and to give a convenient method for finding bases of $\mathscr{F}$ and $\mathscr{T}$. For this, we use spanning trees and their dual, cotrees. But first, we need a crucial theorem that also plays an important role in the theory of flows in networks.

Theorem 6.1. *(Arc Coloring Lemma; Minty [1960]) Let $G = (V, E, s, t)$ be a finite directed graph and assume that the edges of G are colored either in black, red, or green. Pick any edge e and color it black. Then, exactly one of two possibilities may occur:*

(1) There is a simple cycle containing e whose edges are only red or black with all the black edges oriented in the same direction.

(2) There is a simple cocycle containing e whose edges are only green or black with all the black edges oriented in the same direction.

Proof. Let $a = s(e)$ and $b = t(e)$. Apply the following procedure for marking nodes.

Intitially, only b is marked.
while there is some marked node x and some unmarked node y with
    either a black edge, e', with $(x, y) = (s(e'), t(e'))$ or
    a red edge, e', with $(x, y) = \{s(e'), t(e')\}$
    then mark y; $\mathrm{arc}(y) = e'$
endwhile

When the marking algorithm stops, exactly one of the following two cases occurs.

(i) Node a has been marked. Let $e' = \mathrm{arc}(a)$ be the edge that caused a to be marked and let x be the other endpoint of e'. If $x = b$, we found a simple cycle satisfying (i). If not, let $e'' = \mathrm{arc}(x)$ and let y be the other endpoint of e'' and continue in the same manner. This procedure will stop with b and yields the chain C from b to a along which nodes have been marked. This chain must be simple because every edge in it was used once to mark some node (check that the set of edges used for the marking is a tree). If we add the edge e to the chain C, we obtain a simple cycle Γ whose edges are colored black or red and with all edges colored black oriented in the same direction due to the marking scheme. It is impossible to have a cocycle whose edges are colored black or green containing e because it would have been impossible to conduct the marking through this cocycle and a would not have been marked.

(ii) Node a has not been marked. Let Y be the set of unmarked nodes. The set $\Omega(Y)$ is a cocycle whose edges are colored green or black containing e with all black edges in $\Omega^+(Y)$. This cocycle is the disjoint of simple cocycles (by Proposition 6.4) and one of these simple cocycles contains e. If a cycle with black or red edges containing e with all black edges oriented in the same direction existed, then a would have been marked, a contradiction. $\square$

Corollary 6.2. *Every edge of a finite directed graph G belongs either to a simple circuit or to a simple cocircuit but not both.*

Proof. Color all edges black and apply Theorem 6.1. $\square$

Although Minty's theorem looks more like an amusing fact than a deep result, it is actually a rather powerful theorem. For example, we show in Section 6.4 that Minty's theorem can be used to prove the "hard part" of the max-flow min-cut theorem (Theorem 6.7), an important theorem that has many applications. Here are a few more applications of Theorem 6.1.

Proposition 6.6. *Let G be a finite connected directed graph with at least one edge. Then, the following conditions are equivalent.*

(i) G is strongly connected.
(ii) Every edge belongs to some circuit.
(iii) G has no cocircuit.

Proof. $(i) \Longrightarrow (ii)$. If x and y are the endpoints of any edge e in G, as G is strongly connected, there is a simple path from y to x and thus, a simple circuit through e.

$(ii) \Longrightarrow (iii)$. This follows from Corollary 6.2.

$(iii) \Longrightarrow (i)$. Assume that G is not strongly connected and let Y' and Y'' be two strongly connected components linked by some edge e and let $a = s(e)$ and $b = t(e)$, with $a \in Y'$ and $b \in Y''$. The edge e does not belong to any circuit because otherwise a and b would belong to the same strongly connected component. Thus, by Corollary 6.2, the edge e should belong to some cocircuit, a contradiction. $\square$

In order to determine the dimension of the cycle space $\mathcal{T}$, we use spanning trees. Let us assume that G is connected because otherwise the same reasoning applies to the connected components of G. If T is any spanning tree of G, we know from Theorem 3.2, Part (4), that adding any edge $e \in E - T$ (called a *chord* of T) creates a (unique) cycle. We show shortly that the vectors associated with these cycles form a basis of the cycle space. We can find a basis of the cocycle space by considering sets of edges of the form $E - T$, where T is a spanning tree. Such sets of edges are called *cotrees*.

Definition 6.5. Let G be a finite directed connected graph $G = (V, E, s, t)$. A spanning subgraph (V, K, s, t) is a *cotree* iff $(V, E - K, s, t)$ is a spanning tree.

Cotrees are characterized in the following proposition.

Proposition 6.7. *Let G be a finite directed connected graph $G = (V, E, s, t)$. If E is partitioned into two subsets T and K (i.e., $T \cup K = E$; $T \cap K = \emptyset$; $T, K \neq \emptyset$), then the following conditions ar equivalent.*

(1) (V, T, s, t) is tree.
(2) (V, K, s, t) is a cotree.
(3) (V, K, s, t) contains no simple cocycles of G and upon addition of any edge $e \in T$, it does contain a simple cocycle of G.

Proof. By definition of a cotree, (1) and (2) are equivalent, so we prove the equivalence of (1) and (3).

(1) $\Longrightarrow$ (3). We claim that (V, K, s, t) contains no simple cocycles of G. Otherwise, K would contain some simple cocycle $\Gamma(A)$ of G and then no chain in the tree (V, T, s, t) would connect A and $V - E$, a contradiction.

Next, for any edge $e \in T$, observe that $(V, T - \{e\}, s, t)$ has two connected components, say A and B and then $\Omega(A)$ is a simple cocycle contained in $(V, K \cup \{e\}, s, t)$ (in fact, it is easy to see that it is the only one). Therefore, (3) holds

(3) $\Longrightarrow$ (1). We need to prove that (V, T, s, t) is a tree. First, we show that (V, T, s, t) has no cycles. Let $e \in T$ be any edge; color e black; color all edges in $T - \{e\}$ red; color all edges in $K = E - T$ green. By (3), by adding e to K, we find a simple cocycle of black or green edges that contained e. Thus, there is no cycle of red or black edges containing e. As e is arbitrary, there are no cycles in T.

Finally, we prove that (V, T, s, t) is connected. Pick any edge $e \in K$ and color it black; color edges in T red; color edges in $K - \{e\}$ green. Because G has no cocycle of black and green edges containing e, there is a cycle of black or red edges containing e. Therefore, $T \cup \{e\}$ has a cycle, which means that there is a path from any two nodes in T. $\square$

We are now ready for the main theorem of this section.

Theorem 6.2. *Let G be a finite directed graph $G = (V, E, s, t)$, and assume that $|E| = n$, $|V| = m$ and that G has p connected components. Then, the cycle space $\mathscr{F}$ and the cocycle space $\mathscr{T}$ are subspaces of $\mathbb{R}^n$ of dimensions $\dim \mathscr{F} = n - m + p$ and $\dim \mathscr{T} = m - p$ and $\mathscr{T} = \mathscr{F}^\perp$ is the orthogonal complement of $\mathscr{F}$. Furthermore, if $C_1, \ldots, C_p$ are the connected components of G, bases of $\mathscr{F}$ and $\mathscr{T}$ can be found as follows.*

(1) Let $T_1, \ldots, T_p$, be any spanning trees in $C_1, \ldots, C_p$. For each spanning tree T_i form all the simple cycles $\Gamma_{i,e}$ obtained by adding any chord $e \in C_i - T_i$ to T_i. Then, the vectors $\gamma_{i,e} = \gamma(\Gamma_{i,e})$ form a basis of $\mathscr{F}$.
(2) For any spanning tree T_i as above, let $K_i = C_i - T_i$ be the corresponding cotree. For every edge $e \in T_i$ (called a twig), there is a unique simple cocycle $\Omega_{i,e}$ contained in $K_i \cup \{e\}$. Then, the vectors $\omega_{i,e} = \omega(\Omega_{i,e})$ form a basis of $\mathscr{T}$.

Proof. We know from Proposition 6.5 that $\mathscr{F}$ and $\mathscr{T}$ are orthogonal. Thus,

$$\dim \mathscr{F} + \dim \mathscr{T} \le n.$$

Let us follow the procedure specified in (1). Let $C_i = (E_i, V_i)$, be the ith connected component of G and let $n_i = |E_i|$ and $|V_i| = m_i$, so that $n_1 + \cdots + n_p = n$ and $m_1 + \cdots + m_p = m$. For any spanning tree T_i for C_i, recall that T_i has $m_i - 1$ edges and so, $|E_i - T_i| = n_i - m_i + 1$. If $e_{i,1}, \ldots, e_{i,n_i-m_i+1}$ are the edges in $E_i - T_i$, then the vectors

$$\gamma_{i,e_{i,1}}, \ldots, \gamma_{i,e_{i,m_i}}$$

must be linearly independent, because $\gamma_{i,e_{i,j}} = \gamma(\Gamma_{i,e_{i,j}})$ and the simple cycle $\Gamma_{i,e_{i,j}}$ contains the edge $e_{i,j}$ that none of the other $\Gamma_{i,e_{i,k}}$ contain for $k \ne j$. So, we get

$$(n_1 - m_1 + 1) + \cdots + (n_p - m_p + 1) = n - m + p \le \dim \mathscr{F}.$$

Let us now follow the procedure specified in (2). For every spanning tree T_i let $e_{i,1}, \ldots, e_{i,m_i-1}$ be the edges in T_i. We know from Proposition 6.7 that adding any edge $e_{i,j}$ to $C_i - T_i$ determines a unique simple cocycle $\Omega_{i,e_{i,j}}$ containing $e_{i,j}$ and the vectors

$$\omega_{i,e_{i,1}}, \ldots, \omega_{i,e_{i,m_i-1}}$$

must be linearly independent because the simple cocycle $\Omega_{i,e_{i,j}}$ contains the edge $e_{i,j}$ that none of the other $\Omega_{i,e_{i,k}}$ contain for $k \ne j$. So, we get

$$(m_1 - 1) + \cdots + (m_p - 1) = m - p \le \dim \mathscr{T}.$$

But then, $n \le \dim \mathscr{F} + \dim \mathscr{T}$ and inasmuch as we also have $\dim \mathscr{F} + \dim \mathscr{T} \le n$, we get

$$\dim \mathscr{F} = n - m + p \quad \text{and} \quad \dim \mathscr{T} = m - p.$$

The vectors produced in (1) and (2) are linearly independent and in each case, their number is equal to the dimension of the space to which they belong, therefore they are bases of these spaces. $\quad\square$

Because $\dim \mathscr{F} = n - m + p$ and $\dim \mathscr{T} = m - p$ do not depend on the orientation of G, we conclude that the spaces $\mathscr{F}$ and $\mathscr{T}$ are uniquely determined by G, independently of the orientation of G, up to isomorphism. The number $n - m + p$ is called the *cyclomatic number of G* and $m - p$ is called the *cocyclomatic number of G*.

Remarks:

1. Some authors, including Harary [15] and Diestel [9], define the vector spaces $\mathscr{F}$ and $\mathscr{T}$ over the two-element field, $\mathbb{F}_2 = \{0, 1\}$. The same dimensions are obtained for $\mathscr{F}$ and $\mathscr{T}$ and $\mathscr{F}$ and $\mathscr{T}$ still orthogonal. On the other hand, because $1 + 1 = 0$, some interesting phenomena happen. For example, orientation is irrelevant, the sum of two cycles (or cocycles) is their symmetric difference, and the space $\mathscr{F} \cap \mathscr{T}$ is **not** necessarily reduced to the trivial space (0). The

space $\mathscr{F} \cap \mathscr{T}$ is called the *bicycle space*. The bicycle space induces a partition of the edges of a graph called the *principal tripartition*. For more on this, see Godsil and Royle [12], Sections 14.15 and 14.16 (and Chapter 14).

2. For those who know homology, of course, $p = \dim H_0$, the dimension of the zero-th homology group and $n - m + p = \dim H_1$, the dimension of the first homology group of G viewed as a topological space. Usually, the notation used is $b_0 = \dim H_0$ and $b_1 = \dim H_1$ (the first two *Betti numbers*). Then the above equation can be rewritten as

$$m - n = b_0 - b_1,$$

which is just the formula for the *Euler–Poincaré characteristic*.

Fig. 6.3 Enrico Betti, 1823–1892 (left) and Henri Poincaré, 1854–1912 (right)

Figure 6.4 shows an unoriented graph (a cube) and a cocycle Ω, which is also a cycle Γ, shown in thick lines (i.e., a bicycle, over the field $\mathbb{F}_2$). However, as we saw in the example from Figure 6.2, for any orientation of the cube, the vectors γ and ω corresponding to Γ and Ω are different (and orthogonal).

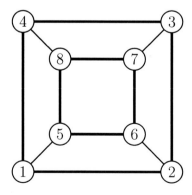

Fig. 6.4 A bicycle in a graph (a cube)

Let us illustrate the procedures for constructing bases of $\mathscr{F}$ and $\mathscr{T}$ on the graph G_8. Figure 6.5 shows a spanning tree T and a cotree K for G_8.

We have $n = 7; m = 5; p = 1$, and so, dim $\mathscr{F} = 7 - 5 + 1 = 3$ and dim $\mathscr{T} = 5 - 1 = 4$. If we successively add the edges e_2, e_6, and e_7 to the spanning tree T, we get the three simple cycles shown in Figure 6.6 with thicker lines.

If we successively add the edges e_1, e_3, e_4, and e_5 to the cotree K, we get the four simple cocycles shown in Figures 6.7 and 6.8 with thicker lines.

Given any node $v \in V$ in a graph G for simplicity of notation let us denote the cocycle $\Omega(\{v\})$ by $\Omega(v)$. Similarly, we write $\Omega^+(v)$ for $\Omega^+(\{v\})$; $\Omega^-(v)$ for $\Omega^-(\{v\})$, and similarly for the the vectors $\omega(\{v\})$, and so on. It turns our that vectors of the form $\omega(v)$ generate the cocycle space and this has important consequences.

Proposition 6.8. *Given any finite directed graph* $G = (V, E, s, t)$ *for every cocycle* $\Omega = \Omega(Y)$ *we have*

$$\omega(Y) = \sum_{v \in Y} \omega(v).$$

Consequently, the vectors of the form $\omega(v)$, *with* $v \in V$, *generate the cocycle space* $\mathscr{T}$.

Proof. For any edge $e \in E$ if $a = s(e)$ and $b = t(e)$, observe that

$$\omega(v)_e = \begin{cases} +1 & \text{if } v = a \\ -1 & \text{if } v = b \\ 0 & \text{if } v \neq a, b. \end{cases}$$

As a consequence, if we evaluate $\sum_{v \in Y} \omega(v)$, we find that

$$\left(\sum_{v \in Y} \omega(v) \right)_e = \begin{cases} +1 & \text{if } a \in Y \text{ and } b \in V - Y \\ -1 & \text{if } a \in V - Y \text{ and } b \in Y \\ 0 & \text{if } a, b \in Y \text{ or } a, b \in V - Y, \end{cases}$$

which is exactly $\omega(Y)_v$. $\quad \square$

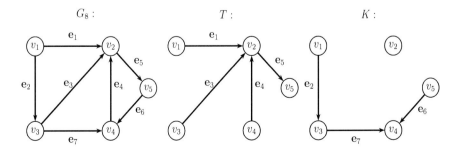

Fig. 6.5 Graph G_8; A spanning tree, T; a cotree, K

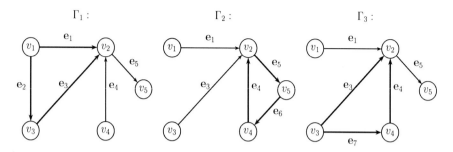

Fig. 6.6 A cycle basis for G_8

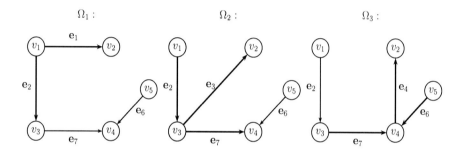

Fig. 6.7 A cocycle basis for G_8

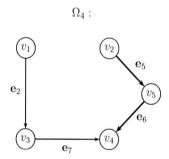

Fig. 6.8 A cocycle basis for G_8 (continued)

Proposition 6.8 allows us to characterize flows (the vectors in $\mathscr{F}$) in an interesting way which also reveals the reason behind the terminology.

Theorem 6.3. *Given any finite directed graph $G = (V,E,s,t)$ a vector $f \in \mathbb{R}^n$ is a flow in $\mathscr{F}$ iff*

$$\sum_{e \in \Omega^+(v)} f(e) - \sum_{e \in \Omega^-(v)} f(e) = 0, \quad \text{for all} \quad v \in V. \tag{†}$$

Proof. By Theorem 6.2, we know that $\mathscr{F}$ is the orthogonal complement of $\mathscr{T}$. Thus, for any $f \in \mathbb{R}^n$, we have $f \in \mathscr{F}$ iff $f \cdot \omega = 0$ for all $\omega \in \mathscr{T}$. Moreover, Proposition 6.8 says that $\mathscr{T}$ is generated by the vectors of the form $\omega(v)$, where $v \in V$ so $f \in \mathscr{F}$ iff $f \cdot \omega(v) = 0$ for all $v \in V$. But (†) is exactly the assertion that $f \cdot \omega(v) = 0$ and the theorem is proved. $\square$

Equation (†) justifies the terminology of "flow" for the elements of the space $\mathscr{F}$. Indeed, a *flow* f in a (directed) graph $G = (V,E,s,t)$, is defined as a function $f \colon E \to \mathbb{R}$, and we say that a flow is *conservative* (Kirchhoff's first law) iff for every node $v \in V$, the total flow $\sum_{e \in \Omega^-(v)} f(e)$ coming into the vertex v is equal to the total flow $\sum_{e \in \Omega^+(v)} f(e)$ coming out of that vertex. This is exactly what equation (†) says.

We can also characterize tensions as follows.

Theorem 6.4. *Given any finite simple directed graph $G = (V,E,s,t)$ for any $\theta \in \mathbb{R}^n$ we have:*

(1) The vector θ is a tension in $\mathscr{T}$ iff for every simple cycle $\Gamma = \Gamma^+ \cup \Gamma^-$ we have

$$\sum_{e \in \Gamma^+} \theta(e) - \sum_{e \in \Gamma^-} \theta(e) = 0. \tag{*}$$

(2) If G has no parallel edges (and no loops), then $\theta \in \mathbb{R}^n$ is a tension in $\mathscr{T}$ iff the following condition holds. There is a function $\pi \colon V \to \mathbb{R}$ called a "potential function", such that

$$\theta(e) = \pi(t(e)) - \pi(s(e)), \tag{**}$$

for every $e \in E$.

Proof. (1) The equation (*) asserts that $\gamma(\Gamma) \cdot \theta = 0$ for every simple cycle Γ. Every cycle is the disjoint union of simple cycles, thus the vectors of the form $\gamma(\Gamma)$ generate the flow space $\mathscr{F}$ and by Theorem 6.2, the tension space $\mathscr{T}$ is the orthogonal complement of $\mathscr{F}$, so θ is a tension iff (*) holds.

(2) Assume a potential function $\pi \colon V \to \mathbb{R}$ exists, let $\Gamma = (v_0, e_1, v_1, \ldots, v_{k-1}, e_k, v_k)$, with $v_k = v_0$, be a simple cycle, and let $\gamma = \gamma(\Gamma)$. We have

$$\gamma_1 \theta(e_1) = \pi(v_1) - \pi(v_0)$$
$$\gamma_2 \theta(e_2) = \pi(v_2) - \pi(v_1)$$
$$\vdots$$
$$\gamma_{k-1} \theta(e_{k-1}) = \pi(v_{k-1}) - \pi(v_{k-2})$$

$$\gamma_k \theta(e_k) = \pi(v_0) - \pi(v_{k-1})$$

and we see that when we add both sides of these equations that we get $(*)$:

$$\sum_{e \in \Gamma^+} \theta(e) - \sum_{e \in \Gamma^-} \theta(e) = 0.$$

Let us now assume that $(*)$ holds for every simple cycle and let $\theta \in \mathscr{T}$ be any tension. Consider the following procedure for assigning a value $\pi(v)$ to every vertex $v \in V$, so that $(**)$ is satisfied. Pick any vertex v_0, and assign it the value, $\pi(v_0) = 0$.

Now, for every vertex $v \in V$ that has not yet been assigned a value, do the following.

1. If there is an edge $e = (u,v)$ with $\pi(u)$ already determined, set

$$\pi(v) = \pi(u) + \theta(e);$$

2. If there is an edge $e = (v,u)$ with $\pi(u)$ already determined, set

$$\pi(v) = \pi(u) - \theta(e).$$

At the end of this process, all the nodes in the connected component of v_0 will have received a value and we repeat this process for all the other connected components. However, we have to check that each node receives a unique value (given the choice of v_0). If some node v is assigned two different values $\pi_1(v)$ and $\pi_2(v)$ then there exist two chains σ_1 and σ_2 from v_0 to v, and if C is the cycle $\sigma_1 \sigma_2^R$, we have

$$\gamma(C) \cdot \theta \neq 0.$$

However, any cycle is the disjoint union of simple cycles, so there would be some simple cycle Γ with

$$\gamma(\Gamma) \cdot \theta \neq 0,$$

contradicting $(*)$. Therefore, the function π is indeed well-defined and, by construction, satisfies $(**)$. $\square$

Some of these results can be improved in various ways. For example, flows have what is called a "conformal decomposition."

Definition 6.6. Given any finite directed graph $G = (V, S, s, t)$, we say that a flow $f \in \mathscr{F}$ has a *conformal decomposition* iff there are some cycles $\Gamma_1, \ldots, \Gamma_k$ such that if $\gamma_i = \gamma(\Gamma_i)$, then

$$f = \alpha_1 \gamma_1 + \cdots + \alpha_k \gamma_k,$$

with

1. $\alpha_i \geq 0$, for $i = 1, \ldots, k$.

2. For any edge, $e \in E$, if $f(e) > 0$ (respectively, $f(e) < 0$) and $e \in \Gamma_j$, then $e \in \Gamma_j^+$ (respectively, $e \in \Gamma_j^-$).

Proposition 6.9. *Given any finite directed graph $G = (V, S, s, t)$ every flow $f \in \mathcal{F}$ has some conformal decomposition. In particular, if $f(e) \geq 0$ for all $e \in E$, then all the Γ_js are circuits.*

Proof. We proceed by induction on the number of nonzero components of f. First, note that $f = 0$ has a trivial conformal decomposition. Next, let $f \in \mathcal{F}$ be a flow and assume that every flow f' having at least one more zero component than f has some conformal decomposition. Let $\overline{G}$ be the graph obtained by reversing the orientation of all edges e for which $f(e) < 0$ and deleting all the edges for which $f(e) = 0$. Observe that $\overline{G}$ has no cocircuit, as the inner product of any simple cocircuit with any nonzero flow cannot be zero. Hence, by the corollary to the coloring lemma, $\overline{G}$ has some circuit C and let Γ be a cycle of G corresponding to C. Let

$$\alpha = \min\{ \min_{e \in \Gamma^+} f(e), \ \min_{e \in \Gamma^-} -f(e)\} \geq 0.$$

Then, the flow

$$f' = f - \alpha \gamma(\Gamma)$$

has at least one more zero component than f. Thus, f' has some conformal decomposition and, by construction, $f = f' + \alpha \gamma(\Gamma)$ is a conformal decomposition of f.
□

We now take a quick look at various matrices associated with a graph.

6.2 Incidence and Adjacency Matrices of a Graph

In this section, we are assuming that our graphs are finite, directed, without loops, and without parallel edges.

Definition 6.7. Let $G = (V, E)$ be a graph and write $V = \{v_1, \ldots, v_m\}$ and $E = \{e_1, \ldots, e_n\}$. The *incidence matrix* $D(G)$ of G is the $m \times n$-matrix whose entries d_{ij} are

$$d_{ij} = \begin{cases} +1 & \text{if } v_i = s(e_j) \\ -1 & \text{if } v_i = t(e_j) \\ 0 & \text{otherwise.} \end{cases}$$

Remark: The incidence matrix actually makes sense for a graph G with parallel edges but without loops.

For simplicity of notation and when no confusion is possible, we write D instead of $D(G)$.

Because we assumed that G has no loops, observe that every column of D contains exactly two nonzero entries, $+1$ and -1. Also, the ith row of D is the vector $\omega(\mathbf{v}_i)$ representing the cocycle $\Omega(\mathbf{v}_i)$. For example, the incidence matrix of the graph G_8 shown again in Figure 6.9 is shown on the next page.

The incidence matrix D of a graph G represents a linear map from $\mathbb{R}^n$ to $\mathbb{R}^m$ called the *incidence map* (or *boundary map*) and denoted by D (or ∂). For every $e \in E$, we have

$$D(\mathbf{e_j}) = s(\mathbf{e_j}) - t(\mathbf{e_j}).$$

Here is the incidence matrix of the graph G_8:

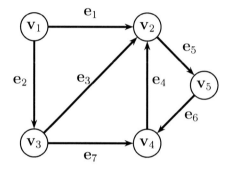

Fig. 6.9 Graph G_8

$$D = \begin{pmatrix} 1 & 1 & 0 & 0 & 0 & 0 & 0 \\ -1 & 0 & -1 & -1 & 1 & 0 & 0 \\ 0 & -1 & 1 & 0 & 0 & 0 & 1 \\ 0 & 0 & 0 & 1 & 0 & -1 & -1 \\ 0 & 0 & 0 & 0 & -1 & 1 & 0 \end{pmatrix}.$$

Remark: Sometimes it is convenient to consider the vector space $C_1(G) = \mathbb{R}^E$, of all functions $f \colon E \to \mathbb{R}$, called the *edge space of G* and the vector space $C_0(G) = \mathbb{R}^V$, of all functions $g \colon V \to \mathbb{R}$, called the *vertex space of G*. Obviously, $C_1(G)$ is isomorphic to $\mathbb{R}^n$ and $C_0(G)$ is isomorphic to $\mathbb{R}^m$. The transpose $D^\top$ of D is a linear map from $C_0(G)$ to $C_1(G)$ also called the *coboundary map* and often denoted by δ. Observe that $\delta(Y) = \Omega(Y)$ (viewing the subset, $Y \subseteq V$, as a vector in $C_0(G)$).

The spaces of flows and tensions can be recovered from the incidence matrix.

Theorem 6.5. *Given any finite graph G if D is the incidence matrix of G and $\mathscr{F}$ and $\mathscr{T}$ are the spaces of flows and tensions on G, then*

(1) $\mathscr{F} = \mathrm{Ker}\, D.$
(2) $\mathscr{T} = \mathrm{Im}\, D^\top.$

Futhermore, if G has p connected components and m nodes, then

$$\text{rank } D = m - p.$$

Proof. We already observed that the ith row of D is the vector $\omega(\mathbf{v}_i)$ and we know from Theorem 6.3 that $\mathscr{F}$ is exactly the set of vectors orthogonal to all vectors of the form $\omega(\mathbf{v}_i)$. Now, for any $f \in \mathbb{R}^n$,

$$Df = \begin{pmatrix} \omega(\mathbf{v}_1) \cdot f \\ \vdots \\ \omega(\mathbf{v}_m) \cdot f \end{pmatrix},$$

and so, $\mathscr{F} = \text{Ker } D$. The vectors $\omega(\mathbf{v}_i)$ generate $\mathscr{T}$, therefore the rows of D generate $\mathscr{T}$; that is, $\mathscr{T} = \text{Im } D^{\top}$.

From Theorem 6.2, we know that

$$\dim \mathscr{T} = m - p$$

and inasmuch as we just proved that $\mathscr{T} = \text{Im } D^{\top}$, we get

$$\text{rank } D = \text{rank } D^{\top} = m - p,$$

which proves the last part of our theorem. $\square$

Corollary 6.3. *For any graph $G = (V, E, s, t)$ if $|V| = m$, $|E| = n$ and G has p connected components, then the incidence matrix D of G has rank n (i.e., the columns of D are linearly independent) iff $\mathscr{F} = (0)$ iff $n = m - p$.*

Proof. By Theorem 6.3, we have $\text{rank } D = m - p$. So, $\text{rank } D = n$ iff $n = m - p$ iff $n - m + p = 0$ iff $\mathscr{F} = (0)$ (because $\dim \mathscr{F} = n - m + p$). $\square$

The incidence matrix of a graph has another interesting property observed by Poincaré. First, let us define a variant of triangular matrices.

Definition 6.8. An $n \times n$ (real or complex) matrix $A = (a_{ij})$ is said to be *pseudo-triangular and nonsingular* iff either

(i) $n = 1$ and $a_{11} \neq 0$.
(ii) $n \geq 2$ and A has some row, say k, with a unique nonzero entry $a_{h,k}$ such that the submatrix B obtained by deleting the hth row and the kth column from A is also pseudo-triangular and nonsingular.

It is easy to see that a matrix defined as in Definition 6.8 can be transformed into a usual triangular matrix by permutation of its columns.

Proposition 6.10. *(Poincaré, 1901) If D is the incidence matrix of a graph, then every square $k \times k$ nonsingular submatrix,*[2] *B of D is pseudo-triangular. Consequently,* $\det(B) = +1, -1,$ *or* 0, *for any square $k \times k$ submatrix B of D.*

Proof. We proceed by induction on k. The result is obvious for $k = 1$.

Next, let B be a square $k \times k$-submatrix of D which is nonsingular, not pseudo-triangular and yet, every nonsingular $h \times h$-submatrix of B is pseudo-triangular if $h < k$. We know that every column of B has at most two nonzero entries (because every column of D contains two nonzero entries: $+1$ and -1). Also, as B is not pseudo-triangular (but nonsingular) every row of B contains at least two nonzero elements. But then, no row of B may contain three or more elements, because the number of nonzero slots in all columns is at most $2k$ and by the pigeonhole principle, we could fit $2k + 1$ objects in $2k$ slots, which is impossible. Therefore, every row of B contains exactly two nonzero entries. Again, the pigeonhole principle implies that every column also contains exactly two nonzero entries. But now, the nonzero entries in each column are $+1$ and -1, so if we add all the rows of B, we get the zero vector, which shows that B is singular, a contradiction. Therefore, B is pseudo-triangular.

The entries in D are $+1, -1, 0$, therefore the above immediately implies that $\det(B) = +1, -1,$ or 0 for any square $k \times k$ submatrix B of D. $\square$

A square matrix such as A such that $\det(B) = +1, -1,$ or 0 for any square $k \times k$ submatrix B of A is said to be *totally unimodular*. This is a very strong property of incidence matrices that has far-reaching implications in the study of optimization problems for networks.

Another important matrix associated with a graph is its adjacency matrix.

Definition 6.9. Let $G = (V, E)$ be a graph with $V = \{\mathbf{v}_1, \ldots, \mathbf{v}_m\}$. The *ajacency matrix* $A(G)$ *of* G is the $m \times m$-matrix whose entries a_{ij} are

$$a_{ij} = \begin{cases} 1 & \text{if } (\exists e \in E)(\{s(e), t(e)\} = \{\mathbf{v}_i, \mathbf{v}_j\}) \\ 0 & \text{otherwise.} \end{cases}$$

When no confusion is possible, we write A for $A(G)$. Note that the matrix A is symmetric and $a_{ii} = 0$. Here is the adjacency matrix of the graph G_8 shown in Figure 6.9:

$$A = \begin{pmatrix} 0 & 1 & 1 & 0 & 0 \\ 1 & 0 & 1 & 1 & 1 \\ 1 & 1 & 0 & 1 & 0 \\ 0 & 1 & 1 & 0 & 1 \\ 0 & 1 & 0 & 1 & 0 \end{pmatrix}.$$

We have the following useful relationship between the incidence matrix and the adjacency matrix of a graph.

---

[2] Given any $m \times n$ matrix $A = (a_{ij})$, if $1 \leq h \leq m$ and $1 \leq k \leq n$, then a $h \times k$-submatrix B of A is obtained by picking any k columns of A and then any h rows of this new matrix.

Proposition 6.11. *Given any graph G if D is the incidence matrix of G, A is the adjacency matrix of G, and Δ is the diagonal matrix such that $\Delta_{ii} = d(\mathbf{v}_i)$, the degree of node $\mathbf{v}_i$, then*

$$DD^\top = \Delta - A.$$

Consequently, $DD^\top$ is independent of the orientation of G and $\Delta - A$ is symmetric positive, semi-definite; that is, the eigenvalues of $\Delta - A$ are real and nonnegative.

Proof. It is well known that $DD^\top_{ij}$ is the inner product of the ith row d_i, and the jth row d_j of D. If $i = j$, then as

$$d_{ik} = \begin{cases} +1 & \text{if } s(\mathbf{e}_k) = \mathbf{v}_i \\ -1 & \text{if } t(\mathbf{e}_k) = \mathbf{v}_i \\ 0 & \text{otherwise} \end{cases}$$

we see that $d_i \cdot d_i = d(\mathbf{v}_i)$. If $i \neq j$, then $d_i \cdot d_j \neq 0$ iff there is some edge $\mathbf{e}_k$ with $s(\mathbf{e}_k) = \mathbf{v}_i$ and $t(\mathbf{e}_k) = \mathbf{v}_i$, in which case, $d_i \cdot d_j = -1$. Therefore,

$$DD^\top = \Delta - A,$$

as claimed. Now, $DD^\top$ is obviously symmetric and it is well known that its eigenvalues are nonnegative (e.g., see Gallier [11], Chapter 12). $\square$

Remarks:

1. The matrix $L = DD^\top = \Delta - A$, is known as the *Laplacian (matrix)* of the graph, G. Another common notation for the matrix $DD^\top$ is Q. The columns of D contain exactly the two nonzero entries, $+1$ and -1, thus we see that the vector $\mathbf{1}$, defined such that $\mathbf{1}_i = 1$, is an eigenvector for the eigenvalue 0.

2. If G is connected, then D has rank $m - 1$, so the rank of $DD^\top$ is also $m - 1$ and the other eigenvalues of $DD^\top$ besides 0 are strictly positive. The smallest positive eigenvalue of $L = DD^\top$ has some remarkable properties. There is an area of graph theory overlapping (linear) algebra, called *spectral graph theory* that investigates the properties of graphs in terms of the eigenvalues of its Laplacian matrix but this is beyond the scope of this book. Some good references for algebraic graph theory include Biggs [3], Godsil and Royle [12], and Chung [6] for spectral graph theory.

   One of the classical and surprising results in algebraic graph theory is a formula that gives the number of spanning trees $\tau(G)$ of a connected graph G in terms of its Laplacian $L = DD^\top$. If J denotes the square matrix whose entries are all 1s and if $\mathrm{adj}\, L$ denotes the adjoint matrix of L (the transpose of the matrix of cofactors of L), that is, the matrix given by

   $$(\mathrm{adj}\, L)_{ij} = (-1)^{i+j} \det L(j,i),$$

   where $L(j,i)$ is the matrix obtained by deleting the jth row and the ith column of L, then we have

$$\operatorname{adj} L = \tau(G)J.$$

We also have

$$\tau(G) = m^{-2}\det(J+L),$$

where m is the number of nodes of G.

3. As we already observed, the incidence matrix also makes sense for graphs with parallel edges and no loops. But now, in order for the equation $DD^\top = \Delta - A$ to hold, we need to define A differently. We still have the same definition as before for the incidence matrix but we can define the new matrix $\mathscr{A}$ such that

$$\mathscr{A}_{ij} = |\{e \in E \mid \{s(e), t(e)\} = \{\mathbf{v}_i, \mathbf{v}_j\}\}|;$$

that is, $\mathscr{A}_{ij}$ is the number of parallel edges between $\mathbf{v}_i$ and $\mathbf{v}_j$. Then, we can check that

$$DD^\top = \Delta - \mathscr{A}.$$

4. There are also versions of the adjacency matrix and of the incidence matrix for undirected graphs. In this case, D is no longer totally unimodular.

6.3 Eulerian and Hamiltonian Cycles

In this short section, we discuss two classical problems that go back to the very beginning of graph theory. These problems have to do with the existence of certain kinds of cycles in graphs. These problems come in two flavors depending on whether the graphs are directed but there are only minor differences between the two versions and traditionally the focus is on undirected graphs.

The first problem goes back to Euler and is usually known as the *Königsberg bridge problem*. In 1736, the town of Königsberg had seven bridges joining four areas of land. Euler was asked whether it were possible to find a cycle that crossed every bridge exactly once (and returned to the starting point).

Fig. 6.10 Leonhard Euler, 1707–1783

The graph shown in Figure 6.11 models the Königsberg bridge problem. The nodes A, B, C, D correspond to four areas of land in Königsberg and the edges to the seven bridges joining these areas of land.

In fact, the problem is unsolvable, as shown by Euler, because some nodes do not have an even degree. We now define the problem precisely and give a complete solution.

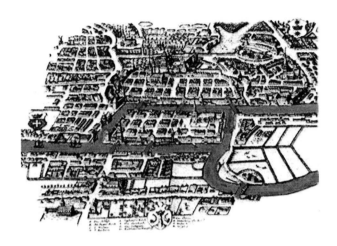

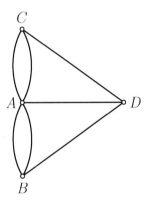

Fig. 6.11 The seven bridges of Königsberg and a graph modeling the Königsberg bridge problem

Definition 6.10. Given a finite undirected graph $G = (V, E)$ (respectively, a directed graph $G = (V, E, s, t)$) a *Euler cycle* (or *Euler tour*), (respectively, a *Euler circuit*) is a cycle in G that passes through every node and every edge (exactly once); (respectively, a circuit in G that passes through every node and every edge (exactly once)). The *Eulerian cycle (resp. circuit) problem* is the problem: given a graph G, is there a Eulerian cycle (respectively, circuit) in G?

Theorem 6.6. *(1) An undirected graph $G = (V, E)$ has a Eulerian cycle iff the following properties hold.*

(a1) The graph G is connected.
(b1) Every node has even degree.

   (2) A directed graph $G = (V, E, s, t)$ has a Eulerian circuit iff the following properties hold.

(a2) The graph G is strongly connected.
(b2) Every node has the same number of incoming and outgoing edges; that is,
   $d^+(v) = d^-(v)$, *for all $v \in V$.*

Proof. We prove (1) leaving (2) as an easy exercise (the proof of (2) is very similar to the proof of (1)). Clearly, if a Euler cycle exists, G is connected and because every edge is traversed exactly once, every node is entered as many times as it is exited so the degree of every node is even.

For the converse, observe that G must contain a cycle as otherwise, being connected, G would be a tree but we proved earlier that every tree has some node of degree 1. (If G is directed and strongly connected, then we know that every edge belongs to a circuit.) Let Γ be any cycle in G. We proceed by induction on the number of edges in G. If G has a single edge, clearly $\Gamma = G$ and we are done. If G has no loops and G has two edges, again $\Gamma = G$ and we are done. If G has no loops and no parallel edges and if G has three edges, then again, $\Gamma = G$. Now, consider the induction step. Assume $\Gamma \neq G$ and consider the graph $G' = (V, E - \Gamma)$. Let $G_1, \ldots, G_p$ be the connected components of G'. Pick any connected component G_i of G'. Now, all nodes in G_i have even degree, G_i is connected and G_i has strictly fewer edges than G so, by the induction hypothesis, G_i contains a Euler cycle Γ_i. But then Γ and each Γ_i share some vertex (because G is connected and the G_i are maximal connected components) and we can combine Γ and the Γ_is to form a Euler cycle in G. $\square$

There are iterative algorithms that will find a Euler cycle if one exists. It should also be noted that testing whether a graph has a Euler cycle is computationally quite an easy problem. This is not so for the Hamiltonian cycle problem described next.

A game invented by Sir William Hamilton in 1859 uses a regular solid dodecahedron whose 20 vertices are labeled with the names of famous cities. The player is challenged to "travel around the world" by finding a circuit along the edges of the dodecahedron that passes through every city exactly once.

In graphical terms, assuming an orientation of the edges between cities, the graph D shown in Figure 6.14 is a plane projection of a regular dodecahedron and we want to know if there is a Hamiltonian cycle in this directed graph (this is a directed version of the problem).

Finding a Hamiltonian cycle in this graph does not appear to be so easy. A solution is shown in Figure 6.15 below.

Definition 6.11. Given any undirected graph G (respectively, directed graph G) a *Hamiltonian cycle* in G (respectively, *Hamiltonian circuit* in G) is a cycle that passes

Fig. 6.12 William Hamilton, 1805–1865

Fig. 6.13 A Voyage Round the World Game and Icosian Game (Hamilton)

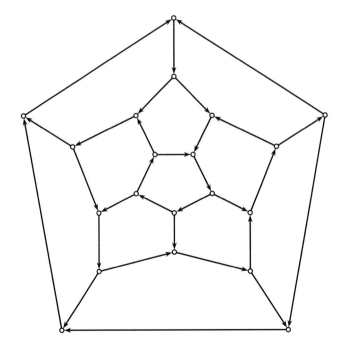

Fig. 6.14 A tour "around the world"

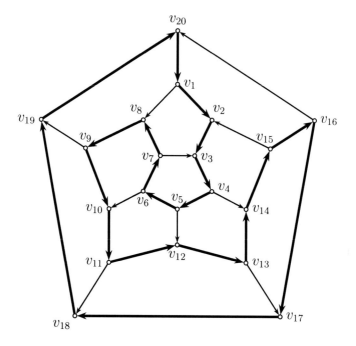

Fig. 6.15 A Hamiltonian cycle in D

though every vertex of G exactly once (respectively, a circuit that passes though every vertex of G exactly once). The *Hamiltonian cycle (respectively, circuit) problem* is to decide whether a graph G has a Hamiltonian cycle (respectively, Hamiltonian circuit).

Unfortunately, no theorem analogous to Theorem 6.6 is known for Hamiltonian cycles. In fact, the Hamiltonian cycle problem is known to be NP-complete and so far, appears to be a computationally hard problem (of exponential time complexity). Here is a proposition that may be used to prove that certain graphs are not Hamiltonian. However, there are graphs satisfying the condition of that proposition that are not Hamiltonian (e.g., *Petersen's graph*; see Problem 6.10).

Proposition 6.12. *If a graph $G = (V, E)$ possesses a Hamiltonian cycle then, for every nonempty set S of nodes, if $G\langle V - S\rangle$ is the induced subgraph of G generated by $V - S$ and if $c(G\langle V - S\rangle)$ is the number of connected components of $G\langle V - S\rangle$, then*

$$c(G\langle V - S\rangle) \leq |S|.$$

Proof. Let Γ be a Hamiltonian cycle in G and let $\widetilde{G}$ be the graph $\widetilde{G} = (V, \Gamma)$. If we delete k vertices we can't cut a cycle into more than k pieces and so

$$c(\widetilde{G}\langle V - S\rangle) \le |S|.$$

However, we also have

$$c(G\langle V - S\rangle) \le c(\widetilde{G}\langle V - S\rangle),$$

which proves the proposition. $\square$

6.4 Network Flow Problems; The Max-Flow Min-Cut Theorem

The network flow problem is a perfect example of a problem that is important practically but also theoretically because in both cases it has unexpected applications. In this section, we solve the network flow problem using some of the notions from Section 6.1. First, let us describe the kinds of graphs that we are dealing with, usually called *networks* (or *transportation networks* or *flow networks*).

Definition 6.12. A *network* (or *flow network*) is a quadruple $N = (G, c, v_s, s_t)$, where G is a finite digraph $G = (V, E, s, t)$ without loops, $c \colon E \to \mathbb{R}_+$ is a function called a *capacity function* assigning a *capacity* $c(e) > 0$ (or *cost* or *weight*) to every edge $e \in E$, and $v_s, v_t \in V$ are two (distinct) distinguished nodes.[3] Moreover, we assume that there are no edges coming into v_s ($d_G^-(v_s) = 0$), which is called the *source* and that there are no outgoing edges from v_t ($d_G^+(v_t) = 0$), which is called the *terminal* (or *sink*).

An example of a network is shown in Figure 6.16 with the capacity of each edge within parentheses.

Intuitively, we can think of the edges of a network as conduits for fluid, or wires for electricity, or highways for vehicle, and so on, and the capacity of each edge is the maximum amount of "flow" that can pass through that edge. The purpose of a network is to carry "flow", defined as follows.

Definition 6.13. Given a network $N = (G, c, v_s, v_t)$ a *flow in N* is a function $f \colon E \to \mathbb{R}$ such that the following conditions hold.

(1) (Conservation of flow)

$$\sum_{t(e)=v} f(e) = \sum_{s(e)=v} f(e), \quad \text{for all } v \in V - \{v_s, v_t\}$$

(2) (Admissibility of flow)

$$0 \le f(e) \le c(e), \quad \text{for all } e \in E$$

---

[3] Most books use the notation s and t for v_s and v_t. Sorry, s and t are already used in the definition of a digraph.

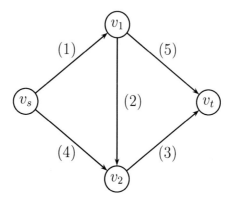

Fig. 6.16 A network N

Given any two sets of nodes $S, T \subseteq V$, let

$$f(S,T) = \sum_{\substack{e \in E \\ s(e) \in S, t(e) \in T}} f(e) \quad \text{and} \quad c(S,T) = \sum_{\substack{e \in E \\ s(e) \in S, t(e) \in T}} c(e).$$

When $S = \{u\}$ or $T = \{v\}$, we write $f(u,T)$ for $f(\{u\},T)$ and $f(S,v)$ for $f(S,\{v\})$ (similarly, we write $c(u,T)$ for $c(\{u\},T)$ and $c(S,v)$ for $c(S,\{v\})$). The *net flow out of S* is defined as $f(S,\overline{S}) - f(\overline{S},S)$ (where $\overline{S} = V - S$). The *value* $|f|$ *(or $v(f)$) of the flow f* is the quantity

$$|f| = f(v_s, V - \{v_s\}).$$

We can now state the following.

Network Flow Problem: Find a flow f in N for which the value $|f|$ is maximum (we call such a flow a *maximum flow*).

Figure 6.17 shows a flow in the network N, with value $|f| = 3$. This is not a maximum flow, as the reader should check (the maximum flow value is 4).

Remarks:

1. For any set of edges $\mathscr{E} \subseteq E$ let

$$f(\mathscr{E}) = \sum_{e \in \mathscr{S}} f(e)$$
$$c(\mathscr{E}) = \sum_{e \in \mathscr{S}} c(e).$$

Then, note that the net flow out of S can also be expressed as

$$f(\Omega^+(S)) - f(\Omega^-(S)) = f(S,\overline{S}) - f(\overline{S},S).$$

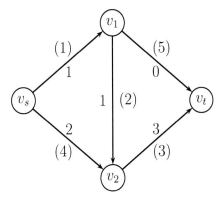

Fig. 6.17 A flow in the network N

Now, recall that $\Omega(S) = \Omega^+(S) \cup \Omega^-(S))$ is a cocycle (see Definition 6.3). So if we define the value $f(\Omega(S))$ of the cocycle $\Omega(S)$ to be

$$f(\Omega(S)) = f(\Omega^+(S)) - f(\Omega^-(S)),$$

the net flow through S is the value of the cocycle, $\Omega(S)$.

2. By definition, $c(S, \overline{S}) = c(\Omega^+(S))$.
3. Because G has no loops, there are no edges from u to itself, so

$$f(u, V - \{u\}) = f(u, V)$$

and similarly,

$$f(V - \{v\}, v) = f(V, v).$$

4. Some authors (e.g., Wilf [22]) do not require the distinguished node v_s to be a source and the distinguished node v_t to be a sink. This makes essentially no difference but if so, the value of the flow f must be defined as

$$|f| = f(v_s, V - \{v_s\}) - f(V - \{v_s\}, v_s) = f(v_s, V) - f(V, v_s).$$

Intuitively, because flow conservation holds for every node except v_s and v_t, the net flow $f(V, v_t)$ into the sink should be equal to the net flow $f(v_s, V)$ out of the source v_s. This is indeed true and follows from the next proposition.

Proposition 6.13. *Given a network $N = (G, c, v_s, v_t)$ for any flow f in N and for any subset $S \subseteq V$, if $v_s \in S$ and $v_t \notin S$, then the net flow through S has the same value, namely $|f|$; that is,*

$$|f| = f(\Omega(S)) = f(S, \overline{S}) - f(\overline{S}, S) \leq c(S, \overline{S}) = c(\Omega^+(S)).$$

In particular,

$$|f| = f(v_s, V) = f(V, v_t).$$

Proof. Recall that $|f| = f(v_s, V)$. Now, for any node $v \in S - \{v_s\}$, because $v \neq v_t$, the equation

$$\sum_{t(e)=v} f(e) = \sum_{s(e)=v} f(e)$$

holds and we see that

$$|f| = f(v_s, V) = \sum_{v \in S} (\sum_{s(e)=v} f(e) - \sum_{t(e)=v} f(e))$$

$$= \sum_{v \in S} (f(v, V) - f(V, v)) = f(S, V) - f(V, S).$$

However, $V = S \cup \overline{S}$, so

$$\begin{aligned}|f| &= f(S, V) - f(V, S) \\ &= f(S, S \cup \overline{S}) - f(S \cup \overline{S}, S) \\ &= f(S, S) + f(S, \overline{S}) - f(\overline{S}, S) - f(S, S) \\ &= f(S, \overline{S}) - f(\overline{S}, S),\end{aligned}$$

as claimed. The capacity of every edge is nonnegative, thus it is obvious that

$$|f| = f(S, \overline{S}) - f(\overline{S}, S) \leq f(S, \overline{S}) \leq c(S, \overline{S}) = c(\Omega^+(S)),$$

inasmuch as a flow is admissible. Finally, if we set $S = V - \{v_t\}$, we get

$$f(S, \overline{S}) - f(\overline{S}, S) = f(V, v_t)$$

and so, $|f| = f(v_s, V) = f(V, v_t)$. $\square$

Proposition 6.13 shows that the sets of edges $\Omega^+(S)$ with $v_s \in S$ and $v_t \notin S$, play a very special role. Indeed, as a corollary of Proposition 6.13, we see that the value of any flow in N is bounded by the capacity $c(\Omega^+(S))$ of the set $\Omega^+(S)$ for any S with $v_s \in S$ and $v_t \notin S$. This suggests the following definition.

Definition 6.14. Given a network $N = (G, c, v_s, v_t)$, a *cut separating* v_s *and* v_t, for short a v_s-v_t-*cut*, is any subset of edges $\mathscr{C} = \Omega^+(W)$, where W is a subset of V with $v_s \in W$ and $v_t \notin W$. The *capacity of a* v_s-v_t-*cut*, $\mathscr{C}$, is

$$c(\mathscr{C}) = c(\Omega^+(W)) = \sum_{e \in \Omega^+(W)} c(e).$$

Remark: Some authors, including Papadimitriou and Steiglitz [18] and Wilf [22], define a v_s-v_t-cut as a pair $(W, \overline{W})$, where W is a subset of V with with $v_s \in W$ and $v_t \notin W$. This definition is clearly equivalent to our definition above, which is due to Sakarovitch [21]. We have a slight preference for Definition 6.14 because it places the emphasis on edges as opposed to nodes. Indeed, the intuition behind v_s-v_t-cuts is that any flow from v_s to v_t must pass through some edge of any v_s-v_t-cut. Thus, it is not surprising that the capacity of v_s-v_t-cuts places a restriction on how much flow can be sent from v_s to v_t.

We can rephrase Proposition 6.13 as follows.

Proposition 6.14. *The maximum value of any flow f in the network N is bounded by the minimum capacity $c(\mathscr{C})$ of any v_s-v_t-cut $\mathscr{C}$ in N; that is,*

$$\max |f| \leq \min c(\mathscr{C}).$$

Proposition 6.14 is half of the so-called *max-flow min-cut theorem*. The other half of this theorem says that the above inequality is indeed an equality. That is, there is actually some v_s-v_t-cut $\mathscr{C}$ whose capacity $c(\mathscr{C})$ is the maximum value of the flow in N.

A v_s-v_t-cut of minimum capacity is called a *minimum v_s-v_t-cut*, for short, a *minimum cut*.

An example of a minimum cut is shown in Figure 6.18, where

$$\mathscr{C} = \Omega^+(\{v_s, v_2\}) = \{(v_s v_1), (v_2 v_t)\},$$

these two edges being shown as thicker lines. The capacity of this cut is 4 and a maximum flow is also shown in Figure 6.18.

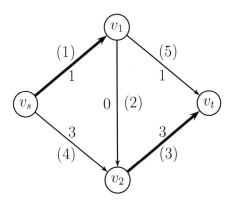

Fig. 6.18 A maximum flow and a minimum cut in the network N

What we intend to do next is to prove the celebrated "max-flow, min-cut theorem" (due to Ford and Fulkerson, 1957) and then to give an algorithm (also due to Ford and Fulkerson) for finding a maximum flow, provided some reasonable assumptions on the capacity function. In preparation for this, we present a handy trick (found both in Berge [1] and Sakarovitch [21]); the *return edge*.

Recall that one of the consequences of Proposition 6.13 is that the net flow out from v_s is equal to the net flow into v_t. Thus, if we add a new edge e_r called the *return edge* to G, obtaining the graph $\widetilde{G}$ (and the network $\widetilde{N}$), we see that any flow f in N satisfying condition (1) of Definition 6.13 yields a genuine flow $\widetilde{f}$ in $\widetilde{N}$ (a flow according to Definition 6.4, by Theorem 6.3), such that $f(e) = \widetilde{f}(e)$ for every edge of G and $\widetilde{f}(e_r) = |f|$. Consequently, the network flow problem is equivalent to finding a (genuine) flow in $\widetilde{N}$ such that $\widetilde{f}(e_r)$ is maximum. Another advantage of this formulation is that all the results on flows from Section 6.1 can be applied directly to $\widetilde{N}$. To simplify the notation, as $\widetilde{f}$ extends f, let us also use the notation f for $\widetilde{f}$. Now, if D is the incidence matrix of $\widetilde{G}$ (again, we use the simpler notation D instead of $\widetilde{D}$), we know that f is a flow iff

$$Df = 0.$$

Therefore, the network flow problem can be stated as a *linear programing problem* as follows:

$$\text{Maximize } z = f(e_r)$$

subject to the linear constraints

$$Df = 0$$
$$0 \leq f$$
$$f \leq c,$$

where we view f as a vector in $\mathbb{R}^{n+1}$, with $n = |E(G)|$.

Consequently, we obtain the existence of maximal flows, a fact that is not immediately obvious.

Proposition 6.15. *Given any network $N = (G, c, v_s, v_t)$, there is some flow f of maximum value.*

Proof. If we go back to the formulation of the max-flow problem as a linear program, we see that the set

$$C = \{x \in \mathbb{R}^{n+1} \mid 0 \leq x \leq c\} \cap \operatorname{Ker} D$$

is compact, as the intersection of a compact subset and a closed subset of $\mathbb{R}^{n+1}$ (in fact, C is also convex) and nonempty, as 0 (the zero vector) is a flow. But then, the projection $\pi : x \mapsto x(e_r)$ is a continuous function $\pi : C \to \mathbb{R}$ on a nonempty compact, so it achieves its maximum value for some $f \in C$. Such an f is a flow on $\widetilde{N}$ with maximal value. $\square$

Now that we know that maximum flows exist, it remains to prove that a maximal flow is realized by some minimal cut to complete the max-flow, min-cut theorem of Ford and Fulkerson. This can be done in various ways usually using some version of an algorithm due to Ford and Fulkerson. Such proofs can be found in Papadimitriou and Steiglitz [18], Wilf [22], Cameron [5], and Sakarovitch [21].

Fig. 6.19 Delbert Ray Fulkerson, 1924–1976

Sakarovitch makes the interesting observation (given as an exercise) that the arc coloring lemma due to Minty (Theorem 6.1) yields a simple proof of the part of the max-flow, min-cut theorem that we seek to establish. (See [21], Chapter 4, Exercise 1, page 105.) Therefore, we choose to present such a proof because it is rather original and quite elegant.

Theorem 6.7. *(Max-Flow, Min-Cut Theorem (Ford and Fulkerson)) For any net-work $N = (G, c, v_s, v_t)$, the maximum value $|f|$ of any flow f in N is equal to the minimum capacity $c(\mathscr{C})$ of any v_s-v_t-cut $\mathscr{C}$ in N.*

Proof. By Proposition 6.14, we already have half of our theorem. By Proposition 6.15, we know that some maximum flow, say f, exists. It remains to show that there is some v_s-v_t-cut $\mathscr{C}$ such that $|f| = c(\mathscr{C})$.

We proceed as follows.

Form the graph $\widetilde{G} = (V, E \cup \{e_r\}, s, t)$ from $G = (V, E, s, t)$, with $s(e_r) = v_t$ and $t(e_r) = v_s$. Then, form the graph, $\widehat{G} = (V, \widehat{E}, \widehat{s}, \widehat{t})$, whose edges are defined as follows.

(a) $e_r \in \widehat{E}$; $\widehat{s}(e_r) = s(e_r), \widehat{t}(e_r) = t(e_r)$.
(b) If $e \in E$ and $0 < f(e) < c(e)$, then $e \in \widehat{E}$; $\widehat{s}(e) = s(e), \widehat{t}(e) = t(e)$.
(c) If $e \in E$ and $f(e) = 0$, then $e \in \widehat{E}$; $\widehat{s}(e) = s(e), \widehat{t}(e) = t(e)$.
(d) If $e \in E$ and $f(e) = c(e)$, then $e \in \widehat{E}$, with $\widehat{s}(e) = t(e)$ and $\widehat{t}(e) = s(e)$.

In order to apply Minty's theorem, we color all edges constructed in (a), (c), and (d) in black and all edges constructed in (b) in red and we pick e_r as the distinguished edge. Now, apply Minty's lemma. We have two possibilities:

1. There is a simple cycle Γ in $\widehat{G}$, with all black edges oriented the same way. Because e_r is coming into v_s, the direction of the cycle is from v_s to v_t, so $e_r \in \Gamma^+$. This implies that all edges of type (d), $e \in \widehat{E}$, have an orientation consistent with the direction of the cycle. Now, Γ is also a cycle in $\widetilde{G}$ and, in $\widetilde{G}$, each edge $e \in E$ with $f(e) = c(e)$ is oriented in the inverse direction of the cycle; that is, $e \in \Gamma^-$ in $\widetilde{G}$. Also, all edges of type (c), $e \in \widehat{E}$, with $f(e) = 0$, are

oriented in the direction of the cycle; that is, $e \in \Gamma^+$ in $\widetilde{G}$. We also have $e_r \in \Gamma^+$ in $\widetilde{G}$.

We show that the value of the flow $|f|$ can be increased. Because $0 < f(e) < c(e)$ for every red edge, $f(e) = 0$ for every edge of type (c) in Γ^+, $f(e) = c(e)$ for every edge of type (d) in Γ^-, and because all capacities are strictly positive, if we let

$$\delta_1 = \min_{e \in \Gamma^+} \{c(e) - f(e)\}$$

$$\delta_2 = \min_{e \in \Gamma^-} \{f(e)\}$$

and

$$\delta = \min\{\delta_1, \delta_2\},$$

then $\delta > 0$. We can increase the flow f in $\widetilde{N}$, by adding δ to $f(e)$ for every edge $e \in \Gamma^+$ (including edges of type (c) for which $f(e) = 0$) and subtracting δ from $f(e)$ for every edge $e \in \Gamma^-$ (including edges of type (d) for which $f(e) = c(e)$) obtaining a flow f' such that

$$|f'| = f(e_r) + \delta = |f| + \delta > |f|,$$

as $e_r \in \Gamma^+$, contradicting the maximality of f. Therefore, we conclude that alternative (1) is impossible and we must have the second alternative.

2. There is a simple cocycle $\Omega_{\widehat{G}}(W)$ in $\widehat{G}$ with all edges black and oriented in the same direction (there are no green edges). Because $e_r \in \Omega_{\widehat{G}}(W)$, either $v_s \in W$ or $v_t \in W$ (but not both). In the second case ($v_t \in W$), we have $e_r \in \Omega_{\widehat{G}}^+(W)$ and $v_s \in \overline{W}$. Then, consider $\Omega_{\widehat{G}}^+(\overline{W}) = \Omega_{\widehat{G}}^-(W)$, with $v_s \in \overline{W}$. Thus, we are reduced to the case where $v_s \in W$.

If $v_s \in W$, then $e_r \in \Omega_{\widehat{G}}^-(W)$ and because all edges are black, $\Omega_{\widehat{G}}(W) = \Omega_{\widehat{G}}^-(W)$, in $\widehat{G}$. However, as every edge $e \in \widehat{E}$ of type (d) corresponds to an inverse edge $e \in E$, we see that $\Omega_{\widehat{G}}(W)$ defines a cocycle, $\Omega_{\widetilde{G}}(W) = \Omega_{\widetilde{G}}^+(W) \cup \Omega_{\widetilde{G}}^-(W)$, with

$$\Omega_{\widetilde{G}}^+(W) = \{e \in E \mid s(e) \in W\}$$

$$\Omega_{\widetilde{G}}^-(W) = \{e \in E \mid t(e) \in W\}.$$

Moreover, by construction, $f(e) = c(e)$ for all $e \in \Omega_{\widetilde{G}}^+(W)$, $f(e) = 0$ for all $e \in \Omega_{\widetilde{G}}^-(W) - \{e_r\}$, and $f(e_r) = |f|$. We say that the edges of the cocycle $\Omega_{\widetilde{G}}(W)$ are *saturated*. Consequently, $\mathscr{C} = \Omega_{\widetilde{G}}^+(W)$ is a v_s-v_t-cut in N with

$$c(\mathscr{C}) = f(e_r) = |f|,$$

establishing our theorem.   □

It is interesting that the proof in part (1) of Theorem 6.7 contains the main idea behind the algorithm of Ford and Fulkerson that we now describe.

The main idea is to look for a (simple) chain from v_s to v_t so that together with the return edge e_r we obtain a cycle Γ such that the edges in Γ satisfy the following properties:

(1) $\delta_1 = \min_{e \in \Gamma^+} \{c(e) - f(e)\} > 0$.
(2) $\delta_2 = \min_{e \in \Gamma^-} \{f(e)\} > 0$.

Such a chain is called a *flow augmenting chain*. Then, if we let $\delta = \min\{\delta_1, \delta_2\}$, we can increase the value of the flow by adding δ to $f(e)$ for every edge $e \in \Gamma^+$ (including the edge e_r, which belongs to Γ^+) and subtracting δ from $f(e)$ for all edges $e \in \Gamma^-$. This way, we get a new flow f' whose value is $|f'| = |f| + \delta$. Indeed, $f' = f + \delta\gamma(\Gamma)$, where $\gamma(\Gamma)$ is the vector (flow) associated with the cycle Γ. The algorithm goes through rounds each consisting of two phases. During phase 1, a flow augmenting chain is found by the procedure *findchain*; During phase 2, the flow along the edges of the augmenting chain is increased using the function *changeflow*.

During phase 1, the nodes of the augmenting chain are saved in the (set) variable Y, and the edges of this chain are saved in the (set) variable $\mathscr{E}$. We assign the special capacity value ∞ to e_r, with the convention that $\infty \pm \alpha = \alpha$ and that $\alpha < \infty$ for all $\alpha \in \mathbb{R}$.

procedure *findchain*(N: network; e_r: edge; Y: node set; $\mathscr{E}$: edge set; δ: real; f: flow)
  begin
    $\delta := \delta(v_s) := \infty$; $Y := \{v_s\}$;
    while $(v_t \notin Y) \wedge (\delta > 0)$ **do**
      if there is an edge e with $s(e) \in Y$, $t(e) \notin Y$ and $f(e) < c(e)$ **then**
        $Y := Y \cup \{t(e)\}$; $\mathscr{E}(t(e)) := e$; $\delta(t(e)) := \min\{\delta(s(e)), c(e) - f(e)\}$
      else
        if there is an edge e with $t(e) \in Y$, $s(e) \notin Y$ and $f(e) > 0$ **then**
          $Y := Y \cup \{s(e)\}$; $\mathscr{E}(s(e)) := e$; $\delta(s(e)) := \min\{\delta(t(e)), f(e)\}$
        else $\delta := 0$ (no new arc can be traversed)
        endif
      endif
    endwhile;
    if $v_t \in Y$ **then** $\delta := \delta(v_t)$ **endif**
  end

Here is the procedure to update the flow.

procedure *changeflow*(N: network; e_r: edge; $\mathscr{E}$: edge set; δ: real; f: flow)
  begin
    $u := v_t$; $f(e_r) := f(e_r) + \delta$;
    while $u \neq v_s$ **do** $e := \mathscr{E}(u)$;
      if $u = t(e)$ **then** $f(e) := f(e) + \delta$; $u := s(e)$;

>   **else** $f(e) := f(e) - \delta; u = t(e)$
>   **endif**
>   **endwhile**
> **end**

Finally, the algorithm *maxflow* is given below.

procedure *maxflow*(N: network; e_r: edge; Y: set of nodes; $\mathcal{E}$: set of edges; f: flow)
  begin
    for each $e \in E$ **do** $f(e) := 0$ **enfdor**;
    repeat until $\delta = 0$
      findchain$(N, e_r, Y, \mathcal{E}, \delta, f)$;
      if $\delta > 0$ **then**
        changeflow$(N, e_r, \mathcal{E}, \delta, f)$
      endif
    endrepeat
  end

The reader should run the algorithm *maxflow* on the network of Figure 6.16 to verify that the maximum flow shown in Figure 6.18 is indeed found, with $Y = \{v_s, v_2\}$ when the algorithm stops.

The correctness of the algorithm *maxflow* is easy to prove.

Theorem 6.8. *If the algorithm maxflow terminates and during the last round through findchain the node v_t is not marked, then the flow f returned by the algorithm is a maximum flow.*

Proof. Observe that if Y is the set of nodes returned when *maxflow* halts, then $v_s \in Y$, $v_t \notin Y$, and

1. If $e \in \Omega^+(Y)$, then $f(e) = c(e)$, as otherwise, procedure *findchain* would have added $t(e)$ to Y.
2. If $e \in \Omega^-(Y)$, then $f(e) = 0$, as otherwise, procedure *findchain* would have added $s(e)$ to Y.

But then, as in the end of the proof of Theorem 6.7, we see that the edges of the cocycle $\Omega(Y)$ are saturated and we know that $\Omega^+(Y)$ is a minimal cut and that $|f| = c(\Omega^+(Y))$ is maximal. $\square$

We still have to show that the algorithm terminates but there is a catch. Indeed, the version of the Ford and Fulkerson algorithm that we just presented may not terminate if the capacities are irrational. Moreover, in the limit, the flow found by the algorithm may not be maximum. An example of this bad behavior due to Ford and Fulkerson is reproduced in Wilf [22] (Chapter 3, Section 5). However, we can prove the following termination result which, for all practical purposes, is good enough, because only rational numbers can be stored by a computer.

Theorem 6.9. *Given a network N if all the capacities are multiples of some number* λ *then the algorithm, maxflow, always terminates. In particular, the algorithm maxflow always terminates if the capacities are rational (or integral).*

Proof. The number δ will always be a multiple of λ, so $f(e_r)$ will increase by at least λ during each iteration. Thus, eventually, the value of a minimal cut, which is a multiple of λ, will be reached. $\square$

If all the capacities are integers, an easy induction yields the following useful and nontrivial proposition.

Proposition 6.16. *Given a network N if all the capacities are integers, then the algorithm maxflow outputs a maximum flow* $f: E \to \mathbb{N}$ *such that the flow in every edge is an integer.*

Remark: Proposition 6.16 only asserts that some maximum flow is of the form $f: E \to \mathbb{N}$. In general, there is more than one maximum flow and other maximum flows may not have integer values on all edges.

Theorem 6.9 is good news but it is also bad news from the point of view of complexity. Indeed, the present version of the Ford and Fulkerson algorithm has a running time that depends on capacities and so, it can be very bad.

There are various ways of getting around this difficulty to find algorithms that do not depend on capacities and quite a few researchers have studied this problem. An excellent discussion of the progress in network flow algorithms can be found in Wilf [22] (Chapter 3).

A fairly simple modification of the Ford and Fulkerson algorithm consists in looking for flow augmenting chains of shortest length. To explain this algorithm we need the concept of *residual network*, which is a useful tool in any case. Given a network $N = (G, c, s, t)$ and given any flow f, the *residual network* $N_f = (G_f, c_f, v_f, v_t)$ is defined as follows.

1. $V_f = V$.
2. For every edge, $e \in E$, if $f(e) < c(e)$, then $e^+ \in E_f$, $s_f(e^+) = s(e)$, $t_f(e^+) = t(e)$ and $c_f(e^+) = c(e) - f(e)$; the edge e^+ is called a *forward edge*.
3. For every edge, $e \in E$, if $f(e) > 0$, then $e^- \in E_f$, $s_f(e^-) = t(e)$, $t_f(e^-) = s(e)$ and $c_f(e^-) = f(e)$; the edge e^- is called a *backward edge* because it has the inverse orientation of the original edge, $e \in E$.

The capacity $c_f(e^\varepsilon)$ of an edge $e^\varepsilon \in E_f$ (with $\varepsilon = \pm$) is usually called the *residual capacity* of e^ε. Observe that the same edge e in G, will give rise to two edges e^+ and e^- (with the same set of endpoints but with opposite orientations) in G_f if $0 < f(e) < c(e)$. Thus, G_f has at most twice as many edges as G. Also, note that every edge $e \in E$ which is *saturated* (i.e., for which $f(e) = c(e)$), does not survive in G_f.

Observe that there is a one-to-one correspondence between (simple) flow augmenting chains in the original graph G and (simple) flow augmenting paths in G_f.

Furthermore, in order to check that a simple path π from v_s to v_t in G_f is a flow augmenting path, all we have to do is to compute

$$c_f(\pi) = \min_{e^\varepsilon \in \pi}\{c_f(e^\varepsilon)\},$$

the *bottleneck* of the path π. Then, as before, we can update the flow f in N to get the new flow f' by setting

$$
\begin{aligned}
f'(e) &= f(e) + c_f(\pi), &&\text{if } e^+ \in \pi \\
f'(e) &= f(e) - c_f(\pi) &&\text{if } e^- \in \pi, \\
f'(e) &= f(e) &&\text{if } e \in E \text{ and } e^\varepsilon \notin \pi,
\end{aligned}
$$

for every edge $e \in E$. Note that the function $f_\pi : E \to \mathbb{R}$, defined by

$$
\begin{aligned}
f_\pi(e) &= c_f(\pi), &&\text{if } e^+ \in \pi \\
f_\pi(e) &= -c_f(\pi) &&\text{if } e^- \in \pi, \\
f_\pi(e) &= 0 &&\text{if } e \in E \text{ and } e^\varepsilon \notin \pi,
\end{aligned}
$$

is a flow in N with $|f_\pi| = c_f(\pi)$ and $f' = f + f_\pi$ is a flow in N, with $|f'| = |f| + c_f(\pi)$ (same reasoning as before). Now, we can repeat this process. Compute the new residual graph $N_{f'}$ from N and f', update the flow f' to get the new flow f'' in N, and so on.

The same reasoning as before shows that if we obtain a residual graph with no flow augmenting path from v_s to v_t, then a maximum flow has been found.

It should be noted that a poor choice of augmenting paths may cause the algorithm to perform a lot more steps than necessary. For example, if we consider the network shown in Figure 6.20, and if we pick the flow augmenting paths in the residual graphs to be alternatively (v_s, v_1, v_2, v_t) and (v_s, v_2, v_1, v_t), at each step, we only increase the flow by 1, so it will take 200 steps to find a maximum flow.

One of the main advantages of using residual graphs is that they make it convenient to look for better strategies for picking flow augmenting paths. For example, we can choose a simple flow augmenting path of shortest length (e.g., using breadth-first search). Then, it can be shown that this revised algorithm terminates in $O(|V| \cdot |E|)$ steps (see Cormen et al. [7], Section 26.2, and Sakarovitch [21], Chapter 4, Exercise 5). Edmonds and Karp designed an algorithm running in time $O(|E| \cdot |V|^2)$ based on this idea (1972), see [7], Section 26.2. Another way of selecting "good" augmenting paths, the *scaling max-flow algorithm*, is described in Kleinberg and Tardos [16] (see Section 7.3).

Here is an illustration of this faster algorithm, starting with the network N shown in Figure 6.16. The sequence of residual network construction and flow augmentation steps is shown in Figures 6.21–6.23. During the first two rounds, the augmented path chosen is shown in thicker lines. In the third and final round, there is no path from v_s to v_t in the residual graph, indicating that a maximum flow has been found.

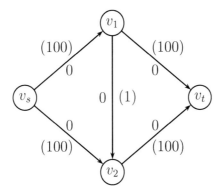

Fig. 6.20 A poor choice of augmenting paths yields a slow method

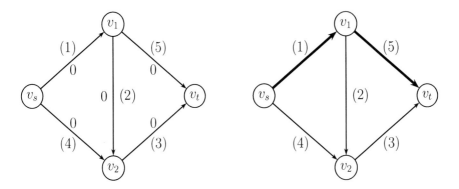

Fig. 6.21 Construction of the residual graph N_f from N, round 1

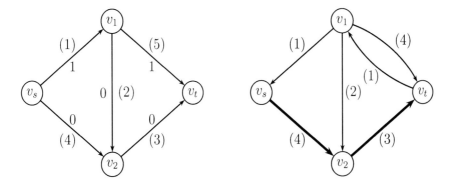

Fig. 6.22 Construction of the residual graph N_f from N, round 2

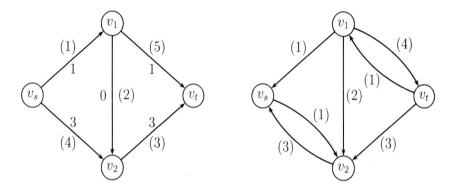

Fig. 6.23 Construction of the residual graph N_f from N, round 3

Another idea originally due to Dinic (1970) is to use *layered networks*; see Wilf [22] (Sections 3.6–3.7) and Papadimitriou and Steiglitz [18] (Chapter 9). An algorithm using layered networks running in time $O(V^3)$ is given in the two references above. There are yet other faster algorithms, for instance "preflow-push algorithms" also called "preflow-push relabel algorithms," originally due to Goldberg. A *preflow* is a function $f \colon E \to \mathbb{R}$ that satisfies Condition (2) of Definition 6.13 but which, instead of satisfying Condition (1), satisfies the inequality

$(1')$ (Nonnegativity of net flow)

$$\sum_{s(e)=v} f(e) \geq \sum_{t(e)=v} f(e) \quad \text{for all } v \in V - \{v_s, v_t\};$$

that is, the net flow out of v is nonnegative. Now, the principle of all methods using preflows is to augment a preflow until it becomes a maximum flow. In order to do this, a labeling algorithm assigning a *height* is used. Algorithms of this type are discussed in Cormen et al. [7], Sections 26.4 and 26.5 and in Kleinberg and Tardos [16], Section 7.4.

The max-flow, min-cut theorem (Theorem 6.7) is a surprisingly powerful theorem in the sense that it can be used to prove a number of other results whose original proof is sometimes quite hard. Among these results, let us mention the *maximum matching problem* in a bipartite graph, discussed in Wilf [22] (Sections 3.8), Cormen et al. [7] (Section 26.3), Kleinberg and Tardos [16] (Section 7.5), and Cameron [5] (Chapter 11, Section 10), finding the *edge connectivity* of a graph, discussed in Wilf [22] (Sections 3.8), and a beautiful *theorem of Menger* on edge-disjoint paths and *Hall's Marriage Theorem*, both discussed in Cameron [5] (Chapter 11, Section 10). More problems that can be solved effectively using flow algorithms, including image segmentation, are discussed in Sections 7.6–7.13 of Kleinberg and Tardos [16]. We only mention one of Menger's theorems, as it is particularly elegant.

Fig. 6.24 Karl Menger, 1902–1985

Theorem 6.10. *(Menger) Given any finite digraph G for any two nodes v_s and v_t, the maximum number of pairwise edge-disjoint paths from v_s to v_t is equal to the the minimum number of edges in a v_s-v_t-separating set. (A a v_s-v_t-separating set in G is a set of edges C such every path from v_s to v_t uses some edge in C.)*

It is also possible to generalize the basic flow problem in which our flows f have the property that $0 \leq f(e) \leq c(e)$ for every edge $e \in E$, to *channeled flows*. This generalization consists in adding another capacity function $b \colon E \to \mathbb{R}$, relaxing the condition that $c(e) > 0$ for all $e \in E$, and in allowing flows such that condition (2) of Definition 6.13 is replaced by the following.

(2′) (Admissibility of flow)

$$b(e) \leq f(e) \leq c(e), \quad \text{for all } e \in E$$

Now, the "flow" $f = 0$ is no longer necessarily admissible and the channeled flow problem does not always have a solution. However, it is possible to characterize when it has a solution.

Theorem 6.11. *(Hoffman) A network $N = (G, b, c, v_s, v_t)$ has a channeled flow iff for every cocycle $\Omega(Y)$ of G we have*

$$\sum_{e \in \Omega^-(Y)} b(e) \leq \sum_{e \in \Omega^+(Y)} c(e). \tag{†}$$

Observe that the necessity of the condition of Theorem 6.11 is an immediate consequence of Proposition 6.5. That it is sufficient can be proved by modifying the algorithm *maxflow* or its version using residual networks. The principle of this method is to start with a flow f in N that does not necessarily satisfy Condition (2′) and to gradually convert it to an admissible flow in N (if one exists) by applying the method for finding a maximum flow to a modified version $\widetilde{N}$ of N in which the capacities have been adjusted so that f is an admissible flow in $\widetilde{N}$. Now, if a flow f in N does not satisfy Condition (2′), then there are some *offending edges* e for which either $f(e) < b(e)$ or $f(e) > c(e)$. The new method makes sure that at the end of every (successful) round through the basic *maxflow* algorithm applied to the modified network $\widetilde{N}$ some offending edge of N is no longer offending.

Let f be a flow in N and assume that $\tilde{e}$ is an offending edge (i.e., either $f(e) < b(e)$ or $f(e) > c(e)$). Then, we construct the network $\tilde{N}(f, \tilde{e})$ as follows. The capacity functions $\tilde{b}$ and $\tilde{c}$ are given by

$$\tilde{b}(e) = \begin{cases} b(e) & \text{if } b(e) \le f(e) \\ f(e) & \text{if } f(e) < b(e) \end{cases}$$

and

$$\tilde{c}(e) = \begin{cases} c(e) & \text{if } f(e) \le c(e) \\ f(e) & \text{if } f(e) > c(e). \end{cases}$$

We also add one new edge $\tilde{e}_r$ to N whose endpoints and capacities are determined by:

1. If $f(\tilde{e}) > c(\tilde{e})$, then $s(\tilde{e}_r) = t(\tilde{e}), t(\tilde{e}_r) = s(\tilde{e}), \tilde{b}(\tilde{e}_r) = 0$ and $\tilde{c}(\tilde{e}_r) = f(\tilde{e}) - c(\tilde{e})$.
2. If $f(\tilde{e}) < b(\tilde{e})$, then $s(\tilde{e}_r) = s(\tilde{e}), t(\tilde{e}_r) = t(\tilde{e}), \tilde{b}(\tilde{e}_r) = 0$ and $\tilde{c}(\tilde{e}_r) = b(\tilde{e}) - f(\tilde{e})$.

Now, observe that the original flow f in N extended so that $f(\tilde{e}_r) = 0$ is a channeled flow in $\tilde{N}(f, \tilde{e})$ (i.e., Conditions (1) and (2') are satisfied). Starting from the new network $\tilde{N}(f, \tilde{e})$ apply the max-flow algorithm, say using residual graphs, with the following small change in 2.

1. For every edge $e \in \tilde{E}$, if $f(e) < \tilde{c}(e)$, then $e^+ \in \tilde{E}_f, s_f(e^+) = s(e), t_f(e^+) = t(e)$ and $c_f(e^+) = \tilde{c}(e) - f(e)$; the edge e^+ is called a *forward edge*.
2. For every edge $e \in \tilde{E}$, if $f(e) > \tilde{b}(e)$, then $e^- \in \tilde{E}_f, s_f(e^-) = t(e), t_f(e^-) = s(e)$ and $c_f(e^-) = f(e) - \tilde{b}(e)$; the edge e^- is called a *backward edge*.

Now, we consider augmenting paths from $t(\tilde{e}_r)$ to $s(\tilde{e}_r)$. For any such simple path π in $\tilde{N}(f, \tilde{e})_f$, as before we compute

$$c_f(\pi) = \min_{e^\varepsilon \in \pi}\{c_f(e^\varepsilon)\},$$

the *bottleneck* of the path π, and we say that π is a flow augmenting path iff $c_f(\pi) > 0$. Then, we can update the flow f in $\tilde{N}(f, \tilde{e})$ to get the new flow f' by setting

$$f'(e) = f(e) + c_f(\pi) \quad \text{if } e^- \in \pi,$$
$$f'(e) = f(e) - c_f(\pi) \quad \text{if } e^- \in \pi,$$
$$f'(e) = f(e) \qquad\qquad \text{if } e \in \tilde{E} \text{ and } e^\varepsilon \notin \pi,$$

for every edge $e \in \tilde{E}$.

We run the flow augmenting path procedure on $\tilde{N}(f, \tilde{e})$ and f until it terminates with a maximum flow $\hat{f}$. If we recall that the offending edge is $\tilde{e}$, then there are four cases:

1. $f(\tilde{e}) > c(\tilde{e})$.

    a. When the max-flow algorithm terminates, $\hat{f}(\tilde{e}_r) = \tilde{c}(\tilde{e}_r) = f(\tilde{e}) - c(\tilde{e})$. If so, define $\hat{f}$ as follows.

$$\widehat{f}(e) = \begin{cases} \widetilde{f}(\widetilde{e}) - \widetilde{f}(\widetilde{e}_r) & \text{if } e = \widetilde{e} \\ \widetilde{f}(e) & \text{if } e \neq \widetilde{e}. \end{cases} \quad (*)$$

It is clear that $\widehat{f}$ is a flow in N and $\widehat{f}(\widetilde{e}) = c(\widetilde{e})$ (there are no simple paths from $t(\widetilde{e})$ to $s(\widetilde{e})$). But then, $\widetilde{e}$ is not an offending edge for $\widehat{f}$, so we repeat the procedure of constructing the modified network, *etc.*

b. When the max-flow algorithm terminates, $\widetilde{f}(\widetilde{e}_r) < \widetilde{c}(\widetilde{e}_r)$. The flow $\widehat{f}$ defined in $(*)$ above, is still a flow but the max-flow algorithm must have terminated with a residual graph with no flow augmenting path from $s(\widetilde{e})$ to $t(\widetilde{e})$. Then, there is a set of nodes Y with $s(\widetilde{e}) \in Y$ and $t(\widetilde{e}) \notin Y$. Moreover, the way the max-flow algorithm is designed implies that

$$\widehat{f}(\widetilde{e}) > c(\widetilde{e})$$
$$\widehat{f}(e) = \widetilde{c}(e) \geq c(e) \quad \text{if } e \in \Omega^+(Y) - \{\widetilde{e}\}$$
$$\widehat{f}(e) = \widetilde{b}(e) \leq b(e) \quad \text{if } e \in \Omega^-(Y).$$

As $\widehat{f}$ also satisfies $(*)$ above, we conclude that the cocycle condition $(\dagger)$ of Theorem 6.11 fails for $\Omega(Y)$.

2. $f(\widetilde{e}) < b(\widetilde{e})$.

a. When the max-flow algorithm terminates, $\widetilde{f}(\widetilde{e}_r) = \widetilde{c}(\widetilde{e}_r) = b(\widetilde{e}) - f(\widetilde{e})$. If so, define $\widehat{f}$ as follows.

$$\widehat{f}(e) = \begin{cases} \widetilde{f}(\widetilde{e}) + \widetilde{f}(\widetilde{e}_r) & \text{if } e = \widetilde{e} \\ \widetilde{f}(e) & \text{if } e \neq \widetilde{e}. \end{cases} \quad (**)$$

It is clear that $\widehat{f}$ is a flow in N and $\widehat{f}(\widetilde{e}) = b(\widetilde{e})$ (there are no simple paths from $s(\widetilde{e})$ to $t(\widetilde{e})$). But then, $\widetilde{e}$ is not an offending edge for $\widehat{f}$, so we repeat the procedure of constructing the modified network, and so on.

b. When the max-flow algorithm terminates, $\widetilde{f}(\widetilde{e}_r) < \widetilde{c}(\widetilde{e}_r)$. The flow $\widehat{f}$ defined in $(**)$ above is still a flow but the max-flow algorithm must have terminated with a residual graph with no flow augmenting path from $t(\widetilde{e})$ to $s(\widetilde{e})$. Then, as in the case where $f(\widetilde{e}) > c(\widetilde{e})$, there is a set of nodes Y with $s(\widetilde{e}) \in Y$ and $t(\widetilde{e}) \notin Y$ and it is easy to show that the cocycle condition $(\dagger)$ of Theorem 6.11 fails for $\Omega(Y)$.

Therefore, if the algorithm does not fail during every round through the max-flow algorithm applied to the modified network $\widetilde{N}$, which, as we observed, is the case if Condition $(\dagger)$ holds, then a channeled flow $\widehat{f}$ will be produced and this flow will be a maximum flow. This proves the converse of Theorem 6.11.

The max-flow, min-cut theorem can also be generalized to channeled flows as follows.

Theorem 6.12. *For any network* $N = (G, b, c, v_s, v_t)$, *if a flow exists in* N, *then the maximum value* $|f|$ *of any flow* f *in* N *is equal to the minimum capacity* $c(\Omega(Y)) = c(\Omega^+(Y)) - b(\Omega^-(Y))$ *of any* v_s-v_t-*cocycle in* N *(this means that* $v_s \in Y$ *and* $v_r \notin Y$).

If the capacity functions b and c have the property that $b(e) < 0$ and $c(e) > 0$ for all $e \in E$, then the condition of Theorem 6.11 is trivially satisfied. Furthermore, in this case, the flow $f = 0$ is admissible, Proposition 6.15 holds, and we can apply directly the construction of the residual network N_f described above.

A variation of our last problem appears in Cormen et al. [7] (Chapter 26). In this version, the underlying graph G of the network N, is assumed to have no parallel edges (and no loops), so that every edge e can be identified with the pair (u, v) of its endpoints (so, $E \subseteq V \times V$). A flow f in N is a function $f: V \times V \to \mathbb{R}$, where it is not necessarily the case that $f(u, v) \geq 0$ for all (u, v), but there is a capacity function $c: V \times V \to \mathbb{R}$ such that $c(u, v) \geq 0$, for all $(u, v) \in V \times V$ and it is required that

$$f(v, u) = -f(u, v) \quad \text{and}$$
$$f(u, v) \leq c(u, v),$$

for all $(u, v) \in V \times V$. Moreover, in view of the skew symmetry condition $(f(v, u) = -f(u, v))$, the equations of conservation of flow are written as

$$\sum_{(u,v) \in E} f(u, v) = 0,$$

for all $u \neq v_s, v_t$.

We can reduce this last version of the flow problem to our previous setting by noticing that in view of skew symmetry, the capacity conditions are equivalent to having capacity functions b' and c', defined such that

$$b'(u, v) = -c(v, u)$$
$$c'(u, v) = c(u, v),$$

for every $(u, v) \in E$ and f must satisfy

$$b'(u, v) \leq f(u, v) \leq c'(u, v),$$

for all $(u, v) \in E$. However, we must also have $f(v, u) = -f(u, v)$, which is an additional constraint in case G has both edges (u, v) and (v, u). This point may be a little confusing because in our previous setting, $f(u, v)$ and $f(v, u)$ are independent values. However, this new problem is solved essentially as the previous one. The construction of the residual graph is identical to the previous case and so is the flow augmentation procedure along a simple path, *except that* we force $f_\pi(v, u) = f_\pi(u, v)$ to hold during this step. For details, the reader is referred to Cormen et al. [7], Chapter 26.

More could be said about flow problems but we believe that we have covered the basics satisfactorily and we refer the reader to the various references mentioned in this section for more on this topic.

6.5 Matchings, Coverings, Bipartite Graphs

In this section, we will deal with finite undirected graphs. Consider the following problem. We have a set of m machines, $M_1, \ldots, M_m$, and n tasks, $T_1, \ldots, T_n$. Furthermore, each machine M_i is capable of performing a subset of tasks $S_i \subseteq \{T_1, \ldots, T_n\}$. Then, the problem is to find a set of assignments $\{(M_{i_1}, T_{j_1}), \ldots, (M_{i_p}, T_{j_p})\}$, with $\{i_1, \ldots, i_p\} \subseteq \{1, \ldots, m\}$ and $\{j_1, \ldots, j_p\} \subseteq \{1, \ldots, n\}$, such that

(1) $T_{j_k} \in S_{i_k}, \quad 1 \le k \le p$.
(2) p is maximum.

The problem we just described is called a *maximum matching problem*. A convenient way to describe this problem is to build a graph G (undirected), with $m + n$ nodes partitioned into two subsets X and Y, with $X = \{x_1, \ldots, x_m\}$ and $Y = \{y_1, \ldots, y_n\}$, and with an edge between x_i and y_j iff $T_j \in S_i$, that is, if machine M_i can perform task T_j. Such a graph G is called a *bipartite graph*. An example of a bipartite graph is shown in Figure 6.25.

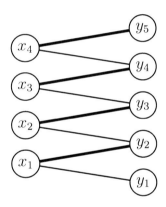

Fig. 6.25 A bipartite graph G and a maximum matching in G
.

Now, our matching problem is to find an edge set of maximum size M, such that no two edges share a common endpoint or, equivalently, such that every node belongs to at most one edge of M. Such a set of edges is called a *maximum matching* in G. A maximum matching whose edges are shown as thicker lines is shown in Figure 6.25.

Definition 6.15. A graph $G = (V, E, st)$ is a *bipartite graph* iff its set of edges V can be partitioned into two nonempty disjoint sets V_1, V_2, so that for every edge $e \in E$, $|st(e) \cap V_1| = |st(e) \cap V_2| = 1$; that is, one endpoint of e belongs to V_1 and the other belongs to V_2.

Note that in a bipartite graph, there are no edges linking nodes in V_1 (or nodes in V_2). Thus, there are no loops.

Remark: The *complete bipartite graph* for which $|V_1| = m$ and $|V_2| = n$ is the bipartite graph that has all edges (i, j), with $i \in \{1, \ldots, m\}$ and $j \in \{1, \ldots, n\}$. This graph is denoted $K_{m,n}$. The complete bipartite graph $K_{3,3}$ plays a special role; namely, it is not a planar graph, which means that it is impossible to draw it on a plane without avoiding that two edges (drawn as continuous simple curves) intersect. A picture of $K_{3,3}$ is shown in Figure 6.26.

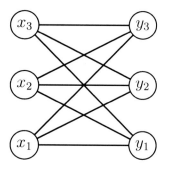

Fig. 6.26 The bipartite graph $K_{3,3}$

The maximum matching problem in a bipartite graph can be nicely solved using the methods of Section 6.4 for finding max-flows. Indeed, our matching problem is equivalent to finding a maximum flow in the network N constructed from the bipartite graph G as follows.

1. Add a new source v_s and a new sink v_t.
2. Add an oriented edge (v_s, u) for every $u \in V_1$.
3. Add an oriented edge (v, v_t) for every $v \in V_2$.
4. Orient every edge $e \in E$ from V_1 to V_2.
5. Define the capacity function c so that $c(e) = 1$, for every edge of this new graph.

The network corresponding to the bipartite graph of Figure 6.25 is shown in Figure 6.27.

Now, it is very easy to check that there is a matching M containing p edges iff there is a flow of value p. Thus, there is a one-to-one correspondence between maximum matchings and maximum integral flows. As we know that the algorithm *maxflow* (actually, its various versions) produces an integral solution when run on the zero flow, this solution yields a maximum matching.

The notion of graph coloring is also important and has bearing on the notion of bipartite graph.

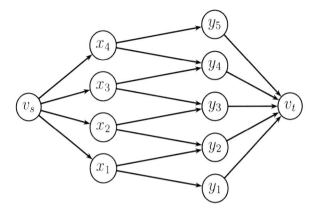

Fig. 6.27 The network associated with a bipartite graph

Definition 6.16. Given a graph $G = (V, E, st)$, a *k-coloring* of G is a partition of V into k pairwise disjoint nonempty subsets $V_1, \ldots, V_k$ so that no two vertices in any subset V_i are adjacent (i.e., the endpoints of every edge $e \in E$ must belong to V_i and V_j, for some $i \neq j$). If a graph G admits a k-coloring, we say that that G is k-*colorable*. The *chromatic number* $\gamma(G)$ (or $\chi(G)$) of a graph G is the minimum k for which G is k-colorable.

Remark: Although the notation $\chi(G)$ for the chromatic number of a graph is often used in the graph theory literature, it is an unfortunate choice because it can be confused with the Euler characteristic of a graph (see Theorem 6.20). We use the notation $\gamma(G)$. Other notations for the chromatic number include $v(G)$ and $\mathrm{chr}(G)$.

The following theorem gives some useful characterizations of bipartite graphs. First, we must define the incidence matrix of an unoriented graph G. Assume that G has edges $\mathbf{e}_1, \ldots, \mathbf{e}_n$ and vertices $\mathbf{v}_1, \ldots, \mathbf{v}_m$. The *incidence matrix A of G* is the $m \times n$ matrix whose entries are given by

$$a_{ij} = \begin{cases} 1 & \text{if } \mathbf{v}_i \in st(\mathbf{e}_j) \\ 0 & \text{otherwise.} \end{cases}$$

Note that, unlike the incidence matrix of a directed graph, the incidence matrix of an undirected graph only has nonnegative entries. As a consequence, these matrices are not necessarily totally unimodular. For example, the reader should check that for any simple cycle C of odd length, the incidence matrix A of C has a determinant whose value is ± 2. However, the next theorem shows that the incidence matrix of a bipartite graph is totally unimodular and in fact, this property characterizes bipartite graphs.

In order to prove part of the next theorem we need the notion of distance in a graph, an important concept in any case. If G is a connected graph, for any two nodes u and v of G, the length of a chain π from u to v is the number of edges in π

and the *distance* $d(u,v)$ from u to v is the minimum length of all path from u to v. Of course, $u = v$ iff $d(u,v) = 0$.

Theorem 6.13. *Given any graph $G = (V,E,st)$ the following properties are equivalent.*

(1) G is bipartite.
(2) $\gamma(G) = 2$.
(3) G has no simple cycle of odd length.
(4) G has no cycle of odd length.
(5) The incidence matrix of G is totally unimodular.

Proof. The equivalence $(1) \Longleftrightarrow (2)$ is clear by definition of the chromatic number.

$(3) \Longleftrightarrow (4)$ holds because every cycle is the concatenation of simple cycles. So, a cycle of odd length must contain some simple cycle of odd length.

$(1) \Longrightarrow (4)$. This is because the vertices of a cycle belong alternatively to V_1 and V_2. So, there must be an even number of them.

$(4) \Longrightarrow (2)$. Clearly, a graph is k-colorable iff all its connected components are k-colorable, so we may assume that G is connected. Pick any node v_0 in G and let V_1 be the subset of nodes whose distance from v_0 is even and V_2 be the subset of nodes whose distance from v_0 is odd. We claim that any two nodes u and v in V_1 (respectively, V_2) are not adjacent. Otherwise, by going up the chains from u and v back to v_0 and by adding the edge from u to v, we would obtain a cycle of odd length, a contradiction. Therefore, G, is 2-colorable.

$(1) \Longrightarrow (5)$. Orient the edges of G so that for every $e \in E$, $s(e) \in V_1$ and $t(e) \in V_2$. Then, we know from Proposition 6.10 that the incidence matrix D of the oriented graph G is totally unimodular. However, because G is bipartite, D is obtained from A by multiplying all the rows corresponding to nodes in V_2 by -1 and so, A is also totally unimodular.

$(5) \Longrightarrow (3)$. Let us prove the contrapositive. If G has a simple cycle C of odd length, then we observe that the submatrix of A corresponding to C has determinant ± 2. $\square$

We now define the general notion of a matching.

Definition 6.17. Given a graph $G = (V,E,st)$ a *matching* M in G is a subset of edges so that any two distinct edges in M have no common endpoint (are not adjacent) or equivalently, so that every vertex $v \in E$ is incident to at most one edge in M. A vertex $v \in V$ is *matched* iff it is incident to some edge in M and otherwise it is said to be *unmatched*. A matching M is a *perfect matching* iff every node is matched.

An example of a perfect matching $M = \{(ab),(cd),(ef)\}$ is shown in Figure 6.28 with the edges of the matching indicated in thicker lines. The pair $\{(bc),(ed)\}$ is also a matching, in fact, a maximal matching (no edge can be added to this matching and still have a matching).

It is possible to characterize maximum matchings in terms of certain types of chains called *alternating chains* defined below.

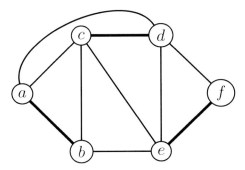

Fig. 6.28 A perfect matching in a graph

Definition 6.18. Given a graph $G = (V, E, st)$ and a matching M in G, a simple chain is an *alternating chain* w.r.t. M iff the edges in this chain belong alternately to M and $E - M$.

Theorem 6.14. *(Berge) Given any graph* $G = (V, E, st)$ *a matching* M *in* G *is a maximum matching iff there are no alternating chains w.r.t.* M *whose endpoints are unmatched.*

Proof. First, assume that M is a maximum matching and that C is an alternating chain w.r.t. M whose enpoints u and v are unmatched. As an example, consider the alternating chain shown in Figure 6.29, where the edges in $C \cap M$ are indicated in thicker lines.

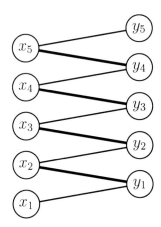

Fig. 6.29 An alternating chain in G

We can form the set of edges

$$M' = (M - (C \cap M)) \cup (C \cap (E - M)),$$

which consists in deleting the edges in M from C and adding the edges from C not in M. It is immediately verified that M' is still a matching but $|M'| = |M| + 1$ (see Figure 6.29), contradicting the fact that M is a maximum matching. Therefore, there are no alternating chains w.r.t. M whose endpoints are unmatched.

Conversely, assume that G has no alternating chains w.r.t. M whose endpoints are unmatched and let M' be another matching with $|M'| > |M|$ (i.e., M is not a maximum matching). Consider the spanning subgraph H of G, whose set of edges is

$$(M - M') \cup (M' - M).$$

As M and M' are matchings, the connected components of H are either isolated vertices, or simple cycles of even length, or simple chains, and in these last two cases, the edges in these cycles or chains belong alternately to M and M'; this is because $d_H(u) \leq 2$ for every vertex $u \in V$ and if $d_H(u) = 2$, then u is adjacent to one edge in M and one edge in M'.

Now, H must possess a connected component that is a chain C whose endpoints are in M', as otherwise we would have $|M'| \leq |M|$, contradicting the assumption $|M'| > |M|$. However, C is an alternating chain w.r.t. M whose endpoints are unmatched, a contradiction.   □

A notion closely related to the concept of a matching but, in some sense, dual, is the notion of a *line cover*.

Definition 6.19. Given any graph $G = (V, E, st)$ without loops or isolated vertices, a *line cover* (or *line covering*) of G is a set of edges $\mathscr{C} \subseteq E$ so that every vertex $u \in V$ is incident to some edge in $\mathscr{C}$. A *minimum line cover* $\mathscr{C}$ is a line cover of minimum size.

The maximum matching M in the graph of Figure 6.28 is also a minimum line cover. The set $\{(ab), (bc), (de), (ef)\}$ is also a line cover but it is not minimum, although minimal. The relationship between maximum matchings and minimum line covers is given by the following theorem.

Theorem 6.15. *Given any graph $G = (V, E, st)$ without loops or isolated vertices, with $|V| = n$, let M be a maximum matching and let $\mathscr{C}$ be a minimum line cover. Then, the following properties hold.*

(1) If we associate with every unmatched vertex of V some edge incident to this vertex and add all such edges to M, then we obtain a minimum line cover, $\mathscr{C}_M$.
(2) Every maximum matching M' of the spanning subgraph $(V, \mathscr{C})$ is a maximum matching of G.
(3) $|M| + |\mathscr{C}| = n$.

Proof. It is clear that $\mathscr{C}_M$ is a line cover. As the number of vertices unmatched by M is $n - 2|M|$ (as each edge in M matches exactly two vertices), we have

$$|\mathscr{C}_M| = |M| + n - 2|M| = n - |M|. \qquad (*)$$

Furthermore, as $\mathscr{C}$ is a minimum line cover, the spanning subgraph $(V, \mathscr{C})$ does not contain any cycle or chain of length greater than or equal to 2. Consequently, each edge $e \in \mathscr{C} - M'$ corresponds to a single vertex unmatched by M'. Thus,

$$|\mathscr{C}| - |M'| = n - 2|M'|;$$

that is,

$$|\mathscr{C}| = n - |M'|. \qquad (**)$$

As M is a maximum matching of G,

$$|M'| \leq |M|$$

and so, using $(*)$ and $(**)$, we get

$$|\mathscr{C}_M| = n - |M| \leq n - |M'| = |\mathscr{C}|;$$

that is, $|\mathscr{C}_M| \leq |\mathscr{C}|$. However, $\mathscr{C}$ is a minimum matching, so $|\mathscr{C}| \leq |\mathscr{C}_M|$, which proves that

$$|\mathscr{C}| = |\mathscr{C}_M|.$$

The last equation proves the remaining claims. $\square$

There are also notions analogous to matchings and line covers but applying to vertices instead of edges.

Definition 6.20. Let $G = (V, E, st)$ be any graph. A set $U \subseteq V$ of nodes is *independent* (or *stable*) iff no two nodes in U are adjacent (there is no edge having these nodes as endpoints). A *maximum independent set* is an independent set of maximum size. A set $\mathscr{U} \subseteq V$ of nodes is a *point cover* (or *vertex cover* or *transversal*) iff every edge of E is incident to some node in $\mathscr{U}$. A *minimum point cover* is a point cover of minimum size.

For example, $\{a, b, c, d, f\}$ is a point cover of the graph of Figure 6.28. The following simple proposition holds.

Proposition 6.17. *Let $G = (V, E, st)$ be any graph, U be any independent set, $\mathscr{C}$ be any line cover, $\mathscr{U}$ be any point cover, and M be any matching. Then, we have the following inequalities.*

(1) $|U| \leq |\mathscr{C}|$.
(2) $|M| \leq |\mathscr{U}|$.
(3) U is an independent set of nodes iff $V - U$ is a point cover.

Proof. (1) Because U is an independent set of nodes, every edge in $\mathscr{C}$ is incident with at most one vertex in U, so $|U| \leq |\mathscr{C}|$.

(2) Because M is a matching, every vertex in $\mathscr{U}$ is incident to at most one edge in M, so $|M| \leq |\mathscr{U}|$.

(3) Clear from the definitions. $\square$

It should be noted that the inequalities of Proposition 6.17 can be strict. For example, if G is a simple cycle with $2k + 1$ edges, the reader should check that both inequalities are strict.

We now go back to bipartite graphs and give an algorithm which, given a bipartite graph $G = (V_1 \cup V_2, E)$, will decide whether a matching M is a maximum matching in G. This algorithm, shown in Figure 6.30, will mark the nodes with one of the three tags, $+$, $-$, or 0.

procedure marking$(G, M, mark)$
  begin
    for each $u \in V_1 \cup V_2$ **do** $mark(u) := 0$ **endfor**;
    while $\exists u \in V_1 \cup V_2$ with $mark(u) = 0$ and u not matched by M **do**
      $mark(u) := +$;
      while $\exists v \in V_1 \cup V_2$ with $mark(v) = 0$ and v adjacent to w with $mark(w) = +$ **do**
        $mark(v) := -$;
        if v is not matched by M **then exit** (α)
        ($*$ an alternating chain has been found $*$)
        else find $w \in V_1 \cup V_2$ so that $(vw) \in M$; $mark(w) := +$
        endif
      endwhile
    endwhile;
    for each $u \in V_1$ with $mark(u) = 0$ **do** $mark(u) := +$ **endfor**;
    for each $u \in V_2$ with $mark(u) = 0$ **do** $mark(u) := -$ **endfor** (β)
  end

Fig. 6.30 Procedure *marking*

The following theorem tells us the behavior of the procedure *marking*.

Theorem 6.16. *Given any bipartite graph as input, the procedure marking always terminates in one of the following two (mutually exclusive) situations.*

(a) The algorithm finds an alternating chain w.r.t. M whose endpoints are unmatched.

(b) The algorithm finds a point cover $\mathscr{U}$ with $|\mathscr{U}| = |M|$, which shows that M is a maximum matching.

Proof. Nodes keep being marked, therefore the algorithm obviously terminates. There are no pairs of adjacent nodes both marked $+$ because, as soon as a node is marked $+$, all of its adjacent nodes are labeled $-$. Consequently, if the algorithm ends in (β), those nodes marked $-$ form a point cover.

We also claim that the endpoints u and v of any edge in the matching can't both be marked $-$. Otherwise, by following backward the chains that allowed the marking of u and v, we would find an odd cycle, which is impossible in a bipartite graph. Thus, if we end in (β), each node marked $-$ is incident to exactly one edge in M. This shows that the set $\mathcal{U}$ of nodes marked $-$ is a point cover with $|\mathcal{U}| = |M|$. By Proposition 6.17, we see that $\mathcal{U}$ is a minimum point cover and that M is a maximum matching.

If the algorithm ends in (α), by tracing the chain starting from the unmatched node u, marked $-$ back to the node marked $+$ causing u to be marked, and so on, we find an alternating chain w.r.t. M whose endpoints are not matched. $\square$

The following important corollaries follow immediately from Theorem 6.16.

Corollary 6.4. *In a bipartite graph, the size of a minimum point cover is equal to the size of maximum matching.*

Corollary 6.5. *In a bipartite graph, the size of a maximum independent set is equal to the size of a minimum line cover.*

Proof. We know from Proposition 6.17 that the complement of a point cover is an independent set. Consequently, by Corollary 6.4, the size of a maximum independent set is $n - |M|$, where M is a maximum matching and n is the number of vertices in G. Now, from Theorem 6.15 (3), for any maximum matching M and any minimal line cover $\mathcal{C}$ we have $|M| + |\mathcal{C}| = n$ and so, the size of a maximum independent set is equal to the size of a minimal line cover. $\square$

We can derive more classical theorems from the above results.
Given any graph $G = (V, E, st)$ for any subset of nodes $U \subseteq V$, let

$$N_G(U) = \{v \in V - U \mid (\exists u \in U)(\exists e \in E)(st(e) = \{u, v\})\},$$

be the set of *neighbours* of U, that is, the set of vertices *not* in U and adjacent to vertices in U.

Theorem 6.17. *(König (1931)) For any bipartite graph $G = (V_1 \cup V_2, E, st)$ the maximum size of a matching is given by*

$$\min_{U \subseteq V_1} (|V_1 - U| + |N_G(U)|).$$

Proof. This theorem follows from Corollary 6.4 if we can show that every minimum point cover is of the form $(V_1 - U) \cup N_G(U)$, for some subset U of V_1. However, a moment of reflection shows that this is indeed the case. $\square$

Theorem 6.17 implies another classical result:

Theorem 6.18. *(König–Hall) For any bipartite graph $G = (V_1 \cup V_2, E, st)$ there is a matching M such that all nodes in V_1 are matched iff*

$$|N_G(U)| \geq |U| \quad \text{for all} \quad U \subseteq V_1.$$

Proof. By Theorem 6.17, there is a matching M in G with $|M| = |V_1|$ iff

$$|V_1| = \min_{U \subseteq V_1} (|V_1 - U| + |N_G(U)|) = \min_{U \subseteq V_1} (|V_1| + |N_G(U)| - |U|),$$

that is, iff $|N_G(U)| - |U| \geq 0$ for all $U \subseteq V_1$. $\square$

Now, it is clear that a bipartite graph has a perfect matching (i.e., a matching such that every vertex is matched, M, iff $|V_1| = |V_2|$ and M matches all nodes in V_1. So, as a corollary of Theorem 6.18, we see that a bipartite graph has a perfect matching iff $|V_1| = |V_2|$ and if

$$|N_G(U)| \geq |U| \quad \text{for all} \quad U \subseteq V_1.$$

As an exercise, the reader should show the following.

Marriage Theorem (Hall, 1935) *Every k-regular bipartite graph with $k \geq 1$ has a perfect matching* (a graph is k-regular iff every node has degree k).

For more on bipartite graphs, matchings, covers, and the like, the reader should consult Diestel [9] (Chapter 2), Berge [1] (Chapter 7), and also Harary [15] and Bollobas [4].

6.6 Planar Graphs

Suppose we have a graph G and that we want to draw it "nicely" on a piece of paper, which means that we draw the vertices as points and the edges as line segments joining some of these points, in such a way that *no two edges cross each other*, except possibly at common endpoints. We have more flexibility and still have a nice picture if we allow each abstract edge to be represented by a continuous simple curve (a curve that has no self-intersection), that is, a subset of the plane homeomorphic to the closed interval $[0, 1]$ (in the case of a loop, a subset homeomorphic to the circle, S^1). If a graph can be drawn in such a fashion, it is called a *planar graph*. For example, consider the graph depicted in Figure 6.31.

If we look at Figure 6.31, we may believe that the graph G is not planar, but this is not so. In fact, by moving the vertices in the plane and by continuously deforming some of the edges, we can obtain a planar drawing of the same graph, as shown in Figure 6.32.

However, we should not be overly optimistic. Indeed, if we add an edge from node 5 to node 4, obtaining the graph known as K_5 shown in Figure 6.33, it can be proved that there is no way to move the nodes around and deform the edge continuously to obtain a planar graph (we prove this a little later using the Euler formula). Another graph that is nonplanar is the bipartite grapk $K_{3,3}$. The two graphs, K_5 and $K_{3,3}$ play a special role with respect to planarity. Indeed, a famous theorem of Kuratowski says that a graph is planar if and only if it does not contain K_5 or $K_{3,3}$ as a minor (we explain later what a minor is).

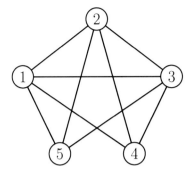

Fig. 6.31 A graph G drawn with intersecting edges

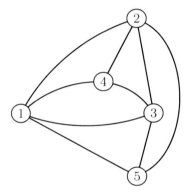

Fig. 6.32 The graph G drawn as a plane graph

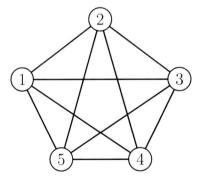

Fig. 6.33 The complete graph K_5, a nonplanar graph

Remark: Given n vertices, say $\{1,\ldots,n\}$, the graph whose edges are all subsets $\{i,j\}$, with $i,j \in \{1,\ldots,n\}$ and $i \neq j$, is the *complete graph on n vertices* and is denoted by K_n (but Diestel uses the notation K^n).

In order to give a precise definition of a planar graph, let us review quickly some basic notions about curves. A *simple curve* (or *Jordan curve*) is any injective continuous function, $\gamma\colon [0,1] \to \mathbb{R}^2$. Because $[0,1]$ is compact and γ is continuous, it is well known that the inverse $f^{-1}\colon \gamma([0,1]) \to [0,1]$ of f is also continuous. So, γ is a homeomorphism between $[0,1]$ and its image $\gamma([0,1])$. With a slight abuse of language we also call the image $\gamma([0,1])$ of γ a simple curve. This image is a connected and compact subset of $\mathbb{R}^2$. The points $a = \gamma(0)$ and $b = \gamma(1)$ are called the *boundaries* or *endpoints* of γ (and $\gamma([0,1])$). The open subset $\gamma([0,1]) - \{\gamma(0),\gamma(1)\}$ is called the *interior* of $\gamma([0,1])$ and is denoted $\overset{\circ}{\gamma}$. A continuous function $\gamma\colon [0,1] \to \mathbb{R}^2$ such that $\gamma(0) = \gamma(1)$ and γ is injective on $[0,1)$ is called a *simple closed curve* or *simple loop* or *closed Jordan curve*. Again, by abuse of language, we call the image $\gamma([0,1])$ of γ a simple closed curve, and so on. Equivalently, if $S^1 = \{(x,y) \in \mathbb{R}^2 \mid x^2 + y^2 = 1\}$ is the unit circle in $\mathbb{R}^2$, a simple closed curve is any subset of $\mathbb{R}^2$ homeomorphic to S^1. In this case, we call $\gamma(0) = \gamma(1)$ the *boundary* or *base point* of γ. The open subset $\gamma([0,1]) - \{\gamma(0)\}$ is called the *interior* of $\gamma([0,1])$ and is also denoted $\overset{\circ}{\gamma}$.

Remark: The notions of simple curve and simple closed curve also make sense if we replace $\mathbb{R}^2$ by any topological space X, in particular, a surface (In this case, a simple (closed) curve is a continuous injective function $\gamma\colon [0,1] \to X$ etc.).

We can now define plane graphs as follows.

Definition 6.21. A *plane graph* is a pair $\mathcal{G} = (V,E)$, where V is a finite set of points in $\mathbb{R}^2$, E is a finite set of simple curves, and closed simple curves in $\mathbb{R}^2$, called *edges* and *loops*, respectively, and satisfying the following properties.

 (i) The endpoints of every edge in E are vertices in V and the base point of every loop is a vertex in V.
 (ii) The interior of every edge contains no vertex and the interiors of any two distinct edges are disjoint. Equivalently, every edge contains no vertex except for its boundaries (base point in the case of a loop) and any two distinct edges intersect only at common boundary points.

We say that G is a *simple plane graph* if it has no loops and if different edges have different sets of endpoints

Obviously, a plane graph $\mathcal{G} = (V,E)$ defines an "abstract graph" $G = (V,E,st)$ such that

(a) For every simple curve γ,

$$st(\gamma) = \{\gamma(0),\gamma(1)\}.$$

(b) For every simple closed curve γ,

$$st(\gamma) = \{\gamma(0)\}.$$

For simplicity of notation, we usually write $\mathscr{G}$ for both the plane graph and the abstract graph associated with $\mathscr{G}$.

Definition 6.22. Given an abstract graph G, we say that G is a planar graph iff there is some plane graph $\mathscr{G}$ and an isomorphism $\varphi\colon G \to \mathscr{G}$ between G and the abstract graph associated with $\mathscr{G}$. We call φ an *embedding of G in the plane* or a *planar embedding of G*.

Remarks:

1. If G is a *simple* planar graph, then by a theorem of Fary, G can be drawn as a plane graph in such a way that the edges are straight line segments (see Gross and Tucker [13], Section 1.6).
2. In view of the remark just before Definition 6.21, given any topological space X for instance, a surface, we can define a graph on X as a pair (V, E) where V is a finite set of points in X and E is a finite set of simple (closed) curves on X satisfying the conditions of Definition 6.21.
3. Recall the *stereographic projection (from the north pole)*, $\sigma_N\colon (S^2 - \{N\}) \to \mathbb{R}^2$, from the sphere, $S^2 = \{(x, y, z) \in \mathbb{R}^3 \mid x^2 + y^2 + z^2 = 1\}$ onto the equatorial plane, $z = 0$, with $N = (0, 0, 1)$ (the north pole), given by

$$\sigma_N(x, y, z) = \left(\frac{x}{1 - z}, \frac{y}{1 - z} \right).$$

   We know that σ_N is a homeomorphism, so if φ is a planar embedding of a graph G into the plane, then $\sigma_N^{-1} \circ \varphi$ is an embedding of G into the sphere. Conversely, if ψ is an embedding of G into the sphere, then $\sigma_N \circ \psi$ is a planar embedding of G. Therefore, a graph can be embedded in the plane iff it can be embedded in the sphere. One of the nice features of embedding in the sphere is that the sphere is compact (closed and bounded), so the faces (see below) of a graph embedded in the sphere are all bounded.
4. The ability to embed a graph in a surface other than the sphere broadens the class of graphs that can be drawn without pairs of intersecting edges (except at endpoints). For example, it is possible to embed K_5 and $K_{3,3}$ (which are known *not* to be planar) into a torus (try it). It can be shown that for every (finite) graph G there is some surface X such that G can be embedded in X. Intuitively, whenever two edges cross on a sphere, by lifting one of the two edges a little bit and adding a "handle" on which the lifted edge lies we can avoid the crossing. An excellent reference on the topic of graphs on surfaces is Gross and Tucker [13].

One of the new ingredients of plane graphs is that the notion of a face makes sense. Given any nonempty open subset Ω of the plane $\mathbb{R}^2$, we say that two points $a, b \in \Omega$ are *arcwise connected*[4] iff there is a simple curve γ such that $\gamma(0) = a$ and

---

[4] In topology, a space is connected iff it cannot be expressed as the union of two nonempty disjoint open subsets. For *open* subsets of $\mathbb{R}^n$, connectedness is equivalent to arc connectedness. So it is legitimate to use the term connected.

$\gamma(1) = b$. Being connected is an equivalence relation and the equivalence classes of Ω w.r.t. connectivity are called the *connected components* (or *regions*) of Ω. Each region is maximally connected and open. If R is any region of Ω and if we denote the closure of R (i.e., the smallest closed set containing R) by $\overline{R}$, then the set $\partial R = \overline{R} - R$ is also a closed set called the *boundary* (or *frontier*) of R.

Now, given a plane graph $\mathscr{G}$ if we let $|\mathscr{G}|$ be the the subset of $\mathbb{R}^2$ consisting of the union of all the vertices and edges of $\mathscr{G}$, then this is a closed set and its complement $\Omega = \mathbb{R}^2 - |\mathscr{G}|$ is an open subset of $\mathbb{R}^2$.

Definition 6.23. Given any plane graph $\mathscr{G}$ the regions of $\Omega = \mathbb{R}^2 - |\mathscr{G}|$ are called the *faces* of $\mathscr{G}$.

As expected, for every face F of $\mathscr{G}$, the boundary ∂F of F is the subset $|\mathscr{H}|$ associated with some subgraph $\mathscr{H}$ of $\mathscr{G}$. However, one should observe that the boundary of a face may be disconnected and may have several "holes". The reader should draw lots of planar graphs to understand this phenomenon. Also, because we are considering finite graphs, the set $|\mathscr{G}|$ is bounded and thus, every plane graph has exactly one unbounded face. Figure 6.34 shows a planar graph and its faces. Observe that there are five faces, where A is bounded by all the edges except the loop around E and the rightmost edge from 7 to 8, B is bounded by the triangle $(4, 5, 6)$ the outside face C is bounded by the two edges from 8 to 2, the loop around node 2, the two edges from 2 to 7, and the outer edge from 7 to 8, D is bounded by the two edges between 7 and 8, and E is bounded by the loop around node 2.

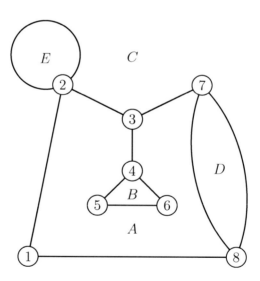

Fig. 6.34 A planar graph and its faces

Remarks:

1. Using (inverse) stereographic projection, we see that all the faces of a graph embedded in the sphere are bounded.
2. If a graph G is embedded in a surface S, then the notion of face still makes sense. Indeed, the faces of G are the regions of the open set $\Omega = S - |G|$.

Actually, one should be careful (as usual) not to rely too much on intuition when dealing with planar graphs. Although certain facts seem obvious, they may turn out to be false after closer scrutiny and when they are true, they may be quite hard to prove. One of the best examples of an "obvious" statement whose proof is much less trivial than one might expect is the Jordan curve theorem which is actually needed to justify certain "obvious" facts about faces of plane graphs.

Fig. 6.35 Camille Jordan, 1838–1922

Theorem 6.19. *(Jordan Curve Theorem) Given any closed simple curve γ in $\mathbb{R}$, the complement $\mathbb{R}^2 - \gamma([0,1])$, of $\gamma([0,1])$ consists of exactly two regions both having $\gamma([0,1])$ as boundary.*

Proof. There are several proofs all using machinery (such as homology or differential topology) beyond the scope of these notes. A proof using the notion of winding number is given in Guillemin and Pollack [14] (Chapter 2, Section 5) and another proof using homology can be found in Munkres [17] (Chapter 4, Section 36). $\quad\square$

Using Theorem 6.19, the following properties can be proved.

Proposition 6.18. *Let $\mathcal{G} = (V, E)$ be any plane graph and let $e \in E$ be any edge of $\mathcal{G}$. Then the following properties hold.*

(1) For any face F of $\mathcal{G}$, either $e \subseteq \partial F$ or $\partial F \cap \overset{\circ}{e} = \emptyset$.
(2) If e lies on a cycle C of $\mathcal{G}$, then e lies on the boundary of exactly two faces of G and these are contained in distinct faces of C.
(3) If e lies on no cycle, then e lies on the boundary of exactly one face of $\mathcal{G}$.

Proof. See Diestel [9], Section 4.2. $\quad\square$

As corollaries, we also have the following.

Proposition 6.19. *Let $\mathcal{G} = (V, E)$ be any plane graph and let F be any face of $\mathcal{G}$. Then, the boundary ∂F of F is a subgraph of $\mathcal{G}$ (more accurately, $\partial F = |\mathcal{H}|$, for some subgraph $\mathcal{H}$ of $\mathcal{G}$).*

Proposition 6.20. *Every plane forest has a single face.*

One of the main theorems about planar graphs is the so-called *Euler formula*.

Theorem 6.20. *(Euler's formula) Let G be any connected planar graph with n_0 vertices, n_1 edges, and n_2 faces. Then, we have*

$$n_0 - n_1 + n_2 = 2.$$

Proof. We proceed by induction on n_1. If $n_1 = 0$, the formula is trivially true, as $n_0 = n_2 = 1$. Assume the theorem holds for any $n_1 < n$ and let G be a connected planar graph with n edges. If G has no cycle, then as it is connected, it is a tree, $n_0 = n + 1$ and $n_2 = 1$, so $n_0 - n_1 + n_2 = n + 1 - n + 1 = 2$, as desired. Otherwise, let e be some edge of G belonging to a cycle. Consider the graph $G' = (V, E - \{e\})$; it is still a connected planar graph. Therefore, by the induction hypothesis,

$$n_0 - (n_1 - 1) + n_2' = 2.$$

However, by Proposition 6.18, as e lies on exactly two faces of G, we deduce that $n_2 = n_2' + 1$. Consequently

$$2 = n_0 - (n_1 - 1) + n_2' = n_0 - n_1 + 1 + n_2 - 1 = n_0 - n_1 + n_2,$$

establishing the induction hypothesis. □

Remarks:

1. Euler's formula was already known to Descartes in 1640 but the first proof by given by Euler in 1752. Poincaré generalized it to higher-dimensional polytopes.
2. The numbers n_0, n_1, and n_2 are often denoted by n_v, n_e, and n_f (v for *vertex*, e for *edge* and f for *face*).
3. The quantity $n_0 - n_1 + n_2$ is called the *Euler–Poincaré characteristic* of the graph G and it is usually denoted by χ_G.
4. If a connected graph G is embedded in a surface (orientable) S, then we still have an Euler formula of the form

$$n_0 - n_1 + n_2 = \chi(S) = 2 - 2g,$$

where $\chi(S)$ is a number depending only on the surface S, called the *Euler–Poincaré characteristic* of the surface and g is called the *genus* of the surface. It turns out that $g \geq 0$ is the number of "handles" that need to be glued to the surface of a sphere to get a homeomorphic copy of the surface S. For more on this fascinating subject, see Gross and Tucker [13].

Fig. 6.36 René Descartes, 1596–1650 (left) and Leonhard Euler, 1707–1783 (right)

It is really remarkable that the quantity $n_0 - n_1 + n_2$ is independent of the way a planar graph is drawn on a sphere (or in the plane). A neat application of Euler's formula is the proof that there are only five regular convex polyhedra (the so-called *platonic solids*). Such a proof can be found in many places, for instance, Berger [2] and Cromwell [8]. It is easy to generalize Euler's formula to planar graphs that are not necessarily connected.

Theorem 6.21. *Let G be any planar graph with n_0 vertices, n_1 edges, n_2 faces, and c connected components. Then, we have*

$$n_0 - n_1 + n_2 = c + 1.$$

Proof. Reduce the proof of Theorem 6.21 to the proof of Theorem 6.20 by adding vertices and edges between connected components to make G connected. Details are left as an exercise. □

Using the Euler formula we can now prove rigorously that K_5 and $K_{3,3}$ are not planar graphs. For this, we need the following fact.

Proposition 6.21. *If G is any simple, connected, plane graph with $n_1 \geq 3$ edges and n_2 faces, then*
$$2n_1 \geq 3n_2.$$

Proof. Let $F(G)$ be the set of faces of G. Because G is connected, by Proposition 6.18 (2), every edge belongs to exactly two faces. Thus, if s_F is the number of sides of a face F of G, we have

$$\sum_{F \in F(G)} s_F = 2n_1.$$

Furthermore, as G has no loops, no parallel edges, and $n_0 \geq 3$, every face has at least three sides; that is, $s_F \geq 3$. It follows that

$$2n_1 = \sum_{F \in F(G)} s_F \geq 3n_2,$$

as claimed. □

The proof of Proposition 6.21 shows that the crucial constant on the right-hand side of the inequality is the the minimum length of all cycles in G. This number is called the *girth* of the graph G. The girth of a graph with a loop is 1 and the girth of a graph with parallel edges is 2. The girth of a tree is undefined (or infinite). Therefore, we actually proved the next proposition.

Proposition 6.22. *If G is any connected plane graph with n_1 edges and n_2 faces and G is not a tree, then*

$$2n_1 \geq \text{girth}(G)\, n_2.$$

Corollary 6.6. *If G is any simple, connected, plane graph with $n \geq 3$ nodes then G has at most $3n - 6$ edges and $2n - 4$ faces.*

Proof. By Proposition 6.21, we have $2n_1 \geq 3n_2$, where n_1 is the number of edges and n_2 is the number of faces. So, $n_2 \leq \frac{2}{3}n_1$ and by Euler's formula

$$n - n_1 + n_2 = 2,$$

we get

$$n - n_1 + \frac{2}{3}n_1 \geq 2;$$

that is,

$$n - \frac{1}{3}n_1 \geq 2,$$

namely $n_1 \leq 3n - 6$. Using $n_2 \leq \frac{2}{3}n_1$, we get $n_2 \leq 2n - 4$.   □

Corollary 6.7. *The graphs K_5 and $K_{3,3}$ are not planar.*

Proof. We proceed by contradiction. First, consider K_5. We have $n_0 = 5$ and K_5 has $n_1 = 10$ edges. On the other hand, by Corollary 6.6, K_5 should have at most $3 \times 5 - 6 = 15 - 6 = 9$ edges, which is absurd.

Next, consider $K_{3,3}$. We have $n_0 = 6$ and $K_{3,3}$ has $n_1 = 9$ edges. By the Euler formula, we should have

$$n_2 = 9 - 6 + 2 = 5.$$

Now, as $K_{3,3}$ is bipartite, it does not contain any cycle of odd length, and so each face has at least *four* sides, which implies that

$$2n_1 \geq 4n_2$$

(because the girth of $K_{3,3}$ is 4.) So, we should have

$$18 = 2 \cdot 9 \geq 4 \cdot 5 = 20,$$

which is absurd.   □

Another important property of simple planar graph is the following.

Proposition 6.23. *If G is any simple planar graph, then there is a vertex u such that $d_G(u) \leq 5$.*

Proof. If the property holds for any connected component of G, then it holds for G, so we may assume that G is connected. We already know from Proposition 6.21 that $2n_1 \geq 3n_2$; that is,

$$n_2 \leq \frac{2}{3}n_1. \qquad (*)$$

If $d_G(u) \geq 6$ for every vertex u, as $\sum_{u \in V} d_G(u) = 2n_1$, then $6n_0 \leq 2n_1$; that is, $n_0 \leq n_1/3$. By Euler's formula, we would have

$$n_2 = n_1 - n_0 + 2 \geq n_1 - \frac{1}{3}n_1 + 2 > \frac{2}{3}n_1,$$

contradicting $(*)$.   □

Remarkably, Proposition 6.23 is the key ingredient in the proof that every planar graph is 5-colorable.

Theorem 6.22. *(5-Color Theorem) Every planar graph G is 5-colorable.*

Proof. Clearly, parallel edges and loops play no role in finding a coloring of the vertices of G, so we may assume that G is a simple graph. Also, the property is clear for graphs with less than 5 vertices. We proceed by induction on the number of vertices m. By Proposition 6.23, the graph G has some vertex u_0 with $d_G(u) \leq 5$. By the induction hypothesis, we can color the subgraph G' induced by $V - \{u_0\}$ with 5 colors. If $d(u_0) < 5$, we can color u_0 with one of the colors not used to color the nodes adjacent to u_0 (at most 4) and we are done. So, assume $d_G(u_0) = 5$ and let $v_1, \ldots, v_5$ be the nodes adjacent to u_0 and encountered in this order when we rotate counterclockwise around u_0 (see Figure 6.37). If $v_1, \ldots, v_5$ are not colored with different colors, again, we are done.

Otherwise, by the induction hypothesis, let $\{X_1, \ldots, X_5\}$ be a coloring of G' and, by renaming the X_is if necessary, assume that $v_i \in X_i$, for $i = 1, \ldots, 5$. There are two cases.

(1) There is no chain from v_1 to v_3 whose nodes belong alternately to X_1 and X_2. If so, v_1 and v_3 must belong to different connected components of the subgraph H' of G' induced by $X_1 \cup X_2$. Then, we can permute the colors 1 and 3 in the connected component of H' that contains v_3 and color u_0 with color 3.

(2) There is a chain from v_1 to v_3 whose nodes belong alternately to X_1 and X_2. In this case, as G is a planar graph, there can't be any chain from v_2 to v_4 whose nodes belong alternately to X_2 and X_4. So, v_2 and v_4 do not belong to the same connected component of the subgraph H'' of G' induced by $X_2 \cup X_4$. But then, we can permute the colors 2 and 4 in the connected component of H'' that contains v_4 and color u_0 with color 4.   □

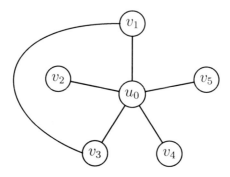

Fig. 6.37 The five nodes adjacent to u_0

Theorem 6.22 raises a very famous problem known as the *four-color problem*: Can every planar graph be colored with four colors?

This question was apparently first raised by Francis Guthrie in 1850, communicated to De Morgan by Guthrie's brother Frederick in 1852, and brought to the attention of a wider public by Cayley in 1878. In the next hundred years, several incorrect proofs were proposed and this problem became known as the *four-color conjecture*. Finally, in 1977, Appel and Haken gave the first "proof" of the four-color conjecture. However, this proof was somewhat controversial for various reasons, one of the reasons being that it relies on a computer program for checking a large number of unavoidable configurations. Appel and Haken subsequently published a 741-page paper correcting a number of errors and addressing various criticisms. More recently (1997) a much shorter proof, still relying on a computer program, but a lot easier to check (including the computer part of it) has been given by Robertson, Sanders, Seymour, and Thomas [19]. For more on the four-color problem, see Diestel [9], Chapter 5, and the references given there.

Let us now go back to Kuratowski's criterion for nonplanarity. For this it is useful to introduce the notion of edge contraction in a graph.

Definition 6.24. Let $G = (V, E, st)$ be any graph and let e be any edge of G. The graph obtained by *contracting the edge e into a new vertex v_e* is the graph $G/e = (V', E', st')$ with $V' = (V - st(e)) \cup \{v_e\}$, where v_e is a new node ($v_e \notin V$); $E' = E - \{e\}$; and with

$$st'(e') = \begin{cases} st(e') & \text{if } st(e') \cap st(e) = \emptyset \\ \{v_e\} & \text{if } st(e') = st(e) \\ \{u, v_e\} & \text{if } st(e') \cap st(e) = \{z\} \text{ and } st(e') = \{u, z\} \text{ with } u \neq z \\ \{v_e\} & \text{if } st(e') = \{x\} \text{ or } st(e') = \{y\} \text{ with } st(e) = \{x, y\}. \end{cases}$$

If G is a simple graph, then we need to eliminate parallel edges and loops. In, this case, $e = \{x, y\}$ and $G/e = (V', E', st)$ is defined so that $V' = (V - \{x, y\}) \cup \{v_e\}$, where v_e is a new node and

$$E' = \{\{u,v\} \mid \{u,v\} \cap \{x,y\} = \emptyset\}$$
$$\cup \{\{u,v_e\} \mid \{u,x\} \in E - \{e\} \quad \text{or} \quad \{u,y\} \in E - \{e\}\}.$$

Figure 6.38 shows the result of contracting the upper edge $\{2,4\}$ (shown as a thicker line) in the graph shown on the left, which is not a simple graph.

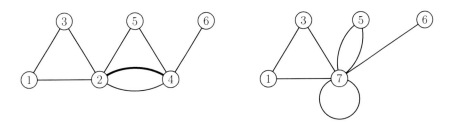

Fig. 6.38 Edge contraction in a graph

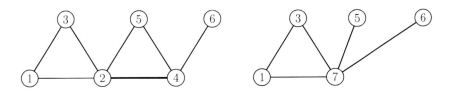

Fig. 6.39 Edge contraction in a simple graph

Observe how the lower edge $\{2,4\}$ becomes a loop around 7 and the two edges $\{5,2\}$ and $\{5,4\}$ become parallel edges between 5 and 7.

Figure 6.39 shows the result of contracting edge $\{2,4\}$ (shown as a thicker line) in the simple graph shown on the left. This time, the two edges $\{5,2\}$ and $\{5,4\}$ become a single edge and there is no loop around 7 as the contracted edge is deleted.

Now, given a graph G we can repeatedly contract edges. We can also take a subgraph of a graph G and then perform some edge contractions. We obtain what is known as a minor of G.

Definition 6.25. Given any graph G, a graph H is a *minor* of G if there is a sequence of graphs $H_0, H_1, \ldots, H_n$ $(n \geq 1)$, such that

(1) $H_0 = G$; $H_n = H$.
(2) Either H_{i+1} is obtained from H_i by deleting some edge or some node of H_i and all the edges incident with this node.
(3) Or H_{i+1} is obtained from H_i by edge contraction,

with $0 \leq i \leq n - 1$. If G is a simple graph, we require that edge contractions be of the second type described in Definition 6.24, so that H is a simple graph.

It is easily shown that the minor relation is a partial order on graphs (and simple graphs). Now, the following remarkable theorem originally due to Kuratowski characterizes planarity in terms of the notion of minor:

Fig. 6.40 Kazimierz Kuratowski, 1896–1980

Theorem 6.23. *(Kuratowski, 1930) For any graph G, the following assertions are equivalent.*

(1) G is planar.
(2) G contains neither K_5 nor $K_{3,3}$ as a minor.

Proof. The proof is quite involved. The first step is to prove the theorem for 3-connected graphs. (A graph, $G = (V,E)$, is *h-connected* iff $|V| > h$ and iff every graph obtained by deleting any set $S \subseteq V$ of nodes with $|S| < h$ and the edges incident to these node is still connected. So, a 1-connected graph is just a connected graph.) We refer the reader to Diestel [9], Section 4.4, for a complete proof. $\square$

Another way to state Kuratowski's theorem involves edge subdivision, an operation of independent interest. Given a graph $G = (V,E,st)$ possibly with loops and parallel edges, the result of subdividing an edge e consists in creating a new vertex v_e, deleting the edge e, and adding two new edges from v_e to the old endpoints of e (possibly the same point). Formally, we have the following definition.

Definition 6.26. Given any graph $G = (V,E,st)$ for any edge $e \in E$, the result of *subdividing the edge e* is the graph $G' = (V \cup \{v_e\}, (E - \{e\}) \cup \{e^1, e^2\}, st')$, where v_e is a new vertex and e^1, e^2 are new edges, $st'(e') = st(e')$ for all $e' \in E - \{e\}$ and if $st(e) = \{u,v\}$ ($u = v$ is possible), then $st'(e^1) = \{v_e, u\}$ and $st'(e^2) = \{v_e, v\}$. If a graph G' is obtained from a graph G by a sequence of edge subdivisions, we say that G' is a *subdivision* of G.

Observe that by repeatedly subdividing edges, any graph can be transformed into a simple graph. Given two graphs G and H, we say that G and H are *homeomorphic*

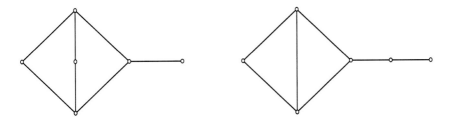

Fig. 6.41 Two homeomorphic graphs

iff they have respective subdivisions G' and H' that are isomorphic graphs. The idea is that homeomorphic graphs "look the same," viewed as topological spaces. Figure 6.41 shows an example of two homeomorphic graphs.

A graph H that has a subdivision H', which is a subgraph of some graph G, is called a *topological minor* of G. Then, it is not hard to show (see Diestel [9], Chapter 4, or Gross and Tucker [13], Chapter 1) that Kuratowski's theorem is equivalent to the statement

A graph G is planar iff it does not contain any subgraph homeomorphic to either K_5 or $K_{3,3}$ or, equivalently, if it has has neither K_5 nor $K_{3,3}$ as a topological minor.

Another somewhat surprising characterization of planarity involving the concept of cycle space over $\mathbb{F}_2$ (see Definition 6.4 and the Remarks after Theorem 6.2) and due to MacLane is the following.

Fig. 6.42 Saunders Mac Lane, 1909–2005

Theorem 6.24. *(MacLane, 1937) A graph G is planar iff its cycle space $\mathscr{F}$ over $\mathbb{F}_2$ has a basis such that every edge of G belongs to at most two cycles of this basis.*

Proof. See Diestel [9], Section 4.4. □

We conclude this section on planarity with a brief discussion of the dual graph of a plane graph, a notion originally due to Poincaré. Duality can be generalized to simplicial complexes and relates Voronoi diagrams and Delaunay triangulations, two very important tools in computational geometry.

Given a plane graph $G = (V, E)$, let $F(G)$ be the set of faces of G. The crucial point is that every edge of G is part of the boundary of at most two faces. A dual graph $G^* = (V^*, E^*)$ of G is a graph whose nodes are in one-to-one correspondence with the faces of G, whose faces are in one-to-one correspondence with the nodes of G, and whose edges are also in one-to-one correspondence with the the egdes of G. For any edge $e \in E$, a dual edge e^* links the two nodes v_{F_1} and v_{F_2} associated with the faces F_1 and F_2 adjacent to e or, e^* is a loop from v_F to itself if e is ajacent to a single face. Here is the precise definition.

Definition 6.27. Let $G = (V, E)$ be a plane graph and let $F(G)$ be its set of faces. A *dual graph* of G is a graph $G^* = (V^*, E^*)$, where

(1) $V^* = \{v_F \mid F \in F(G)\}$, where v_F is a point chosen in the (open) face, F, of G.
(2) $E^* = \{e^* \mid e \in E\}$, where e^* is a simple curve from v_{F_1} to v_{F_2} crossing e, if e is part of the boundary of two faces F_1 and F_2 or else, a closed simple curve crossing e from v_F to itself, if e is part of the boundary of exactly one face F.

(3) For each $e \in E$, we have $e^* \cap G = e \cap G^* = \overset{\circ}{e} \cap \overset{\circ}{e^*}$, a one-point set.

An example of a dual graph is shown in Figure 6.43. The graph G has four faces, a, b, c, d and the dual graph G^* has nodes also denoted a, b, c, d enclosed in a small circle, with the edges of the dual graph shown with thicker lines.

Note how the edge $\{5, 6\}$ gives rise to the loop from d to itself and that there are parallel edges between d and a and between d and c. Thus, even if we start with a simple graph, a dual graph may have loops and parallel edges.

Actually, it is not entirely obvious that a dual of a plane graph is a plane graph but this is not difficult to prove. It is also important to note that a given plane graph G *does not have a unique dual* because the vertices and the edges of a dual graph can be chosen in infinitely different ways in order to satisfy the conditions of Definition 6.27. However, given a plane graph G, if H_1 and H_2 are two dual graphs of G, then it is easy to see that H_1 and H_2 are isomorphic. Therefore, with a slight abuse of language, we may refer to "the" dual graph of a plane graph. Also observe that even if G is not connected, its dual G^* is always connected.

The notion of dual graph applies to a *plane* graph and *not to a planar graph*.

Indeed, the graphs G_1^* and G_2^* associated with two different embeddings G_1 and G_2 of the same abstract planar graph G may **not** be isomorphic, even though G_1 and G_2 are isomorphic as abstract graphs. For example, the two plane graphs G_1 and G_2 shown in Figure 6.44 are isomorphic but their dual graphs G_1^* and G_2^* are not, as the reader should check (one of these two graphs has a node of degree 7 but for the other graph all nodes have degree at most 6).

Remark: If a graph G is embedded in a surface S, then the notion of dual graph also makes sense. For more on this, see Gross and Tucker [13].

In the following proposition, we summarize some useful properties of dual graphs.

Proposition 6.24. *The dual G^* of any plane graph is connected. Furthermore, if G is a connected plane graph, then G^{**} is isomorphic to G.*

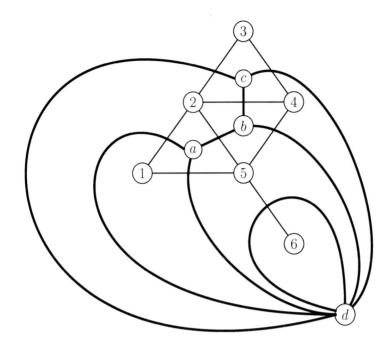

Fig. 6.43 A graph and its dual graph

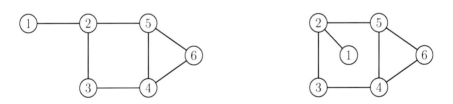

Fig. 6.44 Two isomorphic plane graphs whose dual graphs are not isomorphic

Proof. Left as an exercise. □

With a slight abuse of notation we often write $G^{**} = G$ (when G is connected). A plane graph G whose dual G^* is equal to G (i.e., isomorphic to G) is called *self-dual*. For example, the plane graph shown in Figure 6.45 (the projection of a tetrahedron on the plane) is self-dual.

The duality of plane graphs is also reflected algebraically as a duality between their cycle spaces and their cut spaces (over $\mathbb{F}_2$).

Proposition 6.25. *If G is any connected plane graph G, then the following properties hold.*

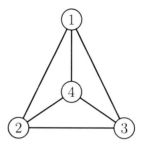

Fig. 6.45 A self-dual graph

(1) A set of edges $C \subseteq E$ is a cycle in G iff $C^* = \{e^* \in E^* \mid e \in C\}$ is a minimal cutset in G^*.

(2) If $\mathscr{F}(G)$ and $\mathscr{T}(G^*)$ denote the cycle space of G over $\mathbb{F}_2$ and the cut space of G^* over $\mathbb{F}_2$, respectively, then the dual $\mathscr{F}^*(G)$ of $\mathscr{F}(G)$ (as a vector space) is equal to the cut space $\mathscr{T}(G^*)$ of G^*; that is,

$$\mathscr{F}^*(G) = \mathscr{T}(G^*).$$

(3) If T is any spanning tree of G, then $(V^*, (E - E(T))^*)$ is a spanning tree of G^* (Here, $E(T)$ is the set of edges of the tree T.)

Proof. See Diestel [9], Section 4.6. □

The interesting problem of finding an algorithmic test for planarity has received quite a bit of attention. Hopcroft and Tarjan have given an algorithm running in linear time in the number of vertices. For more on planarity, the reader should consult Diestel [9], Chapter 4, or Harary [15], Chapter 11.

Besides the four-color "conjecture," the other most famous theorem of graph theory is the *graph minor theorem*, due to Roberston and Seymour and we can't resist stating this beautiful and amazing result. For this, we need to explain what is a *well quasi-order* (for short, a *w.q.o.*). Recall that a partial order on a set X is a binary relation $\leq$, that is reflexive, symmetric, and anti-symmetric. A *quasi-order* (or *preorder*) is a relation which is reflexive and transitive (but not necessarily anti-symmetric). A *well quasi-order* is a quasi-order with the following property.

For every infinite sequence $(x_n)_{n \geq 1}$ of elements $x_i \in X$, there exist some indices i, j, with $1 \leq i < j$, so that $x_i \leq x_j$.

Now, we know that being a minor of another graph is a partial order and thus, a quasi-order. Here is Robertson and Seymour's theorem:

Theorem 6.25. *(Graph Minor Theorem, Robertson and Seymour, 1985–2004) The minor relation on finite graphs is a well quasi-order.*

Remarkably, the proof of Theorem 6.25 is spread over 20 journal papers (under the common title, *Graph Minors*) written over nearly 18 years and taking well over

Fig. 6.46 Paul D. Seymour, 1950– (left) and G Neil Robertson, 1938– (right)

500 pages! Many original techniques had to be invented to come up with this proof, one of which is a careful study of the conditions under which a graph can be embedded in a surface and a "Kuratowski-type" criterion based on a finite family of "forbidden graphs." The interested reader is urged to consult Chapter 12 of Diestel [9] and the references given there.

A precursor of the graph minor theorem is a theorem of Kruskal (1960) that applies to trees. Although much easier to prove than the graph minor theorem, the proof fo Kruskal's theorem is very ingenious. It turns out that there are also some interesting connections between Kruskal's theorem and proof theory, due to Harvey Friedman. A survey on this topic can be found in Gallier [10].

6.7 Summary

This chapter delves more deeply into graph theory. We begin by defining two fundamental vector spaces associated with a finite directed graph G, the *cycle space* or *flow space* $\mathcal{F}(G)$, and the *cocycle space* or *tension space* (or *cut space*) $\mathcal{T}(G)$. These spaces turn out to be orthogonal. We explain how to find bases of these spaces in terms of spanning trees and cotrees and we determine the dimensions of these spaces in terms of the number of edges, the number of vertices, and the number of connected components of the graph. A pretty lemma known as the *arc coloring lemma* (due to Minty) plays a crucial role in the above presentation which is heavily inspired by Berge [1] and Sakarovitch [20]. We discuss the incidence matrix and the adjacency matrix of a graph and explain how the spaces of flows and tensions can be recovered from the incidence matrix. We also define the Laplacian of a graph. Next, we discuss briefly Eulerian and Hamiltonian cycles. We devote a long section to flow problems and in particular to the max-flow min-cut theorem and some of its variants. The proof of the max-flow min-cut theorem uses the *arc-coloring lemma* in an interesting way, as indicated by Sakarovitch [20]. Matchings, coverings, and bipartite graphs are briefly treated. We conclude this chapter with a discussion of planar graphs. Finally, we mention two of the most famous theorems of graph theory: the *four color-conjecture* (now theorem, or is it?) and the *graph minor theorem,* due to Robertson and Seymour.

- We define the *representative vector* of a cycle and then the notion of Γ-*cycle* Γ, *representative vector* of a Γ-cycle $\gamma(\Gamma)$, a Γ-*circuit*, and a *simple* Γ-*cycle*.
- Next, we define a *cocycle* (or *cutset*) Ω, its *representative vector* $\omega(\Omega)$, a *cocircuit*, and a *simple cocycle*.
- We define a *cutset*.
- We prove several characterizations of simple cocycles.
- We prove the fundamental fact that the representative vectors of Γ-cycles and cocycles are *orthogonal*.
- We define the *cycle space* or *flow space* $\mathscr{F}(G)$, and the *cocycle space* or *tension space* (or *cut space*), $\mathscr{T}(G)$.
- We prove a crucial technical result: the *arc coloring lemma* (due to Minty).
- We derive various consequences of the arc-coloring lemma, including the fact that every edge of a finite digraph either belongs to a simple circuit or a simple cocircuit but not both.
- We define a *cotree* and give a useful characterization of them.
- We prove the main theorem of Section 6.1 (Theorem 6.2), namely, we compute the dimensions of the spaces $\mathscr{F}(G)$ and $\mathscr{T}(G)$, and we explain how to compute bases of these spaces in terms of spanning trees and cotrees.
- We define the *cyclomatic number* and the *cocyclomatic number* of a (di)graph.
- We remark that the dimension of $\mathscr{F}(G)$ is the dimension of the *first homology group* of the graph and that the *Euler-Poincaré characteristic* formula is a consequence of the formulae for the dimensions of $\mathscr{F}(G)$ and $\mathscr{T}(G)$.
- We give some useful characterizations of flows and tensions.
- We define the *incidence matrix* $D(G)$ of a directed graph G (without parallel edges or loops).
- We characterize $\mathscr{F}(G)$ and $\mathscr{T}(G)$ in terms of the incidence matrix.
- We prove a theorem of Poincaré about nonsingular submatrices of D which shows that D is *totally unimodular*.
- We define the *adjacency matrix* $A(G)$ of a graph.
- We prove that $DD^\top = \Delta - A$, where Δ is the diagonal matrix consisting of the degrees of the vertices.
- We define $DD^\top$ as the *Laplacian* of the graph.
- The study of the matrix $DD^\top$, especially its eigenvalues, is an active area of research called *spectral graph theory*.
- We define an *Euler cycle* and an *Euler circuit*.
- We prove a simple characterization of the existence of an Euler cycle (or an Euler circuit).
- We define a *Hamiltonian cycle* and a *Hamiltonian circuit*.
- We mention that the Hamiltonian cycle problem is *NP-complete*.
- We define a *network* (or *flow network*), a digraph together with a *capacity function* (or *cost function*).
- We define the notion of *flow*, of *value of a flow*, and state the *network flow problem*.

- We define the notion of v_s-v_t-cut and of *capacity of a v_s-v_t-cut*.
- We prove a basic result relating the maximum value of a flow to the minimum capacity of a v_s-v_t-cut.
- We define a *minimum v_s-v_t-cut* or *minimum cut*.
- We prove that in any network there is a flow of maximum value.
- We prove the celebrated *max-flow min-cut theorem* due to Ford and Fulkerson using the *arc coloring lemma*.
- We define a *flow augmenting chain*.
- We describe the algorithm *maxflow* and prove its correctness (provided that it terminates).
- We give a sufficient condition for the termination of the algorithm *maxflow* (all the capacities are multiples of some given number).
- The above criterion implies termination of *maxflow* if all the capacities are integers and that the algorithm will output some maximum flow with integer capacities.
- In order to improve the complexity of the algorithm *maxflow* we define a *residual network*.
- We briefly discuss faster algorithms for finding a maximum flow. We define a *preflow* and mention "preflow-push relabel algorithms."
- We present a few applications of the max-flow min-cut theorem, such as a theorem due to *Menger* on edge-disjoint paths.
- We discuss *channeled flows* and state a theorem due to Hoffman that characterizes when a channeled flow exists.
- We define a *bottleneck* and give an algorithm for finding a channeled flow.
- We state a *max-flow min-cut theorem* for channeled flows.
- We conclude with a discussion of a variation of the max flow problem considered in Cormen et al. [7] (Chapter 26).
- We define a *bipartite graph* and a *maximum matching*.
- We define the *complete bipartite graphs*, $K_{m,n}$.
- We explain how the *maxflow* algorithm can be used to find a maximum matching.
- We define a *k-coloring* of a graph, when a graph is *k-colorable* and the *chromatic number* of a graph.
- We define the *incidence matrix* of a nonoriented graph and we characterize a bipartite graph in terms of its incidence matrix.
- We define a *matching* in a graph, a *matched vertex*, and a *perfect matching*.
- We define an *alternating chain*.
- We characterize a *maximal matching* in terms of alternating chains.
- We define a *line cover* and a *minimum line cover*.
- We prove a relationship between maximum matchings and minimum line covers.
- We define an *independent* (or *stable*) set of nodes and a *maximum independent set*.
- We define a *point cover* (or *transversal*) and a *minimum point cover*.

- We go back to bipartite graphs and describe a *marking* procedure that decides whether a matching is a maximum matching.
- As a corollary, we derive some properties of minimum point covers, maximum matchings, maximum independent sets, and minimum line covers in a bipartite graph.
- We also derive two classical theorems about matchings in a bipartite graph due to König and König–Hall and we state the *marriage theorem* (due to Hall).
- We introduce the notion of a *planar* graph.
- We define the *complete graph on n vertices* K_n.
- We define a *Jordan curve* (or a *simple curve*), *endpoints* (or *boundaries*) of a simple curve, a *simple loop* or *closed Jordan curve*, a *base point* and the *interior* of a closed Jordan curve.
- We define rigorously a *plane graph* and a *simple plane graph*.
- We define a *planar graph* and a *planar embedding*.
- We define the *stereographic projection* onto the sphere. A graph can be embedded in the plane iff it can be embedded in the sphere.
- We mention the possibility of embedding a graph into a surface.
- We define the *connected components* (or *regions*) of an open subset of the plane as well as its *boundary*.
- We define the *faces* of plane graph.
- We state the *Jordan curve theorem*
- We prove *Euler's formula* for connected planar graphs and talk about the *Euler–Poincaré characteristic* of a planar graph.
- We generalize *Euler's formula* to planar graphs that are not necessarily connected.
- We define the *girth* of a graph and prove an inequality involving the girth for connected planar graphs.
- As a consequence, we prove that K_5 and $K_{3,3}$ are not planar.
- We prove that every planar graph is 5-colorable.
- We mention the *four-color conjecture*.
- We define *edge contraction* and define a *minor* of a graph.
- We state *Kuratowski's theorem* characterizing planarity of a graph in terms of K_3 and $K_{3,3}$.
- We define *edge subdivision* and state another version of *Kuratowski's theorem* in terms of minors.
- We state *MacLane's criterion for planarity* of a graph in terms of a property of its cycle space over $\mathbb{F}_2$.
- We define the *dual graph* of a plane graph and state some results relating the dual and the bidual of a graph to the original graph.
- We define a *self-dual* graph.
- We state a theorem relating the flow and tension spaces of a plane graph and its dual.
- We conclude with a discussion of the *graph minor theorem*.
- We define a *quasi-order* and a *well quasi-order*.
- We state the *graph minor theorem* due to Robertson and Seymour.

Problems

6.1. Recall from Problem 3.8 that an undirected graph G is *h-connected* ($h \geq 1$) iff the result of deleting any $h - 1$ vertices and the edges adjacent to these vertices does not disconnect G. Prove that if G is an undirected graph and G is 2-connected, then there is an orientation of the edges of G for which G (as an oriented graph) is strongly connected.

6.2. Given a directed graph $G = (V, E, s, t)$ prove that a necessary and sufficient condition for a subset of edges $E' \subseteq E$ to be a cocycle of G is that it is possible to color the vertices of G with two colors so that:

1. The endpoints of every edge in E' have different colors.
2. The endpoints of every edge in $E - E'$ have the same color.

Under which condition do the edges of the graph consitute a cocycle? If the graph is connected (as an undirected graph), under which condition is E' a simple cocycle?

6.3. Prove that if G is a strongly connected graph, then its flow space $\mathcal{F}(G)$ has a basis consisting of representative vectors of circuits.
Hint. Use induction on the number of vertices.

6.4. Prove that if the graph G has no circuit, then its tension space $\mathcal{T}(G)$ has a basis consisting of representative vectors of cocircuits.
Hint. Use induction on the number of vertices.

6.5. Let V be a subspace of $\mathbb{R}^n$. The *support* of a vector $v \in V$ is defined by

$$S(v) = \{i \in \{1, \ldots, n\} \mid v_i \neq 0\}.$$

A vector $v \in V$ is said to be *elementary* iff it has minimal support, which means that for any $v' \in V$, if $S(v') \subseteq S(v)$ and $S(v') \neq S(v)$, then $v' = 0$.

(a) Prove that if any two elementary vectors of V have the same support, then they are collinear.

(b) Let f be an elementary vector in the flow space $\mathcal{F}(G)$ of G (respectively, τ be an elementary vector in the tension space, $\mathcal{T}(G)$, of G). Prove that

$$f = \lambda \gamma \ (\textit{respectively}, \ \tau = \mu \omega),$$

with $\lambda, \mu \in \mathbb{R}$ and γ (respectively, ω) is the representative vector of a simple cycle (respectively, of a simple cocycle) of G.

(c) For any $m \times n$ matrix, A, let V be the subspace given by

$$V = \{x \in \mathbb{R}^n \mid Ax = 0\}.$$

Prove that the following conditions are equivalent.

(i) A is totally unimodular.
(ii) For every elementary vector $x \in V$, whenever $x_i \neq 0$ and $x_j \neq 0$, then $|x_i| = |x_j|$.

6.6. Given two $m \times m$ matrices with entries either 0 or 1, define $A + B$ as the matrix whose (i, j)th entry is the Boolean sum $a_{ij} \vee b_{ij}$ and AB as the matrix whose (i, j)th entry is given by

$$(a_{i1} \wedge b_{1j}) \vee (a_{i2} \wedge b_{2j}) \vee \cdots \vee (a_{im} \wedge b_{mj});$$

that is, interpret 0 as **false**, 1 as **true**, $+$ as **or** and $\cdot$ as **and**.

(i) Prove that

$$A_{ij}^k = \begin{cases} 1 & \text{iff there is a path of length } k \text{ from } v_i \text{ to } v_j \\ 0 & \text{otherwise.} \end{cases}$$

(ii) Let

$$B^k = A + A^2 + \cdots + A^k.$$

Prove that there is some k_0 so that

$$B^{n+k_0} = B^{k_0},$$

for all $n \geq 1$. Describe the graph associated with B^{k_0}.

6.7. Let G be an undirected graph known to have an Euler cycle. The principle of *Fleury's algorithm* for finding an Euler cycle in G is the following.

1. Pick some vertex v as starting point and set $k = 1$.
2. Pick as the kth edge in the cycle being constructed an edge e adjacent to v whose deletion does not disconnect G. Update G by deleting edge e and the endpoint of e different from v and set $k := k + 1$.

Prove that if G has an Euler cycle, then the above algorithm outputs an Euler cycle.

6.8. Recall that K_m denotes the (undirected) complete graph on m vertices.

(a) For which values of m does K_m contain an Euler cycle?

Recall that $K_{m,n}$ denotes the (undirected) complete bipartite graph on $m + n$ vertices.

(b) For which values of m and n does $K_{m,n}$ contain an Euler cycle?

6.9. Prove that the graph shown in Figure 6.47 has no Hamiltonian.

6.10. Prove that the graph shown in Figure 6.48 and known as *Petersen's graph* satisfies the conditions of Proposition 6.12, yet this graph has no Hamiltonian.

6.11. Prove that if G is a simple undirected graph with n vertices and if $n \geq 3$ and the degree of every vertex is at least $n/2$, then G is Hamiltonian (this is known as *Dirac's Theorem*).

6.12. Find a minimum cut separating v_s and v_t in the network shown in Figure 6.49:

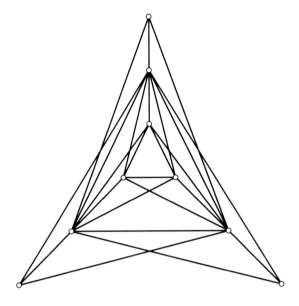

Fig. 6.47 A graph with no Hamiltonian

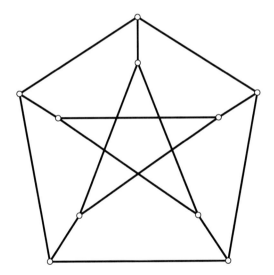

Fig. 6.48 Petersen's graph

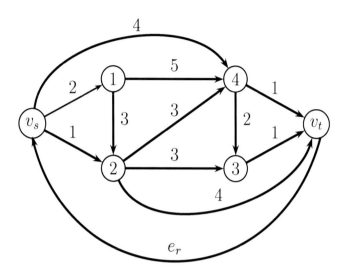

Fig. 6.49 A netwok

6.13. Consider the sequence (u_n) defined by the recurrence

$$u_0 = 0$$
$$u_1 = \frac{\sqrt{5}-1}{2}$$
$$u_{n+2} = -u_{n+1} + u_n.$$

If we let $r = u_1 = (\sqrt{5}-1)/2$, then prove that

$$u_n = r^n.$$

Let $S = \sum_{k=0}^{\infty} r^n = 1/(1-r)$. Construct a network (V, E, c) as follows.

- $V = \{v_s, v_t, x_1, x_2, x_3, x_4, y_1, y_2, y_3, y_4\}$
- $E_1 = \{e_1 = (x_1, y_1), e_2 = (x_2, y_2), e_3 = (x_3, y_3), e_4 = (x_4, y_4)\}$
- $E_2 = \{(v_s, x_i), (y_i, v_t), 1 \leq i \leq 4\}$
- $E_3 = \{(x_i, y_j), (y_i, y_j), (y_i, x_j), 1 \leq i, j \leq 4, i \neq j\}$
- $E = E_1 \cup E_2 \cup E_3 \cup \{(v_t, v_s)\}$
- $c(e) = r^{i-1}$ iff $e = e_i \in E_1$, else $c(e) = S$ iff $e \in E - E_1$.

Prove that it is possible to choose at every iteration of the Ford and Fulkerson algorithm the chains that allow marking v_t from v_s in such a way that at the kth iteration the flow has value $\delta = r^{k-1}$. Deduce from this that the algorithm does not terminate and that it converges to a flow of value S even though the capacity of a minimum cut separating v_s from v_t is $4S$.

6.14. Let $E = \{e_1, \ldots, e_m\}$ be a finite set and let $S = \{S_1, \ldots, S_n\}$ be a family of finite subsets of E. A set $T = \{e_{i_1}, \ldots, e_{i_n}\}$ of distinct elements of E is a *transversal* for S (also called a *system of distinct representatives* for S) iff

$$e_{i_j} \in S_j, \quad j = 1, \ldots, n.$$

Hall's theorem states that the family S has a transversal iff for every subset $I \subseteq \{1, \ldots, n\}$ we have

$$|I| \leq \left| \bigcup_{i \in I} S_i \right|.$$

(a) Prove that the above condition is necessary.

(b) Associate a bipartite graph with S and T and use Theorem 6.18 to prove that the above condition is indeed sufficient.

6.15. Let G be a directed graph without any self-loops or any cicuits (G is acyclic). Two vertices u, v, are *independent* (or *incomparable*) iff they do not belong to any path in G. A set of paths (possibly consisting of a single vertex) *covers* G iff every vertex belongs to one of these paths.

Dilworth's theorem states that in an acyclic directed graph, there is some set of pairwise independent vertices (an antichain) and a covering family of pairwise (vertex-)disjoint paths whose cardinalities are the same.

Two independent vertices can't belong to the same path, thus it is clear that the cardinality of any antichain is smaller than or equal to the cardinality of a path cover. Therefore, in Dilworth's theorem, the antichain has maximum size and the covering family of paths has minimum size.

Given a directed acyclic graph $G = (V, E)$ as above, we construct an undirected bipartite graph $H = (V_1 \cup V_2, E_H)$ such that:

- There are bijections, $h_i \colon V_i \to V$, for $i = 1, 2$.
- There is an edge, $(v_1, v_2) \in E_H$, iff there is a path from $h_1(v_1)$ to $h_2(v_2)$ in G.

(a) Prove that for every matching U of H there is a family $\mathscr{C}$ of paths covering G so that $|\mathscr{C}| + |U| = |V|$.

(b) Use (a) to prove Dilworth's theorem.

6.16. Let $G = (V, E)$ be an undirected graph and pick $v_s, v_t \in V$.

(a) Prove that the maximum number of pairwise edge-disjoint chains from v_s to v_t is equal to the minimum number of edges whose deletion yields a graph in which v_s and v_t belong to disjoint connected components.

(b) Prove that the maximum number of pairwise (intermediate vertex)-disjoint chains from v_s to v_t is equal to the minimum number of vertices in a subset U of V so that in the subgraph induced by $V - U$, the vertices v_s and v_t belong to disjoint connected components.

Remark: The results stated in (a) and (b) are due to Menger.

6.17. Let $G = (V, E)$ be any undirected graph. A subset $U \subseteq V$ is a *clique* iff the subgraph induced by U is complete.

Prove that the cardinality of any matching is at most the number of cliques needed to cover all the vertices in G.

6.18. Given a graph $G = (V, E)$ for any subset of vertices $S \subseteq V$ let $p(S)$ be the number of connected components of the subgraph of G induced by $V - S$ having an odd number of vertices.

(a) Prove that if there is some $S \subseteq V$ such that $p(S) > |S|$, then G does not admit a perfect matching.

(b) From now on, we assume that G satisfies the condition

$$p(S) \leq |S|, \quad \text{for all } S \subseteq V \tag{C}$$

Prove that if Condition (C) holds, then G has an even number of vertices (set $S = \emptyset$) and that $|S|$ and $p(S)$ have the same parity. Prove that if the condition

$$p(S) < |S|, \quad \text{for all } S \subseteq V \tag{C'}$$

is satisfied, then there is a perfect matching in G containing any given edge of G (use induction of the number of vertices).

(c) Assume that Condition (C) holds but that Condition (C') does not hold and let S be maximal so that $p(S) = |S|$.

Prove that the subgraph of G induced by $V - S$ does not have any connected component with an even number of vertices.

Prove that there cannot exist a family of k connected components of the subgraph of G induced by $V - S$ connected to a subset T of S with $|T| < k$. Deduce from this using the theorem of König–Hall (Theorem 6.18) that it is possible to assign a vertex of S to each connected component of the subgraph induced by $V - S$.

Prove that if Condition (C) holds, then G admits a perfect matching. (This is a theorem due to Tutte.)

6.19. The *chromatic index* of a graph G is the minimum number of colors so that we can color the edges of G in such a way that any two adjacent edges have different colors. A simple unoriented graph whose vertices all have degree 3 is called a *cubic* graph.

(a) Prove that every cubic graph has an even number of vertices. What is the number of edges of a cubic graph with $2k$ vertices? Prove that for all $k \geq 1$, there is at least some cubic graph with $2k$ vertices.

(b) Let G be a cubic bipartite graph with $2k$ vertices. What is the number of vertices in each of the two disjoint classes of vertices making G bipartite? Prove that all $k \geq 1$; there is at least some cubic bipartite graph with $2k$ vertices.

(c) Prove that the chromatic index of Petersen's graph (see Problem 6.10) is at least four.

(d) Prove that if the chromatic index of a cubic graph $G = (V, E)$ is equal to three, then

(i) G admits a perfect matching, $E' \subseteq E$.
(ii) Every connected component of the partial graph induced by $E - E'$ has an even number of vertices.

Prove that if Conditions (i) and (ii) above hold, then the chromatic index of G is equal to three.

(e) Prove that a necessary and sufficient condition for a cubic graph G to have a chromatic index equal to three is that G possesses a family of disjoint even cycles such that every vertex of G belongs to one and only one of these cycles.

(f) Prove that Petersen's graph is the cubic graph of chromatic index 4 with the minimum number of vertices.

6.20. Let $G = (V_1 \cup V_2, E)$ be a *regular* bipartite graph, which means that the degree of each vertex is equal to some given $k \geq 1$ (where V_1 and V_2 are the two disjoint classes of nodes making G bipartite).

(a) Prove that $|V_1| = |V_2|$.

(b) Prove that it is possible to color the edges of G with k colors in such a way that any two edges colored identically are not adjacent.

6.21. Prove that if a graph G has the property that for G itself and for all of its partial subgraphs, the cardinality of a minimum point cover is equal to the cardinality of a maximum matching (or, equivalently, the cardinality of a maximum independent set is equal to the cardinality of a minimum line cover), then G is bipartite.

6.22. Let $G = (V_1 \cup V_2, E)$ be a bipartite graph such that every vertex has degree at least 1. Let us also assume that no maximum matching is a perfect matching. A subset $A \subseteq V_1$ is called a *basis* iff there is a matching of G that matches every node of V_1 and if A is maximal for this property.

Prove that if A is any basis, then for every $v' \notin A$ we can find some $v'' \in A$ so that

$$(A \cup \{v'\}) - \{v''\}$$

is also a basis.

(b) Prove that all bases have the same cardinality.

Assume some function $l : V_1 \to \mathbb{R}_+$ is given. Design an algorithm (similar to Kruskal's algorithm) to find a basis of maximum weight, that is, a basis A, so that the sum of the weights of the vertices in A is maximum. Justify the correctness of this algorithm.

6.23. Prove that every undirected graph can be embedded in $\mathbb{R}^3$ in such a way that all edges are line segments.

6.24. A finite set $\mathscr{T}$ of triangles in the plane is a *triangulation* of a region of the plane iff whenever two triangles in $\mathscr{T}$ intersect, then their intersection is either a common edge or a common vertex. A triangulation in the plane defines an obvious plane graph.

Prove that the subgraph of the dual of a triangulation induced by the vertices corresponding to the bounded faces of the triangulation is a forest (a set of disjoint trees).

6.25. Let $G = (V, E)$ be a connected planar graph and set

$$\chi_G = v - e + f,$$

where v is the number of vertices, e is the number of edges, and f is the number of faces.

(a) Prove that if G is a triangle, then $\chi_G = 2$.

(b) Explain precisely how χ_G changes under the following operations:

1. Deletion of an edge e belonging to the boundary of G.
2. Contraction of an edge e that is a bridge of G.
3. Contraction of an edge e having at least some endpoint of degree 2.

Use (a) and (b) to prove Euler's formula: $\chi_G = 2$.

6.26. Prove that every simple planar graph with at least four vertices possesses at least four vertices of degree at most 5.

6.27. A simple planar graph is said to be *maximal* iff adding some edge to it yields a nonplanar graph. Prove that if G is a maximal simple planar graph, then:

(a) G is 3-connected.

(b) The boundary of every face of G is a cycle of length 3.

(c) G has $3v - 6$ edges (where $|V| = v$).

6.28. Prove Proposition 6.24.

6.29. Assume $G = (V, E)$ is a connected plane graph. For any dual graph $G^* = (V^*, E^*)$ of G, prove that

$$|V^*| = |E| - |V| + 2$$
$$|V| = |E^*| - |V^*| + 2.$$

Prove that G is a dual of G^*.

6.30. Let $G = (V, E)$ be a finite planar graph with $v = |V|$ and $e = |E|$ and set

$$\rho = 2e/v, \quad \rho^* = 2e/f.$$

(a) Use Euler's formula ($v - e + f = 2$) to express e, v, f in terms of ρ and ρ^*. Prove that

$$(\rho - 2)(\rho^* - 2) < 4.$$

(b) Use (a) to prove that if G is a simple graph, then G has some vertex of degree at most 5.

(c) Prove that there are exactly five regular convex polyhedra in $\mathbb{R}^3$ and describe them precisely (including their number of vertices, edges, and faces).

(d) Prove that there are exactly three ways of tiling the plane with regular polygons.

References

1. Claude Berge. *Graphs and Hypergraphs*. Amsterdam: Elsevier North-Holland, first edition, 1973.
2. Marcel Berger. *Géométrie 1*. Nathan, 1990. English edition: Geometry 1, Universitext, New York: Springer Verlag.
3. Norman Biggs. *Algebraic Graph Theory*, volume 67 of *Cambridge Tracts in Mathematics*. Cambridge, UK: Cambridge University Press, first edition, 1974.
4. Béla Bollobas. *Modern Graph Theory*. GTM No. 184. New York: Springer Verlag, first edition, 1998.
5. J. Cameron, Peter. *Combinatorics: Topics, Techniques, Algorithms*. Cambridge, UK: Cambridge University Press, first edition, 1994.
6. Fan R. K. Chung. *Spectral Graph Theory*, vol. 92 of *Regional Conference Series in Mathematics*. Providence, RI: AMS, first edition, 1997.
7. H. Cormen, Thomas, E. Leiserson, Charles, L. Rivest, Ronald, and Clifford Stein. *Introduction to Algorithms*. Cambridge, MA: MIT Press, second edition, 2001.
8. Peter Cromwell. *Polyhedra*. Cambridge, UK: Cambridge University Press, first edition, 1994.
9. Reinhard Diestel. *Graph Theory*. GTM No. 173. New York: Springer Verlag, third edition, 2005.
10. Jean Gallier. What's so Special about Kruskal's Theorem and the Ordinal Γ_0? *Annals of Pure and Applied Logic*, 53:199–260, 1991.
11. Jean H. Gallier. *Geometric Methods and Applications, for Computer Science and Engineering*. TAM, Vol. 38. New York: Springer, first edition, 2000.
12. Chris Godsil and Gordon Royle. *Algebraic Graph Theory*. GTM No. 207. New York: Springer Verlag, first edition, 2001.
13. Jonathan L. Gross, and Thomas W. Tucker. *Topological Graph Theory*. New York: Dover, first edition, 2001.
14. Victor Guillemin and Alan Pollack. *Differential Topology*. Englewood Cliffs, NJ: Prentice Hall, first edition, 1974.
15. Frank Harary. *Graph Theory*. Reading, MA: Addison Wesley, first edition, 1971.
16. Jon Kleinberg and Eva Tardos. *Algorithm Design*. Reading, MA: Addison Wesley, first edition, 2006.
17. James R. Munkres. *Elements of Algebraic Topology*. Reading, MA: Addison-Wesley, first edition, 1984.
18. Christos H. Papadimitriou and Kenneth Steiglitz. *Combinatorial Optimization. Algorithms and Complexity*. New York: Dover, first edition, 1998.
19. N. Robertson, D. Sanders, P.D. Seymour and R. Thomas. The four-color theorem. *J. Combin. Theory B*, 70:2-44, 1997.
20. Michel Sakarovitch. *Optimisation Combinatoire, Méthodes mathématiques et algorithmiques. Graphes et Programmation Linéaire*. Paris: Hermann, first edition, 1984.
21. Michel Sakarovitch. *Optimisation Combinatoire, Méthodes mathématiques et algorithmiques. Programmation Discréte*. Paris: Hermann, first edition, 1984.
22. Herbert S. Wilf. *Algorithms and Complexity*. Wellesley, MA: A K Peters, second edition, 2002.

Symbol Index

J. Gallier, *Discrete Mathematics*, Universitext,
DOI 10.1007/978-1-4419-8047-2, © Springer Science+Business Media, LLC 2011

Index

J. Gallier, *Discrete Mathematics,* Universitext,
DOI 10.1007/978-1-4419-8047-2, © Springer Science+Business Media, LLC 2011

Printed by Publishers' Graphics LLC
BT20121014.19.20.2